THE ROOTS OF CONSCIOUS-NESS

THE ROOTS OF CONSCIOUSNESS

THE CLASSIC ENCYCLOPEDIA OF CONSCIOUSNESS STUDIES REVISED AND EXPANDED

JEFFREY MISHLOVE, PH.D.

WITH AN APPENDIX BY SAUL-PAUL SIRAG

Marlowe & Company

The Roots of Consciousness:
Psychic Exploration Through History, Science and Experience

Published 1975, Revised Edition 1993

Published and distributed by
Marlowe & Company
632 Broadway, Seventh Floor
New York, New York 10012

ISBN 1-56924-747-1
LC Number 92-72319
Printed in the United States of America

Objects pictured on the dustjacket courtesy of Peace of Mind Bookstore, Country Charm Antiques, The Antiquary, Don Wheeler, Alice L. Price, Armin L. Saeger, Jean Marie Dennison, Edwin L. Wade, Carol Haralson, Sally Dennison, Michael Hightower, and Paulette Millichap.

Design: Carol Haralson
Production: Pakaáge
Dustjacket photograph: Don Wheeler

To my beloved wife Janelle,
my parents Rose and Hyman, and
my stepson Lewis, a source of pride and joy,
with whom I have shared
many moments of precious awareness.

Acknowledgments

THIS BOOK ONLY EXISTS AS A PRODUCT OF THE ENTIRE COMMUNITY of individuals who have inspired me and whose names can generally be found in the footnotes and text. For each I have a unique feeling of gratitude and appreciation.

I would particularly like to acknowledge Arthur Bloch, James P. Driscoll, Gertrude Schmeidler, and Saul-Paul Sirag — for their friendship and criticism. Additionally, special assistance has been received from John Beloff, Adrian Boshier, William Braud, Stephen Braude, Irvin Child, Henry Dakin, Douglas Dean, Brenda Dunne, Jule Eisenbud, Martin Gardner, Patric Geisler, Bernard Grad, Keith Harary, James Harder, David Hoffman, Albert Krueger, Angela Longo, Ted Mann, James McClenon, Rammurti Mishra, Robert Morris, Thelma Moss, Carroll Nash, A. R. G. Owen, Ted Owens, John Palmer, Harold Puthoff, Dean Radin, K. Ramakrishna Rao, Kathlyn Rhea, William Roll, Milan Ryzl, Jack Sarfatti, Helmut Schmidt, Berthold Schwarz, Ray Stanford, Charles Tart, William Tiller, Jessica Utts, Larissa Vilenskaya, Graham Watkins, Rhea White and Arthur Young and the S. F. Theosophical Library. I would like to thank all of them for the generosity of their time and resources.

Additionally, I am beholden to over a hundred other individuals who have freely revealed to me many of their deepest thoughts in the course of the *Thinking Allowed* and *InnerWork* video interviews which serve as a general background to the revised edition.

CONTENTS

SECTION THREE

THE SCIENTIFIC EXPLORATION OF CONSCIOUSNESS

SECTION FOUR
THEORIES OF CONSCIOUSNESS

APPENDIX:

FOREWORD TO THE REVISED EDITION

JEFFREY MISHLOVE ENTERED MY LIFE IN A MOST UNANTICIPATED AND BIZARRE WAY over a decade ago when he was a graduate student at the University of California, Berkeley and I a faculty member. But that initial meeting was only the beginning, foreshadowing the myriad connections that began to unfold, almost as the unraveling of a destiny that had been decided and known for a long time.

I was sitting in my office at the University at the end of office hours talking to a woman undergraduate student with whom I particularly enjoyed a rapport. Our discussion of history had inexplicably shifted from the Cold War to Carl Jung and synchronicity, a subject we surprisingly discovered was of mutual interest. As we enthusiastically pursued the topic, a young man appeared at the opened door, saying he wanted to talk to me. Already thoroughly caught up in the present discourse, I asked him if he could come back. He said he would wait. It occurred to me that perhaps he had a quick question which could be dealt with immediately; I asked him how I could help him. He said he was a Ph.D. candidate and wanted to talk to me about serving on his dissertation committee. What specialty in history was he researching, I asked. Not history, but an interdisciplinary degree, he responded.

I felt already overloaded with the work I had undertaken with graduate students in my own field, diplomatic history, so I suggested he look for another faculty member who might be more appropriate for his work. In a tone that was quiet but conclusive and considered, he told me he thought I might be that person. How very curious! This exchange was taking place in an increasingly electrifying atmosphere, laden with a compelling momentum which charged the whole room. It was unprecedented in my experience, yet there was a fascination and a sense of familiarity I could not place.

Why have you come to me? I asked the question aloud. He replied that the graduate division had just required him to add a member of the history department to his dissertation committee; he had briefly talked to a secretary who mentioned my name. Then he added to my astonishment that he had been walking down the hall and felt drawn to my office. At this point my undergraduate student began to grow pale. What are you working on? I asked, by now expecting no ordinary answer. Psychic phenomena and consciousness, he explained, moving toward my desk. My student got up, turned to me, and whispered that this experience was too overwhelming for her. She left.

The familiar stranger began to talk about his work, ideas, and an impasse on the current committee. I confessed to my own more than casual interest in his subject and that I had for years taught a core course at M.I.T., "Consciousness and Society." I was spellbound as we talked for what seemed like a few minutes, but, as darkness fell, I realized it was a couple of hours. We would have to continue our talk later, we realized. Yes, I would indeed serve on his committee for a number of reasons, some understood by me at the time. But, I asked, as he was leaving, almost as an afterthought, what is your name? Who are you? I got out a piece of paper I had never used. He replied, Jeffrey Mishlove.

Jeffrey Mishlove! I was stunned. I opened my mouth to tell him, but then I decided to wait. We both were already late for other engagements, and I wanted to sort through the many threads of my reaction. He started to spell his name, but I stopped him. Oh yes, I know. . .

A year before, a particularly dear relative had been dying thousands of miles away and I keenly experienced pain and turmoil. I wanted to do something, but felt helpless. I didn't know why, but I instinctively reached for a book on consciousness I had bought. When originally acquired, it had fascinated me, excited my interest, and profoundly yet simply put before me a new dimension to explore. In my distress, I opened it randomly and found before me the pages of healing exercises practiced for centuries. I read them, and undertook to use them. I experienced immediate release and a flood of understanding overwhelmed me. As a result, I felt a need to get in touch with the author. Opening the book *The Roots of Consciousness*, I looked at the name Jeffrey Mishlove. He seemed to live around Berkeley. I decided that someday I would look him up. Instead, he found me.

It soon became clearer why. His Ph.D. was no ordinary undertaking. It would be the first one designated in parapsychology through the interdisciplinary program which allows the student to construct an innovative major while the university supervises it. His program was very carefully controlled and had a dissertation committee with considerably more members on it than in a regular department. It was a very exacting group of experts in and outside his field, some of whom had been arbitrarily assigned.

From the very start it was glaringly clear to me there was something quite amiss. Jeffrey encountered inexplicable and groundless problems I had never seen crop up on dissertation. Jeffrey's work on a parapsychological degree seemed to have run headlong into invisible barriers and obstacles that always seemed to arise whenever he got close to finishing his degree. The trouble seemed not to be Jeffrey but the prejudice and bias of a certain element which, in an academically unprofessional manner, simply vetoed his work. The detractor who did not even hold professorial rank took particular delight in repeatedly requiring additional and irrelevant tasks. Then, when they were completed, he rejected the results repeatedly over the protestations of the faculty members on the committee. So this was to be the fate of a pioneer like Jeffrey Mishlove! Finally in a showdown over this obstructionism, I asked this antagonist if there were any results the candidate could ever produce from his requests that he would accept. The game was up; no, there were not. He triumphantly announced that he would never agree to a degree in this field. At this point, in addition to my work on the committee on the context and methodology of using historical materials, I undertook to bring this fact to the attention of the more responsible and objective authorities. Throughout this process I felt constantly a powerful clarifying and directing force. Could it be the same one that led Jeffrey into my office or led me to join his committee? Synchronicity?

I have also been aware of a profound interaction between our lives after that experience. It is not dependent on actual contact. Ever since Jeffrey's degree, I have found myself expressing my energies increasingly in the psychic realm. The most important occasion was in 1986 when my father, who is precious to me, suffered a stroke following heart by-pass surgery, and remained in what his doctors called an alpha coma. His body could never function on its own and there was less than a one percent chance of any reversal. I went to his bedside in Cincinnati, hoping to help. Something very powerful took over. I let the years of friendship with Jeffrey speak and fill me with a sublime and concentrated confidence. I knew in a deep an unshakeable way that, despite the medical odds, my father could heal. I spent three months with him, a good part of which was in surgical intensive care during his coma. I talked to him, telling him he was getting better and encouraging him to heal. I used some of the techniques I had learned from Jeffrey's work. I felt the strength from Jeff's previous encouragement for me to follow my instincts. He had rooted in me the desire to use my own sense of inner knowledge telling me what I must do. My father made a miraculous recovery. After three months, he was in rehabilitation and I left him to return to Berkeley the day after we celebrated his eighty-third birthday, a date he had confided to a family member he had, in a dream, seen inscribed on his gravestone.

Jeffrey has, since that fateful day in my office a decade ago, taken his work in many significant directions which come together in this new edition of *The Roots of Consciousness.* His original pathbreaking work has undergone a significant and fundamental metamorphosis through major revision, updating, and synthesis of materials and issues. As important as this work was when it appeared in 1979, the book in its current edition brings us new dimensions. It not only reflects the enormous change that the field has undergone, but also bears testimony to Jeffrey's own personal evolution within it. Thus this edition presents us with an essentially new work, one that speaks to wider audiences and to the expanded concerns of new generations.

A more comprehensive and interpretive section entitled *folklore* supplements the section called *history.* Folklore is an apt term for what Jeffrey has accomplished here, in his digesting, assimilating, and expressing in modern terms the experiences, mythology, and psychic information of the past. Integrated with his own experience and work, this book reveals more of Jeffrey as well as his assessment of what is valid and what makes sense.

For me, there emerged a dialogue dialectically interacting between what at times has been seen as mutually exclusive points of view — those stemming from scientific skeptics and those from the researchers on the frontier. In this dialogue, the reader can find essentially a map of how far the field and Jeffrey have progressed toward a reconciliation of the two. It presents a wonderful courtship of the two camps, in which neither has prevailed, but in which both have given each its best. Jeffrey gives the skeptics voice, in the scientific section, for their best, most sophisticated criticisms, and the researchers their most sophisticated and articulate responses. Together they have shared with each other, in Jeffrey's elucidation, a joint enterprise — even when unwitting — with one dimension enriching the other.

As this manuscript has richly rewarded me, I know that it too will bring to the reader who seeks knowledge and growth rich sustenance for the mind and spirit. In it and from it one can eagerly expect to find new paths.

Diane Shaver Clemens
Professor of American Diplomatic History

Introduction to the Revised Edition

The Roots of Consciousness is a look at the way history, folklore and science shapes our understanding of psychic capacities. The original edition was published in 1975, while I was still a graduate student at the University of California, Berkeley, working in an individual, interdisciplinary doctoral program in parapsychology. It is, in part, a personal book containing descriptions of significant events in my own life. It is also personal because in the field of consciousness exploration there are so many competing interpretations that any telling of the story — even in strictly scientific terms — contains many individual choices.

I might have, for example, written an account from the perspective of a proponent for a particular viewpoint regarding the existence or non-existence of psychic functioning. In so doing, my goal would not be to sift through competing claims to arrive at a balanced and truthful account. Rather, I would be interested in persuading you that my version of reality is superior to those of my opponents.

If I were a skeptical debunker I would rail against magical thinking and would argue that every purported psychic event is the result of human error, folly or fraud.

This view, which is not uncommon in academic circles, has an ancient history and a marvelously fascinating folklore whose heroes are enlightened philosophers — people who have struggled mightily to break free from the shackles of superstition. By popping the illusory bubbles of myth and magic, such heroes can presumably guide humankind toward an age of rational enlightenment. Within the perspective of this folklore, anyone attesting to such events as telepathy, clairvoyance, precognition or psychokinesis is to be considered either suffering pitiable delusion or perpetrating contemptible fraud.

While I doubt that all "skeptics" will feel comfortable with this book, I have become convinced, over the past fifteen years since I wrote the original edition, that the point of view debunkers represent deserves greater respect. True, debunkers often argue from a materialistic, positivistic or scientistic ideology. Their thinking is as colored by their world view as that of other ideologues or "true believers."(The mechanisms by which this can occur are detailed explicitly in Section III.) However, thoughtless dismissal of either true believers or true skeptics sometimes results from a defensive reaction which generally serves no other purpose than to protect our own views from too sharp an outside challenge. As the original edition of *The Roots of Consciousness* was widely used as a college text, I am grateful for the opportunity to inject more critical thinking into the revised edition.

On the other hand, an exploration of consciousness might hardly be thought of as complete without an enumeration of the many inner realms of the mind explored by cultists and occultists, mystics and metaphysicians, witches and warlocks, poets and prophets, seers and saints, spirits and spiritualists, scientists and pseudoscientists of all stripes. Were I to write from the perspective of a New Age proponent, I would not fail to sympathetically treat such important terrain in the geography of consciousness as human beings who are the embodiment of deities, the hierarchy of spiritual beings and planes of non-human existence, the healing power of crystals and pyramids, and the worldwide confluence of prophecies regarding the future of the human race. In so doing, I would find no need to refer respectfully to the arguments of those who challenge my perspective.

Time and space do not permit me to enumerate all the many threads and nuances implicit in the two possible scenarios presented above. Nor do I wish simply to elaborate on all the possibilities. We all possess different genetic patterns, fingerprints and personal histories. Similarly, each of us is the creator of our own unique perspective about the power and creativity of our thoughts and desires. While I have sought to present a balanced viewpoint, I realize that many other knowledgeable persons hold perspectives about consciousness quite different from my own that they also believe to be appropriately balanced!

An author's goal of objectivity suggests that we can be neutral judges, evaluating the world around us as if we were not ourselves part of it, as if we were not players with a stake in the world game. To the degree that I subscribe to this goal (which, I hope, is substantial), I see myself as an impartial observer, accurately and fairly setting down the perspectives of believers and their critics. Yet, consciousness is a unique topic in that it is subjective, that it is direct, that we are it.[1] Thus, while subscribing to the goal of objectivity, I wish to challenge the "myth of objectivity" which holds that we can accurately and fairly describe the world about us without reference to our own selves, our beliefs and attitudes.

Our idealized image of objectivity (especially in science) receives its most severe challenge from neither mystics nor psychics, but from the growing critical literature within the philosophy and sociology of science itself. For an overview, I recommend Michael J. Mahoney's book, *Scientist as Subject: The Psychological Imperative.*[2] Dr. Mahoney persuasively argues that the "storybook image" of the scientist — to which most scientists apparently subscribe — is, in fact, continually contradicted by the empirical evidence. The *actual* behavior of scientists suggests an image that, in practice, overlaps much more with occultism — in both the positive and negative senses in which this might be taken.

Harry Collins and Trevor Pinch in *Frames of Meaning* specifically claim that "radical cultural discontinuities" exist within the scientific community itself. Such cultural differences, they maintain, make it impossible to rationally settle the dispute as to the existence of human psychic abilities.[3] The eminent philosopher of science, Paul Feyerabend, in *Science in a Free Society,* goes even further and argues that major advances in science *necessarily* require the violation of normal scientific rules and standards.[4]

In describing the history, folklore and science of consciousness, I will not pretend to be simply a disinterested observer and student of consciousness, but a *participant* as well. My entire slant is colored by my own experiences. Let me clearly warn you that, while I have done my best to honestly and accurately present all the following material, I had better — due to the possibility of numerous cognitive pitfalls (to be detailed in Section III) — make no further claim that I have demonstrated the truth of any particular version of reality. The purpose of this revised edition of *The Roots of Consciousness* is simply to provide an entry into the language, concepts and assumptions implicit in a sophisticated world view that allows for the possibility of psychic functioning. I am more interested in readers understanding and appreciating this world view than in accepting or following it as "the truth."

One stylistic model which I am setting for myself (and which I hope to attain from time to time) has been called *meaningful thinking* by Sigmund Koch in his presidential address to the Divisions of General Psychology and of Philosophical Psychology of theAmerican Psychological Association. Koch describes *meaningful* thinking in terms which may seem more familiar to mystics, poets and occultists than to scientists and scholars:

> In meaningful thinking, the mind caresses, flows joyously into, over, around, the relational matrix defined by the problem, the object. There is a merging of person and object or problem. Only the problem or object, its terms and relations, exist. And *these* are real in the fullest, most vivid, electric, undeniable way. It is a fair descriptive generalization to say that meaningful thinking is ontologistic in some primitive, accepting, artless, unselfconscious sense.[5]

We are all ourselves

"Why am I me?" The chills and sensations of first being conscious of myself being conscious of myself are still vivid in my memory. I was a ten-year-old child then, sitting alone in my parents' bedroom, touching my own solid consciousness and wondering at it. I was stepping through the looking glass seeing myself being myself seeing myself being myself...tasting infinity in a small body. I could be anybody. But I happen to be *me.* Why not someone else? And if I were someone else, could I not still be *me*?

What does it mean to be an individual being? How is it possible that *I exist?* How is it I am able to sense myself? What is the *self* I sense I *am*?

How is it I am able to be conscious? What does it mean to exercise consciousness?

Does conscious awareness naturally emerge from the complex structure of physical atoms, molecules, cells and organs that compose my body? Does consciousness reside somehow or emerge from the higher structure of my brain and nervous system? And, if so, how does that occur? What is it about the structure of my nervous system that allows me to discover myself as a human being? How can a brain formulate questions? Are thoughts and questions even *things* in the same sense that neurons and brains are *things*?

As conscious beings, do we possess spirits and souls? Are we sparks of the divine fire?

How close are we to understanding the origins of the universe, of life, of consciousness? Is it possible to answer questions such as: Who are we? What does it mean to be human? What is the ultimate nature of matter? Of mind?

In our time, the spiritual and material views seem quite divergent. In a way they both ring true. They each speak to part of our awareness. And, for many if not most people, they each, by themselves, leave us unsatisfied.

We have myths and stories. We have world views, paradigms, constructs and hypotheses. We have competing dogmas, theologies and sciences. Do we have understanding? Can an integration of our scientific knowledge with the spiritual insights of humanity bring greater harmony to human civilization?

We go about our business. We build cities and industries. We engage in buying and selling. We have families and raise children. We affiliate with religious teachings or other traditions.

We sometimes avoid confronting the

deep issues of being because there we feel insecure, even helpless. And like a mirror of our inner being, society reflects our tension.

Yet the mystery of being continues to rear its head. It will not go away. As we face ecological disaster, nuclear war, widespread drug addiction, global inhumanity, we are forced to notice the consequences of our lives in ever greater detail. Are not these horrendous situations the products of human consciousness and behavior? Can we any longer continue to address the major political, technological and social issues of our time without also examining the roots of our consciousness and our behavior?

Can we reconcile our spiritual and material natures? Can we discover a cultural unity underlying the diverse dogmas, religions, and political systems on our planet?

This book suggests that we have that potential. It details the progress of some who have dared to probe the roots of being. Let us now begin the journey of discovery together.

Introduction to the Original Edition

The title for *The Roots of Consciousness* was inspired by a statement by cosmologist Arthur M. Young who cautioned against seeking only the flowers of consciousness. Although flowers provide moments of pleasure and delight, they are forgotten after they wilt and die.

The flowers of consciousness are the exquisitely intriguing foliage blooming in psychology's borderland — telepathy, clairvoyance, precognition, psychokinesis, astral projection and other potential powers seemingly latent within us. These things may seem strange to Western humanity's current ways of thinking — or even non-existent. However, it is my intention in this book to explore the possibility that they are rooted in the essential core of our existence, in our cultural history and in our scientific knowledge.

For me, the exploration of consciousness really has its origins in sparks of wonderment at my own existence which recur many times in different ways. These are the simplest experiences underlying the science of consciousness — a newly emerging discipline which, like music, art, medicine or physical education, involves intense personal commitment as well as objective understanding.

One of most profound speculations on the origins of consciousness occurs in a hymn from the *Rig Veda*, written over three thousand years ago, in which the sages search their hearts for the personal, social and cosmic origins of being:

> Neither not being nor being was there at that time; there was no air-filled space nor was there sky which is beyond it. What enveloped all? Under whose protection? What was the unfathomable deep water?
>
> Neither was death there, nor even immortality at that time; there was no distinguishing mark of day and night. That One breathed without wind in its own special manner. Other than It, indeed, and beyond, there did not exist anything whatsoever.
>
> In the beginning there was darkness concealed in darkness; all this was an indistinguishable flood of water. That which, possessing life-force, was enclosed by the vacuum, the One, was born from the power of heat from its austerity.
>
> Upon It rose up, in the beginning, desire, which was the mind's first seed. Having sought in their hearts, the wise ones discovered, through deliberation, the bond of being and non-being.
>
> Right across was their dividing line extended. Did the below exist then, was there the above? There were the seed planters, there were the great forces of expansion. Below there was self-impulse, above active imparting.
>
> Who knows it for certain; who can proclaim it here; namely, out of what it was born and wherefrom this creation issued? The gods appeared only later — after the creation of the world. Who knows, then, out of what it has evolved?
>
> Wherefrom this creation has issued, whether He has made it or whether He has not — He who is the superintendent of this world in the highest heaven — He alone knows, or, perhaps, even He does not know.[6]

Let us examine the fundamental origins of being. There is the void, the absolute, darkness concealed in darkness, the unknown, that which is beyond. About the unknown void little can be said, although we say that this void permeates everything — including the most solid-appearing objects.

From this, according to the myth, through the power of heat was born the One. Regarding heat, the origin of the One, physics can shed some light. Heat is transmitted by particle-waves called photons in the infrared area of the electromagnetic spectrum. All electromagnetic interactions from radio waves and light to cosmic rays are mediated by photons. This idea perhaps correlates with other creation myths, including the version in Genesis, which states that light was the first manifestation from the void. Photons are the basic quantum unit of the action of electromagnetic radiation. They have no mass and no charge. They travel through space at 186,000 miles per second with no loss in energy until they collide with other particles.

Imagine that you are a photon traveling through space at the speed of light. If you were to look at your watch, you would, according to the theory of relativity, discover that time was standing still. Hence you could travel to the very edges of the known universe without aging a single day, although, to an observer on earth, it would take you three billion years to get there. Thus photons, tiny particle-waves with no mass, no charge, no time, neither matter nor antimatter, but with a unit spin, constitute one of the basic units of action in physics.

In the seventeenth century, the *principle of least action* was discovered to be true of light — and subsequently found to apply to almost all physical phenomena. This principle states that light always follows the path that gets it to its destination in the shortest possible time. Photons are also subject to Heisenberg's *principle of uncertainty* or indeterminacy. It is impossible to predict the destination of any given photon, which has led physicists to describe these wave-particles as "packets of uncertainty." Any given uncertainty packet is theoretically located everywhere in the universe, with the probability densities being greater for some particular space-time coordinates.

This idea of unpredictability represents a breakdown of the nineteenth-century notion of a mechanical determinism governing all of nature — including human consciousness. Now scientists are beginning to see that the process of observation itself influences the universe. Because this is problematic in the context of current theories, physicists are beginning to search for a new understanding.

Photons, like all known physical phenomena, pulsate or vibrate. Planck's law states that the photon's energy is directly proportional to the frequency of its vibration. The constant of proportionality between the energy of photons and the frequency of vibration is known as *Planck's Constant*, or *h*, which is a very important unit in describing wholeness as well as indeterminacy in physics. It, like the other constants in physics, is suggestive of the Pythagorean notion that the underpinnings of the physical universe are mathematical in nature.

A photon, when it is annihilated, is able to create particles of matter and antimatter which have both mass, charge and time, such as *electrons* and *positrons*, or *protons* and *antiprotons*. It is tempting to suggest that these particles with their charges represent the principle of desire or attraction, which in the *Vedic* myth arose from the One.

Protons and electrons, of course, are attracted to each other and form the basic constituents of atoms and molecules. Electrons have a negative charge and protons, which are much heavier, have a positive charge.

Positrons and antiprotons are particles of what is called antimatter. In an atom of antimatter, the light weight positrons orbit around a nucleus that contains negatively charged antiprotons. When particles of matter and antimatter come into contact with each other, they are annihilated and photons are produced. Physicists suggest that particles of antimatter move backward through time — like a movie played backward.

As the extreme energy of the photons becomes somewhat solidified in the form of the mass of protons and electrons and their antimatter counterparts, the amount of free energy these particles possess is accordingly reduced. Also the amount of indeterminacy, or unpredictability of these particles, while still great, is less that of the photon. According to Heisenberg's uncertainty principle, the product of uncertainty of the position and momentum of these particles is equal to or greater than Planck's Constant, h. In other words, the more certain you are of the position of a particle, the less certain you can be of its momentum, and vice versa.

Of course, protons and electrons combine to form atoms which have even greater mass and less indeterminacy. There are 106 known different kinds of atoms whose identity is determined by the number of charged particles that have come together. These atoms compose the elements of the periodic table that combine chemically into molecules to form all of the substances we experience in our day-to-day living. The indeterminacy exhibited by atoms and molecules is limited to the amounts of energy they can absorb and release and the times at which they release and absorb this energy, which is in the form of photons. This indeterminacy though very small is real. Consequently, scientists no longer believe that atoms and molecules generally behave exactly like the predictable billiard balls of nineteenth-century physics.

The particles, atoms and molecules of the universe combine to form the stars and the planets and newly discovered fantastic structures in outer space whose origins and properties are still mysterious to us. The expansion of the universe, the nature of the black holes and the nature of quasars all imply notions of time, space and matter foreign to the rules of classical Newtonian physics that generally apply in daily life.

The small amount of uncertainty that remains in molecules may play an important role in the curious growth displayed by properties polymers. These long molecular chains —such as rubber, cellulose and nylon — seem to anticipate the growth of

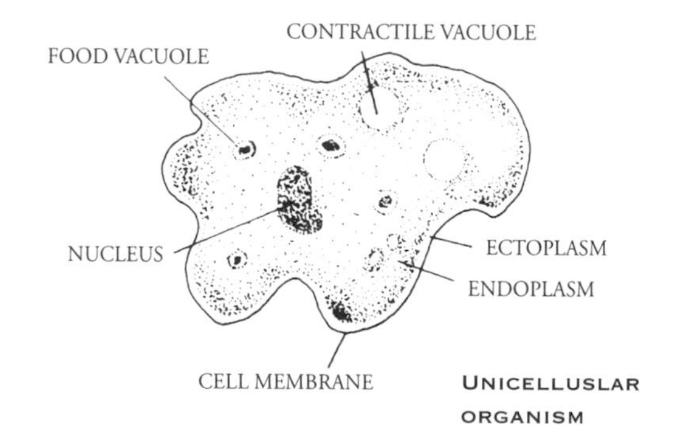

Unicelluslar organism

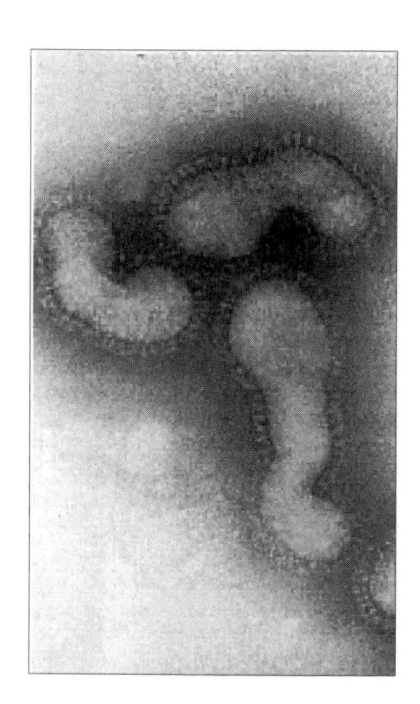

Influenza Virus

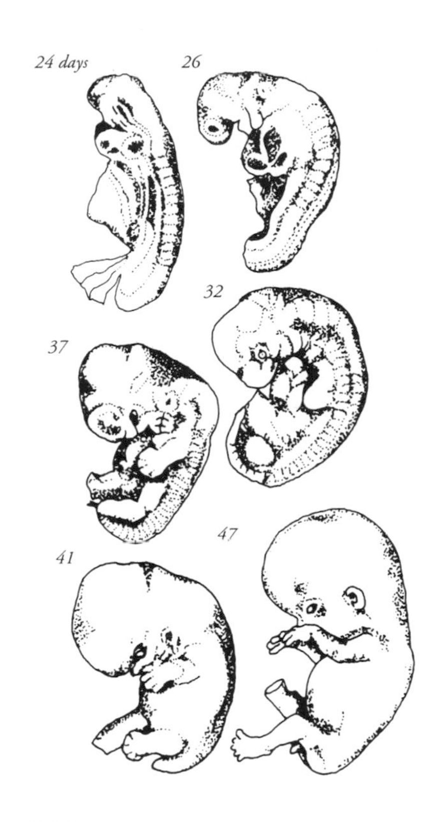

Stages in the growth of a human foetus.

cellular life itself. Functional polymers such as proteins are the primary constituents of animal life, and the proteins actin and myosin, the primary ingredients of muscle tissue, exhibit properties of animal mobility.[7]

Perhaps the most important molecule of all is deoxyribosenucleic acid, or DNA. These complex molecules contain in their double-helix structure the information necessary for living cells to grow and function. Some scientists say that these molecules contain within their structure all of the information necessary for the complete development of an organism, such as the human being. This has yet to be proven; however, a single DNA molecule can store much more information than is contained in this book. When cells divide, DNA molecules also divide in two and are able to reproduce themselves. DNA molecules are able to transcribe the information they contain onto other molecules of ribonucleic acid or RNA.

Certain very complex molecules of DNA or RNA combined with protein, called viruses, actually seem to be alive and can reproduce themselves when they are inside of another living organism. However, these viruses are quite inert in the free state.

Although science has not yet filled in all of the links in the process of evolution, one can sense a naturalness in the emergence of cellular life from a sea of complex molecules. Unicellular organisms constitute the majority of living creatures on earth. Microscopic structures within these cells, called organelles, perform the digestive, respiratory, metabolic and reproductive functions of the organism. While the cells are said to be alive (although this is perhaps questionable in the case of the virus), the *organelles* themselves are not.

When similar kinds of cells group together, *colonies* are formed — such as fungi or algae or sponges. Several different types of cells coming together lead to the formation of different tissues within each organism. In more complex creatures, these tissues join further into units called *organs.* The most complex organisms contain not only many tissues and organs, but groups of organs may also form one or more structural organizations called *organ systems.*

Ontogeny recapitulates phylogeny. This means that the growth of any individual organism, from conception, follows the same pattern of development as the evolution of that species. Thus, when you were an embryo you passed through all the stages of growth in a nine-month period that led to humanity's three-billion-year evolution from a single-celled organism.

Beyond organisms of all kinds, higher levels of life may be distinguished in the form of social groupings such as *families' hives, tribes, societies, populations, species*, and local *communities.* The sum of all living communities represents the ecological system of the planet, which in turn interacts with the solar system, galaxy and, indeed, the known universe.

Human cultural history goes back about five thousand years; however, the existence of the *homo sapiens* species can be traced back at least 500,000 years. Thus in some sense it can be said that humanity has evolved through qualitative changes of consciousness during the life of our species. Human beings seem to exhibit free will, a phenomenon reminiscent of the unpredictability displayed by photons and subatomic wave-particles.

From the evolution of the universe we have just briefly traced, there seems to emerge patterns, pulsations, vibrations and cycles. The loss of uncertainty, the entrapment of spirit in matter as we descend from the photon to the molecule, seems to be balanced by the increased freedom, the rise of matter into spirit, as we ascent through the plant, animal and human kingdoms. The orchestration of the universe is a most complex and subtle symphony. This book, this moment, is a small part of the harmony of the universe as it comes to understand itself.

THE ROOTS OF CONSCIOUS-NESS

The History of Consciousness Exploration

Shamanistic Traditions

The peoples of prehistoric times and primitive cultures have laid the groundwork for modern consciousness exploration. Our knowledge of these groups comes from archeological or anthropological observation. In some cases, researchers have lived for long periods in the wilderness with primitive peoples. We will find that return to the wilderness has been used throughout history to explore the deeper layers of the psyche.

The way of perceiving the world that emphasizes the existence of spirits, ghosts, and gods who interact with men and inhabit objects is called *animism.* Animism characterizes virtually all primitive and ancient cultures.[1] In many languages, the word for spirit is also the word for breath — which leaves the body at death. Spirits could occupy the bodies of living men and animals causing either illness or insanity, but they often imparted higher wisdom. Psychic powers were ascribed to aid from such spirits. Also commonly found in primitive cultures is the correlative belief in a general spiritual force, or *mana*, permeating all of nature.

Evidence from cave art, dating back at least thirty thousand years, suggests that caves were used for magical ritual purposes. In certain cases it must have been necessary to crawl for hours through the caves in order to reach the locale of the artwork and related artifacts. It may be that solitude inside such a cavern was an initiation technique used to explore the inner realms of being. Markings on antlers and bones indicate that people made notations of the phases of the moon as long as 30,000 years ago and suggest that the cave rituals and other cultural practices had a seasonal or periodical orientation.[2] It has been suggested that prehistoric people may have been sensitive to different phases in the lunar cycle as special times for meditation.[3] The monumental

Stonehenge, built in prehistoric England, is oriented towards equinoxes, solstices and lunar eclipses during the equinox and suggests similar usage.[4]

The leaders in such practices and rituals were called shamans by anthropologists. They were the earliest professionals. They mediated between the inner life of the tribe and its external affairs. They presided at all "rites of passage" such as births, puberty initiations, marriages, and deaths as well as all "rites of intensification" which attempt to strengthen the tribe's relation with powerful natural forces in times of crisis such as famine, storm, and epidemic.

The role of the shaman varied from culture to culture and with different circumstances. In some cultures shamanic ideology, technique and ritual dominated social interaction while in others it constituted a secondary influence. The word *shaman* itself has a Sanskrit origin, and means ascetic.

The shaman's power essentially lies in mastering the ecstatic techniques of dreams, visions, and trances. Ecstasy in its original sense meant an altered state of consciousness with an awareness of the single emotion rapture. The shaman also mastered the traditional mythology, genealogy, belief system and secret language of the tribe as well as its healing methods. The youth who are called to be shamans attract attention through their love for solitude, desire to roam in the woods or in unfrequented places, visions, and spontaneous song-making. Sometimes they enter trance-like states which make them unconscious. These signs are regarded with pleasure and awe by the tribespeople who generally believe that the soul of the shaman is being carried away by spirits to a place where he is instructed, sometimes by his shaman ancestors, in the secrets of the profession.

In some cultures the behavior of the prospective shaman is described in terms that seem to indicate psychopathology. However, it is precisely because they succeed in curing themselves that these individuals become shamans. Often a crisis bordering on madness is provoked in future shamans by the sudden announcement to others in the tribe that they have been chosen by the spirits for this profession. In other cases this initiatory sickness is induced by the use of drugs or fasting and other austerities. Regardless of the means, the symbolic pattern of death and rebirth common to all initiation rites will be reenacted.

The initiatory rituals peculiar to Siberian and central Asian shamanism include a ritual series of waking dreams. During this ritual Siberian shamans maintain that they "die" and lie inanimate for from three to seven days in a tent or other solitary place:

> Imagine having your body dismembered by demons or ancestral spirits; your bones cleaned, the flesh scraped off, the body fluids thrown away, and your eyes torn from their sockets, but set aside so that you may watch the entire procedure. It is only after such a purgative experience that shamans are said to obtain the powers of shamanistic healing. Then they are given new flesh and the spirits instruct them in magical arts. They experience the gods of the heavens; they learn to find the souls of sick men who have wandered or been carried by demons away from their bodies. They learn how to guide the soul of the dead to their new abode; and they add to their knowledge by regular association with higher beings.

Shamans are said to "die" and return to life many times. They know how to orient themselves in the unknown regions they enter during their ecstasy. They learn to explore the new planes of existence their experiences disclose.[5]

It is standard anthropological knowledge that shamanistic systems are similar in places like Tierra del Fuego, at the tip of South America, or in Lapland in northern Europe, or in Siberia or Southern Africa. The Australian aboriginals have the same system, basically, and they were separated thirty or forty thousand years from other peoples.

A unique approach to shamanism is that of Michael Harner, author of *The Jivaro, Hallucinogens and Shamanism*, and *The Way of the Shaman*. Dr. Harner is a former professor of anthropology at the New School for Social Research, and is currently acting as the director of the Center for Shamanic Studies. He is actively involved in teaching Westerners how to live and practice as shamanic healers. In my *Thinking Allowed* interview with him, he responded to an old anthropological position that shamans were mentally imbalanced:

> There was this tendency to feel that shamans were psychotic individuals — in other words, crazy — but they had the good fortune to live in crazy cultures, i.e., cultures other than our own, which of course is very sane. There also was the point of view that they were fakers — that when they claimed to go into a trance, which is the word they often used in the literature, that they could not possibly be going into the trance and having these experiences they claimed to do. I have run into ethnologists from Germany, from Russia, who . . . stuck pins or burning embers under the skin of shamans while they were in a trance, to see if they were really faking or not. There was this kind of skepticism.

I responded by citing Mircea Eliade who wrote that even if shamans were crazy, it was uncanny how they could dance all night long, and maintain this incredible level of energy. Harner's reply took us right to what he maintains is the heart of shamanism:

> The person who is doing this work is drawing upon an experience of power far beyond himself or herself. Shamanic ecstasy, where one is having ineffable experiences, make living very worthwhile. Such experiences are connected with helping others and working in harmony with nature.
>
> The world of the shaman is one of cosmic unity and a sense of love. In shamanic ecstasy, tears of joy exist. It is the same world of the Christian mystics of the medieval times. It is the same world of the great Eastern saints.[6]

MICHAEL HARNER (FROM *THE WAY OF THE SHAMAN* VIDEOTAPE, COURTESY THINKING ALLOWED PRODUCTIONS)

Evidence suggests to us that ancient shamans possessed a very detailed knowledge about the use of a wide variety of mind-altering drugs. The earliest religious literature of India points to prehistoric use of a mythical, or at least undiscovered, drug called soma for inducing contact with nature's innermost forces. The primitive tribes of Central and South America are known for their ritualistic use of drugs such as yage, peyote and a number of others for the purpose of inducing ecstatic experiences. At times ecstasy is induced through drum rhythms and night-long dancing.[7]

There is also reason to believe that ancient shamans engaged in practices that could be considered the prototypes of modern systems of yoga and meditation.

It was commonly thought that while in these various altered states of consciousness the shaman had the ability to diagnose diseases, see into the future, see objects at a great distance, walk over hot coals, and speak to the spirits of the dead.

Accounts of this type are all too common in the stories researchers and explorers bring back.[8] However, a most promising line of research into the nature of such oddities was for a time taken up by one unusual young man, unfortunately now deceased. Adrian Boshier, of the Museums of Science and Man in Johannesburg, South Africa, used an approach that combines living off the land in the African wilds while receiving initiations as a *sangoma* or witch doctor with the objective work of a scientist studying other shamans.

In his explorations, Boshier discovered 112 previously unknown prehistoric cave paintings whose ritual function and value has been preserved in the secret oral traditions of the witch doctors whose friendship and trust he has cultivated. He discussed one such encounter:

> Upon arrival at a village some ten miles away, we were directed to one of the mud huts where we found the woman, Makosa, sitting on the floor amongst bones, dice and shells — her instruments of divination. Completely unperturbed by the arrival of our party, she did not even look up, but continued studying and rethrowing the bones. Eventually she spoke. "One of you is here to ask me questions, he has a head full of questions, he is not a man of this land, but comes from over the big water." Then ignoring the others, she looked directly at me and asked, "What do you want?" I chided her and in the traditional manner told her to inquire this of her spirits. Again she picked up her bones, blew on them, and cast them down. She repeated this process three times, studying carefully the pattern between each throw. After some time she picked up a small knuckle bone and said that this bone represented me. It was the bone of the impala. The impala ram is an animal that lives with its herd most of the time, she told me, but periodically it leaves its group and goes off into the wilds by itself. It always returns to its herd, but again it must leave to wander alone. "This is you," she said, "you live with your people, but sometimes you go into the bush alone. You go out to learn, living in the wild places, the mountain, the desert.... This is your life's

Boshier and a few of the forty sangoma of the Komi tribe in North Western Transvaal, attending an initiation ceremony, January 1975 (courtesy Adrian K. Boshier).

A female trainee (on right with bladders in hair) is divining the whereabouts of hidden objects to prove her powers. The sangoma on the left who is testing her is one that Boshier had been studying for several years. The bladders in the hair of the initiate have been taken from sacrificial animals and have been filled with the breath of her teacher. Today this girl is being "reborn" as a sangoma. Swaziland 1972 (courtesy Adrian K. Boshier).

During the ceremonies at which animals are sacrificed to the ancestral spirits each sangoma is possessed by those spirits who talk through them — thanking the host, praising the spirits, giving advice, etc. The woman in the right foreground is in the midst of this activity. Jackson, Transvaal, 1971 (courtesy Adrian K. Boshier).

A sangoma studying her bones (taula) of divination. Her patients look on as she studies the pattern of the bones, shakes the rattle in her hand and calls upon her ancestors to help her "see." The taula are composed of ivory tablets with symbols engraved upon them, seashells, knucklebones, tortoise shells, seeds and coins (courtesy Adrian K. Boshier).

work. What you learn is what the spirits are teaching you. This is the only way."

The old woman continued throwing the bones and revealing personal details concerning my life, which were absolutely accurate....Such an existence taught me much about the country and its wildlife, but probing the customs and beliefs of the people proved to be a far more difficult and lengthy undertaking, for the historians and spiritual leaders of African society are the witchdoctors — people like Makosa who tell only so much, whose revelations are very limited to the uninitiated.[9]

These innate visions, so familiar to primitive peoples, are the heritage of humanity and have been preserved in various forms within all cultures.

Ancient Mesopotamia

In ancient Mesopotamia the art of divination received more intense, sustained interest than in any other known civilization. Reading omens was particularly important since every event was thought to have a personal meaning to the observer. This attitude was scientific in that it stressed minute observation and description of phenomena. However, causality was not an important notion — all events were seen as communications from the divine.[10]

The Mesopotamians were specialists in the arts of prescience, predicting the future from the livers and intestines of slaughtered animals, from fire and smoke, and from the brilliancy of precious stones; they foretold events from the murmuring of springs and from the shape of plants. Trees spoke to them, as did serpents, "wisest of all animals." Monstrous births of animals and of men were believed to be portents, and dreams always found skillful interpreters.

Atmospheric signs, rain, clouds, wind and lightning were interpreted as forebodings; the cracking of furniture and wooden panels foretold future events. Flies and other insects, as well as dogs, were the carriers of occult messages.[11]

Babylonian Winged God Marduk

Mesopotamia was noted throughout the ancient world for its magi — men and women for whom nothing was accidental. They also saw a unity in nature and harmony in the universe which bound together all objects and all events. The Assyrians made accurate observations of stellar movements and developed mathematical formulas to predict heavenly events. Omens were often interpreted through paranomastic relations — puns and plays on word — between the ominous portent and its consequence.

Dream Portents

The idea is expressed in Mesopotamian literature, that the soul, or some part of it, moves out from the body of the sleeping person and actually visits the places and persons the dreamer "sees" in his sleep. Sometimes the god of dreams is said to carry the dreamer.

In times of crisis, ancient kings, priests, or heros would spend the night in the inner room of the sanctuary of a god. After due ritual preparation, the god would appear to the dreamer and give him a very clear and literal message which would require no further interpretation.

The Assyrian king, Assurbanipal, 668-626 B.C., recounted this incident in an ancient dream-book:

> The army saw the river Idid'e which was at that moment a raging torrent, and was afraid of crossing. But the goddess Ishtar who dwells in Arbela let my army have a dream in the midst of the night addressing them as follows: "I shall go in front of Assurbanipal, the king whom I have created myself." The army relied upon this dream and crossed safely the river Idid'e.

This dream seems to have been reported simultaneously by many sleepers.[12]

Bad dreams dealing with sexual life or tabooed relationships were thought of as diseases caused by evil demons rising from the lower worlds to attack people. Their contents were rarely mentioned for fear of causing increased entanglements. One intriguing technique used to obviate the consequences of such a dream was a practice of telling the dream to a lump of clay that is then dissolved in water.

Ancient India

Contemporary Hindu culture originated primarily with the Aryans who invaded India about 1500 B.C. bringing with them the Sanskrit language and the *Vedic* religion. However, for at least a thousand years prior to this invasion, there existed a culture in India about which we know very little.

The cities of the Indus river valley left no large monuments and although they did have a written language, it has not yet been translated. From fragmentary evidence that does remain, scholars conclude this early culture contained within it many elements later incorporated into the Hindu religion.[13]

The Language of Consciousness

The exploration of consciousness has developed to a remarkable degree in the Hindu culture. In fact, the Sanskrit language has shown itself to be sufficiently precise in describing the subtleties of consciousness exploration that many Sanskrit words, with no adequate English equivalents, have become commonplace in our own contemporary culture. Consider for example these terms:

ahimsa: The doctrine of non-violence toward sentient beings.

akasha: The ether; primordial substance that pervades the entire universe; the substratum of both mind and matter. All thoughts, feelings, or actions are recorded within it.

asanas: Postures used to stimulate flow of life-force through the body and to aid meditation.

atman: The human soul or spirit — the essence of the inner being.

Brahman: Hindu god who represents the highest principle in the universe; the essence that permeates all existence. *Brahman* is the same as *atman* in the philosophy of the *Upanishads.*

dharma: One's personal path in life, the fulfillment of which leads to a higher state of consciousness.

dhyana: The focusing of attention on a particular spiritual idea in continuous meditation.

guna: A cosmic force or quality. Hindu cosmology maintains the universe is composed of three such qualities: *satvic,* meaning pure or truthful; *rajasic,* meaning rich or royal; and *tomasic* meaning rancid or decaying.

Ishwara: Personal manifestation of the supreme; cosmic self; cosmic consciousness.

karma: The principle by which all of our actions will affect our future circumstances, either in the present or in future lifetimes.

mandala: Images used to meditate upon.

mantras: Syllables, inaudible or vocalized, that are repeated during meditation.

maya: The illusions the physical world generates to ensnare our consciousness.

moksha: The attainment of liberation from the worldly life.

nirvana: The transcendental state that is beyond the possibility of full comprehension or expression by the ordinary being enmeshed in the concept of selfhood.

ojas: Energy developed by certain yogic practices that stimulates endocrine activity within the body.

prana: Life energy that permeates the atmosphere, enters the human being through the breath, and can be directed by thought.

pranayama: Yogic exercises for the regulation of the breath flow.

sadhanas: Spiritual disciplines. Practical means for the attainment of a spiritual goal.

samadhi: State of enlightenment of superconsciousness. The union of the individual consciousness with cosmic consciousness.

samsara: The phenomena of the senses. Attachment to *samsara* leads to rebirth.

siddhis: Powers of the soul and spirit that are the fruits of yogic disciplines.

soma: A plant, probably with psychedelic properties, that was prepared and used in ritual fashion to enable men to communicate with the gods.

tantras: Books dealing with the worship of the female deities and specifying certain practices to attain liberation through sensuality, particularly through the heightened union of male and female energies.

yoga: The Sanskrit word meaning union and refers to various practices designed to attain a state of perfect union between the self and the infinite.[14]

The Discipline of Yoga

The *Yoga Sutras of Patanjali* prescribe a system of eight stages, or limbs, for one's higher development. The first two limbs are known as *yama* and *niyama.* They involve a highly ethical and disciplined lifestyle — control, indifference, detachment, renunciation, charity, celibacy, vegetarianism, cleanliness, and non-violence. The third step involves the development and care of the body through the use of exercises and postures called *asanas.* The fourth stage involves *pranayama* breathing exercises. The next stage, *pratyahara,* involves meditation, by means of which one withdraws consciousness from the senses.

The sixth limb of yoga is called *dharana* which means concentration. An object of contemplation is held fixedly in the mind; it must not be allowed to waver or change its form or color — as it will have a tendency to do. Often the yogi will concentrate on different *chakras* or focal points within the body. Self-analysis is used to observe breaks in concentration. Often he will carry a string of beads and one is pulled over the finger every time a break begins. The next stage of *dhyana* occurs when the sense of separateness of the self from the object of concentration disappears and one experiences a union or oneness with that object. In the final stage, *samadhi,* one experiences an absolute, ecstatic cosmic consciousness. This does not, as some suppose, entail a loss of individuality. "The Drop is not poured into the Ocean; the Ocean is poured into the Drop." The self and the entire universe are simultaneously experienced.[15]

> When one is so far advanced that every shadow and every echo has disappeared, so that one is entirely quiet and firm, this is refuge within the cave of energy, where all that is miraculous returns to its roots. One does not alter the place, but the place divides itself. This is incorporeal space where a thousand and ten thousand places are one place. One does not alter the time, but the time divides itself. This is immeasurable time when all the aeons are like a moment. (*The Secret of the Golden Flower,* translated by Richard Wilhelm).

In the past decades, Western scientists have begun to study the abilities yoga practitioners can achieve. Body functions such as heartbeat, temperature, and brainwaves, that which had been previously thought of as totally autonomic, have been shown to be under the conscious control of some yogis. This research has paved the way for the emerging science of consciousness which is discussed in Sections III and IV.

Ancient China

Taoism

Lao Tsu was born in China in 601 B.C. He held a number of public offices during his long life and was the curator of the royal library in Loyang. As an old man he retired from government service and traveled on buffaloback to the region of the Gobi desert. Because of a boundary warden's plea, the sage paused long enough to inscribe a short record of his teachings. Leaving this record, he resumed his journey to an unknown destination and was seen no more. His small book, the *Tao Te Ching*, is one of the world's great religious classics. Within a matter of centuries, Lao Tsu was worshipped as a god and his masterpiece was engraved on stone at the capitol of every Chinese state. Is it possible to describe the way of the *Tao* in this book? Though thousands of volumes have been written about it, Lao Tsu himself states that "The Tao that can be put into words is not the eternal Tao." We can only try to outline and summarize:

There is a perfect balance that lies within each individual, and following this balance requires neither cunning nor striving. A story is told of King Wen's wonderful cook who has used the same knife to cut meat for over twenty years without every having to sharpen his knife, because he spontaneously moves his blade through the meat precisely and without effort or hacking. This expression of the *Tao* as effortless attainment is referred to as *wu-wei.*

Wu-wei has been translated to mean noninterference, non-doing, action without deeds, or actionless activity. The Taoists did not mean that one should never act, but that one should be fluid and changing so as to always adjust one's self to circumstances.[16] This is why it is said:

> What is of all things most yielding (water)
> Can overwhelm that which is of all things most hard (rock).
> Being substanceless it can enter even where there is no space.
> That is how I know the value of action that is actionless.
> But that there can be teaching without words,
> Value in action that is actionless,
> Few indeed can understand.[17]

Like the yogis in India, the Taoists developed exercises enabling them to gain conscious control over internal states. Disciplined breathing constituted one basis of these exercises. Meditation was also used in order to separate the spirit from the body and travel independently of it as well as to maintain an eternal calm in the midst of changing conditions.[18]

Qi energy was said to be the product of the two great forces of *yin* and *yang* which run through the entire universe. All phenomenal manifestation exists as a tension between these two principles, which operate on all levels.

Yin	**Yang**
feminine	masculine
negative	positive
moon	sun
darkness	light
yielding	aggressive
wetness	dryness
left side	right side
heart	fire
lungs	intestines
spleen	stomach
kidneys	bladder
autumn	spring
winter	summer

Ancient Greece

Mystery Traditions

An important part of ancient Greek culture were the mystery cults, into which many Greek philosophers were initiated. These cults developed impressive rituals that involved fasting, purification, song and dance, the use of mythology and poetry. It is said that Greek drama developed from these rituals. Many authors have written about the enormous impact these initiations have had upon their understanding — although the specific nature of the rituals has still remained a secret. Doubtless, profound states of consciousness developed.

The rituals of the Eleusinian mysteries dealt with the myth of Persephone, the daughter of Ceres, who was abducted by Pluto, Lord of the Underworld, who forced her to become his queen. Ceres entered the Underworld in search of her daughter and at her request Pluto agreed to allow Persephone to live in the upper world half of the year if she would stay with him in the darkness of Hades for the remaining half. Of this ritual, Manley Palmer Hall states:

> It is probable that the Eleusinians realized the soul left the body during sleep, or at least was made capable of leaving by the special training which undoubtedly they were in a position to give. Thus Persephone would remain as the queen of Pluto's realm during the waking hours, but would ascend to the spiritual worlds during periods of sleep. The initiate was taught how to intercede with Pluto to permit Persephone (the initiate's soul) to ascend from the darkness of his material nature into the light of understanding.

The fact that initiates maintained they had conquered the fear of death leads one to surmise that these rituals — akin, perhaps, to the cave rituals of prehistoric humans — developed a state we now call the out-of-body experience.

Pythagoras, Plato and other Greek philosophers were said to have been initiated into these cults — which are said to have originated in Egypt.[19] The tradition of Hermetic mysticism also claims an origin in the legendary Egyptian-Greek god-sage *Hermes Trismegistus.*

HERMES TRISMEGISTUS

Oracles

The first recorded controlled parapsychological experiment took place in ancient Greece during the sixth century B.C. Greece was at that time famous for its oracles which were generally connected with the temples of the various gods. Generally these oracles operated through a priestess or medium who went into a trance or became possessed by the god of the oracle and uttered prophetic words which were then interpreted by the priests. Their enormous prestige and political influence was attested to by kings and generals who would consult with these oracles before making major decisions.

Herodotus, the Father of History, reports that the King of Lydia, Croesus, wishing to test the different oracles, sent messengers to those of Aba, Miletus, Dodona, Delphi, Amphiaraus, Trophonius, and Jupiter Ammon. His idea was by this means to choose the best of them to consult about his proposed campaign against the Persians. On the hundredth day after their departures all of the messengers were to simultaneously ask the oracles to tell them what Croesus was doing at that very moment. Accordingly, on the day appointed, when the emissaries had entered the temple of Delphi, even before they had time to utter their mandate, which had been kept secret, the priestess said in verse:

I count the grains of sand on the ocean shore
I measure the ocean's depths
I hear the dumb man
I likewise hear the man who keeps silence.
My senses perceive an odor as when one cooks
together the flesh of the tortoise and the lamb.
Brass is on the sides and beneath;
Brass also covers the top.

This reply was committed to writing and rushed back to Croesus who received the lines of the priestess with utmost veneration. On the appointed day he had sought for something impossible to guess: having caused a tortoise and lamb to be cut into pieces, he had had them cooked together in a brass pan upon which he had afterwards placed a lid of the same metal. The oracle of Amphiaraus also proved lucid in this experiment; others were less definite. The presents that Croesus sent to Delphi were of incalculable value. A detailed list may be found in Herodotus.[20]

Out of this cultural milieu developed a philosophical tradition that was *hylozoistic*, conceiving of nature as animated or alive; *ontological*, inquiring into the very essence of things; and *monistic*, seeking to find a single principle to explain all phenomena. Now, we will explore the theories of mind and consciousness promulgated by the ancient philosophers. Note that their teachings consistently emphasized a unity between the goals of philosophy and the practices of living. Such a unity of thought and action is sadly deemphasized in the contemporary quibbling of much modern academic philosophy.

Pythagoras

Pythagoras — a great mystic who is also regarded as the father of the Western scientific tradition — did not begin to teach until the age of forty. Until that time he studied in foreign countries with the resolution to submit to all of his teachers and make himself a master of their secret wisdom. We are told that the Egyptian priests with whom he studied were jealous of admitting a foreigner into their secrets. They baffled him as long as they could, sending him from one temple to another. However, Pythagoras endured until he was rewarded accordingly for his patience. Later on he was no less strict in dealing with his own disciples.

Pythagoras proposed philosophy as a means of spiritual purification. He suggested the heavenly destiny of the soul and the possibility of its rising to union with the divine. He claimed that he could personally hear the "music of the heavens," and he generally allowed his followers to believe that he was the Hyperborean Apollo and that he assumed a human form in order to invite men to approach him. When one examines the scope of his profound scientific and philosophical contributions, these metaphors seem almost apt. Twenty-five hundred years later we can recognize Pythagoras as a being of rare visionary power.[21]

Scholars suggest that he incorporated the ecstatic practices of the Orphic mystery cult, known for its raw emotional manifestations — which he attempted to reform with his emphasis on knowledge and science as a path to salvation.[22] During the novitiate period of five years, a vow of strict silence was required. Only then were disciples permitted to participate in intellectual discourse with the master. Members of the Pythagorean brotherhood held their goods in common, made fidelity chief virtue and held that the best should rule.

That Pythagoras was indeed a shaman is a credible notion. Among the many threads woven into the fabric of his philosophy are the shamanistic cult practices of Thracian medicine men. He also claimed to have full memory of the many forms he had taken in past incarnations.

The Pythagorean brotherhood formed a direct link between the mystery school tradition and the development of Greek philosophy. Pythagoras is credited with inventing the Western musical scale and the theorem by which the length of the hypotenuse of a right triangle is derived. His philosophy was looked upon as mystical since he stressed the harmonious development of the soul within humanity. (He also taught the doctrine of the transmigration of the soul through successive incarnations.) Relationships between all things could be expressed by numbers which took on qualitative properties that were analogous to the qualitative differences found in musical harmonies. The Pythagoreans devoted themselves to studying the countless peculiarities discoverable in numbers, and ascribed these to the universe at large.

Numbers were thus seen as a principle that linked the symbolic properties of the mind with the mechanisms of the universe.[23] It is precisely this Pythagorean notion that forms the backbone of the theoretical models of consciousness which will be presented in Section IV and, in greater detail, as a work-in-progress in the Appendix. For the most ancient of all philosophical-scientific traditions relating consciousness and reality may yet prove the most fruitful. If so, it would not be the first time that Pythagorean concepts have demonstrated extraordinarily penetrating insights. Pythagoras was the first to suggest that the earth was a spherical planet orbiting the sun. The origin of human consciousness was understood in the context of this astronomy.

One of the most influential concepts developed by the Pythagoreans was the notion of the *harmony of the spheres*, which related the inner states of the mind to a contemplation of these celestial spheres.

> Before the beginning of reflective thought, man feels, in various contexts, an involvement. He unconsciously arranges the multiplicity of phenomena into a restricted number of schemata. It is the business of reflection, when it begins, to raise these transitory insights into the realm of

PYTHAGORAS

consciousness, to name them, and to assimilate them to one another. This is how the world becomes comprehensible. In myth and ritual man tries to make these realizations present and clear, to assure himself that, in spite of all confusion and all the immediate threats of his environment, everything is "in order." It is in such a prescientific conception of order that the idea of cosmic music has its roots; and number speculation springs from the same soil.

But relationships that usually have their effect unconsciously, or only enter consciousness as the result of slow and patient reflection, become immediate, overwhelming experiences in ecstasy. The soul that in ecstasy, or dream, or trance, travels to heaven, hears there the music of the universe, and its mysterious structure immediately becomes clear to him. The incomparable and supernatural sound is part of the same thing as the incomparable beauty and colorfulness of other worlds. If Pythagoras was something like a shaman, who in ecstasy made contact with worlds "beyond," then the tradition that he personally heard the heavenly music surely preserves something of truth.[24]

Democritus

In contrast to Pythagoras, Democritus, the originator of atomic theory, was more concerned with the substance of the universe than with its form. He maintained that the soul is composed of the finest, roundest, most nimble and fiery atoms. These atoms cannot be seen visually, but can be perceived in thought. At death, Democritus maintained in a book called *Chirokmeta*, that soul molecules detach themselves from the corpse, thus giving rise to specters. Through this theory, Democritus also attempted to explain dreams, prophetic visions, and the gods.

Democritus held that objects of all sorts, and especially people continually emitted what he termed *images* — particles on the atomic level that carried representations of the mental activities, thoughts, characters and emotions of the persons who originated them. "And thus charged, they have the effect of living agents: by their impact they could communicate and transmit to their recipients the opinions, thoughts, and impulses of their senders, when they reach their goal with the images intact and undistorted." The images "which leap out from persons in an excited and inflamed condition," yield, owing to their high frequency and rapid transit, especially vivid and significant representations.[25]

Socrates

Socrates, who left no writings of his own, remains an enigma. Regarded by many as possibly the wisest man who ever lived, he has developed a more recent reputation as something of a protofascist, rabble rouser.[26] I see him as a role model for the exploration of consciousness presented herein because he spoke freely about his experiences. His honesty was profoundly respected by all who knew him — in fact, his life was a model of integrity which inspired many philosophers. The Socratic injunction of *Know Thyself* provides a basic impulse for the work in this book.

Socrates' life bridged the gap between the spirit and the intellect. Most people would agree that he is one of the greatest philosophers; yet he never left any writings whatsoever. Jacob Needleman, professor of philosophy at San Francisco State University and author of *The Heart of Philosophy,* expressed the essence of Socrates' life and teaching in a *Thinking Allowed* television interview with me:

> He was a philosopher in the original sense of the term, which is a lover of wisdom. That is what the word means — to love, to seek wisdom. Wisdom is a state of the whole human being. A person who is wise not only knows the truth, but can live it.

JACOB NEEDLEMAN (FROM *SPIRITUALITY AND THE INTELLECT* VIDEOTAPE, COURTESY THINKING ALLOWED PRODUCTIONS).

> The philosopher Socrates sought to be wise, not simply to know facts and propositions and ideas.
>
> We know he stood for this inquiry, the development of the soul, the self — what we would call the deep self today. He was always trying, according to legend, to engage people in this kind of exchange which I call real inquiry. That was his life, and that was his only aim. People came to him. Many people were shocked by it, offended because he blasted their opinions away. He made them see they did not know what they thought they knew. That is the precondition for real learning.[27]

Socrates, who himself was apparently gifted with precognitive perception, attributed his abilities to the aid of a personal *daemon*, which then meant demigod and not (evil) demon. In the *Theagetes*, Plato makes Socrates say:

> By favour of the Gods, I have, since my childhood, been attended by a semi-divine being whose voice from time to time dissuades me from some undertaking, but never directs me what I am to do. You know Charmides the son of Glaucon. One day he told me that he intended to compete at the Nemean games. I tried to turn Charmides from his design, telling him, "While you were speaking, I heard the divine voice. Go not to Nemea." He would not listen. Well, you know he has fallen.

In his *Apology for Socrates*, Xenophon attributes to him these words:

> This prophetic voice has been heard by me throughout my life: it is certainly more trustworthy than omens from the flight or entrails of birds: I call it a God or daemon. I have told my friends the warnings I have received, and up to now the voice has never been wrong.

Plato

Plato, a student of Socrates, was greatly concerned with exploring consciousness. Plato maintains that the world of ideas is just as real as the world of objects, and that through ideas humanity attains consciousness of the absolute.

He compares our human condition to that of slaves enchained in a cave where one can only see the shadows of people and objects passing by outside. Eventually we come to accept these shadows as reality itself, while the source of the shadows is ignored. Such is our ignorance of the spiritual world in which our own ideas have their being.[28]

PLATO

Plato also makes a distinction between mere augury, which tries to comprehend the workings of God through a rational process, and genuine prophecy — such as that instanced by Socrates — which utilizes inner voices, out-of-body experiences, inspired states, dreams and trances.

Like Pythagoras, Plato admits the pre-existence of the *nous*, or divine soul of humanity, which chooses the existence for which it must incarnate. It survives the death of the body, and if it has not attained sufficient perfection to merit endless bliss, it must be subjected to new tests by reincarnating in order to attain further progress and perfection. In many other respects, particularly his emphasis on mathematics, Plato can be viewed as teaching in the Pythagorean tradition.[29]

In the last chapter of his masterpiece, *The Republic*, Plato vividly describes a vision of life after death attained by a young man named Er who was wounded in battle and thought to be dead. While this story is clearly augmented by the moralistic philosophy and cosmological views prevalent in his culture, Plato's account carries a ring that echoes the after-death mythology of many different visionaries, and was very likely derived from the lore of the mystery cults into which he was initiated.

After the souls of the dead have received a thousand years of reward or punishment for the deeds of their previous lives, they are brought to a place where they choose their next incarnations. Reincarnation is necessary for the development of consciousness in order to...

> know what the effect of beauty is when combined with poverty or wealth in a particular soul, and what are the good and evil consequences of noble and humble birth, of private and public station, of strength and weakness, of cleverness and dullness, and of all the natural and acquired gifts of the soul.[30]

IAMBLICHUS, TEACHER OF THE PHILOSOPHER PORPHERY

Plato taught that before each incarnation, the soul enters into a forgetfulness of what has gone before. The purpose of human learning and philosophy is, then, to reawaken in the soul remembrance of the eternal, spiritual realm of pure forms and ideas.[31]

Aristotle

Plato's greatest student, Aristotle, is noted for having turned away from the inner world of spiritual ideals Plato loved toward a more rational and scientific philosophy. Instead of describing the spiritual world as having greater reality than the physical, he describes an entelechy or vital force urging the organism toward self-fulfillment. He describes this urge as the ultimate and immortal reality of the body. Aristotle also recognized in the stars embodied deities, beings of superhuman intelligence.

Neoplatonism

The Neoplatonic school, centered in Alexandria, combined mystical elements found in Judaism with Greek philosophy. Philosophers within this tradition sought to explain the world as an emanation from a transcendent God who was both the source and goal of all being. Philo Judaeus (30 B.C. - 50 A.D.) described a process of mediation between God and humanity, in which the Jewish notions of angels and demons was equated with the world-soul or realm of ideas of the Greeks. Philo advocated using forms of asceticism in order to free oneself from the grip of sensory reality and enter into communion with spiritual reality.

Similar doctrines were taught by Plotinus who insisted that union with God cannot be realized by thought even freed from the senses. This experience is possible only in a state of ecstasy in which the soul totally transcends its own thought, loses itself in the being of God and becomes one with divinity.

The way in which the neo-Platonists probed into the magical workings of nature is reflected in the questions posed by the philosopher Porphery to his teacher Iamblichus:

> . . . granted that there are Gods. But I inquire what the peculiarities are of each of the more excellent genera, by which they are separated from each other; and whether we must say that the cause of the distinction between them is from their energies, or their passive motions, or from things that are consequent, or from their different arrangement with respect to bodies; as, for instance, from the arrangement of the Gods with reference to ethereal, but of daemons to aerial, and of souls to terrestrial bodies?...
>
> I likewise ask concerning the mode of divination, what it is, and what the quality by which it is distinguished?...
>
> The ecstasy, also, of the reasoning power is the cause of divination, as is likewise the mania which happens in diseases, or mental aberration, or a sober and vigilant condition, or suffusions of the body, or the imaginations excited by diseases, or an ambiguous state of mind, such as that which takes place between a sober condition and ecstasy, or the imaginations artificially procured by enchantment.
>
> What also is the meaning of those mystic narrations which say that a certain divinity is unfolded into the light from the mire, that he is seated above the lotus, that he sails in a ship, and that he changes his forms every hour, according to the signs of the zodiac? For thus they say he presents himself to view, and thus ignorantly adapt the peculiar passion of their own imagination to the God himself. But if these things are asserted symbolically, being symbols of the powers of this divinity, I request an interpretation of these symbols.[32]

These questions were all directed toward the operation of certain theurgical (magical) rites designed to evoke the powers of the gods to aid the philosophers.

The neoplatonic philosophers did not consider themselves the originators of a new school of thought. Rather, they felt they were carrying on the tradition of Plato, however, they developed a more active mysticism than is found in Plato himself, a mysticism which was to carry a great influence on later hermeticists, alchemists, and cabalists.

Ancient Rome

The Romans, like the Greeks, were fascinated with stories of the marvelous. Pliney the Elder asserts that he collected twenty thousand theurgical incidents taken from the writings of a hundred different authors. Historians such as Herodotus, Tacitus, Suetonius, Plutarch, and Titus Livius relate many such incidents in the lives of prominent men.

Cicero (106-43 B.C.), known as Rome's greatest orator, wrote a book, *Divination*, in which he discusses the evidence pro and con for the accuracy of predictions. He attempts to take an impartial philosophic stance to the evidence. Inquiring into the nature of fate, he asks what use predictions are if the foretold events cannot be changed. How does free will fit into this picture? Cicero implies that for some events fate and determinism rule, while for other categories of events men exercise an amount of free will.

The use of puns in prophecy finds a striking illustration in a story related by Tacitus about Vespasian before he attained the throne of the Roman Empire. Disturbed by several miraculous healings which occurred in his presence, Vespasian had decided to consult the oracle.

> Entering the temple, he ordered everyone to leave. Suddenly, while his attention was turned to the god, he noticed behind him one of the principle Egyptian priests named Basilides, whom he knew to be several days' journey from Alexandria, and ill in bed at the time. Leaving the temple he went out into the streets and enquired if Basilides had not been seen in the city; finally he sent horsemen to the place where this priest lived, and learned that at the time he saw him Basilides was eighty miles away. Then he was forced to admit that he had really been favored with a vision: the word Basilides (from a Greek word for king) meant that he would attain to empire.[33]

The great historian Plutarch (born 47 A.D.) held that the human soul had a natural faculty for divination and added that it must be exercised at favorable times and in favorable bodily states. He described the daemon of Socrates as an intelligent light that resonated with Socrates because of his inner light. Plutarch also viewed such spirits as the mediators between God and men.

Plutarch reports for example that Calpurnia, Caesar's fourth wife, dreamed she saw her husband's blood spilled on the eve of the fatal Ides of March. A comet and other portents also forewarned of Caesar's death.

The philosopher, statesman, and playwright Seneca (4 B.C. -65 A.D.) saw the emerging scientific outlook as a plausible substitute for existing religions and as a basis for a moral philosophy. He accepted divination in all of its forms, but stressed the personal inner growth of the scientist:

> Those secrets [of nature] open not promiscuously nor to every comer. They are remote of access, enshrined in the inner sanctuary.[34]

Seneca was an extraordinary historical figure who was largely responsible for introducing Stoic philosophy to the Roman world. I personally view his moral outlook as particularly appropriate for our own times. Like Socrates, he exemplified the principle that a deep understanding of consciousness emerges as much from self-development as from logical inquiry — or perhaps even more so.

STATUE OF SENECA IN CORDOBA, SPAIN (COURTESY JANELLE BARLOW)

Pliny the Elder wrote his *Natural History* in 77 A.D. It contains thirty-seven books investigating most of the ancient arts and sciences. In this work, Pliny posits a stance which is often found in the writings of scholars since his time. He brands the magi of his day as fools and impostors. Yet nonetheless he also deems certain "magical" procedures to be proved by experience. He recognizes the importance of the right spirit in science and offers frequent advice on chastity, virginity, nudity and fasting. Sometimes he urges physicians to be totally silent. He also speaks about sympathies and antipathies between various material objects, and uses this idea as the basis for many medical treatments. He acknowledges that the positions of the sun and the moon are important for such treatments.

One of the fathers of medicine, Galen (born 129 A.D.) began to study philosophy but turned to medicine at age seventeen because of a dream his father had. He innovated many medical practices and left us about twenty volumes of medical treatises, averaging one thousand pages each. He was the first to recognize the physiological symptoms of emotional states, such as the quickening of the heart beat of those in love. He refused to accept supernatural influences in medicine and felt that all his remedies were shown by experiment and experience and were naturally understandable.

Galen recognized the value of using dreams for diagnosing illness as well as for predicting the future. He accepted the doctrine of occult virtues in medicine that were the property of the substance as a whole and not any part of it that might be isolated. These virtues were discovered through contemplation on a given substance.

In 150 B.C. the Romans passed a law declaring that no important resolution could be adopted without consulting the augers.

THE LEGENDARY APOLLONIUS OF TYANA

Apollonius of Tyana

Approximately 217 A.D., Philostratus composed the *Life of Apollonius* at the request of Julia Domna, the learned wife of the emperor Septimus Severus who possessed documents belonging to Damis of Ninevah, a disciple and companion of Apollonius. Philostratus used the will and epistles of Apollonius and also personally took the trouble to visit the cities and temples Apollonius had frequented in his lifetime about a hundred years earlier.

Apollonius was a Pythagorean philosopher whose miracles in raising the dead and healing the sick have been compared to those performed by Christ. During his travels he associated with the Brahmins of India and also the Persian magi. In Rome he was arrested and tried before the emperor Domition for sorcery, because he had managed to predict the plague at Ephesus. He claimed that it was merely his moderate diet that kept his senses clear and enabled him to see the present and the future with an unclouded vision. Philostratus implied that

Apollonius managed to inexplicably vanish from the courtroom.

Apollonius believed that health and purity were a prerequisite for divination. His life was also guided by dreams, and he would interpret the dreams of others as well. He would not sacrifice animals, but he enlarged his divinatory powers during his sojourn among the Arab tribes, by learning to understand the language of animals and listen to the birds — for animals and birds seemed to predict the future. He would also observe smoke rising from burning incense.

His ability to detect and deal with demons is illustrated in the story of a lamia (which has become immortalized in a poem by Keats) or evil demon which he disposed of through his penetrating insight. In fact, he was held with such awe by his disciples they believed him to be a daemon or demigod.

Ancient Hebrews and Early Christians

One encounters many instances of higher communication in the Bible. Dream interpretation is common. God speaks to men directly and also through angels and at times appears in burning bushes and whirlwinds. Prophets communicate with God. Joseph, for example, interprets dreams that foretell the future. Miracles of a wide variety abound in the works attributed to Moses. Of particular interest is the Ark of the Covenant, a device through which the God of the Hebrews spoke to his people.

Judaism was neither a nature religion of the type that focused primarily on the changing of seasons and ensuring fertility, nor was it primarily a religion of mystical union through contemplation. There were elements of natural and mystical religion in Judaism, but it was primarily a historical religion in which God interacted with and shaped the destiny of the Hebrew nation.

Prophecy

Nevertheless, there are many instances in Judaism of prayer as an altered state of consciousness, of healing, glossolalia or speaking in tongues, revelry, fasting, retreating to the wilderness, and possibly the use of mind-altering drugs as anointing oils, as well as esoteric communities such as the Essenes and various schools of prophecy.[35] An important branch of Jewish mysticism was based on attaining the vision of the throne of God as described in the first chapter of the *Book of Ezekiel.* The cabalistic tradition in Judaism is based on the ascent of consciousness

The Vision of Ezekiel

through various stages to the ultimate vision of the throne of glory — and beyond all vision to union with God.[36]

The Teachings of Jesus

Whether or not one is a Christian, there is little room for doubt that few persons in the ancient world have exerted a greater influence on humankind than Jesus Christ. To some extent it would be more accurate to say that we have been influenced by the myth or the story or the archetype of Jesus Christ — for there is great dispute among scholars as to what the actual person, Jesus, taught. Yet, for almost two thousand years, each generation has sought to find wisdom in the life of Jesus Christ. In so doing, each generation has uniquely contributed to our picture of the Western spiritual quest.

The gospels of Matthew, Mark, Luke and John were all written several generations after the death and supposed resurrection of Jesus Christ. In the King James *red letter edition* which highlights those statements directly attributed to Christ, one can find a teaching of depth and wisdom which places an emphasis on love as central to the inner spiritual life. The early Christians, until the time Christianity was accepted by the Roman emperor Constantine, were strictly pacifists.

The teachings of Jesus can be seen as a form of *bhakti yoga* — attainment through love and devotion to a master. By stating that "the Kingdom of Heaven is within you," Jesus was, perhaps, relating spiritual salvation to the primordial traditions of psychological growth. The main message of Jesus is that of living a life of virtue, obeying certain precepts — particularly the gentle virtues of love and kindness, the virtues of the heart. For example, when asked what the greatest of the commandments was, Jesus stated:

> Thou shalt love the Lord thy God with all thy heart, and with all thy soul, and with all thy mind. This is the first and great commandment. And the second is like unto it, Thou shalt love thy neighbor as thyself. On these two commandments hang all the law and the prophets.[37]

This excerpt from an apocryphal gospel of Mark quoted in a letter from an early Church father, Clement of Alexandria, provides a glimpse, independent of the Bible, into the powers of spirit within the Christian context:

> And they came into Bethany. And a certain woman whose brother had died was there. And, coming, she prostrated herself before Jesus and says to him, "Son of David,/ have mercy on me." But the disciples rebuked her. And Jesus, being angered, went off with her into the garden where the tomb was, and /straightaway a great cry was heard from the tomb. And going near, Jesus rolled away the stone from the door of the tomb. And straightaway, going in where the youth was, he stretched forth his hand and raised him, seizing his hand. But the youth, looking upon him, loved him / began to beseech him that he might be with him. And going out of the tomb they came into the house of the youth, for he was rich. And after six days Jesus told him what to do and in the evening the youth comes to him, wearing a linen cloth over his naked body. And he remained with him that night / for Jesus taught him the mystery of the kingdom of God.[38]

This passage suggests that Jesus administered some sort of nocturnal initiation ritual, reminiscent of the Greek and Egyptian mysteries. Clement, who lived in the second century, was instrumental in integrating these pagan mysteries into the framework of a Christian spiritual life.

In Christianity one finds an emphasis on the *gifts of the spirit.* Jesus is noted for miraculous psychic feats: healing the sick, raising the dead, walking on water, multiplying loaves and fishes. Christ encouraged his believers to accept the possibility of certain behaviors attributable to spiritual power:

> And these signs shall follow them that believe; In my name shall they cast out devils; they shall speak with new tongues; They shall take up serpents; and if they drink any deadly thing, it shall not hurt them; they shall lay hands on the sick, and they shall recover.[39]

One of my favorite New Testament passages is in the first epistle of Paul the apostle to the Corinthians. Here he enumerates the gifts of the spirit:

> For to one is given by the Spirit the word of wisdom; to another the word of knowledge by the same Spirit; To another faith by the same Spirit; to another the gifts of healing by the same Spirit; To another the working of miracles; to another prophecy; to another discerning of spirits; to another divers kinds of tongues; to another the interpretation of tongues.[40]

Paul continues to point out that the gifts of the spirit are available to all for the benefit of all, whether or not they are Christian:

> For as the body is one, and hath many members, and all the members of that one body, being many, are one body: so also is Christ. For by one Spirit are we all baptized into one body, whether we be Jews or Gentiles, whether we be bond or free; and have been all made to drink into one Spirit.[41]

In an elegant passage, Paul then emphasizes the prime importance of Christian love which while translated as "charity" in the King James version is referred to as "love" in other versions, such as the Gideon Bible:

> Even though I speak in human and angelic language and have no love, I am as noisy brass or a clashing cymbal. And although I have the prophetic gift and see through every secret and through all that may be known, and have sufficient faith for the removal of mountains, but I have no love, I am nothing. And though I give all my belongings to feed the hungry and surrender my body to be burned, but I have no love, I am not in the least benefitted.[42]
>
> Love never fails. As for prophesyings, they will pass away; as for tongues, they

THE TEMPTATION OF ST. ANTHONY

> will cease; as for knowledge, it will lose its meaning. For our knowledge is fragmentary and so is our prophesying. But when the perfect is come then the fragmentary will come to an end.[43]
>
> There remain then, faith, hope, love, these three; but the greatest of these is love.[44]
>
> Make love your great quest; then desire spiritual gifts, and especially that you may prophesy.[45]

Christian Saints

In his *Dialogues*, Gregory the Great, who was Pope of the Church from 590-604 A.D., described many marvelous wonders he had learned about Italy's saintly men of God either by personal experience or through trustworthy witnesses.

Talking to animals, raising the dead, and stopping avalanches were all recorded in his time. St. Augustine of Hippo, who died in 430 A.D., claimed to attain his knowledge through a series of contemplative glimpses of supramundane reality which set the tone of Christian mysticism since his time:

> My mind withdrew its thoughts from experience, extracting itself from the contradictory throng of sensuous images, that it might find out what light was wherein it was bathed. And thus with the flash on one hurried glance, it attained to the vision of That Which Is. And then at last I saw Thy invisible things understood by means of the things that are made, but I could not sustain my gaze: my weakness was dashed back, and I was relegated to my ordinary experience, bearing with me only a loving memory, and as it were the fragrance of those desirable meats on which as yet I was not able to feed.[46]

The early saints of the church achieved their status through popular veneration. However, by the thirteenth century, within the Roman Catholic Church, the sole right to canonize was reserved to the papacy. Recognition of a saint required evidence of the holiness of his or her life and of miracles obtained through the saint's intercession before God.

In Catholic theology, the saint does not himself or herself work miracles or answer prayers. Miracles are said to be the work of God and the saint intercedes with God to grant the petitions of the faithful.

The modern process of canonization takes the form of a law suit in which the pope is the final judge. The case for the candidate is presented by the "postulator" and it is the duty of the "promoter of the faith," popularly known as "the Devil's Advocate," to draw attention to weak points in the case. Two lower ranks than saints have been instituted, those of "venerable" and "blessed."

The Monastic Tradition

The early Christian monastics, particularly in Egypt, practiced a number of austerities such as fasting, solitude, self-mortification, celibacy, sitting on poles, sleep-deprivation, etc.[47] Theirs was a mystical philosophy of the inner vision of Christ, and often many varieties of demons came to tempt them from the purity of their path.

Mystics such as St. Anthony, when confronted by tormenting demons, recognized them as intangible thoughtforms who could not do harm (similar to the teachings of *The Tibetan Book of the Dead*).

The desert became so crowded with solitary monks that they founded their own communities. These monasteries were a mystical source of the power of the Church for many centuries. In fact, the practice of holy penitence eventually led to the penitentiaries of our "modern" penal system. The first institutions of this sort were operated by the Catholic Church in Europe and by the Quakers in the United States.

Hermeticism

The writers of the *Corpus Hermetica* attributed their philosophy to a legendary figure named *Hermes Trismegistus* (thrice-great). This figure is associated with the god Thoth or Hermes who was alleged to have given the Egyptians their knowledge of the arts and sciences. The main dictum of the hermetic tradition as found in the *Emerald Tablet* in essence states, "As Above, So Below," meaning that humanity will find within itself the nature of the entire universe.[48]

Professor Wayne Shumaker, a historian of the occult sciences, uses an analogy to elucidate Hermeticism:

> Again and again we are told the whole world is alive. "If therefore the world is always a living animal was, and is, and will be — nothing in the world is mortal. Since every single part, such as it is, is always living and is in a world which is always one and is always a living animal, there is no place in the world for death" (Ascl. 29). When in our own day, C. S. Lewis' fictive character Ransom first traveled through space, the word "space," the reader is told, was "a blasphemous libel for this empyrean ocean of radiance in which they swam. He could not call it 'dead'; he felt life pouring toward him from it every minute." The Hermetic universe was similarly vitalistic, permeated with life. So is the universe of the low savage, the *Naturmensch*; but long before the second and third centuries of our era the primitive belief had been rationalized.[49]

The notion that humanity can discover the nature of God and the universe by looking within itself has been called The Perennial Philosophy by Aldous Huxley:

> *Philosophia perennis* — the phrase was coined by Leibnitz; but the thing — the metaphysic that recognizes a divine Reality substantial to the world of things and lives and minds; the psychology that finds in the soul something similar to, or even identical with, divine Reality; the ethic that places man's final end in the knowledge of the immanent and transcendent Ground of all being — the thing is immemorial and universal. Rudiments of the Perennial Philosophy may be found among the traditionary lore of primitive peoples in every region of the world, and in its fully developed form it has a place in every one of the higher religions. A version of this Highest Common Factor in all preceding and subsequent theologies was first committed to writing more than twenty-five centuries ago, and since that time the inexhaustible theme has been treated again and again, from the standpoint of every religious tradition and in all the principle languages of Asia and Europe.[50]

Huston Smith (From the *Primordial Tradition* videotape, courtesy Thinking Allowed Productions

Huston Smith, emeritus professor of philosophy and religion at MIT and Syracuse University, is currently affiliated with the Graduate Theological Union in Berkeley, California. Dr. Smith is the author of *The Religions of Man*, a classic book which has sold over two million copies, as well as *Beyond the Post-Modern Mind*, and *Forgotten Truth*. In a *Thinking Allowed* interview, he described his impressions of the hermetic thread that runs through all religious and spiritual traditions:

> The primordial tradition is intended to have the ring of being timeless — and, I would add, spaceless as well — because it was not only always, but everywhere. The

space-time world, fits into the primordial tradition but does not exhaust it. There are reaches beyond the physical.

> The primordial tradition sees us as situated in a world where the order of measure is quality. It is fifty percent happiness, fifty percent sorrow; fifty percent knowledge, fifty percent being in the dark about things. It is again situated midway between what in the traditional cosmologies are shown as the heavens, which are incomparably better, and the hells, which are also, alas, incomparably worse. But the interesting point is the difference in quality, whereas science gives us almost the same structure, but in quantitative rather than qualitative terms. Science everywhere can pick up quantitative distinctions, but qualitative ones, like beauty and spirit, really slip through its nets like the sea slips through the nets of fishermen.
>
> All the virtues keep lock step as they advance. So greater wisdom and greater power and greater beauty and greater bliss — they all rise concomitantly towards the goal wherein they lose their distinctness and completely merge in what is beyond words at that point.[51]

Not all scholars accept this notion. In a recent anthology on mystical traditions, an assortment of noted scholars argued that we must recognize the enormous cultural and language differences that separate each form of mysticism. Philosopher Steven Katz summarizes this perspective:

> "God" can be "God," "Brahman" can be "Brahman," and *nirvana* can be *nirvana* without any reductionist attempt to equate the concept of "God" with that of "Brahman," or "Brahman" with *nirvana.*[52]

Islamic Exploration of Consciousness

Theories of Occult Radiation

One of the greatest Arab occult scholars was Ya'kub ibn Sabbah al Kindi who died in 873 A.D. and is simply known as Alkindi. He translated the works of Aristotle and other Greeks into Arabic, and wrote books about philosophy, politics, mathematics, medicine, music, astronomy and astrology. He developed his own very detailed philosophy based on the concept of the radiation of forces or rays from everything in the world. Fire, color and sound were common examples of this radiation.

Alkindi was quite careful to distinguish between radiation that could be observed through the science of physics — due to the action of objects upon one another by contact — and radiation of a more hidden interaction, over a distance, which sages perceive inwardly. Radiant interactions were for him the basis of astrology. Human imagination was capable of forming concepts and then emitting rays that were able to affect exterior objects. Alkindi claimed that frequent experiments have proven the potency of words when uttered in exact accordance with the imagination and intention.

Favorable astrological conditions were capable of heightening these "magical" effects. Furthermore, the rays emitted by the human mind and voice became the more efficacious for moving matter if the speaker had his mind fixed upon the names of God or some powerful angel. Such an appeal to higher powers was not necessary, however, when the person was attuned to the harmony of nature (or in Chinese terms, the Tao). Alkindi also advocated the use of magical charms and words:

> The sages have proved by frequent experiments that figures and characters inscribed by the hand of man on various materials with intention and due solemnity of place and time and other circumstances have the effect of motion upon external objects.[53]

He further recognized that humanity's psychic vision is heightened when the soul dismisses the senses and employs the formative or imaginative virtues of the mind. This happens naturally in sleep. Unfortunately, the details of the experimental techniques of Alkindi and his associates have not been handed down. Nevertheless, he does deserve credit as an important pioneer.

One of the most sophisticated critics of psychic phenomena, a contemporary of Alkindi, was Costa ben Luca of Baalbek who wrote an important work on magic called *The Epistle Concerning Incantations, Adjurations and Suspensions from the Neck.* In this document he strongly asserts that the state of one's consciousness will have an effect on the body. If one believes that a magical ritual or incantation will help, one will at least benefit by his or her own confidence. Similarly, if a person is afraid that magic is being used against him, he may fret himself into illness. Ben Luca did not accept the notion of the occult virtues of stars or demons but did admit that strange phenomena were possible and would one day be understood. He listed a number of ancient magical techniques and maintained that these were useful in treating people who felt that they were enchanted.[54]

Although both Alkindi and ben Luca lived in Arab countries and wrote in Arabic, neither of them was Moslem. Like Judaism and Christianity, Islam was essentially an historical religion with primary emphasis on the law. Yet within Islam the perennial philosophy was maintained by the Sufi mystics who were often persecuted.

Medieval Exploration of Consciousness

Maimonides

Moses Maimonides (1135-1204 A.D.), the greatest Jewish medieval philosopher, lived in Cairo although he was born in Spain. There he was the chief physician to the vizier of the Sultan. Continuing the allegorical method of Philo, and well steeped in the Cabala, he attempted to reconcile Jewish thought with Greek philosophy. He held that the celestial bodies were living, animated beings and that the heavenly spheres were conscious and free. In his *Guide to the Perplexed,* written in Arabic in 1190, he stated that all philosophers are agreed that the inferior world, of earthly corruption and degeneration, is ruled by the natural virtues and influences of the more refined celestial spheres. He even felt every human soul has its origin in the soul of the celestial sphere.

Maimonides believed in a human faculty of natural divination and that in some men "imagination and divination are so strong that they correctly forecast" the greater part of future events. Nevertheless, he upheld human free will and human responsibility for our actions.

While he did believe in angels, he would not accept the existence of demons — saying that evil was mere privation. Alleged cases of possession by demons were diagnosed by him as simple melancholy. In accordance with Mosaic law, he accepted injunctions against the occult practices of idolatry and magic. Yet he maintained that any practices known to have a natural cause or proved efficacious by experience, as in the use of medicinal charms, were permissible. This differentiation between demonic and natural magic was to be emphasized by scholars for several centuries.

Today we have little difficulty in finding fault with the scientific methodology of even the greatest thinkers of these times, and there is no doubt in my mind that the professed magi of medieval and renaissance times were often the gullible dupes of many superstitious fallacies. However, magic was also the art of bringing divine life into physical manifestation. We can see throughout cultural history that the magi were artists who were able to infuse a delicately balanced state of consciousness into their lives and work — one that opened the intuition to the deepest levels of being and then exposed the insights attained to intellectual scrutiny and carefully controlled craftsmanship. It is precisely a process of this sort that underlies all genius. As history unfolds we shall cite other examples in which the development of this creative state of consciousness is clearly linked to esoteric or spiritual practices.

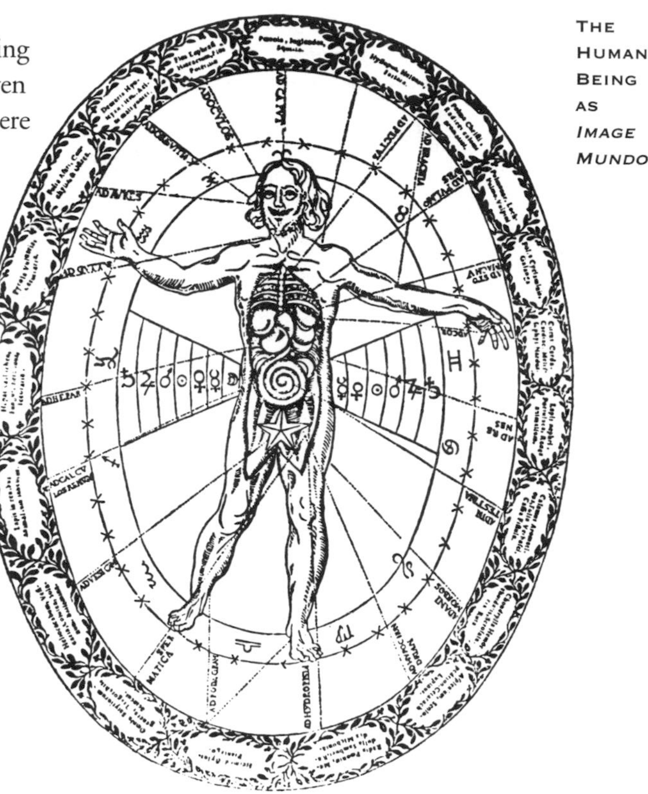

The Human Being as *Image Mundo*

Albertus Magnus

Albertus Magnus

The leading figure in thirteenth-century learning was Albertus Magnus, a Dominican friar who was finally canonized as a saint in 1931. Albertus, who has left us eight books on physics, six on psychology, eight on astronomy, twenty-six on zoology, seven on botany, five on minerals, one on geography, and three on life in general, was strongly influenced by Aristotle. Believing that God acts through natural causes in natural phenomena, he conducted experiments in the field of animal behavior and thus became an important forerunner of modern experimental science. He was known to have had miraculous visions since childhood.

He was also an ardent philosopher of magic and expressed a very positive attitude toward the magi of the Bible as "masters who philosophize about the universe and…search the future in stars." This view still persists in the Roman Catholic Church.

For Albertus, heaven and the stars were the mediums between the primal cause, or Aristotle's prime mover, and matter. All things produced in nature or in art are influenced by celestial virtues. The human being is an *image mundo*, or image of the universe, similar in conception to the hermetic notion of human as a microcosm. His natural magic thus made use of nature and the stars. It included astrology to find a favored hour for beginning a contemplated act, or an act of contemplation. And Albertus was clearly interested in the transmutation of metals as well as the use of psychic abilities to find metals within the earth. Towards this last end, he recommended employing potions to clog and stupefy the senses, thereby producing visions. He also advocated dream interpretation, the use of herbs and magical stones, animal potions and images engraved on gems. When these practices did not work, Albertus maintained that the defects were not to be found in the science of natural magic but in the souls of those who abused it.[55]

Renaissance Explorations

Cornelius Agrippa

As for the art of invoking spirits, Cornelius Agrippa, a magus whose influence was considerable in his day, has left us a description:

> If you would call any evil Spirit to the Circle it first behooveth us to consider and to know his nature, to which of the planets he agreeth, and what offices are distributed to him from the planet.
>
> This being known, let there be sought out a place fit and proper for his invocation, according to the nature of the planet, and the quality of the offices of the same Spirit, as near as the same may be done.
>
> For example, if his power be over the sea, rivers or floods, then let a place be chosen on the shore, and so of the rest.
>
> In like manner, let there be chosen a convenient time, both for the quality of the air — which should be serene, quiet and fitting for the spirits to assume bodies — and for the quality and nature of the planet, and so, too, of the Spirit: to wit, on his day, ignoring the time wherein he ruleth, whether it be fortunate or unfortunate, day or night, as the stars and Spirits do require.

These things being considered, let there be a circle framed at the place elected, as well as for the defense of the invocant as for the confirmation of the spirit. In the Circle itself there are to be written the general Divine names, and those things which do yield defense unto us; the Divine names which do rule the said planet, with the offices of the spirit himself; and the names, finally, of the good Spirits which bear rule and are able to bind and constrain the Spirit which we intend to call.

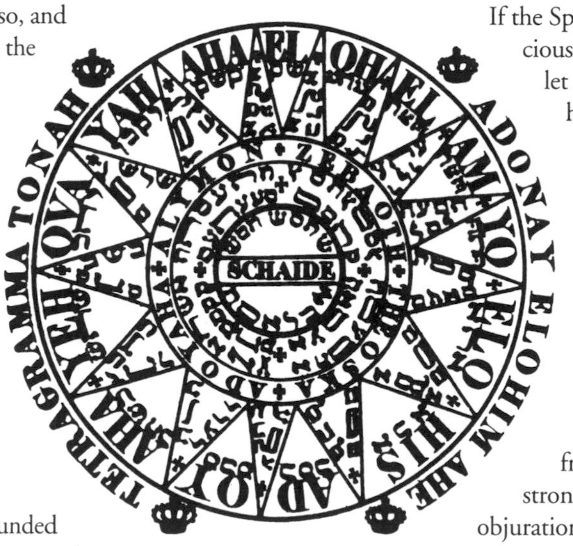

If we would further fortify our Circle, we may add characters and pentacles to the work. So also, and within or without the Circle, we may frame an angular figure, inscribed with such numbers as are congruent among themselves to our work. Moreover, the operator is to be provided with lights, perfumes, unguents and medicines compounded according to the nature of the Planet and Spirit, which do partly agree with the Spirit, by reason of their natural and celestial virtue, and partly are exhibited to the Spirit for religious and superstitious worship.

The operator must also be furnished with holy and consecrated things, necessary as well for the defense of the invocant and his followers as to serve for bonds which shall bind and constrain the Spirits.

Such are holy papers, lamens, pictures, pentacles, swords, sceptres, garments of convenient matter and color, and things of like sort.

When all these are provided, the master and his fellows being in the Circle, and all those things which he useth, let him begin to pray with a loud voice and convenient gesture and countenance. Let him make an oration unto God, and afterwards entreat the good Spirits. If he will read any prayers, psalms or gospels for his defence, these should take the first place.

Thereafter, let him begin to invocate the Spirit which he desireth, with a gentle and loving enchantment to all the coasts of the world, commemorating his own authority and power. Let him then rest a little, looking about him to see if any Spirit do appear, which if he delay, let him repeat his invocation as before, until he hath done it three times.

If the Spirit be still pertinacious and will not appear, let him begin to conjure him with the Divine Power, but in such a way that all the conjurations and commemorations do agree with the nature and offices of the spirit himself.

Reiterate the same three times, from stronger to stronger, using objurations, contumelies, cursings, punishments, suspensions from his office and power and the like.

After all these courses are finished, again cease a little, and if any Spirit shall appear, let the invocant turn towards him, and receive him courteously and, honestly entreating him, let him require his name. Then proceeding further, let him ask whatsoever he will.

But if in anything the Spirit shall show himself obstinate and lying, let him be bound by convenient conjurations, and if you still doubt of any lie, make outside the Circle, with the consecrated Sword, the figure of a triangle or pentacle, and compel the Spirit to enter it. If you would have any promise confirmed by oath, stretch the sword out of the Circle, and swear the Spirit by laying his hand upon the sword.

Then having obtained of the Spirit that which you desire, or being otherwise contented, license him to depart with courteous words, giving command unto him that he do no hurt.

If he will not depart, compel him by powerful conjurations and, if need require, expel him by exorcism and by making contrary fumigations.

When he is departed, go not out of the Circle, but stay, making prayer for your defense and conservation, and giving thanks unto God and the good angels. All these things being orderly performed, you may depart.

But if your hopes are frustrated, and no Spirit will appear, yet for this do not despair but, leaving the Circle, return again at other times, doing as before.[56]

Occult scholarship attempted to systematize everything from tastes, smells, colors, and body parts, to herbs, charms, spirits and dreams. It was an imaginative effort based primarily on introspection and reflection, but without proper standards of measurement and adequate means of correcting error. Nevertheless, deep levels of the psyche were involved in this effort to condense esoteric knowledge into meaningful symbols. This in-depth study of the intuitive and emotional connections between consciousness and the external world has a built-in difficulty in that the exact conditions necessary to create subtle intuitions and visions do not readily repeat themselves.

PARACELSUS

Paracelsus

Foremost among the occult scientists of his age was Phillipus Aureolus Theophrastus Bombastus von Hohenheim otherwise known as Paracelsus. He was born in Switzerland in 1493 and spent his entire life wandering throughout Europe and acquiring a great reputation for medical ability, unorthodox views, and a testy personality. For example, he was known to have publicly burned established medical texts. It is very difficult to distinguish his work from that of his students, interpreters, translators and editors. Very little of his writing was published in his own lifetime, and few of his original manuscripts survive today. His German writings were only noticed for their originality about twenty years after his death when scholars saw in him an alternative to stale medieval and Latin learning.[57]

Today he is recognized as the first modern medical scientist, as the precursor of microchemistry, antisepsis, modern wound surgery, and homeopathy. He wrote the first comprehensive work on the causes, symptoms and treatment of syphilis. He proposed that epileptics should be treated as sick persons and not as lunatics possessed by demons. He studied bronchial illnesses in mining districts and was one of the first people to recognize the connection between an industrial environment and certain types of disease. Notwithstanding this accurate scientific bent, his work is in close accord with the mystical alchemical tradition.

He wrote on furies in sleep, on ghosts appearing after death, on gnomes in mines and underground, of nymphs, pygmies, and

Magical Salamander from Paracelsus' *Auslegung von 30 magischen Figuren*

magical salamanders. His world view was animistic. Invisible forces were always at work and the physician had to constantly be aware of this fourth dimension in which he was moving. He utilized various techniques for divination and astrology as well as magical amulets, talismans, and incantations. He believed in a vital force radiating around every person like a luminous sphere and acting at a distance. He is also credited with the early use of what we now know as hypnotism. He believed that there was a star in each human being.[58]

John Dee

Another important occult scholar was John Dee (1527-1608 A.D.) who was one of the most celebrated and remarkable men of the Elizabethan age. His world was half magical and half scientific; he was noted as a philosopher, mathematician, technologist, antiquarian, as well as a teacher and astrologer. He was the first Englishman to encourage the founding of a royal library. He personally owned the largest library in sixteenth-century England, which contained over four thousand volumes. He held a large influence over the intellectual life of his times. He wrote the preface for the first English translation of Euclid and is given credit for the revival of mathematical learning in renaissance England. According to Lynn Thorndike in *The History of Magic and Experimental Science*:

John Dee and Edmund Kelly evoking a spirit.

For John Dee the world was a lyre from which a skillful player could draw new harmonies. Every thing and place in the world radiated force to all other parts and received rays from them. There were also relations of sympathy and antipathy between things. Species, both spiritual and natural, flowed off from objects with light or without it, impressing themselves not only on the sight but on the other senses, and especially coalescing in our imaginative spirit and working marvels in us. Moreover, the human soul and specific form of every thing has many more and more excellent virtues and operations than has the human body or the matter of the thing in question. Similarly the invisible rays of the planets or their secret influence surpass their sensible rays or light.[59]

He maintained that these invisible influences could be made manifest through the art of crystal gazing, which involved entering into a trance-like consciousness. He conducted many experiments in which he claimed to be in contact with angels through the use of a medium.

Dee's philosophy was embodied in a work entitled *Hieroglyphic Monad Explained Mathematically, Cabalistically and Anagogically.* This book, which served as an important foundation of the Rosicrucian movement, attempted to synthesize and condense all of the then-current mystical traditions within the symbolism characterizing the planet Mercury.

Queen Elizabeth herself was very taken with Dee's ideas. She appointed him as her court philosopher and astrologer, and asked for personal instruction into the abstruse symbolic meanings of his book. Nevertheless, he was still a very controversial figure because of his reputation as a conjurer. Dee lost favor with the court when James ascended to the throne.[60]

The Rosicrucians

This same fusion of world views is to be found in the teachings of the Rosicrucian movement, which caused quite a public stir in seventeenth-century England, France, Italy, and Germany. Only a limited number of men, most notably John Dee's student Robert Fludd, openly identified themselves as Rosicrucians. Most of the manifestos that caused a great uproar were published anonymously. Emphasizing earlier notions common to hermeticism, alchemy and the Cabala, the Rosicrucian documents proclaimed the existence of a hidden brotherhood of scholars and explorers who were united in teaching the deepest mysteries of nature, free from religious and political prejudice.

The following excerpt is taken from the last paragraph of *Fame of the Fraternity of the Rosie Cross* — an early manifesto first printed in 1614 and translated into English by Thomas Vaughan in 1652:

> And although at this time we make no mention either of our names, or meetings, yet nevertheless every ones opinion shal assuredly come to our hands, in what language soever it be; nor any body shal fail, who so gives but his name to speak with some of us, either by word of mouth, or else if there be some lett in writing. And this we say for a truth, That whosoever shal earnestly, and from his heart, bear affection unto us, it shal be beneficial to him in goods, body and soul; but he that is false-hearted, or only greedy of riches, the same first of all shal not be able in any manner of wise to hurt us, but bring himself to utter ruine and destruction. Also our building (although one hundred thousand people had very near seen and beheld the same) shal for ever remain untouched, undestroyed, and hidden to the wicked world, *sub umbra alarum tuarum Jehova.*[61,62]

At this same time Sir Francis Bacon (1561-1626) in England was also calling for a brotherhood that would foster the "advancement of learning." His effort ultimately led to the founding of the Royal Society in 1660. During his association with King James in England, Bacon was careful never to publicly connect himself with the Rosicrucians or any other occult movements. However, in a work published after his death, *The New Atlantis*, he describes his own version of a utopian society, revealing his sympathies and possible connection with this movement, and the *Invisible College.*

There are serious thinkers today who believe Bacon to have been a spiritual adept of the highest rank — founder of the Rosicrucians, secret author of the works attributed to William Shakespeare, the prime mover behind the English Renaissance, a man who contributed thousands of words to the English language and who first articulated the spiritual ideals upon which the United States of America was founded.

In *The New Atlantis*, the governor of the invisible island of which Bacon writes, describes the preeminent reason for the greatness of his society:

> It was the erection and institution of an order, or society, which we call Salomon's House; the noblest foundation, as we think, that was ever upon the earth, and the lantern of this kingdom. It is dedicated to the study of the works and creatures of God. Some think it beareth the founder's name a little corrupted, as if it should be Solamona's House. But the records write it as it is spoken. So as I take it to be denominate of the king of the Hebrews, which is famous with you, and no stranger to us; for we have some parts of his works which with you are lost; namely, that Natural History which he wrote of all plants, from the Cedar of Libanus to the moss that groweth out of the wall; and of all things that have life and motion. This maketh me think that our king finding himself to symbolize, in many things, with that king of the Hebrews (which lived many years before him) honoured him with the title of this foundation. And I am the rather induced to be of this opinion, for that I find in ancient records, this order or society is sometimes called Salomon's House, and sometimes the College of the Six Days' Works; whereby I am satisfied that our excellent king had

THE ROSICRUCIAN INVISIBLE COLLEGE

> learned from the Hebrews that God had created the world, and all that therein is, within six days: and therefore he instituted that house, for the finding out of the true nature of all things (whereby God ought have the more glory in the workmanship of them, and men the more fruit in the use of them), did give it also that second name....[63]

The *Invisible College* was a foundation of the Rosicrucian teaching. It was a building with wings which existed nowhere and yet united the entire secret movement. The high initiates of the society, the R. C. Brothers, said to be invisible, were able to teach their knowledge of a higher social and scientific order to worthy disciples who themselves became invisible. The symbolism of the *Invisible College* is very complex and further complicated by the social furor that resulted from it. As adventurers and scholars desiring a new social order sought to make contact with the fabled R. C. Brothers, an increasing public outcry resulted in witch hunts and persecutions.[64]

In one sense, the *Invisible College* refers to that type of teaching and inspiration that occurs to one in dreams. An allegory, written in 1651 by Thomas Vaughan, is quite suggestive of this theory:

> There is a mountain situated in the midst of the earth or center of the world, which is both small and great. It is soft, also above measure hard and stony. It is far off and near at hand, but by the providence of God invisible. In it are hidden the most ample treasures, which the world is not able to value. This mountain — by envy of the devil, who always opposes the glory of God and the happiness of man — is compassed about with very cruel beasts and ravening bird — which make the way thither both difficult and dangerous. And therefore until now — because the time is not yet come — the way thither could not be sought after nor found out. But now at last the way is to be found by those that are worthy — but

THE INVISIBLE MAGICAL MOUNTAIN

> nonetheless by every man's self-labor and endeavors.
>
> To this mountain you shall go in a certain night — when it comes — most long and most dark, and see that you prepare yourself by prayer. Insist upon the way that leads to the Mountain, but ask not of anywhere the way lies. Only follow your Guide, who will offer himself to you and will meet you in the way. But you are not to know him. This Guide will bring you to the Mountain at midnight, when all things are silent and dark. It is necessary that you arm yourself with a resolute and heroic courage, lest you fear those things that will happen, and so fall back. You need no sword nor any other bodily weapons; only call upon God sincerely and heartily.
>
> When you have discovered the Mountain the first miracle that will appear is this: A most vehement and very great wind will shake the Mountain and shatter the rocks to pieces. You will be encountered also by lions and dragons and other terrible beasts; but fear not any of these things. Be resolute

> and take heed that you turn not back, for your Guide — who brought you thither — will not suffer any evil to befall you. As for the treasure, it is not yet found, but it is very near. . . .
>
> After these things and near the daybreak there will be a great calm, and you will see the Day-star arise, the dawn will appear, and you will perceive a great treasure. The most important thing in it and the most perfect is a certain exalted Tincture, with which the world — if it served God and were worthy of such gifts — might be touched and turned into most pure gold.
>
> This Tincture being used as your guide shall teach you will make you young when you are old, and you will perceive no disease in any part of your bodies. By means of this Tincture also you will find pearls of an excellence which cannot he imagined. But do not you arrogate anything to yourselves because of your present power, but be contented with what your Guide shall communicate to you. Praise God perpetually for this His gift, and have a special care that you do not use it for worldly pride, but employ it in such works as are contrary to the world. Use it rightly and enjoy it as if you had it not. Likewise live a temperate life and beware of all sin. Otherwise your Guide will forsake you and you will be deprived of this happiness. For know of a truth: whosoever abuses this Tincture and does not live exemplarly, purely and devoutly before men, will lose this benefit and scarcely any hope will be left of recovering it afterward.[65]

In another sense, the Invisible College referred to an influential, though hidden, political, artistic, and scientific movement which included Francis Bacon and other notable renaissance figures dedicated to the teachings of the perennial philosophy. For example, there is evidence connecting Robert Boyle, who developed the laws relating the pressure of a gas at a fixed temperature to the inverse of its volume, with the College. Sir Isaac Newton also indicated an awareness of this movement.

THE AGE OF ENLIGHTENMENT

Descartes and Mind-Body Dualism

Rene Descartes (1596-1650), certainly not an occult scholar or even a sympathizer, nevertheless attributed all of his philosophic ideas to images that appeared to him either in dreams or when he was in the hypnogogic state just before awakening.[66] (In fact, he had to "prove his visibility" to keep from being associated with the Invisible College. The association of creativity with dreaming apparently gave rise to public speculation about an actual college, perhaps diabolical, that dreamers visited in their sleep.)

Mind-body *dualism* was first formally stated in modern philosophy by Descartes. His famous *cogito ego sum* ("I think therefore I am") implies that only through personal consciousness can one be certain of one's own existence. He identified consciousness with mind or soul, which to him was a substance as real and as concrete as the substance he called body. Descartes defined body as extended (space-filling), physical material and defined mind as "thinking thing" (*res cogitans*), which was unextended (did not take up space) and was not made of any physical material, but was purely spiritual. He also posited that these two substances mutually affect each other, giving the name *interactionism* to his position.

Leibnitz and Monadology

Carrying on the Pythagorean-Platonic doctrine of universal harmony, Gottfried Wilhelm Leibnitz, who with Isaac Newton was the co-inventor of calculus, developed an elegant grand philosophy based on the concept of an evolving unit of consciousness called the monad. Monads for Leibnitz are the most fundamental metaphysical points which have always existed and can never be destroyed.

GOTTFRIED WILHELM LEIBNITZ

Leibnitz felt that all matter is alive and animated throughout with monads. The monad is the principle of continuity between the physical and the psychological realms. The same principle that expresses itself within our minds is active in inanimate matter, in plants, and in animals. Thus the nature of the monad is best understood by studying the spiritual and psychic forces within ourselves.

Monads themselves vary in the amount of consciousness or clarity of their perceptions. Certain physical facts, such as the principle of least action, indicated to Leibnitz an intelligence within the most basic particles in creation. On the other hand, the findings of psychology have indicated that there are areas of the mind that are unconscious in their nature. In the lowest monads everything is obscure and confused, resembling sleep. While in humanity, consciousness attains a state of *apperception* — a reflexive knowledge of the self.

Every monad discovers its nature from within itself. It is not determined from without; there are no windows through which anything can enter; all of its experience already exists within each monad.

Both organisms and inorganic bodies are composed of monads, or centers of force, but the organism contains a central monad or "soul" which is the guiding principle of the other monads within its body. Inorganic bodies are not centralized in this way, but consist of a mere mass or aggregation of monads. The higher the organism, the more well-ordered will be its system of monads.

Every monad has the power to represent the entire universe within itself. It is a world in miniature, a microcosm, a "living mirror of the universe." Yet each monad has its own unique point of view, with its own characteristic degree of clarity. The higher the monad, the more distinctly it perceives and expresses the world; the monads with which it is most closely associated constitute its own body, and these it represents most clearly. Leibnitz stated:

> Every body feels everything that occurs in the entire universe, so that anyone who sees all could read in each particular thing that which happens everywhere else and, besides all that has happened and will happen, perceiving in the present that which is remote in time and space.[67]

The monads form a graduated progressive series from the lowest to the highest. There is a continuous line of infinitesimal gradation from the dullest piece of inorganic matter to God, the monad of all other monads — just as the soul is the presiding monad over the other monads within the human body. There is a parallelism between mental and physical states here. The body is the material expression of the soul. However, while the body operates according to the deterministic laws of cause and effect, the soul acts according to the teleological principle of final causes towards its ultimate evolution. These two realms are in harmony with each other.

Idealism

Another important philosopher of consciousness in this period was Bishop George Berkeley (1685-1753), after whom the City of Berkeley, California — where this book has been written — was named. Berkeley was a strict *idealist* who tried to demonstrate that the only things we ever experience are the perceptions, thoughts and feelings within our own minds. There is no need to ever assume that anything material exists whatsoever. The external world of physics is for all we know a figment of the imagination. Look about you. Everything that you see or sense in any way is simply a sensation in your mind. Is this a book you are reading? Did it take raw materials to produce? That was all somehow an illusion.

But, you will say, there must be some cause of the thoughts and sensations in our minds. For Berkeley this cause is one undivided active spirit which produces these effects upon our consciousness. Although we cannot perceive this spirit itself (any more than we could perceive such nonsense as matter), we still have some notion of it, some apprehension of the greater reality beyond us. You see, we all exist in the mind of God. As the Hindus would say, this is all simply Shiva's dream.

Berkeley's philosophical conclusions have never gained ascendancy within Western culture. Nor have they been logically discredited. The logical cohesion of the idealist philosophy as developed in Western culture by Berkeley and others (notably Hegel) and as developed by the Hindu and Buddhist philosophers of Asia has been unrivaled. Yet logical cohesion is not the sole criterion of a philosophical theory.

When Samuel Johnson (who wrote the first great English-language dictionary) was asked about the theories of Berkeley, his response was simply to kick a stone. "That's an end to that," he proclaimed. His response has been typical of the strong gut reaction against the idealist position. Yet, on one very important point, there is no difference between the idealist and the materialist position. In either case, the ultimate nature of reality is both unknown and unknowable. Whether one describes reality as mind or as matter is more of a social convention, or a foundation of belief, than the product of a well-reasoned inquiry. The Pythagorean position that there may be a mathematical unification offers more promise.

Sir Isaac Newton

Sir Isaac Newton (1642-1727) is generally regarded as one of the greatest scientists who ever lived. He discovered the binomial theorem, invented differential calculus, made the first calculations of the moon's attraction by the earth, described the laws of motion of classical mechanics, and formulated the theory of universal gravitation.

BISHOP GEORGE BERKELEY

He was very careful not to publish anything not firmly supported by experimental proofs or geometrical demonstrations — thus he exemplified and ushered in the Age of Reason. Newton, who is generally thought of as the archetypal materialist scientist, was astounded by the startling, and contradictory, nature of his own theories.

> That gravity should be innate, inherent and essential to matter, so that one body may act upon another at a distance through a vacuum, without the mediation of anything else, by which their force and action may be conveyed to one another....[68]

Here he has framed a major problem that remained unsolved until Einstein developed his theory of General Relativity. The problem continues to remind us of the incompleteness of a "common-sense," materialistic viewpoint. Gravity is such a common effect that it is taken for granted. Nevertheless, Einstein's understanding of gravity, while it solved Newton's problem of "action at a distance," requires that we accept that space itself is *curved.* The quest for a completeness in science, articulated by Sir Isaac Newton, has now found its expression in the search for a "grand unified field theory" in physics. This theory, which will be elaborated further in Section IV and in the Appendix, in its most current form echoes the Pythagorean principle of a mathematical structure underlying all of reality — consciousness and matter.

However, if we look at Newton's own personal notes and diaries, over a million words in his own handwriting, a startlingly different picture of the man emerges. Sir Isaac Newton was an alchemist. He devoted himself to such endeavors as the transmutation of metals, the philosopher's stone, and the elixir of life. Lord Keynes describes this work in the Royal Society's *Newton Tercentenary Celebrations* of 1947:

> His deepest instincts were occult, esoteric, semantic — with profound shrinking from the world...a wrapt, consecrated solitary, pursuing his studies by intense introspection, with a mental endurance perhaps never equalled.[69]

He attempted to discover the secrets of the universe in apocalyptic writings like the *Book of Revelation* or in occult interpretations of the measurements of Solomon's temple. But Lord Keynes even maintained that there was a magical quality to his scientific thought as well — that he solved a problem intuitively and dressed it up in logical proofs afterwards. Columbia University historian Lynn Thorndike feels that one can safely go further than Lord Keynes and compares Newton's method of scientific discovery "to that of a medium coming out of a trance."[70]

Newton used the term *ether* following Descartes to refer to a hypothetical substance that permeated the entire universe and was responsible for gravitation and electromagnetism as well as sensations and nervous stimuli. He felt that this ether itself was the living spirit, although he recognized that sufficient experimental proof did not exist in his own time.

It was only in the twentieth century that scientists actually discarded the concept of *ether,* although the term is still used pervasively in occult and spiritual circles. The elucidation of the field that unifies both psychical and physical phenomena is still one of the greatest challenges facing scientific research.

Newton normally spelled the word Nature with a capital and regarded her as a Being or at least a wonderful mechanism second only to God. Newton described his conception of God as:

> Creator and governor of this mechanistic universe, who first created the fermental aether and its principles of action, and

> then assigned to a lesser power, Nature, the duty of forming and operating the perceptible mechanical universe.

Like most men at the close of the seventeenth century, Newton still believed in the existence of animal spirits in the human body. He described them as of an ethereal nature and subtle enough to flow through animal voices as freely as the magnetic effluvia flow through glass. For him, all animal motions resulted from this spirit flowing into the motor nerves and moving the muscles by inspiration.

His followers, however, emphasized his mechanistic view of the universe to the exclusion of his religious and alchemical views. In a sense, their action ushered in a controversy that has existed ever since. Since Newton's time, all hypotheses suggesting the presence of a force that transcended time or space were ironically considered to be in violation of Newton's Laws — even though Newton himself realized that his laws were not sacrosanct!

THE FOLKLORE
OF CONSCIOUSNESS EXPLORATION

ASTROLOGY

THE ART AND SCIENCE OF ASTROLOGY, which bridges the ancient and modern worlds, is a very problematic subject. On the one hand, it has been justly castigated by modern scientists as a throwback to superstitious thinking. On the other hand, it is an expression of humanity's perennial yearning for a comprehensive understanding of our unity with the cosmos.

I am partially sympathetic with the skeptics who decry the current interest in astrology as a symptom of scientific and rational illiteracy. The major claims of traditional horoscope interpretation lack scientific support.

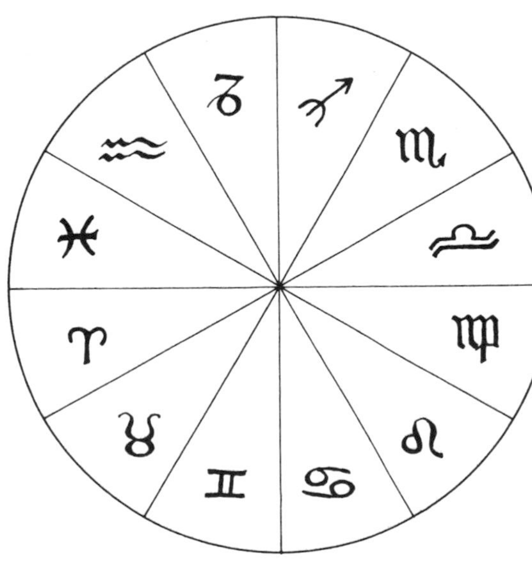

The mystery deepens as I (and, no doubt, you also) find numerous examples in my own personal life which seem to validate the horoscope. Quite honestly, while this leads me to suspect that there is more to astrology than I appreciate, it should also lead me to further grasp the extent to which my own well-educated mind is subject to mundane folly and fallibility. To compound matters further, some general astrological claims as well as some very specific claims have received clear-cut experimental support. On a day-to-day basis, this tension is both exciting and uncomfortable.

Some astrological correlations are self-evident to everybody. The changing of the seasons which is correlated with the movement of the sun through the zodiac. The influence of the moon upon the earth's tides, as well as the apparent correlation of the lunar cycle with female menstruation, is well known.

In fact, our language is peppered with astrological innuendos. When we call an unpredictable person *mercurial*, we are referring to the planet, and Roman god, Mercury. If a lustful person is described as *venial*, the reference is to the planet, and ancient Roman goddess, Venus. A *lunatic* is someone who has fallen under the influence of the moon, or the Roman goddess personifying the moon (sometimes identified with Diana). A person who studies the *martial* arts is engaged in a practice under the influence of the planet, and Roman god, Mars. When we describe a magnanimous person as *jovial*, we are making reference to the planet, and Roman god, Jupiter. To ancient peoples all of these things must have seemed obvious.

The social critic and historian, Theodore Roszak, in attempting to understand the persistence of astrology, writes:

> The essential teaching of astrology, reaching back to the ancient worship of the stars, was that of spiritual communion between the human and the heavenly....The modern fascination with astrology — even in its crudest forms — stems from a growing nostalgia for that older, more unified sense of nature in which the sun, moon and stars were experienced as a vast network of living consciousness. For a growing number of people, the rich imagery of these old traditions has become a more inspirational way of talking about emotions, values, motivations and goals than conventional psychiatry. The astrological universe is, after all, the universe of Greco-Roman myth, of Dante, Chaucer, Shakespeare, Milton, Blake. It has poetry and philosophy built into it.[1]

THEODORE ROSZAK (FROM *THE CULT OF INFORMATION* VIDEOTAPE, COURTESY THINKING ALLOWED PRODUCTIONS)

Some "skeptics" carry their arguments too far. In attacking astrology's quasi-scientific claims, they often overlook its cultural relevance as a symbolic language. In 1975, at the time of the original publication of *The Roots of Consciousness,* over 180 leading scientists — including eighteen Nobel Prize winners — signed a public letter proclaiming that astrology made invalid and unsupportable scientific claims.

Apparently the effort did little to stem the "rising tide of irrationality" which is the bane of those who proclaim themselves to be "skeptics." Two years later a Gallup poll reported that more than thirty million Americans believed in astrology.

When a representative of the BBC wanted to interview some of these eminent scientists, they declined with the remark that they had never studied astrology and had no idea of its details. The inappropriateness of such learned scientists attempting to combat astrology by using the weight of their academic reputations has been criticized by the eminent philosopher of science, Paul Feyerabend. Feyerabend suggests that the fifteenth-century Roman Catholic Church made a more cogent argument against witchcraft (in the classic *Malleus Malificarium*) than the skeptics were able to make against the underlying principles of astrology. He poses an interesting rhetorical question: "Why 186 signatures if one has arguments?"[2]

Where does one draw the line? Can both the proponents and skeptics of astrology be wrong? Feyerabend sums up his position very eloquently:

> Modern astrology is in many respects similar to early mediaeval astronomy: it inherited interesting and profound ideas, but it distorted them, and replaced them by caricatures more adapted to the limited understanding of its practitioners. The

> caricatures are not used for research; there is no attempt to proceed into new domains and to enlarge our knowledge of extra-terrestrial influences; they simply serve as a reservoir of naive rules and phrases suited to impress the ignorant. Yet this is not the objection raised by our scientists. They do not criticize the air of stagnation that has been permitted to obscure the basic assumptions of astrology, they criticize these basic assumptions themselves and in the process turn their own subjects into caricatures. It is interesting to see how closely both parties approach each other in ignorance, conceit and the wish for easy power over minds.[3]

If we can sort out the valid threads from the superstition in the fabric of astrology, perhaps we will discover potentially useful clues to understanding the nature of consciousness itself. In fact, a great deal of Pythagorean wisdom — which is largely lost to modern thinking — is likely to be embodied in astrology. Perhaps no one has done more to excavate this understanding than Arthur M. Young, whose work will be described later in this discussion of astrology and again in Section IV. When asked, one evening, why he found astrology of relevance today, Young stated it very simply, "Astrology connects us to the realm of mythos."

Ptolmaic Astrology

Ptolemy (110-151 A.D.) who lived in Alexandria during the neoplatonic era was the ablest mathematician and closest scientific observer of his day. He made important contributions to trigonometry and cartography and while his geocentric theory was proven erroneous, it was well substantiated on the basis of the existing scientific evidence and in fact held for over one thousand years. He also formulated the principle (later known as Occam's razor) that one should always use the simplest hypothesis consistent with the facts. He wrote twenty-one books on geography, mathematics and astronomy, as well as four on control of human life by the stars. These books, known as the *Tetrabiblos* have formed the substantial base of all Western astrology since his day, although few modern astrologers have ever read Ptolemy or understood the caution he urged.

Ptolemy contended that a certain force is diffused from the heavens over all things on earth. Yet he recognized that most applications of astrology were still hypothetical. He also acknowledged that it was easier to predict events affecting large areas, whole peoples or cities, rather than individuals. But astrology was not to be rejected simply because it was difficult to apply and only partially accurate any more than one would reject the art of navigation because ships are frequently wrecked.

Kepler and Astrology

In Germany at the beginning of the seventeenth century, astrology itself was becoming very controversial. In 1610, the great astronomer Johannes Kepler published a work that attempted to intervene in a public conflict between a pastor who issued prognostications and a physician who had attacked astrology. The title was:

> *A Warning to Sundry Theologians, Medical Men and Philosophers...that They, while very Properly Overthrowing Stargazing Superstition, do not Chuck out the Baby with the Bathwater and thereby Unwittingly Injure Their Profession.*

Kepler's reasoned attitude toward astrology was to try and determine precisely the extent and manner of the influence of the heavenly bodies upon the earth and its inhabitants. He was very concerned with revising and reforming the traditional rules of astrology in accordance with his own observations.

For example, he condemned the general run of astrological predictions, maintaining

JOHANNES KEPLER

that only one in a hundred was accurate. He further argued that the division of the zodiac into twelve signs was also completely arbitrary and irrelevant (a claim that, as we shall see, has been disputed by Arthur M. Young). Nevertheless, he felt that these practices could be kept simply as a matter of convenience.

He emphasized the importance of the aspects, or angular relations between the different planets — in fact, he added several aspects of his own invention to those traditionally used by astrologers. Modern experimental workers in astrology have taken a similar position.

In 1619, Kepler published *Harmonice Mundi* in which he described the results of twenty years of astrological observations. He maintained that the degree of harmony of the rays descending from the heavenly bodies to the earth was a function of their angular relationships. He described the similarities between planetary aspects and musical consonance and the effects different configurations of the planets exerted on the emotional and mental lives of animals and humans. He also observed the relationship of the weather to the planetary aspects. Of particular importance to him was the influence of the sun and the moon upon the earth. All bodily fluids waxed and waned with the moon, which was therefore very influential in the treatment of diseases. And he felt that the nature of planetary influences was revealed by their colors.

Kepler reaffirmed the importance of the positions of the planets at the moment of birth:

> In general there is no expedite and happy genesis unless the rays and qualities of the planets meet in apt, and indeed geometric, agreement.

He even thought that sons, particularly the first-born, were often born under horoscopes similar to their parents. This curious hypothesis is also borne out by some current research which will be mentioned in the next chapter.

Kepler, of course, is best known to modern science for his laws of planetary motions and his support of the heliocentric theory of the solar system. He also felt that the sun itself was the soul of the whole universe, presiding over the movements of the stars, the regeneration of the elements, and the conservation of animals and plants. To him the earth was like an animal:

> ...[with] the twin faculty of attracting sea waters into the secret seats of concoction, and of expelling the vapors which have been thus concocted. By its perception of the celestial aspects it is stimulated and excited to excrete these vapors with a pleasure akin to that which an animal feels in the ejaculation of its semen. Man, too, is not merely a rational being but is endowed with a natural faculty like that of the earth for sensing celestial configurations, "without discourse, without learning, without progress, without even being aware of it."[4]

Kepler developed his theories on the basis of explorations into the dimly lit archetypal regions of the mind as surely as on his mathematical observations of the planetary motions. He was clearly a student in the tradition of earlier mystic-scientists such as Pythagoras and Paracelsus.[5]

Astrology in Contemporary Times

It is on this level that we need to consider some of the more extreme findings of astrology. Electromagnetic radiations and solar storm activity can certainly account for certain mass phenomena. But we cannot expect electromagnetic effects to bear much relationship to the individual horoscope. Nevertheless, data of this sort exists. Most of the significant studies are the result of experimentation by Michel Gauquelin in France which began in the 1940s and continues to the present day. His results do not vindicate some widely used horoscope claims such as the reality of zodiacal influences, the meaning of the twelve astrological "houses," or the role of the "aspects" between planets. (In fact, there is really no good data supporting the astrological value of sun signs, moon signs, or any zodiacal signs at all.) What Gauquelin was able to discover was a weak relationship between the position of planets relative to the horizon and success in certain professions. This branch of study is sometimes termed *neo-astrology.* While, with a few notable exceptions, Gauquelin did not predict his results in advance, his findings are consistent with the astrological interpretations of Ptolemy and Kepler. These results are independent of normal seasonal or circadian factors.

For example, of 3,647 famous doctors and scientists he studied, 724 were born with Mars just above the eastern horizon or at the mid-heaven, directly overhead. Since you would only expect 626 men in this sample to have Mars rising or culminating, this effect has a probability of only 1 in 500,000 of having occurred by chance. In the same group of individuals, an unusually high number, statistically speaking, were born with Saturn rising or culminating.

In a group of 3,438 famous soldiers, Jupiter or Mars was frequently found in the sectors following their rise or culmination.

In a group of 1,409 actors, Jupiter appears to be more frequent after its rise and superior culmination. In a group of 1,352 writers, the moon appears more frequently in the key positions after the rise and superior culmination.[6]

In a group of 2,088 sports champions, Mars dominates, being recorded 452 times rising or culminating instead of the expected 358. The probability that this effect could have happened by chance was only one in five million. Furthermore, this finding was repeated independently of Gauquelin by a Belgian committee of scientists studying para-conceptual phenomena.

The treatment of Michel Gauquelin's astrology research remains one of the sadder chapters in the history of skeptical efforts to debunk astrology. A *Fate* magazine article titled "sTarbaby" by Dennis Rawlins claimed that the Mars effect was also independently uncovered in a study by the Committee for the Scientific Investigation of the Claims of the Paranormal (CSICOP) — a group normally devoted to skeptical debunking. Rawlins, a staunch skeptic himself, described in detail how CSICOP chairman Paul Kurtz obfuscated the evidence in order to avoid the implications of this embarassing finding.[7] CSICOP, for its part, has attacked the Rawlins report, claiming that the astrology study was not one of its official activities.

Interestingly, Gauquelin's data is significant only when dealing with prominent individuals. No correlations were found for individuals who did not achieve notable success in their professions. Furthermore, the correlations do not seem to apply for individuals whose birth was artificially induced at a certain time in order to conform with the schedule requirements of the attending physician. Only if the birth was allowed to occur at its own speed were the astrological correlates significant.[8] Since an increasing proportion of modern births are artificially induced (i.e., fewer hospital births take place at 1:20 A.M., such as mine did), the actual application and opportunity for continued replication of Gauquelin's "neo-astrology" is extremely limited.

Although scientists find Gauquelin's findings very disquieting, increasingly sophisticated analysis seems to confirm, rather than discomfirm, certain of the original results. For example, in a 1986 study, the German researcher, Suitbert Ertel, reported:

> A reanalysis of Gauquelin professional data using alternative procedures of statistical treatment supports previous Gauquelin results. Frequency deviations from chance expectancy along the scale of planetary sectors differ markedly between professions.[9]

Gauquelin himself is very clear that his findings may be interpreted as completely disproving the claims of conventional astrology.[10] It is hard to think that his findings have much relevance to the work of any given astrological interpreter of charts. Even the most significant correlations applied to less than one-fifth of the population in his sample. It is a big jump from a few weak, albeit significant, correlations to the interpretations of thousands of factors on an individual chart. Astrologers have never been able to explain this process to my satisfaction, and they often are in marked disagreement with each other. The strongest key to astrology seems to be meanings implicit in the mythological symbolism of the astronomical names. This is particularly evident when we think of soldiers being influenced by Mars and Jupiter.

CARL GUSTAV JUNG

Using an independent approach, the great Swiss psychiatrist Carl Jung also conducted studies in astrology. In an analysis of 483 pairs of charts from married couples chosen at random, Jung found a preponderance of aspects that would signify such a relationship. Most notably the moon of the woman was found to be in conjunction with either the sun, the moon, or the ascendant of her husband. A control group of more than thirty-two thousand unmarried pairs of charts did not show these aspects. Statistical analysis indicated that the results obtained could have happened by chance only one time in ten thousand.[11]

Jung maintained that this amazing finding could be attributed to the role ideas, thoughtforms, and archetypes play in organizing events. He labeled this process *synchronicity*, an acausal connecting principle.

An interesting study recently conducted in Germany used horoscopes of "accident-prone persons" — a group of car drivers who stand out for having been involved in above-average numbers of accidents. Aggressive and risky driving styles are among the characteristic features of accident-prone persons. Current astrology ascribes this kind of behavior to the planet Mars and Uranus which are expected to turn up in dominant positions in their horoscopes. For the purpose of an empirical investigation of this hypothesized relationship, several study groups formed by the "German Society of Astrologers, Inc." were asked to distinguish one or two accident-prone persons, on the basis of their horoscopes, from between one and four persons who were not prone to accidents. Five of six accident-prone persons were recognized as such; in each case, the solution was found by a cooperative attempt of all participants. Because of the small number of tasks, this positive result will require further replication before it can be considered reliable.[12]

In 1959, the American psychologist Vernon Clark conducted a few simple experiments designed to test the accuracy of astrologers' claims. These studies are often cited by the proponents of conventional horoscope interpretation. He collected horoscope charts showing planetary positions at birth, from people in ten different professions. Half were men and half were women. All were between forty-five and sixty years of age. The horoscopes were given to twenty different astrologers together with a separate list of professions; the astrologers were asked to match them up. The same information was given to another group of psychologists and social workers who knew little about astrology. Oddly enough, the astrologers were able to complete their task with a statistically significant accuracy — although they still made many mistakes. The psychiatrists and social workers scored no better than chance expectations.

In another experiment, Clark took ten pairs of horoscopes, matched for sex and year of birth. To each pair he attached a list of dates showing important events such as marriage, children, new jobs, and death. The astrologers were asked to state which of the two horoscopes in each pair predicted the pattern of events listed. Three of the astrologers got all ten right and the rest scored significantly above chance — with a probability of one in a hundred.

In a third test, the astrologers were given ten pairs of charts with no other data to work with. They were asked to determine which of the two charts indicated that the individual was a victim of cerebral palsy. Again the astrologers scored well above chance expectations.[13]

One might attribute the astrologers' success to ESP. But, for those who are unwilling to credit the validity of horoscopes, there are other interpretations. For example, it is unclear under what experimental conditions these studies were conducted. It may be the case that Clark preselected horoscopes which would match the textbook interpretations which are taught to astrologers. Unless the horoscopes were selected independently (using "double blind" experimental procedures) the research would not have been valid. This criticism was applied by Michel Gauquelin himself with regard to an ostensibly successful study reported by astrologer Joseph Vidmar.[14] A number of other studies, based on Vernon Clark's model have failed to produce positive results.[15]

An English astrologer, Jeff Mayo, conducted an interesting study in which he correlated astrological sun signs with personality measures of introversion and extroversion in over two thousand individuals. The initial results were remarkable. Individuals with "masculine" sun signs (Aries, Gemini, Leo, Libra, Sagitarius and Aquarius) scored higher on extroversion while those with "feminine" sun signs (Taurus, Cancer, Virgo, Scorpio, Capricorn and Pisces) scored higher on introversion. This study was even published in a psychological journal, indicating the co-authorship of Hans J. Eyesenck — one of Britain's greatest psychologists.[16] However, efforts to replicate failed to produce significant results. Eyesenck later concluded that the results were skewed because the individuals taking the personality tests were beginning astrology students whose tests results were biased by their identification with new material in astrology that they had learned. When tests were taken by either advanced astrology students or individuals ignorant of astrology, the correlations failed to hold.[17]

When Geoffrey Dean showed a large group of astrologers what they believed to be British singer Petula Clark's natal chart, they derived descriptions that tended to match her personality. Actually, the chart was that of mass murderer Charles Manson.[18] Another astrology research found, in the popular literature, three publications analysing John Lennon's natal chart, each postdicting the exact time of Lenon's violent death. However, each was derived from a different birth time.[19] Such findings leave little doubt that popular astrologers will interpret a natal chart in a manner that supports their own preconceptions.

In 1985, the *British Journal of Social Psychology* published three studies testing the astrological doctrine of *aspects* — the angular relationship between the earth and the various planets, a standard parameter for interpreting planetary influences in natal astrology. No evidence for the validity of the astrological concepts was found.[20]

As was noted earlier, many popular beliefs relate to the influence of the moon. It is said that more crimes occur during the full moon and that there are more traffic accidents then. In 1976, Berkeley health feminist Louise Lacey published a study suggesting that women might use the lunar cycle for purposes of contraception (which she claimed was influenced by exposure to light).[21] A novel study published in *Social Behavior and Personality* tested the lunar-aggression hypothesis by using the aggressive penalties awarded in ice hockey over a season of competition. Interpersonal aggression was found to be unrelated to lunar cycles. These findings suggested the persistence of lunar beliefs might be related to social factors such as expectations and selective exposure.[22,23]

In 1985, Shawn Carlson, then a UCLA graduate student in physics, published the results of a carefully controlled experiment in the prestigious British science journal, *Nature.*[24] Its purpose was to give recognized astrologers an opportunity to apply their craft in a controlled experimental setting to see if they could obtain accurate information about people they had never met.

Carlson summarizes his experiment:

> Three recognized expert astrologers consented to act as advisors on the experiment's design. They also selected those of their peers they considered sufficiently competent to participate, made "worst case" predictions for how well their colleagues would do, and approved the design as a "fair test" of astrology. Twenty-eight of the ninety astrologers who had been recommended by the advisors agreed to participate.
>
> Volunteer test subjects took a widely used and generally accepted personality test, the California Personality Inventory (CPI), to provide an objective measure of their character traits. The astrologers chose the CPI as the personality test which came closest to describing those character traits/attributes accessible to astrology. A computer constructed a natal chart for each subject.

The experiment consisted of two parts. In the first, the natal charts were divided up amongst the astrologers. For each natal chart, the astrologers wrote personality descriptions covering the material that they felt sure the subjects would be able to recognize as accurate. The subjects received their own personality descriptions and two others chosen at random, and were asked to select the description they felt best fit them. Extensive controls were established to eliminate self-attribution and other possible biases. If astrology does not work, one would expect one third of the subjects to select the correct personality descriptions. The astrologers' own "worst case" prediction was 50% correct.

In the second test, the natal charts were again divided up amongst the astrologers. For each natal chart, the astrologers were given three CPI test results, one of which correctly corresponded to the given natal chart. They were asked to make a first and second place choice as to which CPI came closest to matching each natal chart, and rate how well each fit the natal chart on a one to ten scale. If the natal chart contains no information about the subject, the astrologers had a one in three chance of making a given selection correctly. Their worst case prediction here too was 50% correct.

The subjects failed to select the correct personality descriptions more than one third of the time. However, if people tend not to know themselves very well they would be unable to select the correct descriptions no matter how well astrology worked. Since the CPI descriptions are known to be accurate, each subject was asked to select his or her own CPI from a group of three. Since the subjects were also unable to do this, their failure to select the correct astrological description could not be held against astrology.

However, astrologers' confidence in their abilities comes from their clients recognizing astrologically derived personality descriptions as accurate. The subjects' failure here means either that astrology does not work, or people do not know themselves well enough to recognize the astrologers' descriptions as accurate when reviewed in a controlled setting in which two alternate descriptions are also presented. Either way, these results show that astrologers' faith in their abilities is unfounded.

The astrologers failed to match the correct CPIs to the natal charts. They scored exactly chance and were very far from their "worst case" prediction of 50%. Even worse, the astrologers did no better when they rated a CPI as being a perfect fit to a natal chart. None of the astrologers scored well enough to warrant being retested. Thus, even though the astrologers started off very confident that they would pass this test, they failed.

There are two conclusions which one can draw from Carlson's study. The first is that twenty-eight astrologers were unable to pass what they agreed was a fair test of their abilities. The second, which I regard as most significant, is that the public at large is so lacking in self-awareness that people are unable to recognize their own personality profiles as measured by the California Personality Inventory — one of the most valid and reliable tests known in psychology. (This may, however, also be taken as a sad reflection on the present state of personality assessment.) Before summarizing all of the astrology research, I should point out that a number of other studies fail to support conventional horoscope interpretations.[25,26,27] However, the studies of Gauquelin and of Jung seem to provide weak support for some of the underlying principles of planetary influence. The findings will continue to remain on the fringes of science until many of the speculations are grounded in a mass of incontrovertible data.

There is some suggestion that astrologers (as well as psychics and metaphysical counselors) may play a role similar to that played by psychotherapists. Their clients are individuals with real needs who, nevertheless, do not wish to characterize themselves as having psychological problems. These ideas were documented recently in the *American Journal of Psychotherapy* by extracts from the writings of astrologers as well as by transcripts of visits to astrologers.[28] With regard to this possibility, Shawn Carlson comments:

> If an astrologer is a skilled counselor and a caring person, he/she may be of benefit to his/her clientele. However, most astrologers have no training in counseling. Many studied astrology because its unconventionality appealed to their own unconventional natures. A few are outright charlatans who use astrology as a scam to bilk the gullible. A person with a real problem is, in my opinion, courting disaster in seeking astrological counseling. One would be much safer in the hands of a trained, licensed, respected, reputable counselor.

In order to safeguard the public against astrological hucksters and other "paranormalists," Carlson argues that individuals who charge a fee for the exercise of special occult skills should be required to pass an examination which requires them to demonstrate those skills in a fashion that precludes trickery or self-deception.[29]

Astro-Biology

There is a sense in which many terrestrial phenomena may be influenced by electromagnetic and gravitational effects originating within the solar system. Such influences, do not support traditional Ptolemaic astrology. Nevertheless, they may play a role in understanding possible subtle environmental effects upon consciousness. The correlations that are reported here stand on the edge of scientific respectability and

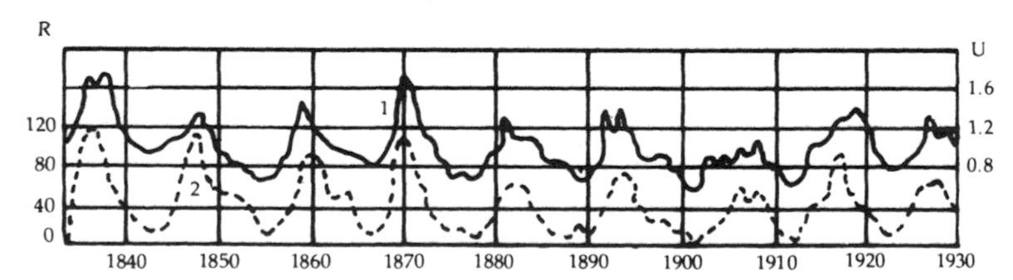

RELATIONSHIP BETWEEN MEAN ANNUAL MAGNETIC ACTIVITY (U) AND NUMBER OF SUNSPOTS (R). (AFTER A. S. PRESSMAN, *ELECTROMAGNETIC FIELDS AND LIFE*. NEW YORK: PLENUM PRESS, 1970)

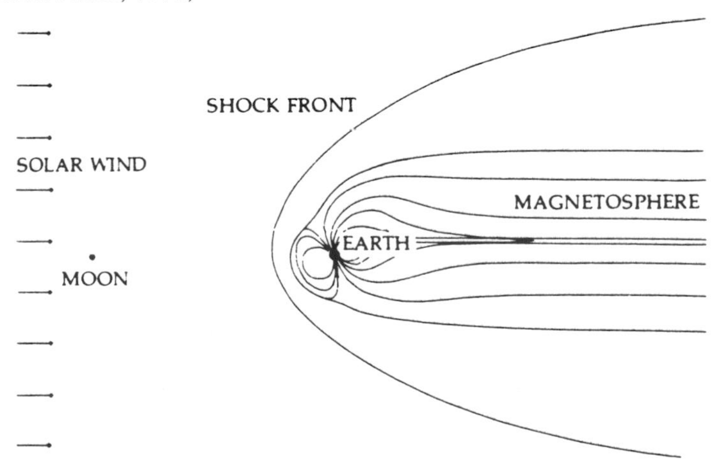

THE MAGNETIC FIELD EXTENDING OUT AROUND THE EARTH, BLOWN BY SOLAR WINDS AWAY FROM THE SUN. FLUCTUATIONS IN SOLAR STORM ACTIVITY FOLLOW A CYCLE AVERAGING 11.2 YEARS AND VARYING FROM ABOUT NINE TO THIRTEEN YEARS IN LENGTH.

must be considered more seriously than other claims in Section II.

The earth's magnetic field changes slightly according to the solar day, the lunar day, and the lunar month. Geomagnetic disturbances are particularly correlated with solar storms discharging large clouds of ionic plasma. These solar eddies generally impinge upon the earth's magnetosphere about two days after the solar flare causing polar lights, radio interference, and compression of the earth's magnetic lines of flux.

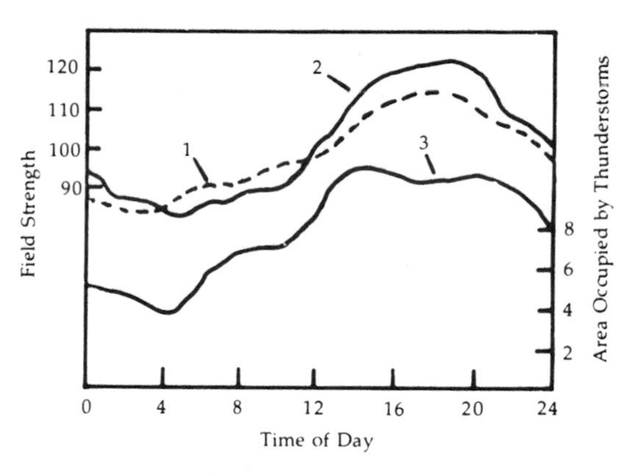

DIURNAL VARIATIONS OF ELECTRIC FIELD OF ATMOSPHERE: 1) OVER THE OCEAN; 2) IN POLAR REGIONS; 3) THUNDERSTORM ACTIVITY OVER WHOLE EARTH'S SURFACE. (AFTER PRESSMAN.)

Scientists have correlated solar storm activity to rates of heart attacks, lung disease, eclampsias, and the activity of microbes. Epidemics of diptheria, typhus, cholera, and smallpox have also been correlated with solar activity. Much of this work was done between the two world wars by the Russian scientist A. L. Tchijewski. In a huge study he drew up lists of wars, epidemics, revolutions, and population movements from 500 B.C. to 1900 A.D. and plotted them against curves of solar activity. He found that 78% of these outbreaks correlated with peaks of solar activity. He also found an amazing assortment of correlating phenomena ranging from locust hordes in Russia to succession of liberal and conservative governments in England from 1830 to 1930. Sturgeon in the Caspian Sea reproduce and then die in masses following cycles of eleven and thirty-three years which occur during periods of many sunspots (solar storms). The great financial crisis of 1929 coincided with a peak in solar activity. Other research has shown correlations between solar activity and the number of road accidents and mining disasters reported. This may be due to delayed or inaccurate human reactions in conjunction with very violent solar activity.[30]

An Italian chemist, Giorgi Piccardi, was asked to figure out why "activated" water dissolves the calcium deposits from a water boiler at certain times and not at others. (Activated water is a vestige of alchemy. A sealed phial containing neon and mercury is moved around in the water until the neon lights up; there is no chemical change in the water; however, the structure of the molecular bonds are altered somewhat.) After years of patient research measuring the rate at

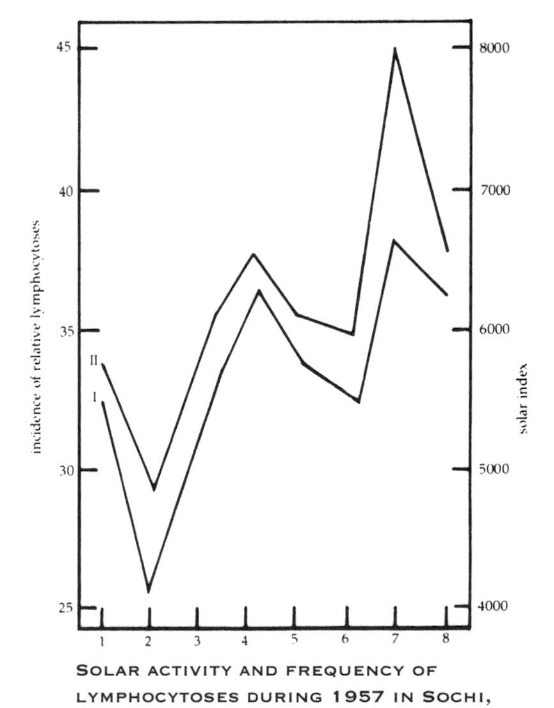

SOLAR ACTIVITY AND FREQUENCY OF LYMPHOCYTOSES DURING 1957 IN SOCHI, USSR. (AFTER PRESSMAN, P. 199.)

which bismuth sulfide becomes a colloid in activated and normal water, Piccardi showed that this colloid-forming rate varies with sunspot activity.[31] A colloid solution is one in which the dissolved particles have large enough molecular weight so that the surface tension of each molecule is of importance in determining the behavior of the solution. Common colloids are glue, gelatin, milk, egg white and blood. (The word colloid is derived from the Greek word *kolla*, meaning glue.) In general colloidal particles are too big to pass through membranes which will pass smaller dissolved molecules. The influence solar activity exerts upon the molecular structure of water is likely to be even more acute in human organisms as the human body temperature is fixed near the limit where changes in the structure of water normally occur.[32]

Not only are inorganic colloidal suspensions affected by solar activity, so is at least one other organic colloid as well — blood. Research by Dr. M. Takata in Japan, since verified in Germany and the Soviet Union, indicates that blood samples showing flocculation (a cloud-like formation) as well as the leucocyte (white blood cell) content of the blood varies in accordance with solar storm activity.[33] This widespread solar influence upon all colloidal substances manifests itself in a wide variety of ways. Individual reaction times, pain felt by amputees, and the number of suicides all reveal a similar variation in response to sunspots. Michel Gauquelin lists numerous ways in which the sunspot cycle affects weather conditions:

> During violent solar eruptions, or at the time when important groups of sunspots move to the sun's central zone, a certain number of disturbances occur in our atmosphere, particularly the *aurorae boreales*, as a result of the greater ionization in the upper atmosphere, and magnetic storms, revealed by violent agitation of compass needles....
>
> The level of Lake Victoria-Nyanza changes in accordance with the rhythm of the sunspots, also the number of icebergs, and famines in India due to lack of rain. The *Bulletin Astronomique de France* brought out a very interesting article on the relation between the activity of sunspots and the quality of Burgundy wines: excellent vintages correspond with periods of maximum solar activity, and bad vintages with periods of minimum activity. Douglas, an American, and Schvedov, a Russian, have observed that the concentric rings formed in the growth of trees have a period of recurrence of eleven years as well. Finally, there is Lury's well-known statistical observation that the number of rabbit skins taken by the Hudson Bay Company follows a curve parallel to that of solar activity.
>
> On this subject, perhaps the most interesting study is that carried out on varves. These say Piccardi, are many-layered deposits of sand or clay which are formed in calm waters, lakes, ponds, swamps, etc., in glacial zones. A varve's thickness depends on the rainfall in a given year. Examination of these fossilized deposits in sedementary rock formed through the geological ages reveals the same inevitable eleven year cycle in the most distant past.[34]

F. A. Brown is an eminent biologist who has advanced the theory that the "biological clock" mechanisms observed in organisms can be explained by animals being sensitive to various subtle environmental factors. In addition to demonstrating the influence magnetic fields exert on a wide variety of living organisms, Brown has also demonstrated that several organisms including the potato, oyster, fiddler crab, and rat modify their behavior according to lunar rhythms. The experimental subjects were enclosed for long periods in hermetically sealed rooms where the light, pressure, temperature, and humidity were carefully kept constant.

He also notes that "fluctuations in intensity of primary cosmic rays entering the earth's atmosphere were dependent upon the strength of geomagnetism. The magnetic field steadily undergoes fluctuations in intensity. When the field is stronger, fewer primary cosmic rays come into the outer atmosphere; when it weakens, more get in."[35] Other researchers have shown influences on the circadian rhythm of electrostatic fields, gamma radiation, x rays, and weak radio waves.[36,37] Recent years have shown an upsurge of interest in the ways in which human activity is affected by the remote environment. Scientists around the world who are doing research in this area have been meeting under the auspices of the International Society for Biometeorology. In 1969, the society created a special study group on the "biological effects of low and high energy particles and of extra-terrestrial factors." On this committee sit such scientists as F. A. Brown, Giorgi Piccardi, and Michel Gauquelin. Many poorly understood phenomena are now coming under the scrutiny of respectable members of the scientific community. Gradually the frontiers of science are being extended into territory that once belonged to mystics and occultists.

Perhaps one could think of objects as complicated concatinations of interpenetrating electromagnetic fields. As far as we know, all objects in our universe, with a temperature above absolute zero, are emitting electromagnetic radiation. In this sense, the alchemical theories of Alkindi are true.

Chronotopology

Dr. Charles Muses — a mathematician, philosopher, and computer scientist — is author of *Destiny and Control in Human Systems* and *The Lion Path.* Muses has developed an approach to astrology couched in the language of systems theory. In a *Thinking Allowed* interview, he describes the theoretical principles underlying the methodology he has come to refer to as *chronotopology,* i.e., studying the structure of time. One has a sense from this discussion that contemporary astrology is, perhaps, a decadent form of what was once a philosophically well-grounded and noble pursuit:

> Muses: The hypothesis of chronotopology is that whether you have pointers of any kind — ionospheric disturbances or planetary orbits — independently of those pointers, time itself has a flux, a wave motion.
>
> Mishlove: In quantum physics there's this notion that the underlying basis for the physical universe are probability waves — nonphysical, nonmaterial waves — underlying everything.
>
> Muses: These waves are standing waves. Actually the wave-particle so-called paradox is not that bad, when you consider that a particle is a wave packet, a packet of standing waves. That's why an electron can go through a plate and leave wavelike things. Actually our bodies are like fountains. The fountain has a shape only because it's being renewed every minute, and our bodies are being renewed. So we are standing waves; we are no exception.
>
> Time is the master control. I will give you an illustration of that. If you take a moment of time, this moment cuts through the entire physical universe as we're talking. It

CHARLES MUSES (FROM *TIME AND DESTINY* VIDEOTAPE, COURTESY THINKING ALLOWED PRODUCTIONS)

holds all of space in itself. But one point of space doesn't hold all of time. In other words, time is much bigger than space.

A line of time is then an occurrence, and a wave of time is a recurrence. And then if you get out from the circle of time, which Nietzsche saw, the eternal recurrence — if you break that, as we know we do, we develop, and then we're on a helix, because we come around but it's a little different each time.

Mishlove: Well, now you're beginning to introduce the notion of symbols — point, line, wave, helix, and so on.

Muses: Yes, the dimensions of time.

Mishlove: Symbols themselves — words, pictures — point to the deeper structure of things, including the deeper structure of time. I gather that you are suggesting the mind is part of a nonphysical, mathematically definable reality that can interface and interact with physical reality, and in which physical reality is embedded.

Muses: There can be some things which are physically effective which are not physical. I can give you an illustration, a very recondite one, but there is the zero-point energy of the vacuum. The vacuum is defined in quantum physics as space devoid of radiation or matter — no energy, no matter. Yet there is an inherent energy in there which can be measured — this is one of the great triumphs of modern physics — and that is physically effective.

Mishlove: The energy of a pure vacuum.

Muses: Yes. Yet it obviously is not a pure vacuum. The so-called savage would say to us, "The room is empty, and the wind is a magic spirit." We know it is air. So we are like the savage in saying that the vacuum is empty. There is something there.[38]

Muses, in effect, has been echoing the ancient claim of the Primordial Tradition that there is a fundamental unity between the universal mind and the cosmos itself — including the unfolding of time. The structure of the relationship between macrocosm and microcosm is expressed in mathematical, scientific and mythological symbols. It is the intuitive grasp of these symbols which is ultimately the goal of astrology.

Such a system must always be larger and more enduring than any rational attempt to understand or contain it. If this perspective of astrology is correct, I would predict that rationalists will forever shun astrology's pseudoscientific face. And, ironically, astrology and other "superstitions" will persist because the yearning human soul can never be content with rational materialism.

Arthur M. Young's "Geometry of Meaning"[39]

A most elegant and brilliant search for the deep unity of myth, mathematics and morphology is Arthur M. Young's derivation of a "geometry of meaning" from the angular relationships that exist between the measure formula of physics.[40] Starting from pure physical and mathematical relationships, Young has built an elegant theoretical model that bears an uncanny isomorphism to the twelve signs of the zodiac.

He begins by plotting the motion of a pendulum (as a representative of simple harmonic motion — the basis of all wave motion) over time on a Cartesian coordinate system. The completion of the swing to the left and its return to A is referred to as a *cycle of action* whose halfway point is C.

ARTHUR M. YOUNG

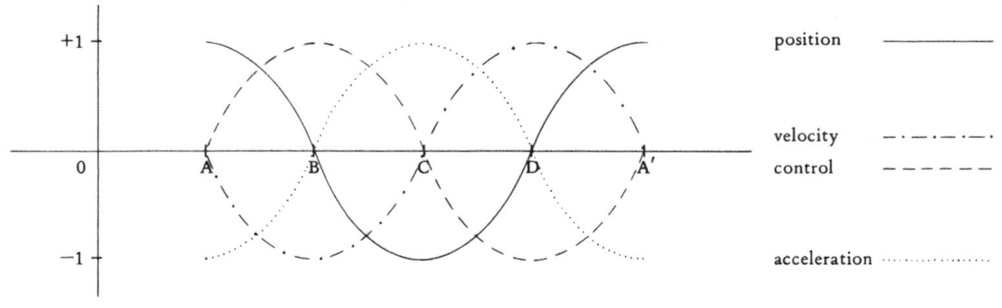

The slope of the following curve represents the rate at which the position of the pendulum or its velocity changes. The rate at which the velocity changes is the acceleration, and the rate at which acceleration changes Young designates as control. These dimensions can be plotted in a likewise fashion.

These curves are all the same shape, but the curve for velocity is displaced one-fourth of the cycle before the position curve. The velocity is at 0 when the position is either +1 or -l. The velocity is at a maximum (in a positive or negative direction) when the position of the pendulum is at 0. The curve for acceleration is displaced one-fourth of the cycle from the curve of velocity, and one-half of a cycle from the curve of position. Thus, when the pendulum's position is 0, and the velocity is at a maximum, the acceleration is 0. The control curve is displaced one-fourth of a cycle from the acceleration curve and one-half of a cycle from the velocity. This cycle of action can be represented as a full circle — with the designations of maximum position, velocity, acceleration, and control falling on four points 90 degrees apart.

The fourth positional derivative repeats the cycle and becomes once more "position" in this scheme. Young maintains that "these four categories of the measure are necessary and sufficient for the analysis of motion of a moving body." He also states that there are fundamental qualitative differences between these measures.

In order to determine position, we make an observation — either visually or by some equally direct process. On the other hand, velocity cannot be observed directly. It must be computed. It can only be known intellectually. We must make two observations of

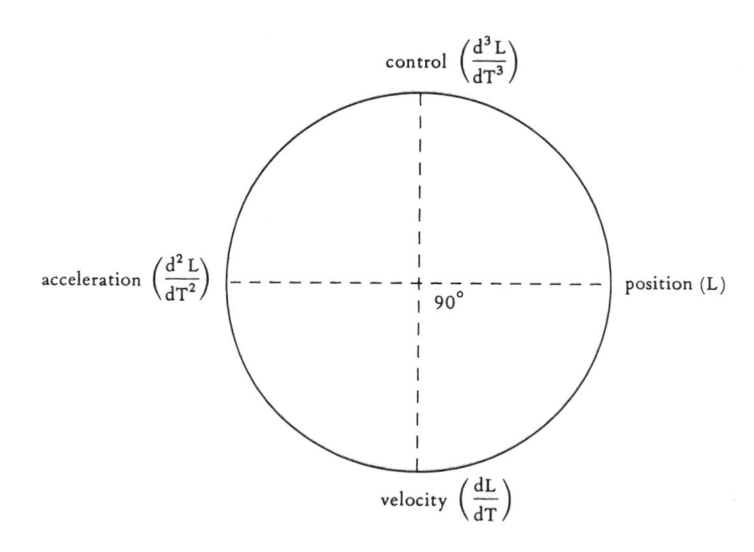

position, determine their difference and divide by the time elapsed, obtaining a ratio. Acceleration can also be computed, but it can be directly and physically experienced by the knower, through feeling, in a way by which the other measures cannot be known. Control, or change in acceleration created by an operator. This element is indeterminate and unknowable to an observer.

The four types of experience derived from the physical measure formulae now become the basis for another cycle of action called the learning cycle. This cycle begins with a spontaneous act, an impulse derived from feeling, and equivalent, Young believes, to acceleration. Unconscious reaction is the second element of this cycle. This reaction is based on habits, instincts, or "programming" in the language of bio-computer theory. It is equivalent to the computations involved in recognizing velocity. The third element of the cycle is observation and is equivalent to position, the observable physical factor. The fourth element of the learning cycle is control.

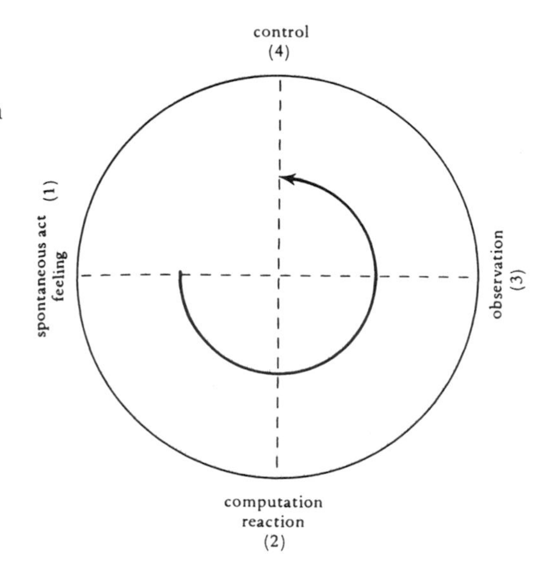

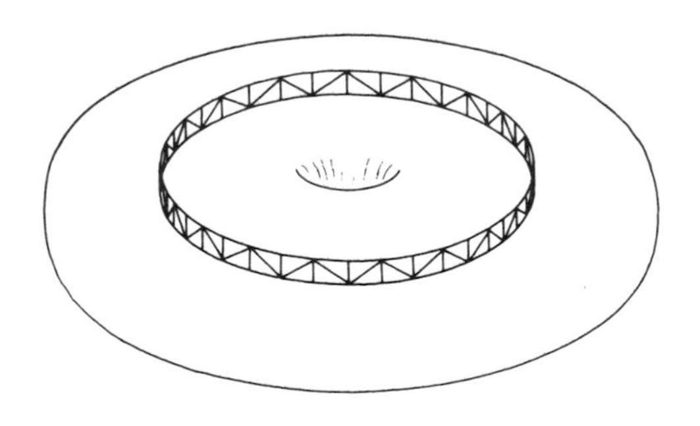

An example of the learning cycle would be a baby reaching out to touch something (spontaneous action). If a hot object like a flame is encountered, the hand is withdrawn (reaction). Then there follows observation upon the event. Further exploration consciously avoids fire until the learned behavior becomes automatically programmed and spontaneity is resumed. The cycle can be diagrammed as shown.

Young derives a formula for the cycle of action or learning that leads to consciousness. Note that the distance from spontaneous action to conscious control is 3/4 of 360 degrees or 3/4 x 2π which equals $3\pi/2$. The common sense view of the universe is to consider it shaped like a sphere extending infinitely in all directions. However, if we multiply the formula for the volume of a sphere ($4\pi r^3/3$) by Young's formula for the cycle leading to consciousness, we obtain the formula for the volume of a torus (donut) with an infinitely small hole ($2\pi^2 r^3$). This is significant for Young in that the formula for the volume of the Einstein-Eddington Universe, the boundary region of the so-called hypersphere is also $2\pi^2 r^3$.

Young sees in the torus topology a possible answer to the philosophical problem of the individual (or part, or microcosm) versus the collective (or whole, or macrocosm). In a toroidal universe, a part can be seemingly separate and yet connected with the rest. If we think of the fence as separating the inner from the outer, the torus provides a paradigm that permits us to see a monad as both separated from the rest of the universe by the fence and still connected with everything else through the core. The core of the torus with its infinitely small hole is for Young a representation of inner consciousness.

Young points out that magnetic fields, vortices, and tornados all have the toroidal form. The vortex is furthermore the only manner in which a fluid can move on itself. Thus, it is a very suitable shape for the universe to have. However, we must bear in mind that the volume of the torus is three dimensional and is something akin to the surface of the four dimensional hypersphere of Eddington and Einstein.

If we expand upon the "geometry of meaning," we can add eight other measure formulae of physics for a total of twelve.

You will recall the logic by which Young determined that length (or position) and its three derivatives divided the circle into four quarters, thus giving the operation of *T* (time) an angular value of 90 degrees. Through a process of trial and error, Young found that by assuming that *M* (mass) has the value of 120 degrees and *L* the value 30 degrees, the measure formula could equally spread around a circle in twelve positions. No proof is given for this assumption. However, when these values are applied to the measure formula and incorporated into the cycle of action, we do get the above, symmetrically elegant, results shown on page 78.

Young then found that he could assign the different astrological signs to the measure formula according to the appropriateness of the physical and astrological symbolism. Acceleration, at the starting point of our learning cycle, is equated with Aries, the first sign of the zodiac. Mass control is translated into the sign of Taurus, the bull. Gemini, the sign of knowledge ruled by Mercury, is equated with the physical measure for power. ("After all," says Young, quoting Francis Bacon, "Knowledge is Power.") The process continues in a way that is rather reasonable from the standpoint of astrology, if not outrageous to physicists. The results are seen on page 86.

The diagram of Young's "Rosetta Stone" is a subject treated at length in his book *The Geometry of Meaning*. Its significance lies beyond the traditional realms of either astrology or physics. It may be a perfectly useless intellectual "bead game." On the other hand, it points toward the unification of symbolic meaning with mathematical manipulation. It provides a comprehensive metaphor with which to describe the processes of consciousness. It is also suggestive of a metasystem within which one can integrate the diverse disciplines of human endeavor. Such synergistic approaches are necessary in order to apply the vast resources of our information explosion to the social problems confronting us.

Young's theory fits within the context of historical cosmological speculations beginning with Pythagoras. In earlier cultures, psi phenomena were generally incorporated in the prevailing world view with a great deal of eloquence, if not rational lucidity. They were thought to stem from the heavenly hierarchies and celestial realms believed to interpenetrate or transcend all gross matter. However, in the attempt to be empirically grounded, modern cosmologists in rejecting supernatural elements have also unthinkingly rejected psi. Consciousness itself has been left out of the scientific picture. Something of an operational substitute has been provided by observable behavior.

Arthur Young has attempted to bridge this gap by developing a modern, scientific meta-theory within which one can account for human consciousness and process, including the data of parapsychology. What such a theory leads to is a holistic view of

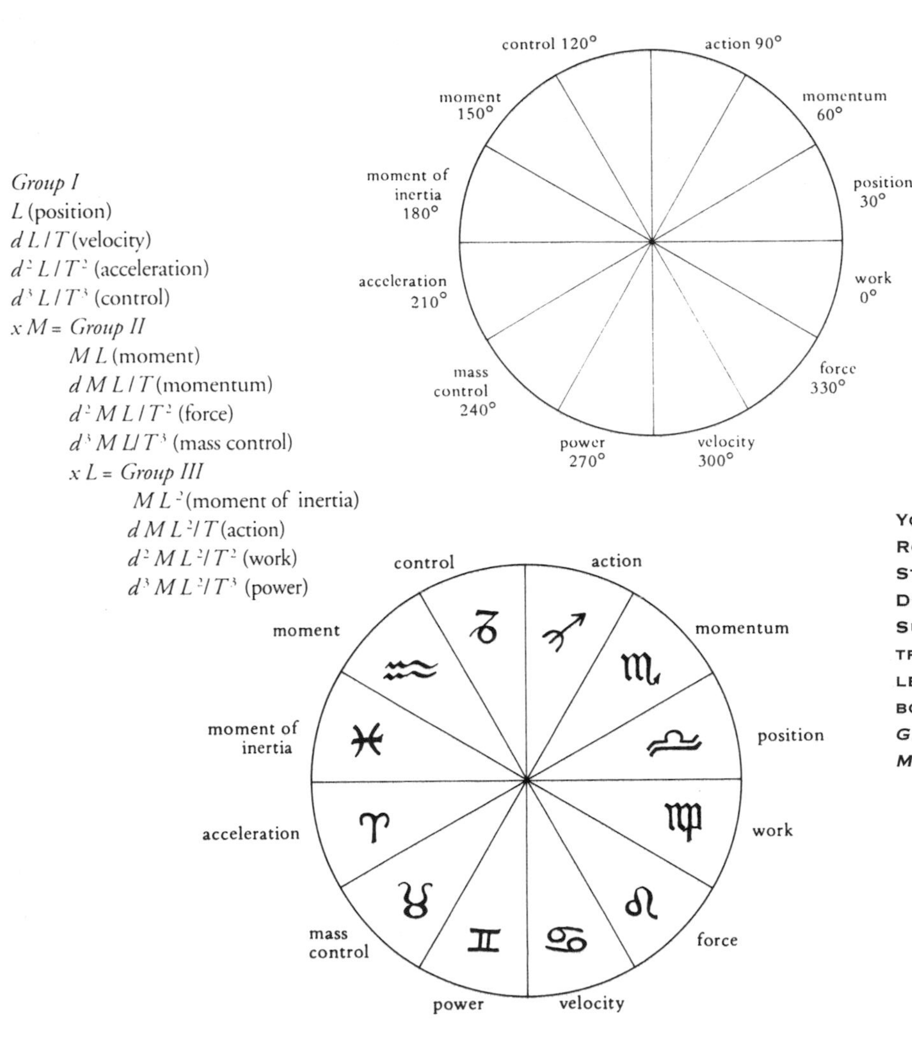

YOUNG'S ROSETTA STONE. A DIAGRAM AND SUBJECT TREATED AT LENGTH IN HIS BOOK *THE GEOMETRY OF MEANING.*

the universe, oneself, and humankind.

Arthur Young is the founder of the Institute for the Study of Consciousness located in Berkeley, California. The focus of the institute is primarily intellectual or gnostic. It operates under Young's precept that a true understanding of the universe can only be grasped through the exercise of one's full human capacities.

Concluding Thoughts on Astrology

How ironic that Arthur Young's *Rosetta Stone* would seem to provide a mathematical/physical foundation for the somewhat discredited mythical signs of the zodiac. On one hand, Young's "geometry of meaning" falls firmly within the Pythagorean tradition. In principle, he seems to be digging into a mathematical structure that offers the potential to unify the mythic-subjective and the scientific-objective aspects of consciousness. This is a quest and a theme to which we will return again and again. However, when one looks at the research in astrology, particularly that of Gauquelin, it seems that the planets and houses fare the best — while little empirical support is offered for the zodiacal signs.

To be sure, Arthur Young himself is not disconcerted by this turn of events. He does not view his work as a theory to be fitted within the structure of science. Rather, he sees it as a paradigm or model which encompasses science and mythology. From Young's perspective, astrology is *not* a science. It is a language beyond science that connects us to the realm of mythos. It's validity is ideosyncratic. It deals with the uniqueness of individuals and their connection to the mysterious realm of the gods.

If we view astrology as an ancient folk tradition, we may say that it is derived from several sources: the deep mathematics of Pythagoras, the numenous realm of mythology, the cycles of the seasons and agriculture, the ecology of consciousness, the uncanny planetary influences documented by Gauquelin and his critics, as well as the human need for certainty in the face of the unknown. One might well also add the human propensity for folly.

In dealing with astrology, as well as in dealing with so much more that will be discussed, the one character trait that will most adequately serve us is the ability to tolerate ambiguity. Of course, as any astrologer can tell you, some of us have this trait in greater proportion than others. (If it didn't come rather naturally to me, I doubt whether I would be able to write this book at all.) As psychologist Charles Tart is fond of saying, "I do not believe in astrology; but then, people of my sign never do."

Tart's statement reflects an attitude of humor and irony which is an essential prerequisite for intelligent, sensitive exploration into the murky waters of consciousness.

Astral Projection and Out-of-Body Experiences

According to tradition, shamans are individuals who, by definition, have overcome the common perceptual mindset of space-time conditioning. Space for them no longer exists. It is sufficient for them to

intensify the sharpness of their interior sight and the intensity of their interior light to be able to penetrate everything. We read in traditional texts and in reports of scholars that shamans bilocate themselves, that they move in space, can go far and return in an instant, that they are clairvoyant, besides being therapists and magicians.

Certain shamanistic conceptions suggest that our past thoughts and our mind have never left our body, for the simple fact that they have *always* been outside our body and our brain. They do not need to emerge from inside, because they have never been inside. The shamanistic initiation makes the trainee simply aware that mind and consciousness are outside and above, and they have always been so. The initiation practices reawaken a corresponding state of consciousness, that gives the living awareness of this vision.[41]

The shamanistic conceptions and the converging descriptions of out-of-body experiences coincide with the notion that mind and consciousness are a *prius* in respect of the brain and the body — in other words, that brain and body evolve out of consciousness.

Astral projection is one of the most salient features of the mystical tradition underlying most of the major religions. In fact, anthropologists have found that OBE beliefs appear in about 95% of the world's cultures.[42]

The idea that consciousness can function independently of and outside of the physical body is found in Egyptian manuscripts that delve in detail into the nature of the *ka*, or double that can separate from the physical body and travel at will. The *ba* is the principle of life that dwells in the *ka*, much like the heart of the physical body. The *klu* is the radiance of the being in eternal life, and *sekhem* is the form through which a person exists in heaven. In addition there is the *ren* or spiritual name of a being.[43]

Allusions to astral projection are particularly prominent in the scripts of *Tantric* Buddhism, a subdivision of *Mahayana* Buddhism found in Tibet and parts of Mongolia. Such experiences are considered to be a mark of his devotion to the Buddha.[44] Pure Land Buddhism in China is a tradition which not only admits to OBEs, but is philosophically grounded upon their reality and accessibility to all people.[45]

The particular notion of *astral projection* can be traced back to Pythagoras' claim to hear the music of the heavens. The Pythagoreans assumed that the distances of the heavenly bodies from the earth somehow corresponded to musical intervals. By allowing one's consciousness, uplifted by philosophy, to rise through these astral spheres one ultimately might attain to union with the divine.

Subsequently such terms as astral projection and out-of-body experience have come to be applied to a wide variety of visionary, mystical and psychic experiences. For example, the experience developed in the mystery traditions which enabled participants to lose their fear of death might be viewed in this way. St. Augustine's visionary experience, described earlier, is another possible instance.

Dante Allegheiri epitomizes the artistic evolution that signaled the end of the Middle Ages and the rise of the Italian Renaissance. His descriptions in the *Divine Comedy* of his own visions into the worlds of hell, purgatory, and paradise enjoy a paramount position in Western poetry. Yet there is a striking similarity in his work to the understanding of the afterworld one finds earlier in Egypt, Tibet, and Plato, as well as later in the visions of Emmanuel Swedenborg. Dante's familiarity with several systems of medieval mysticism leads us to believe that he used dream and reverie states as an inspirational source for his artwork:

> At the hour near morning when the swallow begins her plaintive songs, in remembrance, perhaps, of her ancient woes, and when our mind, more a pilgrim from the flesh and less held by thoughts, is in its visions almost prophetic, I seemed to see in a dream an eagle poised in the sky, with feathers of gold, with open wings and prepared to swoop. And I seemed to be in the place where his own people were left behind by Ganymede when he was caught up to the supreme conclave;....[46, 47]

Broadly speaking, we can define an astral projection, bilocation or out-of-body experience (OBE) as the sensation of observing phenomena from a perspective that does not coincide with the physical body. Often one will experience consciousness being transferred from the physical body to another "astral body," "second body," "etheric body," "double," or "doppelganger." On other occasions, one may experience oneself as a mere point of awareness outside of the physical body. There seem to be several distinct, but related, types of experience lumped together under the general rubric of out-of-body experience. These include (1) lucid dreams where one seems to be conscious within a dream world, (2) clairvoyant awareness of distant locations, (3) the actual sensation of separation from one's physical body, floating above it, and looking down upon the physical form, (4) traveling outside of one's body to different locations in physical time and space, and (5) gliding and flying through the various supersensible "astral" and spiritual planes.

Thousands of OBEs have been reported by individuals of all ages and from all walks of life. Such experiences have played a major role in the shamanistic rites and esoteric schools of many previous cultures. Much occult literature abounds in unsubstantiated claims regarding the vast scientific and historical knowledge that can be imparted to visitors upon the "higher planes." Some of this literature is actually quite valuable because of the systematic explorations conducted by philosophically trained clairvoyants. This body of literature seems to put the OBE into a larger perspective.

Ramacharaka's Theosophical Perspective

A typical description of astral traveling, from the "occult" viewpoint, is provided by Yogi Ramacharaka:

> It is possible for a person to project his astral body, or travel in his astral body, to any point within the limits of the earth's attraction,[48] and the trained occultist may do so at will, under the proper conditions. Others may occasionally take such trips (without knowing just how they do it, and having afterwards, the remembrance of a particular and very vivid dream); in fact many of us do take such trips, when the physical body is wrapped in sleep, and one often gains much information in this way, upon subjects in which he is interested, by holding astral communication with others interested in the same subject, all unconsciously of course. The conscious acquirement of knowledge in this way, is possible only to those who have progressed quite a way along the path of attainment. The trained occultist merely places himself in the proper mental condition, and then wishes himself at some particular place, and his astral travels there with the rapidity of light, or even more rapidly. The untrained occultist, of course, has no such degree of control over his astral body and is more or less clumsy in his management of it. The Astral Body is always connected with the physical body (during the life of the latter) by a thin silk-like, astral thread, which maintains the communication between the two. Were this cord to be severed the physical body would die, as the connection of the soul with it would be terminated....
>
> Perhaps the best way to make plain to you the general aspects and phenomena of the Astral World, would be to describe to you an imaginary trip made by yourself in that world, in the charge of an experienced occultist. We will send you, in imagination, on such a trip, in this lesson, in charge of a competent guide it being presupposed that you have made considerable spiritual progress, as otherwise even the guide could not take you very far, except by adopting heroic and very unusual methods, which he probably would not see fit to do in your case. Are you ready for your trip? Well, here is your guide.
>
> You have gone into the silence, and suddenly become aware of having passed out of your body, and to be now occupying only your astral body. You stand beside your physical body, and see it sleeping on the couch, but you realize that you are connected with it by a bright silvery thread, looking something like a large bit of bright spider-web. You feel the presence of your guide, who is to conduct you on your journey. He also has left his physical body, and is in his astral form, which reminds you of a vapory something, the shape of the human body, but which can be seen through, and which can move through solid objects at will. Your guide takes your hand in his and says, "Come," and in an instant you have left your room and are over the city in which you dwell, floating along like a summer cloud. You begin to fear lest you may fall, and as soon as the thought enters your mind you find yourself

sinking. But your guide places a hand under you and sustains you, saying, "No just realize that you cannot sink unless you fear to — hold the thought that you are buoyant and you will be so." You do so, and are delighted to find that you may float at will, moving here and there in accordance to your wish or desire.

You see great volumes of thought-clouds arising from the city like great clouds of smoke, rolling along and settling here and there. You also see some finer vapory thought-clouds in certain quarters, which seem to have the property of scattering the dark clouds when they come in contact with them. Here and there you see bright thin lines of bright light, like an electric spark, traveling rapidly through space, which your guide tells you are telepathic messages passing from one person to another, the light being caused by the Prana with which the thought is charged. You see, as you descend toward the ground, that every person is surrounded by an egg-shaped body of color, — his aura — which reflects his thought and prevailing mental state, the character of the thought being represented by varying colors. Some are surrounded by beautiful auras, while others have around them a black, smoky aura, in which are seen flashes of red light. Some of these auras make you heart-sick to observe, as they give evidence of such base, gross, and animal thoughts, that they cause you pain, as you have become more sensitive now that you are out of your physical body. But you have not much time to spare here, as your trip is but a short one, and your guide bids you come on.

You do not seem to change your place in space, but a change seems to have come over everything — like the lifting of a gauzy curtain in the pantomime. You no longer see the physical world with its astral phenomena, but seem to be in a new world — a land of queer shapes. You see astral "shells" floating about — discarded astral bodies of those who have shed them as they passed on. These are not pleasant to look upon, and you hurry on with your guide, but before you leave this second ante-room to the real Astral World, your guide bids you relax your mental dependence upon your astral body, and much to your surprise you find yourself slipping out of it, leaving it in the world of shells, but being still connected with it by a silk-like cord, or thread, just as it, in turn, is connected with your physical body, which you have almost forgotten by this time, but to which you are still bound by these almost invisible ties. You pass on clothed in a new body, or rather an inner garment of ethereal matter, for it seems as if you have been merely shedding one cloak, and then another, the YOU part of yourself remains unchanged — you smile now at the recollection that once upon a time you thought that the body was "you." The plane of the "astral shells" fades away, and you seem to have entered a great room of sleeping forms, lying at rest and in peace, the only moving shapes being those from higher spheres who have descended to this plane in order to perform tasks for the good of their humbler brethren. Occasionally some sleeper will show signs of awakening, and at once some of these helpers will cluster around him, and seem to melt away into some other plane with him. But the most wonderful thing about this region seems to be that as the sleeper awakens slowly, his astral body slips away from him just as yours a little before, and passes out of that plane to the place of "shells," where it slowly disintegrates and is resolved into its original elements. This discarded shell is not connected with the physical body of the sleeping soul, which physical body has been buried or cremated, as it is "dead"; nor is the shell connected with the soul which has gone on, as it has finally discarded it and thrown it off. It is different in your case, for you have merely left it in the ante-room, and will return and resume its use, presently.

The scene again changes, and you find yourself in the regions of the awakened souls, through which you, with your guides, wander backward and forward. You notice that as the awakening souls pass along, they seem to rapidly drop sheath after sheath of their mental-bodies (for so these higher forms of ethereal covering are called), and you notice that as you move toward the higher planes your substance becomes more and more etherealized, and that as you return to the lower planes it becomes coarser and grosser, although always far more etherealized than even the astral body, and infinitely finer than the material body. You also notice that each awakening soul is left to finally awaken on some particular plane. Your guide tells you that the particular plane is determined by the spiritual progress and attainment made by the soul in its past lives (for it has had many earthly visits or lives), and that it is practically impossible for a soul to go beyond the plane to which it belongs, although those on the upper planes may freely revisit the lower planes, this being the rule of the Astral World — not an arbitrary law, but a law of nature....[49]

This description bears at least some resemblance to other accounts from such diverse sources as *The Egyptian Book of the Dead*, *The Tibetan Book of the Dead*, Plato's description of *Er*, Dante's *Divine Comedy*, and Swedenborg. Although the social climate in our culture is arriving at a point where it will soon be more prevalent, so far there have been few spiritual visionaries who felt that working with scientists would be a beneficial use of their time. Similarly, very few scientists are interested in working with visionaries. Thus, science has currently little to say about such experiences.

An Accidental Projection

Apparently not everyone who leaves his body is able to travel to the Empyrean heights (if they exist). Many individuals who have been spontaneously thrust outside of their bodies, or who have cultivated the ability to have OBEs at will, have sought a scientific confirmation and understanding of their experiences. These projections often result from hypnosis, anesthesia, drugs, stress, or accidents. A typical accidental projection occurred to a seventy-year-old Wisconsin man:

> He had hitched his team, one wintry day, and gone into the country after a load of firewood. On his return, he was sitting atop the loaded sleigh. A light snow was falling. Without warning, a hunter (who happened to be near the road) discharged his gun at a rabbit. The horses jumped, jerking the sleigh and throwing the driver to the ground head-first.
>
> He said...that no sooner had he landed upon the ground than he was conscious of standing up and seeing another "himself" lying motionless near the road, face down in the snow. He saw the snow falling all about, saw the steam rising from the horses, saw the hunter running toward him. All this was very exact; but his great bemuddlement was that there were *two* of him, for he believed at the time that he was observing all that occurred from another physical body.
>
> As the hunter came near, things seemed to grow dim. The next conscious impression he had was of finding himself upon the ground, with the hunter trying to revive him. What he had seen from his astral body was so real that he could not believe that there were not two physical bodies, and he even went so far as to look for tracks in the snow, in the place where he knew he had been standing.[50]

OBE in a Dream

Projections frequently occur in dreams. A classic example of a dream OBE, was reported in 1863 by Mr. Wilmot of Bridgeport, Connecticut:

> I sailed from Liverpool for New York, on the steamer City of Limerick....On the evening of the second day out,...a severe storm began which lasted for nine days....Upon the night of the eighth day,...for the first time I enjoyed refreshing sleep. Toward morning I dreamed that I

> saw my wife, whom I had left in the U.S., come to the door of the stateroom, clad in her night dress. At the door she seemed to discover that I was not the only occupant in the room, hesitated a little, then advanced to my side, stooped down and kissed me, and quietly withdrew.
>
> Upon waking I was surprised to see my fellow-passenger…leaning upon his elbow and looking fixedly at me. "You're a pretty fellow," he said at length, "to have a lady come and visit you this way." I pressed him for an explanation, and he related what he had seen while wide awake, lying on his berth. It exactly corresponded with my dream….
>
> The day after landing I went to Watertown, Conn., where my children and my wife were…visiting her parents. Almost her first question when we were back alone was, "Did you receive a visit from me a week ago Tuesday?"…."It would be impossible," I said. "Tell me what makes you think so." My wife then told me that on account of the severity of the weather,…she had been extremely anxious about me. On the night mentioned above she had lain awake a long time thinking about me, and about four o'clock in the morning it seemed to her that she went out to seek me….She came at length…to my stateroom. "Tell me," she said, "do they ever have staterooms like the one I saw, where the upper berth extends further back than the under one? A man was in the upper berth looking right at me, and for a moment I was afraid to go in, but soon I went up to the side of your berth, bent down and kissed you, and embraced you, and then went away." The description given by my wife of the steamship was correct in all particulars, though she had never seen it.[51]

Astral projection is so often associated with dreaming that many writers insist that the astral body normally separates from the physical during sleep. Most of us, this theory posits, are not sensitive to the separation and only maintain a vague memory of the experience as a dream.

Conscious Astral Projection

Many techniques for conscious astral projection involve regaining consciousness within the dream state. I would suggest, however, that you not engage in such practice if you often experience great discord within yourself.

Sylvan Muldoon's Method

One technique is offered by Sylvan Muldoon in his book *The Projection of The Astral Body*:

> 1. Develop yourself so that you are enabled to hold consciousness up to the very moment of "rising to sleep." The best way to do this is to hold some member of the physical body in such a position that it will not be at rest, but will be inclined to fall as you enter sleep.
>
> 2. Construct a dream which will have the action of Self predominant. The dream must be of the aviation type, in which you move upward and outward, corresponding to the action of the astral body while projecting. It must be a dream of something which you enjoy doing.
>
> 3. Hold the dream clearly in mind; visualize it as you are rising to sleep; project yourself right into it and go on dreaming.[52]

Through the use of properly applied suggestion, prior to the dream, you will be able to remember yourself in your dream and bring your dream body — or astral body — to full waking consciousness. This technique may require months of gentle persistence.

Muldoon's book, first published in 1929, offers a wealth of information based on the hundreds of out-of-body experiences he had over a period of many years. However, Muldoon's experiences were seldom completely conscious, and never beyond the limits of the immediate earth environment. In one "superconscious" experience, after a lonely evening, he found himself in a strange house, watching a young lady, who happened to be sewing at the time. Six weeks later, he chanced to recognize this woman on the streets of the small Wisconsin town where he lived. Upon his approaching her, she was startled to discover that he was able to accurately describe the inside of her home. She eventually became a very close friend of his and participated with him in a number of projection experiments.

By systematically observing his own condition in the out-of-body state, Muldoon was able to derive some very interesting hypotheses. For example, he made numerous measurements of the "silver cord" connecting the astral and the physical bodies, stating that it varied in thickness from about 1½ inches to about the size of a sewing thread according to the proximity of the astral body to the physical. Muldoon does not tell us how these measurements were made. Presumably they are simply estimates of some sort. At a distance of from eight to fifteen feet, the cord reached its minimum width. It was only after this incident occurred that Muldoon was able to exercise complete control over his astral body. He also noticed that the impulses for the heartbeat and breath seemed to travel from the astral through the cord to the physical body.

> I have tried the experiment many times of holding the breath, while consciously projected, and within cord-activity range. The instant that it is suspended the before-mentioned action of slight expansion and contraction ceases, in the psychic cable, as it likewise does in the physical body; but while the respiration ceases the regular pulsating action [the heartbeat] continues. A deep breath in the astral will produce an identical breath in the physical; a short one will produce a short one; a quick one will produce a quick one, etc.[53]

Muldoon also observed that physical debilities and morbid physical conditions seemed to provide an incentive for projection. He himself was quite frail and sickly during the years when his experiences were most pronounced. It was his hypothesis that the unconscious will — motivated by desires, necessities, or habits that would otherwise have resulted in somnambulism or sleepwalking — led to astral projection for him because of the debility of his body. When he was thirsty at night, for instance, he might find his astral body traveling to the pump for water.

On one occasion it occurred to Muldoon that his heart was beating rather slowly. He went to a doctor who told him that his pulse was only forty-two beats per minute and gave him a cardiac stimulant — strychnine — to correct the condition. For the next two months Muldoon took this stimulant, and during this period he was not able to induce a projection — although during the previous year he had been averaging at least one OBE each week. After he discontinued the medication, he was again able to astral project. He also noticed that if he experienced intense emotions while out of his body, it tended to cause his heart to beat faster. This resulted in his being suddenly "interiorized" again, often against his conscious will. Such sudden interiorization often resulted in painful, sometimes cataleptic, repercussions within the body.

As his health improved, Muldoon's abilities waned and practically disappeared. Eventually he lost all interest in astral projection — after having made the most significant contribution of his time. Since then several other individuals have contributed extensive reports of their own out-of-body experiences.[54,55]

Robert Monroe's Method

Robert A. Monroe, the author of *Journeys Out of the Body*, describes how he visited several medical doctors looking for an explanation of his condition. They could find nothing wrong with him. In fact, Monroe is an excellent example of an individual whose reported experiences could not easily be attributed to defective mental or emotional functioning. A former vice-president of Mutual Broadcasting Corporation, Monroe is now president of two corporations active in cable-vision and electronics. He has produced over six hundred television programs. During the years of his reported OBEs, Monroe has continued to lead an active business and a rewarding family life.

His book documents many dimensions of OBE activity. In what he termed "Locale I" and "Locale II" are found the common experiences of the occult literature — floating outside of one's body within the familiar physical environment and then traveling to the "astral" worlds of heaven and hell complete with spirits and thoughtforms. In "Locale III" Monroe describes his visits to a plane rather parallel to our own. Human beings there lived much as we do, with some rather odd exceptions. They had no electricity or internal combustion motors, yet a rather sophisticated technology was built around a sort of steam power. Their automobiles held a single bench seat large enough for five or six people abreast.[56]

Monroe is currently engaged in a very sophisticated program of training scientists and others to participate in out-of-body experiences.

Robert Crookall's Observations

While personal accounts of this sort are invaluable, they do little to satisfy the scientific need for information that has no possibility of subjective distortion or falsification. A small, but important, step in this direction has been made by the eminent British geologist, Dr. Robert Crookall. Struck by the many independent reports of OBE, Crookall attempted a critical analysis of the data from as many sources as he could possibly collect. By looking at this collection from different perspectives, he was able to discover a number of interesting, and previously undetected, patterns of out-of-body experience.

In his first analysis, Crookall revealed a basic OBE pattern which was scattered among hundreds of cases from many different cultures: The replica body is "born" from the physical body and takes a position above it. At the moment of separation, there is generally a blackout of consciousness "much as the changing of gears in a car causes a momentary break in the transmission of power." Commonly the vacated physical body is seen from the released "double." Sometimes the "silver cord" can be noticed. The experience is generally not frightening. Many different phenomena are viewed after separation and the return of the double follows a reversal of the pattern just indicated. Rapid re-entry can cause shock to the physical body.[57]

In a second analysis, all cases were broken down into two large groupings. One group contained projections resulting naturally and gradually — from illness, exhaustion or sleep. The other included forcible and sudden projections caused by accidents, anesthetics, suffocation, or willful projection. Crookall reports that people who left their bodies in a natural manner enjoyed consciousness of a clear and extensive type, with telepathy, while the consciousness of the forcibly ejected was remarkably restricted and dim, with dreamlike elements. Those who left naturally tended to glimpse bright and peaceful conditions. The forcibly ejected, if not on earth, tended to be in the confused, and semi-dreamlike conditions corresponding to the "Hades" of the ancients. The former met many helpers (including dead friends and relatives), the latter sometimes encountered discarnate would-be hinderers.[58]

A third analysis compared the differences in experiences reported by ordinary people with those of individuals who claim to be psychics. By and large, the psychics reported experiences very much like enforced projections, whereas the non-psychics had experiences of natural projection. He also noted that the psychic and mediumistic people commonly observed a mist or vapor leaving their bodies and forming part of the double. Similar statements are often made by those who observe the permanent release of the double during the process of death.

This suggested to Crookall that the double actually comprises a semi-physical aspect called the vital or etheric body as well as an astral or super-physical Soul Body. If after the projection, the semi-physical body is still attached, the double will be able to move physical objects, cause rappings, etc. However, if the projection occurs in two stages, so that the Soul Body is separate from the vital body, then the Soul Body is free to travel to the higher "paradise" realms. This second stage would be equivalent to discarding the "astral shell" in Ramacharaka's account. In his most recent work, Crookall documents many cases in which the projection experience occurs in two stages.[59]

Contemporary Perspectives About OBEs

In a survey of experimental psi research literature, Carlos Alvarado determined that — in spite of occasional striking results — the evidence for ESP occurrence, as well as for the possible ESP-conducive properties of the OBE, is considered to be weak.[60]

Alvarado also noted that since 1980, the psychological community has been giving increasing attention to the out-of-body experience characterized by surveys of spontaneous OBEs studying psychological and other correlates and aspects of the experience. Psychologists have been able to find very few factors that reliably correlate with the OBE experience in terms of age, sex or religious orientation. However, psychological variables such as absorption, fantasy proneness, locus of control, lucid dreams, dream recall and others, indicate the importance of internal cognitive processes related to the OBE because of the positive relationship with the experience.[61]

For example, in a survey of 321 people by British psychologist Susan Blackmore, 12% reported out of body experiences. Most OBEs occurred when resting but not asleep and lasted one to five minutes. Many details of the OBEs were obtained — the most significant finding being that OBErs were highly likely to report lucid dreams, flying dreams, hallucinations, body image distortions, psychic experiences and beliefs, and mystical experiences.[62]

SUSAN BLACKMORE, AUTHOR OF *BEYOND THE BODY*. LONDON: PALADIN GRAFTON, 1982

Research into the out-of-body experience is at an extremely primitive state. The experience itself has yet to be adequately defined. When groups are asked if they have had an out-of-body experience, the percentage of yes answers varies so widely that it demands an explanation. Gertrude Schmeidler, a psi researcher emeritus from the City College of New York, reports that twelve surveys show a range of yes answers between 4% and 98% when people are asked whether or not they have had an OBE. This analysis indicates that the differences do not depend on the wording of the question or the explanation of it. Perhaps differences in prior exposure to direct suggestion — or to cultural conditioning — can account for such enormous extremes in the groups' reports of OBEs.[63]

A unique perspective on out-of-body experience has been developed by the Jungian analyst Arnold Mindell and his concept of the "dreambody." He views the dreambody as a multi-channeled information center which communicates a message concerning one's life process through dreams, via pain, verbally, by auditory channels, and in body symptoms, and movements. In order to achieve and increase awareness of the patient's life tendency, Mindell advocates amplifying not only dreams but body symptoms. Mindell observes that dreambody awareness increases most dramatically near death. He says that near death, dying people experience their dream bodies as clairvoyant or lucid dreams. They feel that they can go places and often actually hear, see, and feel what is going on at a distance even though their real bodies still lie in bed. He suggests that their dreambody is almost free to do the impossible because their proprioception no longer relates to the pressures, pains, and agonies of their physical body. The dreambody appears to transcend the real body. It is a form of awareness that is independent of the living body. He suggests that death may be the last edge, the one at which we truly begin to live as we are.[64]

Some out-of-body experiencers describe the sensation of possessing and, in some instances, simultaneously occupying a multiple number of "bodies" at varying locations, sometimes in conjunction with the sense of being disembodied. A variant of this experience occurs, during which subjects seem to shift awareness alternately between two or more locations.[65] Researchers tend to think that, if the phenomenology of this experience is to be taken at face value, it could possible be modeled by a hyperdimensional view of consciousness (such as is presented in the Appendix).

Interesting confirmation of the need for a hyperdimensional model comes from J. H. M. Whiteman, a South African mathematician and physicist who has written extensively on his own out-of-body experiences. He maintains that models of OBE, purporting to explain the evidence in terms of conventional psychological, psychiatric, or physical theories, are premature and incomplete. Whiteman feels they seriously misrepresent the states in question. He suggests that little if any real progress in understanding OBEs can be expected in the field so long as *one-space theories* govern research in the subject.[66]

Perhaps the most interesting question about out-of-body experience was posed by British psychologist Susan Blackmore, who asks how "the human information processing system construct[s] the illusion of a separate self in the first place."[67]

Healing

Even before the dawn of civilization, tribal peoples developed shamanistic healing practices of a rather sophisticated nature. Many of these traditions survive today in cultural contexts where scientific medicine is unavailable either because of distance or cost. These modes of healing have been examined throughout the world by Stanley Krippner, professor of psychology and director of the Center for Consciousness Studies at the Saybrook Institute in San Francisco. Krippner has coauthored (with Alberto Villoldo) *Healing States* and *Realms of Healing*. In a *Thinking Allowed* interview, Dr. Krippner summarizes his understanding of shamanistic healing practices involving spiritualistic views:

> Some of the more sophisticated of the Brazilian practitioners have said, "You know, the worst black magic is the black magic we commit against ourselves. It is the sorcery that hurts ourselves when we think negative thoughts, or we hold onto a destructive self concept, or when we allow ourselves to say negative, hostile things about ourselves and the people around us, and those sentences go over and over in our mind. It's no wonder people get stomachaches, backaches and headaches with those negative thought images going around."
>
> To me that sounds like sorcery, and if that can be exorcised, so much the better. If you want to call this a malevolent spirit, fine. If you want to call it negative thinking, fine. But either the spiritist or the psychotherapist, or both of them, really have to approach that negativity and get rid of it if the person's going to recover.

STANLEY KRIPPNER (FROM *PSYCHIC AND SPIRITUAL HEALING* VIDEOTAPE, COURTESY THINKING ALLOWED PRODUCTIONS)

I then asked Krippner if he would extend his pragmatic point of view to situations where a healer is actually using out-and-out fraud, such as some of the cases of alleged psychic surgery, where fraud seems to be used and then people recover. He responded:

> The amazing thing is that there is a history of sleight of hand in shamanism, and sometimes the sleight of hand is used for very benign purposes. In other cases it's used to earn a buck. But sometimes, sleight of hand will be used by the shamans, especially when they do the cupping and sucking routine, and they suck on a person's skin and their mouth fills up with black fluid and they spit it out, and they say, "OK, I've sucked all the poison out of you." Usually, that's tobacco juice, and sometimes the patient knows that's tobacco juice, but that's beside the point. It's the ritual that is so important. The shaman is saying, "I have sucked that poison out of you," OK, the patient is sometimes willing to let go of what has been poisoning him or her, symbolically, and that can be very beneficial from a therapeutic point of view.[68]

Healing Temples

In the Egyptian temples of *Imhotep* an art of healing known as incubation was practiced. It is known that *Imhotep* is the architect who designed the first known pyramid for the Pharaoh *Zosar* around 2700 B.C. He developed a reputation as a magician and healer and some two thousand years later came to be regarded as a god. Exactly how this deification occurred is uncertain. However, people with illnesses would go to his temple and sleep there. It is claimed that he appeared to them in their dreams and healed them. This tradition, which was carried on in the Greek temples with the healing God *Asclepios*, implies at the very least a very practical understanding of hypnotic suggestion — if not actual spiritual healing.

HEALING TEMPLE OF THE GOD ASCLEPIOS ON THE GREEK ISLAND OF KOS, WHERE HIPPOCRATES RECEIVED HIS TRAINING AS A HEALER

An Egyptian text dating from over three thousand years ago, called the *Bent-rosh Stela,* describes the sufferings a princess endured in a profound experience described as "possession by a spirit." The ministrations of the local psychiatrist or "artist of the heart" were insufficient to relieve her suffering. However, an immediate cure was effected upon the arrival of a high priest of *Thoth* who brought with him a great stone statue of the healing moon-deity aspect of the god. His journey had taken nearly a year and a half.[69] These techniques were preserved by the priesthood and also passed through mystery cults associated with the pyramids and sphinx.

It is hard to appreciate the enormous influence of the healing temples which dotted the ancient world and served as both hospitals and centers of learning. Hippocrates, the father of Western medicine, received his training at the temple of Asclepios on the Island of Kos.

An Inner Healing Advisor

I had an opportunity to explore a process very akin to that used in the ancient healing temples, during a videotaped *InnerWork* interview with Dr. Martin Rossman, author of *Healing Yourself With Mental Imagery.*[70,71] According to Dr. Rossman, the "inner healing advisor" can take a wide variety of forms when the mind is prepared through deep relaxation and suggestion (or *hypnosis*). It may appear as an animal, as a wise person, as a spirit or even as a deity. His key directive is that one must accept and attempt to learn from whatever imagery appears.

During our session, while in a trance state, the image appeared of a stately Roman wearing a toga who identified himself as "Seneca." Thinking of the Roman playwright (I had heard on the radio that afternoon), I tried to change the image. If my inner advisor was to be one of the ancients, I thought, I would rather have Demosthenes who, born with a speech impediment, rose to be one of the greatest Greek orators. Although I did try, I found I was unable to change the image. Seneca remained.

Accepting the image at face value as my "inner healing advisor," I entered into a dialogue and asked Seneca how we could work together. The response was, "Study my life." Upon awakening from the hypnotic session, I was struck by the vividness and intensity of the experience. I have subsequently spent a good deal of time studying the remarkable life of this ancient writer (and

MARTIN ROSSMAN, M.D. (FROM *HEALING YOURSELF WITH MENTAL IMAGERY* VIDEOTAPE, COURTESY THINKING ALLOWED PRODUCTIONS)

even arranged a visit to Cordoba, Spain, where Seneca was born). He was not only a playwright, but also a statesman, scholar and philosopher who wrote extensively on matters of health and healing.[72]

Although Seneca lived a productive life until the age of sixty-nine, he frequently suffered from asthma and other diseases. Based on the tradition of Stoicism, Seneca maintained a philosophical balance between an acceptance of fate and a continual search for virtuous self-improvement. He once offered this prescription to a friend who was suffering from illness:

> Call to mind things which you have done that have been upright or courageous; run over in your mind the finest parts that you have played. And cast your memory over the things you have most admired; this is a time for recollecting all those individuals of exceptional courage who have triumphed over pain. . . . . If pain has been conquered by a smile, will it not be conquered by reason?. . . Comforting thoughts contribute to a person's cure; anything which raises one's spirits benefits one physically as well.[73]

Seneca's advice is reminiscent of an old Jewish tradition which prescribes the recitation of the *Song of Solomon* — the most sensuous of biblical texts — as a remedy for ill health.[74] I regard the ostensible contact with Seneca as meaningful from a *deep healing* perspective — one in which healing, to the extent it occurs, is a natural outcome of a balanced, philosophical (i.e., love of wisdom) approach to life. Such a perspective differs in scope and depth from those schools of thought that view healing as a consequence of particular techniques. Christian healing, when practiced within the context of Christian principles as a way of life, is another example of deep healing. Numerous examples of miraculous cures are to be found within the spiritual traditions of all cultures.

Mesmerism

In the eighteenth century, the Age of Reason, techniques for healing through suggestion and consciousness alteration rose to the very forefront of public attention. Franz Anton Mesmer (1734-1815), a Viennese-trained physician who held to the old astrological beliefs, initiated this new era of consciousness exploration.

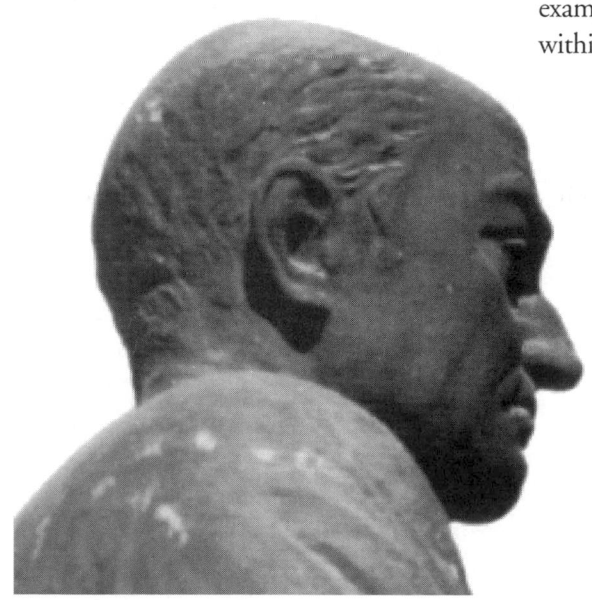

STATUE OF SENECA IN CORDOBA, SPAIN (COURTESY JANELLE BARLOW)

Mesmer hypothesized that the planets affected human beings through an influence resembling magnetism. This was the subject of his doctoral dissertation at the University of Vienna. It was a reasonable hypothesis at the time, as magnetism in Mesmer's day constituted a challenge to physical science — the solution to which might shed light on the mysteries of the psyche. Mesmer, however, ventured beyond mere hypothesis into the realm of pseudoscience when he began treating patients for a variety of ailments by applying magnets to their bodies. Later he stopped using magnets and maintained that any curative influence emanated from the hands and nervous system of the healer. He believed that this influence, which he named *animal magnetism*, could be transmitted to objects held in or stroked by the hand. It could then be discharged to a patient through a suitable conductor. In 1778 he moved to Paris and attracted great notoriety with many patients and pupils among the wealthy classes.

In order to deal with the crowds of patients he attracted, he devised a wooden tub filled with water and iron filings, and containing many bottles of "magnetized water." Iron rods protruded from it that could be applied to diseased or damaged parts of his patients. Clothed in a magician's gown and wand, Mesmer himself moved among the company, making *magnetic passes* over his patients to a background of soft music and mysterious lighting. Many cures of various diseases were noted. Often they were preceded by convulsive movements and rapturous noises. Numbered among Mesmer's acquaintances, and possibly his patients, were Mozart and his family, King Louis XVI and his queen, Marie Antoinette, as well as Empress Maria Theresa of Austria.

In 1784, the French government appointed an official commission to investigate Mesmer. It was composed of several renowned scientists including Antoine Lavoisier, "the founder of modern chemistry" and Benjamin Franklin. They reported to the Academy of Sciences and the Royal Society of Medicine that the "magnetic fluid" was a myth. Although Mesmer's techniques did give rise to certain psychophysiological states that might result in the curing of diseases, they agreed with the assumption that Mesmer played upon the imagination of his subjects. Primarily their report stressed the immorality of the healer making magnetic passes over the bodies of his female patients.[75]

Mesmer's reputation waned. He remained in Paris until 1789 when he fled the revolution. He never saw himself as a fraud. His writings show that he knew his treatments held greater significance than simply curing patients.[76]

Animal Magnetism

The most influential of Mesmer's pupils, the Marquis de Puysegur, discovered that patients could be put by "Magnetization" into a sleep-like *somnambulistic* state in which cures could be effected. Patients in this state showed themselves unusually responsive to the suggestions of the mesmerist, and could be made not only to perform actions, but also to feel emotions or to entertain delusional beliefs. Ordinary senses might be heightened, and other psychic sensitivities seemed to be induced. Of this state, Mesmer writes:

> The somnambulist may perceive the past and the future through an inner sense of his. Man is in contact through his inner sense with the whole of nature and can always perceive the concatenation of cause and effect. Past and future are only different relations of its different parts.[77]

Some patients would diagnose and prescribe for their own ailments, and sometimes they would do this for others using clairvoyance. Puysegur emphasized the importance of the will of the therapist as the directing influence behind the mesmerizing process and claimed the ability to put people into a trance telepathically. Ultimately his theories were no less influential than Mesmer's on the history and development of consciousness.

The rules of magnetizing, based on Puysegur's theories, were published in 1825 by Joseph Philippe Francois Deleuze:

> Cause your patient to sit down in the easiest position possible, and place yourself before him, on a seat a little more elevated, so that his knees may be between yours, and your feet by the side of his. Demand of him in the first place that he give himself up entirely, that he think of nothing, that he do not trouble himself by examining the effects that he experiences, that he banish all fear, and indulge hope, and that he be not disquieted or discouraged if the action of magnetism produces in him temporary pains.
>
> After you have brought yourself to a state of self-collectedness, take his thumbs between your two fingers, so that the inside of your thumbs may touch the inside of his. Remain in this situation five minutes, or until you perceive there is an equal degree of heat between your thumbs and his: that being done, you will withdraw your hands, removing them to the right and left, and waving them so that the interior surface be turned outwards, and raise them to his head; then place them upon his two shoulders, leaving them there about a minute; you will then draw them along the arm to the extremity of the fingers, touching lightly. You will repeat this pass[78] five or six times, always turning your hands and sweeping them off a little, before reascending: you will then place your hands upon the head, hold them there a moment, and bring them down before the face, at the distance of one or two inches, as far as the pit of the stomach: there you will let them remain about two minutes, passing the thumb along the pit of the stomach, and the other fingers down the sides. Then descend slowly along the body as far as the knees, or farther; and, if you can conveniently, as far as the ends of the feet. You may repeat the same processes during the greater part of the sitting. You may sometimes draw nearer to the patient so as to place your hands behind his shoulders, descending slowly along the spine, thence to the hips, and along the thighs as far as the knees, or to the feet. After the first passes you may dispense with putting your hands upon the head, and make the succeeding passes along the arms beginning at the shoulder: or along the body commencing at the stomach.
>
> When you wish to put an end to the sitting, take care to draw towards the extremity of the hands, and towards the extremity of the feet, prolonging your passes beyond these extremities, and shaking your fingers each time. Finally, make several passes transversely before the face, and also before the breast, at the distance of three or four inches: these passes are made by presenting the two hands together and briskly drawing them from each other, as if to carry off the super-abundance of fluid with which the patient may be charged. You see that it is essential to magnetize, always descending from the head to the extremities, and never mounting from the extremities to the head. It is on this account that we turn the hands obliquely when they are raised again from the feet to the head. The descending passes are magnetic, that is, they are accompanied with the intention of magnetizing. The ascending movements are not. Many magnetizers shake their fingers slightly after each pass. This method, which is never injurious, is in certain cases advantageous, and for this reason it is good to get in the habit of doing it.
>
> Although you may have at the close of the sitting taken care to spread the fluid over all the surface of the body, it is proper, in finishing, to make several passes along the legs from the knees to the end of the feet.
>
> This manner of magnetizing by longitudinal passes, directing the fluid

from the head to the extremities, without fixing upon any part in preference to others, is called *magnetizing by the long pass*. . . . It is more or less proper in all cases, and it is requisite to employ it in the first sitting, when there is no special reason for using any other. The fluid is thus distributed into all the organs, and it accumulates naturally in those which have need of it. Besides the passes made at a short distance, others are made, just before finishing, at the distance of two or three feet. They generally produce a calm, refreshing and pleasurable sensation.

There is one more process by which it is very advantageous to terminate the sitting. It consists in placing one's self by the side of the patient, as he stands up, and, at the distance of a foot, making with both hands, one before the body and the other behind, seven or eight passes, commencing above the head and descending to the floor, along which the hands are spread apart. This process frees the head, reestablishes the equilibrium and imparts strength.

When the magnetizer acts upon the patient, they are said *to be in communication* (rapport). That is to say, we mean by the word *communication*, a peculiar and induced condition, which causes the magnetizer to exert an influence upon the patient, there being between them a communication of the vital principle. . . . Ordinarily magnetism acts as well and even better in the interior of the body, at the distance of one or two inches, than by the touch. It is enough at the commencement of the sitting to take the thumbs a moment. Sometimes it is necessary to magnetize at the distance of several feet. Magnetism at a distance is more soothing, and some nervous persons cannot bear any other. . . .

It is by the ends of the fingers, and especially by the thumbs, that the fluid escapes with the most activity. For this reason it is, we take the thumbs of the patient in the first place, and hold them whenever we are at rest. . . .

The processes I have now indicated, are the most regular and advantageous for magnetism by the long pass, but it is far from being always proper, or even possible to employ them. When a man magnetizes a woman, even if it were his sister, it might not be proper to place himself before her in the manner described: and also when a patient is obliged to keep his bed, it would be impossible... .

Let us now consider the circumstances which point out particular processes.

When any one has a local pain, it is natural, after establishing a communication, to carry the magnetic action to the suffering part. It is not by passing the hands over the arms that we undertake to cure a sciatic; it is not by putting the hand upon the stomach that we can dissipate a pain in the knee. Here are some principles to guide us.

The magnetic fluid, when motion is given to it, draws along with it the blood, the humors and the cause of the complaint. For example, if...one has a pain in the shoulder, and the magnetizer makes passes from the shoulder to the end of the fingers, the pain will descend with the hand: it stops sometimes at the elbow, or at the wrist, and goes off by the hands, in which a slight perspiration is perceived....[Magnetism seems to chase away and bear off with it what disturbs the equilibrium, and its action ceases when the equilibrium is restored.] It is useless to search out the causes of these facts, it is sufficient that experience has established them, for us to conduct ourselves accordingly, when we have no reason to do otherwise. . . .

You may be assured that the motions you make externally, will operate sympathetically in the interior of the patient's body, wherever you have sent the fluid into it....

I think it important to combat an opinion which appears to me entirely erroneous, although it is maintained by men well versed in the knowledge of magnetism: viz., that the processes are in themselves *indifferent*; that they serve only to fix the attention, and that the will alone does all....

The processes *are* nothing if they are not in unison with a determined intention. We may even say they are not the *cause* of the magnetic action; but it is indisputable that they are necessary for directing and concentrating and that they ought to be varied according to the end one has in view....Each one might modify the processes according to his own views and practice; but not that he could omit them, or employ them in a manner contrary to the general rules. For example, various magnetizers act equally well by passes, more gentle or more rapid; by contact, or at a distance; by holding the hands to the same place, or by establishing currents. But it is absurd to believe one can cure chilblains on the feet, by placing the hands on the breast.

Persons who are not in the habit of magnetizing, think they ought to exert a great deal of force. For which purpose, they contract their muscles, and make efforts of attention and will. This method is...often injurious. When the will is calm and constant, and the attention sustained by the interest we take in the patient, the most salutary effects ensue, without our giving ourselves the least pain....A person ought not to fatigue himself by magnetic processes: he will experience fatigue enough from the loss of the vital fluid....

It frequently happens that magnetism gradually re-establishes the harmony of the system without producing any sensation, and its influence is perceived only in the restoration of health. In that case you ought to continue zealously to follow the processes I have pointed out, without troubling yourself about the manner in which the magnetism acts, and without seeking for any apparent effect....

There are patients in whom the influence of magnetism is displayed in two or three minutes; others, who do not feel it for a long time. There are some in whom the effects are constantly increasing; others, who experience at the first time all that they will experience in the course of a long treatment....

The effects by which magnetism manifests its action are greatly varied....They change sometimes, in proportion to the change wrought in the malady.

I will now describe the effects which are most commonly exhibited.

The magnetized person perceives a heat escaping from the ends of your fingers, when you pass them at a little distance before the face, although your hands appear cold to him, if you touch him. He feels this heat through his clothes, in some parts, or in all parts of his body before which your hands pass. He often compares it to water moderately warm, flowing over him, and this sensation precedes your hand. His legs become numb, especially if you do not carry your hands as low as his feet; and this numbness ceases when, towards the close, you make passes along the legs to the toes, or below them. Sometimes instead of communicating heat, you communicate cold, sometimes also you produce heat upon one part of the body, and cold upon another. There is often induced a general warmth, and a perspiration more or less considerable. Pain is felt in the parts where the disease is seated. These pains change place, and descend.

Magnetism causes the eyes to be closed. They are shut in such a manner that the patient cannot open them; he feels a calm, a sensation of tranquil enjoyment; he grows drowsy, he sleeps; he wakes when spoken to, or else he wakes of himself at the end of a certain time, and finds himself refreshed. Sometimes he enters into somnambulism, in which state he hears the magnetizer and answers him without awaking....The state of somnambulism...does not take place except in a small number of cases....

Here I ought to observe, that the magnetic sleep is of itself essentially restorative. During this sleep, nature unassisted works a cure; and it is often sufficient to re-establish the equilibrium, and cure nervous complaints.[79]

In 1825, Baron du Potet and Dr. Husson, a physician of the Paris Hospital and a member of the Academy of Medicine, made some remarkable experiments in which he telepathically induced somnambulism on an unaware subject. This report was quite controversial.[80] (This was a precursor of a line of research in telepathic hypnotic induction which has been pursued in the former Soviet Union.)

In 1831 a second government report was issued that was much more favorable to the mesmerists.[81] Students of Mesmer were in practice throughout Europe and by that time most of the characteristic phenomena were well known. These included:

Amnesia
Acceptance of totally delusive beliefs
Clairvoyance
Heightened muscular power
Impersonation with outstanding flair
Improvement of skin complaints and warts
Increased powers of memory
Alleviation of neurotic states
Anesthesia
Sensory acuity
Sensory hallucinations in all senses
Healing ability of self and others
Stopping of bad habits
Production of blisters and marks on the skin
Suppression of physiological responses
Suppression of pain

The majority of those interested in *mesmerism* could be divided into two camps — those who believed in psychic phenomena and accepted some version of the magnetic fluid theory, and those who did not. The first camp evolved into numerous "new thought," spiritualist and Christian Science movements. James Braid (1795-1860) did the most to make mesmerism acceptable to the second group by changing the name to hypnotism and referring to the process strictly as a matter of "suggestion." Modern depth psychiatry and psychoanalysis evolved from Braid's use of hypnosis in treating clinical disorders.

The state of our scientific understanding of healing is still at a very primitive level. It wasn't until 1959, for example, that the American Medical Association officially approved hypnosis as a therapeutic tool.[82]

Holistic Faith Healing

Mental healing today is widely practiced and is gaining increasing attention from researchers. The variety of alternative forms of healing is enormous and overwhelmingly confusing. Such therapies include acupuncture, Alexander technique, aromatherapy, Aston patterning, astrological diagnosis, aura analysis, Auyervedic medicine, Bach flower remedies, Bates method, bioenergetics, biofeedback, biorhythms, charismatic healing, chiropractic, Christian Science, color therapy, copper effects, crystal healing, cupping, diet and health, fasting, Feldenkreis work, herbs, homeopathy, hypnotherapy, iridology, Joh Rei healing, Kirlian diagnosis, laying on of hands, macrobiotics, magnetic passes, meditation, moxibustion, naturopathy, negative ions, osteopathy, past lives therapy, phrenology, polarity balancing, prayer, psychic surgery, psychosomatic healing, pyramid energy, radiesthesia, radionics, reflexology, Reichian therapy, Rikei healing, Rolfing, sexual therapy, shamanic healing, Shiatsu and acupressure, spiritual healing, Tai Chi Ch'uan, toning, and yoga.

Undoubtedly, some of these practices are predicated upon valid insights into the workings of the human body. Others are based upon theories which have virtually no validity or empirical support. The positive results which sometimes stem from the use of such techniques is almost certainly unrelated to the theories that are claimed. Virtually, any method can activate a *placebo effect* which is one of the strongest healing principles known to humanity.

Every drug which is released for medical use in the United States is tested for its efficacy in comparison with a *placebo. Placebo* is a Latin word meaning "I shall please." A placebo is simply a sugar pill, or some other inert substance, which is administered to a test population (called a "control group") under the same environmental and psychological conditions as the drug to be tested. Researchers have been amazed that the positive changes reported from placebo use often rival those of medicinally potent medicines. Studies have even shown that placebos are capable of activating the body's own mechanisms for producing natural painkiller substances called *endorphins.*

Many of the alternative healing approaches, regardless of the merits of their other claims, emphasize creating conditions particularly conducive to the placebo effect. For example, in an article on "healer-healee interactions and beliefs in therapeutic touch," Rene Beck and Erik Peper describe five key concepts as essential to the healing process:

> (1) What we communicate by attitude, act, and word as well as the setting we provide will affect our potential to heal; (2) what the healer believes is important; (3) all persons involved are interconnected — there can be no independent observer; (4) some of the essential qualities may neither be observable nor measurable; and (5) time may contract or expand.[83]

There are many different theories of healing. Harry Edwards, one of the grand old statesmen of British spiritual healers, put the matter rather simply: The healing energies emanate from the nonphysical dimension of spirit. Healing takes place, according to Edwards, when there is a merging of the spiritual and physical aspects of being. Basing his teachings on the somewhat outdated concepts of Rene Descartes, Edwards postulated that this merging of the physical and nonphysical takes place in the pineal gland.[84]

In today's "post-modern" society, there are many practitioners of healing and psychic arts who are not willing to acknowledge the inner dimensions of their work. Captivated by the scientistic mythology of our age, they build mysterious-looking pseudoscientific devices to which they then attribute remarkable healing abilities.

Radionics

Radionics is the term applied to a social movement concerned with diagnosis and healing through the use of complicated devices — or *black boxes* — that (although no one understood how they worked) resulted in miraculous cures. The father of this movement was Dr. Albert Abrams, a professor of pathology at Stanford University's medical school. Basing his discoveries on the philosophy that all matter radiates information that can be detected by his instruments in conjunction with the unconscious reflexes of another human being, Abrams succeeded in attracting a large following and also arousing the unremitting ire of the medical and scientific establishment. Thousands of self-professed healers were effecting cures, making diagnoses, and even removing pests from gardens merely by

RADIONICS "BLACK BOX"

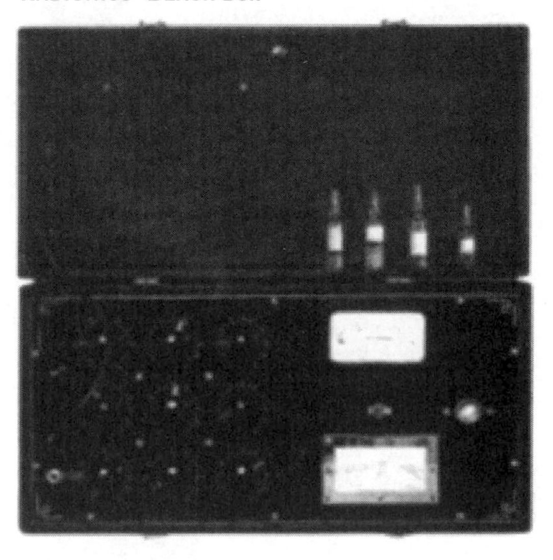

twisting dials, swinging pendulums, or rubbing their fingers across strange devices.[85,86] The following passage describes the use of one such instrument known as the Delawarr machine:

> Suppose that it is required to find out the condition of a patient's liver. We place a bloodspot or saliva sample in one of the two containers at the top of the main panel, according to whether the patient is male or female, and start turning the tuning knob slowly, passing the fingers of the right hand over the rubber detector at the same time with a series of "brushing" strokes until a "stick" is obtained. The patient's bloodspot is then tuned into the set.[87]

The stick refers to a particular rubbing sensation in the finger. The location of the dials when the "stick" occurs, when properly translated is said to indicate the diagnosis of the disease. When the disease is tuned in to the instrument, the cure can be "broadcast" over any distance, to the patient.

Other radionic developments are said to have been even more startling, such as the camera developed by the Los Angeles chiropractor, Ruth Drown. Using nothing but a drop of blood, it is claimed that this camera could take pictures of the organs and tissues of patients — sometimes at a distance of thousands of miles. Drown also claimed to take pictures in "cross-section," a feat that cannot be duplicated even with X rays. While she received a British patent for her apparatus, Drown was persecuted as a charlatan by the FDA.[88]

A story about Drown's ability is told by the cosmologist Arthur M. Young, who invented the Bell helicopter:

> Ruth Drown was truly an angelic sort of a person — if you can imagine an angel in the flesh. She started reeling off these Pythagorean relationships that just made my mind spin....I couldn't keep up with her.
>
> It wasn't on the first occasion, but maybe on the second, that I wanted to put her to a test. I was at that time having a toothache. So I asked her if she would diagnose my condition and take a photograph. But I didn't tell her anything. And she took these photographs that were about eight by ten. It looked like a very detailed picture of teeth!
>
> She put the film in this box, but there were no lenses or anything like that. Whatever this radiation was, it exposed the film. It was not done with light. And she got a photograph of the tooth.
>
> Being scientific in nature, I said, "Now do it again." This was all in the dark. She couldn't see me. So I pressed the tooth hard with my finger to make it hurt more, to see what would happen. The next picture was an enlargement of this same tooth![89]

Today there are two developments affecting the standing of radionics. On the one hand, researchers in a new area dubbed *psychotronics* are taking a serious interest in understanding the possible mechanisms such instrumentation might have.[90,91] In fact, several new radionics devices have recently been manufactured with computerized components. At the time of this writing, there is no clear indication that such new devices represent any genuine advance in the arcane art of radionics. There is no reason to suspect that a major breakthrough is at hand. On the other hand, a number of radionic practitioners and investigators have reported that after becoming proficient in the use of the "black boxes," they were able to obtain the same effects without them.[92,93]

One radionics expert, Frances Farrelly, demonstrated her ability to work without her instrument at the International Conference on Psychotronics in Prague in 1973:

> . . . she was confronted by a professor from the Czechoslovak Academy of Sciences who gave her a chip of mineralized rock and asked her before a large audience if she could state its origin and age. Rubbing the table before her to get a radionic type "stick," Farrelly, after putting a dozen

questions to herself, stated that the mineral in question came from a meteor and was about 3,200,000 years old, answers which exactly matched the most considered conclusions of expert Czech mineralogists.[94]

It was her contention she had learned to "run the instrument in [her] head." Perhaps, then, the "black box" is to radionics what the pencil is to arithmetic — a tool for focusing consciousness within the structure of a disciplined system.

Edgar Cayce

Perhaps the most prominent twentieth-century practitioner of psychic diagnosis and prescription healer was Edgar Cayce (pronounced Casey) who died in 1945. Cayce originally received hypnosis treatment for his own asthma. While in an unconscious trance state, Cayce made diagnoses and prescribed treatments for thousands of individuals.

Although Cayce tried to arrange for his patients to be treated by qualified physicians, like all unlicensed healers, he met with a deeply entrenched opposition from the medical profession. It was, however, a homeopathic doctor, Wesley Ketchum, M.D., who examined the records of Cayce's treatments and made a favorable report, in 1910, to the American Society of Clinical Research at Boston:

> I have used him in about 100 cases and to date have never known of any errors in diagnosis, except in two cases where he described a child in each case by the same name and who resided in the same house as the one wanted. He simply described the wrong person....
>
> The cases I have used him in have, in the main, been the rounds before coming to my attention, and in six important cases which had been diagnosed as strictly surgical he stated that no such condition existed, and outlined treatment which was followed with gratifying results in every case.

EDGAR CAYCE (COURTESY ASSOCIATION FOR RESEARCH AND ENLIGHTENMENT)

Files of over nine thousand readings by Edgar Cayce are kept on record by the Association for Research and Enlightenment in Virginia Beach, Virginia. Since 1931, this organization has attempted to foster scientific investigation of Cayce's healings.

Critics of Edgar Cayce argue that the cures which he outlined while in a trance state seem very much in the homeopathic/naturopathic tradition of Dr. Ketchum himself. A very real possibility exists here that many of Cayce's notions were actually suggested to him while he was in the hypnotized state. Skeptics also point out that no bona fide scientific studies have ever actually verified the efficacy of Cayce's cures.[95] Therefore the massive records housed by the Association for Research and Enlightenment have resulted in almost no scientific advance in our understanding of the healing process. Nevertheless, their contribution to the folklore of healing, and other psychic matters, has been enormous. Over thirty books have been published describing the details of Cayce's prescriptions.

Library of the Association for Research and Enlightenment

Psychic Surgery

Many forms of spiritual healing, including some derived from the New Testament, are based on the notion of driving out evil spirits or demons. This approach is practiced by (otherwise) sophisticated Christian faith healers in the United States as well as by folk healers and "psychic surgeons" in Third World countries such as Brazil and the Philippines. In these countries, healers often appear to remove ugly black tobacco leaves and other unsightly objects from the bodies of the ill. While slight of hand is generally used to create the effect of spiritual intervention, the theoretical premise is that these objects were implanted as the result of dark magic or sorcery.

These cases of psychic surgery center around healers whose talents are said to be guided by powerful spirits. Operations are performed without the benefit of anesthesia or antisepsis, under unsterile conditions, often without even the benefit of a knife. Blood appears. Tissue is removed. And yet, when the procedure is completed there is often no trace of a wound or an opening. As Dr. Andrija Puharich has put it, these operations mimic, yet violate every principle of modern surgery.

One of the most extraordinary accounts in the literature of psychic surgery is one in which psychologist Alberto Villoldo participated as an assistant to Dona Pachita, a Mexican healer. Villoldo describes how Dona Pachita performed surgery while in trance and assuming the personality of Cuahutemoc, an Aztec prince. During this procedure, she appeared to removed a cancerous tumor from the urinary bladder of a Texas woman and then insert a new human bladder purportedly obtained from a local medical school. The operation was performed under unsterilized conditions using a hunting knife. Villoldo describes how, at the instructions of the healer, he actually helped to remove tissues from the woman's abdomen and insert the new bladder — during which time his own finger was accidently cut by the hunting knife.

The session was also witnessed by Gabriel Cousens, an American doctor of medicine, who stated:

> There was no question in my mind that she was opening the skin; there was no question that I was smelling blood. I could see the opening in the abdomen...and I could see Pachita's hands going into the bladder area, into the abdominal cavity. There is no question in my mind that things had been taken out from these incisions and things put back in. I had no doubts that I had seen authentic "psychic surgery."[96]

Approximately ninety minutes after the "operation," Cousens examined the patient and observed:

> The best that I could understand from the English speaking woman was that she had experienced some pain. [She was] still wrapped in sheets and it appeared as if [she was] still experiencing some kind of trauma — psychological or physical, or perhaps both. [She was] dazed; this brought back memories of being in post-operative anesthesia rooms in hospitals where people had not regained full consciousness. [She was] not given medication or drugs of any kind before or during the "operation." I examined the woman from Texas and found there was a scar, and it had apparently "healed" more rapidly than any scar I have ever seen immediately after "surgery." There were no stitches. It looked reddish but not particularly inflamed and as if it were perhaps a month old. I was told that the scar tended to disappear within a few weeks.[97]

This account by Villoldo and Cousens is so extraordinary as to be virtually unbelievable. (In fact, in 1976, I conducted a radio interview with Villoldo in which this episode was discussed. It was the only instance in hundreds of broadcasts that actually provoked a listener to phone in and voice obscenities.) Many readers may feel more comfortable believing that Cousens and Villoldo were deceived (or participating in a deception), rather than accept their account at face value. However, their account was neither the first nor the last in the puzzling area of psychic surgery.

During the period from 1963 to 1968, Andrija Puharich, then a senior medical researcher at New York University, conducted studies in Brazil with the wonder-healer Jose Pedro de Freita — known by his nickname Arigo. Born in 1918, Arigo received four years of education in primary school, worked as a mine laborer, and also owned his own small restaurant. At the age of thirty he entered into a period of severe depression, nightmares, sleep talking and sleepwalking. For two years neither a doctor nor a priest were of any benefit to him whatsoever. Finally a local spiritualist, Senor Olivera, prayed for Arigo and told him that a spirit was trying to work through him.[98]

In 1950, Arigo first achieved unsought fame as a healer, during a visit to the city of Belo Horizonte. There he happened to stay at a hotel where state senator Bittencourt, who had been diagnosed with an inoperable colonic tumor, was also a guest. According to Puharich, Arigo entered Bittencourt's room early one morning, telling him to lie down on the bed. Then he produced a razor and proceeded to remove a tumor from Bittencourt's abdomen. Subsequent diagnosis indicated that the tumor dissipated. Bittencourt's pajamas were torn and an orange-sized piece of tissue was found in the hotel room. While there was blood on his pajamas and body, no scar was found. Eventually Arigo went on to do surgery of a similar kind in public.[99]

Puharich himself claims to have observed over one thousand instances in which Arigo diagnosed and treated patients — with complete accuracy as far as he himself was able to determine.

> We found we were able to verify 550 verdicts, because in those cases we ourselves were able to establish a pretty definite diagnosis of the problem. In the remaining 450 cases, for example in rare blood cases, we could not be certain of our own diagnosis because we lacked available on-the-spot resources to enable us to do so. But of those of which we were certain we did not find a single case in which Arigo was at fault. Every patient was helped and none had post operative complications.[100]

Puharich was further impressed with the accuracy and sophisticated terminology of the medical prescriptions Arigo frequently gave. Puharich claims that the medical team was never able to find a mistake in the medical or registered trade name of a drug prescribed.[101] Thousands of surgical operations were conducted under conditions Puharich described as resembling a train station at rush hour. In order to satisfy his own curiosity, Puharich even allowed himself to be operated on.

Arigo agreed to operate on a benign tumor on his elbow. The scene was a crowded room with some ninety people gathered around as spectators. With a flourish, Arigo asked that someone furnish him with a pocket knife; this was produced. Arigo asked Puharich not to watch the operation so Puharich turned his head toward his cameraman who was filming a motion picture. Within a matter of seconds, Arigo placed the tumor and the pocket knife in Puharich's hands.

In spite of being perfectly conscious, Puharich had not experienced any pain or even any sensation at the surgical site. Yet there was a bleeding incision and a tumor. Knowing that the knife was dirty, that his skin was not cleansed and that Arigo's hands were not clean, Puharich suspected he might get an infection — perhaps even blood poisoning. Nevertheless, the wound healed clean without a drop of pus in three days, according to Puharich, half the time required under normal precautions.[102]

According to Arigo, however, all of this is very simple: "I simply listen to a voice in my right ear and repeat whatever it says. It is always right." Arigo claims this voice belongs to a deceased German medical student, Adolphus Fritz. However, after five years of observation, Puharich still felt unable to arrive at any conclusion as to the reality of "Dr. Fritz" as an independent spirit.

Arigo has been twice put in jail for the illegal practice of medicine. In 1971, he died of an automobile crash. Nevertheless, other healers continue to be active in Brazil where the spiritualist movement has even established its own hospitals.[103]

The psychic surgeons of the Philippines are renowned for their ability to operate without knives, removing tissue, yet leaving no wounds. This phenomena has consistently been extremely controversial and the healers have often been accused of fraud even by those who studied them sympathetically. This undoubtedly places the healers in a questionable legal and moral position; however, it does not answer the scientific question.

Hundreds of home movies have been taken by Americans who went to the Philippines with serious ailments. Medical doctors who have examined these patients before and after their treatment are reportedly often baffled.[104]

In December 1988, my wife Janelle and I were in the Philippines and had the opportunity to witness a psychic surgery session in

ANDRIJA PUHARICH

Manila with Alex Orbito — one of the most prominent and respected healers, who is president of the Philippine Healers Association. Orbito generously allowed us to observe and photograph freely. We witnessed over twenty instances of ostensible psychic surgery over a period of several hours.

At the end of the session, Janelle was convinced that the whole business was easily explained as sleight of hand. She pointed out the many movements that Orbito made where he could have hidden various animal parts that he could have palmed and then later produced as evidence of blood and removed tissues. I found her hypothesis quite acceptable, but not proven. Other observers, such as Canadian psychologist Lee Pulos, have observed and filmed hundreds of sessions with Orbito and are convinced that they are genuine.

Jesus B. Lava, M.D., conducted an interesting study of psychic surgery. Lava was a political activist. Between 1950 and 1954 he served as secretary general of the Communist Party of the Philippines. In 1964 he was captured by the military and detained for ten years as a political prisoner. Following his release, at the age of sixty, he felt that the only thing left for him would be to get involved in something apolitical and, yet, of social relevance. He set up a research and healing center, under the auspices of the Philippine Society for Psychical Research, which allowed him, over a period of several years, to observe the procedures of the native healers and to conduct follow-up studies of his own with patients.

Dr. Lava's report of his findings reads very much like the work of a man who, believing in psychic surgery, wanted to obtain the strongest possible evidence of its success. Yet, as an honest researcher, he was nevertheless forced to admit that his results were quite limited. For example, Lava could find no objective evidence that psychic surgery could extirpate or excise tumors. And he admitted that "fakeries abound." While he did acknowledge that the Philippine methods did "at least give relief to suffering," this was so uneven that it was "not possible to predict the probabilistic results of a particular healing session." Lava also admitted that whenever he was allowed to examine the materials supposedly removed from a patient through psychic surgery they showed "no significant relation to diseased tissues."[105]

Perhaps most significant is Lava's report that healers are somehow able to create incisions in the skin without the use of surgical instruments:

> Another form of energy or force, perhaps, is involved in the instrumentless incision done from a distance of one foot by the mere stroke of the healer's finger, or anybody's finger held by the healer. (This force is capable of passing through the x ray plate interposed between the finger and the skin, without itself being cut.)[106]

Lava's account, if true, would be quite remarkable. Yet in all other respects, the procedure is rather conventional:

> The subsequent procedure, however, follows orthodox surgery. The opening is widened by the fingers and the cyst isolated and pulled out by pincers. Closing is by pressure of the fingers on the skin flaps followed by the application of adhesive tape. After two or three days the wound is completely healed with hardly a scar. There seems to be no susceptibility to infections.[107]

Skeptics who have witnessed this same procedure are convinced that the trick is accomplished through slight of hand with a hidden razor blade.

My own position as regards psychic surgery as well as many other equally unbelievable claims is one of zetetic skepticism. I find no need to jump to a conclusion that the phenomena are either real or false in the absence of clear and convincing evidence one way or the other. I am perfectly willing to tolerate the ambiguity and to accept that I do not know what really happened. If I were seriously ill, I doubt that I would be further harmed by psychic surgery treatments. However, my own choice, and my strongest recommendation to readers, is that such treatment only be obtained while working in close consultation with a medical doctor. The risk in seeking exotic, alternative treatments is that of neglecting or overlooking essential forms of basic health care — both attitudinal and medical interventions.

Delusion and Fraud

There can be little doubt that, whether or not healing actually works, fraud is certainly rampant. One of the best exposés of fraudulent practices is James Randi's book *The Faith Healers*.[108] A practicing magician and a tireless crusader against fraudulent supernatural claims, Randi describes the great lengths to which he and his colleagues have gone in order to uncover many unsavory practices cloaked in the trappings of religion.[109]

While implicitly accepting the possibility of serious psi research into the claims of healers, Randi claims that none of the faith healers he questioned were able to provide any solid evidence of a *single* healing that could be attributed to supernatural intervention. His book leaves little doubt that he uncovered blatant fakery in the ministries of Leroy Jenkins, W. V. Grant and Peter Popoff. He noted, however, that the claims of many faith healers — such as Pat Robertson and Richard Roberts — could not be examined scientifically because they were scaled-down and non-falsifiable. For example, Randi quotes from a Pat Robertson broadcast:

> There is a woman in Kansas City who has sinus. The Lord is drying that up right now. Thank you, Jesus. There is a man with a financial need — I think a hundred thousand dollars. That need is being met right now, and within three days, the money will be supplied through the miraculous power of the Holy Spirit. Thank you, Jesus! There is a woman in Cincinnati with cancer of the lymph nodes. I don't know whether it's been diagnosed yet, but you haven't been feeling well, and the Lord is dissolving that cancer right *now*! There is a lady in Saskatchewan in a wheelchair — curvature of the spine. The Lord is straightening that out right now, and you can stand up and walk![110]

With such claims, Randi claims, the task of evaluation "can be compared to trying to nail a handful of grape jelly to a wall." There is little question, however, that Pat Robertson's followers are convinced and that their conviction has been backed up by generous donations to his evangelical ministry and television broadcast operations.

James Randi

BRENDON O'REGAN (FROM *THE INNER MECHANISMS OF HEALING* VIDEOTAPE, COURTESY THINKING ALLOWED PRODUCTIONS)

Healing at Lourdes

One of the investigators of spontaneous remissions and supposedly miraculous healings is Brendon O'Regan, vice president for research of the Institute of Noetic Sciences, and former research coordinator for Buckminster Fuller at Southern Illinois University (who died in 1992). During a *Thinking Allowed* television interview on "The Mechanisms of Healing," O'Regan provided this vivid description of the Catholic healing shrine at Lourdes:

> The classic assembly of data where spiritual events or deeply psychological events are involved, are the cases of apparent miraculous cures in Lourdes, which have been documented since 1858. The International Medical Commission at Lourdes, which has been in existence in current form since 1947, first asks when a claim is presented, if it could be a remission. If they think that the disease could have healed naturally, they throw it out from further consideration.
>
> One of the interesting cases is, for instance, one that happened in 1976 to a man who had a sarcoma of the pelvis. The bone was literally being eaten away by this cancer, and his hip separated from the pelvis, and he was in a full-body cast and couldn't walk. He had been in the hospital for about a year and was brought to Lourdes and immersed in the water, which is what they do with people.
>
> We have all the x rays at the office. We have been in touch with the Medical Commission on several of these. The doctors and the priests did not even believe that anything had happened to him, even though when he was put in the water he felt this electrical charge run through his body, and immediately regained his appetite. He had been unable to eat from nausea, had gangrene setting in, and was in a debilitated condition. They took the cast off and re-x rayed him and found that the tumor was disappearing, and within two months he was walking. The joint and the socket reconstructed. The bone actually grew back.
>
> I think millions of people have gone to Lourdes. There have been six thousand claims of extraordinary healing, of which only sixty-four have made it through to the status as miracle.
>
> The commission includes people from every medical specialty, from almost every country in Europe. What is also interesting is that they will dismiss any case where the disease is known to have what they call a strong psychological component. So they try to weed out placebo, they try to weed out remission, they try to take out all of the easier cases.[111]

British psychiatrist and psychical researcher, Donald West, examined the records of eleven cases of cures at Lourdes judged by the Roman Catholic Church as "miraculous." Although aware from previous experience how difficult it was to obtain sufficient medical information to be able to draw any conclusions from cases of faith healing, West felt that the situation at Lourdes would be different because "the files at the Lourdes Medical Bureau contain an accumulation of medical data on faith cures that has no parallel elsewhere."[112] Although the cases he reviewed were among the best documented on record, he showed that, even in these cases, crucial information was missing. A similar conclusion was also reached by James Randi, who looked into some Lourdes healings.[113]

Mental Imagery

Of course, it is almost impossible, within a scientific framework, to accept the strict world view of healers who battle illness by invoking the power of Christ, Thoth or Asclepios; by searching for lost souls in the underworld; by fighting evil sorcerers with magical weapons; by enlisting the aid of spirit doctors; or by application of animal magnetism and radionics. Yet, within the past fifteen years, many researchers have recognized a common thread in almost all shamanistic/spiritual healing — the use of mental imagery. Contemporary researchers might well lose their jobs for seriously considering that a shaman became a raven, for example, and flew into the nether realms to fight for the soul of the healee. But to accept that the shaman mediated a healing process through the use of vivid mental imagery is a viable research hypothesis which is bearing fruit.[114,115]

Studies have shown, for example, that transcendental meditators report fewer instances of allergies and infectious diseases than before they began meditating regularly.[116] Other studies conducted at the University of Rochester over a twenty-year period have shown that the presence or absence of cancer in a patient can be predicted on the basis of the feelings of "hopelessness" which the person has toward life in general.[117] Studies of this sort have led to the development of a recognized specialty in psychosomatic medicine. Doctors now realize that the attitudes of their patients are just as significant as the symptoms of their disease.

In Fort Worth, Texas, for example Dr. Carl Simonton — in addition to treating cancer patients with conventional radiation, chemotherapy, and surgery — has also used relaxation and visualization techniques.

> The patient is asked to meditate regularly three times a day for 15 minutes in the morning upon arising, around noon, and at night before going to bed. In the meditation exercise, the first couple of minutes are used to go into a state of relaxation, then once the body is completely relaxed, the patient visualizes a peaceful scene from nature. A minute later, the patient begins the main part of the work of mental imagery. First, he tunes in on the cancer, "sees" it in his mind's eye. Then as Simonton describes it, "he pictures his immune mechanism working the way it's supposed to work, picking up the dead and dying cells." Patients are asked to visualize the army of white blood cells coming in, swarming over the cancer, and carrying off the malignant cells which have been weakened or killed by the barrage of high energy particles of radiation therapy given off by the cobalt machine, the linear accelerator, or whatever the source is. These white cells then break down the malignant cells which are then flushed out of the body. Finally, just before the end of the meditation, the patient visualizes himself well.[118]

The patient is instructed in the general principles of the immune mechanism and is shown photographs of other patients whose visible cancers — such as on the skin and

CARL SIMONTON, M.D. (FROM *THE HEALING PROCESS* VIDEOTAPE, COURTESY THINKING ALLOWED PRODUCTIONS)

mouth — are actually responding to the treatment, getting smaller and disappearing. In a study of 152 patients, Simonton found the greatest success with those who were the most optimistic and committed to full participation in the entire therapeutic process. Furthermore, these patients also showed fewer distressing side effects to the radiation therapy.[119] These results are very encouraging. However, further research and longer follow up studies are necessary before the medical establishment can form a conclusive judgment on the radical possibility of using psychological treatment to cure organic disease.

Omega Seminar Techniques

As an Omega Seminar trainer, I myself have had the opportunity to function as a mental healer of sorts. Founded over thirty years ago by John Boyle, the Omega Seminar teaches a number of powerful techniques for accomplishing goals — predicated upon the assumption that one can access the power of the *superconscious mind* through the use of *positive affirmations.*

In many seminars I have successfully helped participants alleviate discomfort from headaches, sinus congestion, and stomachaches by simply repeating the affirmation, "I CAN" Neck, shoulder and back pain has shown a similar positive response to the affirmation, "I DON'T HAVE TO!" In fact, I have never seen these techniques fail to work — at least temporarily — within the seminar context.

Another technique from the Omega Seminar has consistently worked like magic to alleviate long-term, chronic pain such as that resulting from athletic or automotive injuries. The method is simple. I find a volunteer who is experiencing pain and ask him to visualize the pain as being a certain color. Then I ask him to compare the size of the pain to a fruit (i.e., watermelon, cantaloupe, grapefruit, orange, plum, cherry, raisin). Once he has identified a color and a fruit, I ask him to imagine that the pain is shrinking in size (to a fruit of a slightly smaller size). This generally takes only a few moments. As the pain shrinks in size, I keep asking the volunteer to tell me what color it is (the color generally changes with size). After the pain has been reduced to about the size of a raisin (which can take from five to ten minutes of concentrated visualization), I ask him to shrink it further to the size of a match head — and then to imagine that he is simply flicking it away from his body. Inevitably, when the exercise is finished the pain is gone — at least temporarily.

While these Omega techniques are far from a holistic approach to *deep healing,* they clearly demonstrate the power of suggestion in dealing with pain and the control which we can exert over our state of well-being on a moment-to-moment basis. Many other *affirmations* can be used for long-term health goals.

A model which has been promoted by both medical doctors and scientists, such as Steven Locke at Harvard University, refers to "the healer within." This approach assumes that within each of us is an intelligence which has a deep understanding of our emotional and physical well-being. Using the principles of *psychoneuroimmunology* and *psychoneuroendocrinology,* this intelligence, when activated, is capable of producing dramatic, positive changes in health. A variety of methods can be used to activate the inner healer — including progressive relaxation, autogenic training, biofeedback and hypnosis.[120] This is essentially the perspective Dr. Martin Rossman used in our interview during which I engaged in an internal dialogue with an image of the ancient Roman, Seneca.

The perspective toward healing described by Seneca almost two thousand years ago is not at all unlike that of transpersonal psychologists such as Frances Vaughan, author of *The Inward Arc.* Vaughan suggests that a lifestyle aimed at integrating spiritual, philosophical and psychological teachings can lead to healing and wholeness on any level of consciousness. Such a lifestyle involves a metaphoric awareness of spiritual paths and an awareness of dreams. Healing relationships are described by Vaughan as "an essential aspect of emerging awareness of wholeness and optional relational exchange at any stage of development."[121]

At this point many readers will feel justifiably entitled to ask whether any of this healing, by whatever name and mythology, actually works. A good question! Let's look at one carefully conducted case study of an individual healer.

A Case Study

In 1955, the Institute for Border Areas of Psychology and Mental Hygiene, headed by Prof. Hans Bender in Freiburg, Germany, conducted a thorough study of a mental healer, Dr. (of political science) Kurt Trampler. This study, while not indicative of any paranormal healing, does give us a good picture of the role which psychological factors can play in the healing process. Trampler was seeking to exculpate himself legally as he had been tried and found guilty of violating the statutes governing medical practice. In light of hundreds of testimonials from his patients, the Board of Health ruled that a research study would be of sociological and medical interest.

Trampler's philosophy and methods are not untypical of psychic healers in general. He stresses the need for the patient to establish a "reconnection with the fundamental source of life." In his view, sickness is a "disturbance in man's contact with the higher interrelationships of life." Each treatment session begins with a philosophical discussion of this sort, eloquently delivered in a manner found appealing to an audience of varied backgrounds.

> Trampler then "charges" the patient with his own raised hands, held at some distance. He claims that he can feel the streaming of "an impulse which is transmitted to the patient who then, by some so far unexplained process of a spiritual or energetic nature seems to bring about a change for the better." The patient describes his own sensations during this "atunement." He experiences feelings of warmth and cold, a prickling sensation or a sense of a powerful current. To sustain his therapy Trampler gives the patients sheets of aluminum foil which he has first "charged" in his hand and which upon returning home the patients are to lay on the afflicted spots or spread out under their pillows, or even carry constantly on their persons.
>
> Every evening at a certain hour Trampler tunes in on all his patients. In his preliminary lecture, he gives notice of this "remote treatment" and cites examples of its success.[122]

During a six-month period, 650 patients treated by Trampler were examined intensively by a research team before treatment. Follow-up studies were conducted on 538 of these individuals. Two thirds of these patients were women.[123] As far as educational, occupational, or family background the patients were representative of the population of the area surrounding Freiburg.

A wide variety of maladies was found in this group. Almost 75% of the patients were chronic cases who had been suffering for more than five years from the conditions which prompted them to see the mental healer. Over half of them were simultaneously undergoing medical treatment — which is something that Dr. Trampler encouraged. They had come to the mental healer because other modes of treatment had failed.

Medical evaluation indicated unexpected, objective improvement in 9% of Dr. Trampler's patients. On the other hand, 61% of Trampler's patients had the subjective experience of permanent or temporary improvement in their condition. In fact, 50% of those patients whose condition had objectively worsened nevertheless declared that they were considerably better, at least temporarily. The subjective improvement of the malady seemed to depend very little on the diagnosis or seriousness of the disease. The results indicated that the subjective improvement was chiefly a function of the attitude which the patients had before treatment by Dr. Trampler. Patients with the highest expectations seemed to respond the most.

Oddly enough, the patients who responded the least to Dr. Trampler's methods were more intelligent, imaginative, and self-confident than those who seemed to benefit the most. The patients who experienced the greatest improvement were, however, more relaxed! In no case was Trampler's treatment found to be objectively harmful to the patient![124] Experimental evidence suggestive of psychic healing effects will be further examined in Section III.

Ramacharaka's Healing Exercise

If after having read this far, you are much more willing to explore the potential for healing you may possess yourself, you may find the following passages, written more than eighty years ago by Yogi Ramacharaka, of practical value. Remember *prana* is the Hindu term for the life energy that permeates the atmosphere, enters the human being through the breath, and can be directed by thought:

Pranic Healing

We will first take up a few experiments in Pranic Healing (or "Magnetic Healing," if you prefer the term):

(I) Let the patient sit in a chair, you standing before him. Let your hands hang loosely by your sides, and then swing them loosely to and fro for a few seconds, until you feel a tingling sensation at the tips of your fingers. Then raise them to the level of the patient's head, and sweep them slowly toward his feet, with your palms toward him with fingers outstretched, as if you were pouring force from your finger tips upon him. Then step back a foot and bring up your hands to the level of his head, being sure that your palms face each other in the upward movement, as, if you bring them up in the same position as you swept them down, you would draw back the magnetism you send toward him. Then repeat several times. In sweeping downward, do not stiffen the muscles, but allow the arms and hands to be loose and relaxed. You may treat the affected parts of the body in a similar way, finishing the treatment by saturating the entire body with magnetism. After treating the affected parts, it will be better for you to flick the fingers away from your sides, as if you were throwing off drops of water which had adhered to your fingers. Otherwise you might absorb some of the patient's conditions. This treatment is very strengthening to the patient, and if frequently practiced will greatly benefit him.

In case of chronic or long seated troubles, the trouble may be "loosened up" by making "sideways" passes before the afflicted part, that is by standing before the patient with your hands together, palms touching, and then swinging the arms out sideways several times. This treatment should always be followed by the downward passes to equalize the circulation.

Self-Healing

(II) Lying in a relaxed condition, breathe rhythmically, and command that a good supply of prana be inhaled. With the exhalation, send the prana to the affected part for the purpose of stimulating it. Vary this occasionally by exhaling, with the mental command that the diseased condition be forced out and disappear. Use the hands in this exercise, passing them down the body from the head to the affected part. In using the hands in healing yourself or others always hold the mental image that the prana is flowing down the arm and through the finger tips into the body, thus reaching the affected part and healing it....

A little practice of the above exercise, varying it slightly to fit the conditions of the case, will produce wonderful results. Some Yogis follow the plan of placing both hands on the affected part, and then breathing rhythmically, holding the mental image that they are pumping prana into the diseased organ and part, stimulating it and driving out diseased conditions, as pumping into a pail of dirty water will drive out the latter and fill the bucket with fresh water. This last plan is very effective if the mental image of the pump is clearly held, the inhalation representing the lifting of the pump handle and the exhalation the actual pumping.

Healing Others

The main principle to remember is that by rhythmic breathing and controlled thought you are enabled to absorb a considerable amount of prana, and are also able to pass it into the body of another person, stimulating weakened parts and organs, imparting health and driving out diseased conditions. You must first learn to form such a clear mental image of the desired condition that you will be able to actually feel the influx of prana, and the force running down your arms and out of your finger tips into the body of the patient. Breathe rhythmically a few times until the rhythm is fairly established, then place your hands upon the affected part of the body of the patient, letting them rest lightly over the part. Then follow the "pumping" process described in the preceding exercise and fill the patient full of prana until the diseased condition is driven out. Every once in a while raise the hands and "flick" the fingers as if you were throwing off the diseased condition. It is well to do this occasionally and also to wash the hands after treatment, as otherwise you may take on a trace of the diseased condition of the patient. Also practice the Cleansing Breath several times after the treatment. During the treatment let the prana pour into the patient in one continuous stream, allowing yourself to be merely the pumping machinery connecting the patient with the universal supply of prana, and allowing it to flow freely through you. You need not work the hands vigorously, but simply enough that the prana freely reaches the affected parts. The rhythmic breathing must be practiced frequently during the treatment, so as to keep the rhythm normal and to afford the prana a free passage. It is better to place the hands on the bare skin, but where this is not advisable or possible place them over the clothing. Vary above methods occasionally during the treatment by stroking the body gently and softly with the finger tips, the fingers being kept slightly separated. This is very soothing to the patient. In cases of long standing you may find it helpful to give the mental command in words, such as "get out, get out," or "be strong, be strong," as the case may be, the words helping you to exercise the will more forcibly and to the point. Vary these instructions to suit the needs of the case, and use your own judgement and inventive faculty.

(III) Headaches may be relieved by having the patient sit down in front of you, you standing back of his chair, and passing your hands, fingers down and spread open in double circles over the top of his head, not touching his head, however. After a few seconds you will actually feel the passage of the magnetism from your fingers, and the patient's pain will be soothed.

(IV) Another good method of removing pain in the body is to stand before the patient, and present your palm to the affected part, at a distance of several inches from the body. Hold the palm steady for a few seconds and then begin a slow rotary motion, round and round, over the seat of the pain. This is quite stimulating and tends to restore normal conditions.

(V) Point your forefinger toward the affected part a few inches away from the body, and keeping the finger steadily pointed move the hand around just as if you were boring a hole with the point of the finger. This will often start the circulation at the point affected, and bring about improved conditions.

(VI) Placing the hands on the head of the patient, over the temples and holding them for a time, has a good effect, and is a favorite form of treatment of this kind.

(VII) Stroking the patient's body (over the clothing) has a tendency to stimulate and equalize the circulation, and to relieve congestion.

(VIII) Much of the value of Massage and similar forms of manipulative treatment, comes from the Prana which is projected from the healer into the patient, during the process of rubbing and manipulating. If the rubbing and manipulating is accompanied by the conscious desire of the healer to direct the flow of Prana into the patient, a greatly increased flow is obtained. If the practice is accompanied with Rhythmic Breathing, the effect is much better.

(IX) Breathing upon the affected part, is practiced by many races of people, and is often a potent means of conveying Prana to the affected part. This is often performed by placing a bit of cotton cloth between the flesh of the person and the healer, the breath heating up the cloth and adding the stimulation of warmth in addition to the other effects.

(X) Magnetized water is often employed by "magnetic healers," and many good results are reported to have been obtained in this way. The simplest form of magnetizing water is to hold the glass by the bottom, in the left hand, and then, gathering together the fingers of the right hand, shake them gently over the glass of water just as if you were shaking drops of water into the glass from your fingertips. You may add to the effect afterwards making downward passes over the glass with the right hand, passing the Prana into the water. Rhythmic breathing will assist in the transferring of the Prana into the water. Water thus charged with Prana is stimulating to sick people, or those suffering from weakness, particularly if they sip it slowly holding their mind in a receptive attitude, and if possible forming a mental picture of the Prana from the water being taken up by the system and invigorating them.

Mental Healing

We will now take up a few experiments in the several forms of Mental Healing....

(I) Auto-suggestion consists in suggesting to oneself the physical conditions one wishes to bring about. The auto-suggestions should be spoken (audibly or silently) just as one would speak to another, earnestly and seriously, letting the mind form a mental picture of the conditions referred to in the words. For instance: "*My stomach is strong, strong, strong — able to digest the food given it — able to assimilate the nourishment from the food — able to give me the nourishment which means health and strength to me. My digestion is good, good, good, and I am enjoying and digesting and assimilating my food, converting it into rich red blood, which is carrying health and strength to all parts of my body, building it up and making me a strong man (or woman).*" Similar auto-suggestions, or affirmations, applied to other parts of the body, will work equally good results, the attention and mind being directed to the parts mentioned causing an increased supply of Prana to be sent there, and the pictured condition to be brought about. Enter into the spirit of the auto-suggestions, and get thoroughly in earnest over them, and so far as possible form the mental image of the healthy condition desired. See yourself as you wish yourself to be. You may help the cure along by treating yourself by the methods described in the experiments on Pranic Healing.

(II) Suggestions for healing, given to others, operate on the same principle as do the auto-suggestions just described, except that the healer must impress upon the patient's mind the desired conditions instead of the patient's doing it for himself. Much better results may be obtained where the healer and patient both co-operate in the mental image and when the patient follows the healer's suggestions in his mind, and forms the mental picture implied by the healer's words. The healer suggests that which he wishes to bring about and the patient allows the suggestions to sink into his Instinctive Mind, where they are taken up and afterwards manifested in physical results....

In many cases all that is needed in suggestive treatment, is to relieve the patient's mind of Fear and Worry and depressing thoughts, which have interfered with the proper harmony of the body, and which have prevented the proper amount of Prana from being distributed to the parts. Removing these harmful thoughts is like removing the speck of dust which has caused our watch to run improperly, having disarranged the harmony of the delicate mechanism....

(III) In what is called strictly Mental Healing, the patient sits relaxed and allows the mind to become receptive. The healer then projects to the patient thoughts of a strengthening and uplifting character which, reacting upon the mind of the patient, causes it to cast off its negative conditions and to assume its normal poise and power, the result being that as soon as the patient's mind recovers its equilibrium it asserts itself and starts into operation the recuperative power within the organism of the person, sending an increased supply of Prana to all parts of the body and taking the first step toward regaining health and strength.

...In treating a patient in this way, keep firmly in your mind the thought that physical harmony is being re-established in the patient, and that health is his normal condition and that all the negative thoughts are being expelled from his mind. Picture him as strong and healthy in mind and in body. Picture as existing all the conditions you wish to establish within him. Then concentrate the mind and fairly *dart* into his body, or into the affected part, a strong penetrating thought, the purpose of which is to work the desired physical change, casting out the abnormal conditions and re-establishing normal conditions and functioning. Form the mental image that the thought is fully and heavily charged with Prana and fairly drive it into the affected part by an effort of the will. Considerable practice is usually needed to accomplish this last result, but to some it appears to come without much effort.

(IV) Distant healing, or "absent treatment," is performed in precisely the same way as is the treatment when the patient is present....

Prana colored by the thought of the sender may be projected to persons at a distance, who are willing to receive it, and healing work done in this way. This is the secret of the "absent healing," of which the Western world has heard so much of late years. The thought of the healer sends forth and colors the prana of the sender, and it flashes across space and finds lodgment in the psychic mechanism of the patient. It is unseen, and it passes through intervening obstacles and seeks the person attuned to receive it. In order to treat persons at a distance, you must form a mental image of them until you can feel yourself to be in rapport with them. This is a psychic process dependent upon the mental imagery of the healer. You can feel the sense of rapport when it is established, it manifesting in a sense of nearness. That is about as plain as we can describe it. It may be acquired by a little practice, and some will get it at the first trial. When rapport is established, say mentally to the distant patient, "I am sending you a supply of vital force or power, which will invigorate you and heal you." Then picture the prana as leaving your mind with each exhalation of rhythmic breath, and traveling across space instantaneously and reaching the patient and healing him. It is not necessary to fix certain hours for treatment, although you may do so if you wish. The respective condition of the patient, as he is expecting and opening himself up to your psychic force, attunes him to receive your vibrations whenever you may send them. If you agree upon hours, let him place himself in a relaxed attitude and receptive condition.

Some healers form the picture of the patient sitting in front of them, and then proceed to give the treatment, just as if the patient were really present. Others form the mental image of projecting the thought, picturing it as leaving their mind, and then traversing space entering the mind of the patient. Others merely sit in a passive, contemplative attitude and intently *think* of the patient, without regard to intervening space. Others prefer to have a handkerchief, or some other article belonging to the patient, in order to render more perfect the *rapport* conditions. Any, or all, of these methods are good, the temperament and inclinations of the person causing him to prefer some particular method. But the same principle underlies them all.

A little practice along the lines of the several forms of healing just mentioned, will give the student confidence, and ease in operating the healing power, until he will often radiate healing power without being fully conscious of it. If much healing work is done, and the heart of the healer is in his work, he soon gets so that he heals almost automatically and involuntarily when he comes into the presence of one who is suffering. The healer must, however, guard against depleting himself of Prana, and thus injuring his own health. He should study...methods...of recharging himself, and protecting himself against undue drains upon his vitality. And he should make haste slowly in these matters, remembering that forced growth is not desirable.

This lesson has not been written to advise our students to become healers. They must use their own judgment and intuitions regarding that question....

For ourselves, we cling to the principles of "Hatha Yoga," which teaches the doctrine of preserving health by right living and right thinking, and we regard all forms of healing as things made necessary only by Man's ignorance and disobedience of Natural laws. But so long as man will not live and think properly, some forms of healing are necessary, and therefore the importance of their study.[125,126]

RECHARGING YOURSELF

If you feel that your vital energy is at a low ebb, and that you need to store up a new supply quickly, the best plan is to place the feet close together (side by side, of course) and to lock the fingers of both hands in any way that seems the most comfortable. This closes the circuit, as it were, and prevents any escape of prana through the extremities. Then breathe rhythmically a few times, and you will feel the effect of the recharging.[127]

Deep Healing

The exercises and techniques presented above by Ramacharaka can be very powerful and effective. However, they do not necessarily provide the only or the deepest approach to healing. Another very profound point of view is exemplified in the writings and teachings of Stephen Levine, a poet, an author, a spiritual teacher. Stephen has written numerous books, including *Who Dies?, Meetings on the Edge, Healing into Life and Death,* and (with Ram Dass) *Grist for the Mill.* His perspective is summarized in a *Thinking Allowed* interview:

> The very idea that you are a good person if you heal, makes you a bad person if you die. Who needs to die with a sense of failure? Many have been injured by the idea that you are responsible for your illness. You are not responsible for your illness; you are not responsible for your cancer. You are responsible to your cancer. When we see that we are responsible to our illness, then when pain arises we can send mercy, we can send kindness.
>
> You and I, we're conditioned. We walk

across a room, we stub our toe. What do we do with the pain in our toe? We're conditioned to send hatred into it. We're conditioned to try to exorcise it and…we cut the pain off. In fact, even many meditative techniques for working with pain are to take your awareness, your attention, and put it elsewhere. Just when that throbbing toe is most calling out for mercy, for kindness, for embrace, for softness, it's least available. In some ways it's amazing that anybody heals, considering our conditioning to send hatred into our pain, which is the antithesis of healing.

The way we respond to pain is the way we respond to life. When things aren't the way we want them to be, what do we do? Do we close down, or do we open up to get more of a sense of what's needed in the moment? Our conditioning is to close down — aversion, rejection, put it away, denial. Nothing heals. That is the very basis on which unfinished business accumulates, putting it away — I'm right, they're wrong; no quality of forgiveness. Where can there be healing in that?

We suggest that people treat their illness as though it were their only child, with that same mercy and loving-kindness. If that was in your child's body, you'd caress it, you'd hold it, you'd do all you could to make it well. But somehow when it's in our body we wall it off, we send hatred into it and anger into it. We treat ourselves with so little kindness, so little softness. And there are physical correlations to the difference between softening around an illness — blood flow, availability of the immune system, etcetera — and hardness. You know, if you've got a hard belly and your jaw is tight, and that hardness is around your eyes, it's very difficult for anything to get through.

A lot of healers, if they can't "heal" you, they have no business with you anymore. But when our work is on ourself, then even the teaching of helplessness is honored. Sometimes you can't help everybody, but that doesn't mean anything has to come out of you that limits their access to who you are, to your heart, to your connection with them. If it's work on yourself, they're in the presence of good healing. But all the healers I know who are really phenomenal, who are some of the phenomenal healers, they all say God does it.

When the mind sinks into the heart, and vice versa, there's healing. When we become one with ourselves, there's healing.[128]

STEPHEN LEVINE (FROM *CONSCIOUS LIVING/ CONSCIOUS DYING* VIDEOTAPE, COURTESY THINKING ALLOWED PRODUCTIONS)

SPIRITUAL ANATOMY

Many cultural traditions contain an esoteric thread describing what might be referred to as the anatomy of the human soul. There are the *ka* and *ba* of Egyptian mythology, the *meridians* of acupuncture, the *chakras* and *nadis* of yoga, the *sephirot* of Hebrew cabalistic tradition, and the *etheric* and *astral bodies* of Western esoteric lore.

It is very natural that this should be so since, as master mythologist Joseph Campbell pointed out during a *Thinking Allowed* interview, the mythologies of all cultures are borne of our bodily experiences:

> Fantasy and imagination is a product of the body. The energies that bring forth the fantasies derive from the organs of the body. The organs of the body are the source of our life, and of our intentions for life.

JOSEPH CAMPBELL (FROM *UNDERSTANDING MYTHOLOGY* VIDEOTAPE, COURTESY THINKING ALLOWED PRODUCTIONS)

> They conflict with each other. Among these organs, of course, is the brain. And then you must think of the various impulses that dominate our life system — the erotic impulse; the impulse to conquer, conquest and all that; self preservation; and then certain thoughts that have to do with ideals and things that are held up before us as aims worth living for and giving life its value. All of these different forces come into conflict within us. And the function of mythological imagery is to harmonize and coordinate the energies of our body, so that we will live a harmonious and fruitful life in accord with our society, and with the new mystery that emerges with every new human being — namely, what are the possibilities of this particular human life?[129]

The deeper truth embedded within cultures that emphasize mythological systems of spiritual anatomy is that *the divine is within us.* Cultures that do not emphasize the anatomy of the soul tend to be those which view the deity as external and apart from the human being.

One might say that our spiritual bodies are made of thought itself. Of course, from the perspective of psychic folklore, thought is tangible — almost solid — and certainly very potent. As we journey through the lore of spiritual anatomy, it is appropriate that we begin by examining the role of thought itself. The concept of *thoughtforms* provides an excellent vehicle for the journey — for in many systems and teachings thought, itself, is very spiritual in nature.

Thoughtforms

Descriptions of thoughtforms and the mental body come from Theosophists Annie Besant and C. W. Leadbeater who were both very influential in the shaping of modern psychic folklore:

> The mental body is an object of great beauty, the delicacy and rapid motion of its particles giving it an aspect of living iridescent light, and this beauty becomes an extraordinarily radiant and entrancing loveliness and the intellect becomes more highly evolved and is employed chiefly on pure and sublime topics. Every thought gives rise to a set of correlated vibrations in the matter of this body, accompanied with a marvelous display of color, like that in the spray of a waterfall as the sunlight strikes it, raised to the *n*th degree of color and vivid delicacy. The body under this impulse throws off a vibrating portion of itself, shaped by the nature of the vibration — as figures are made by sand on a disk vibrating to a musical note — and this gathers from the surrounding atmosphere matter like itself in fineness from the elemental essence of the mental world. We have then a thought-form pure and simple, and it is a living entity of intense activity animated by the one idea that generated it. If made of finer kinds of matter, it will be of great power and energy, and may be used as a most potent agent when directed by a strong and steady will....
>
> Each definite thought produces a double effect — a radiating vibration and a floating form. The thought itself appears first to clairvoyant sight as a vibration in the mental body, and this may be either simple or complex....
>
> If a man's thought or feeling is directly connected with someone else, the resultant thought-form moves toward that person

and discharges itself upon his astral and mental bodies. If the man's thought is about himself, or is based upon a personal feeling, as the vast majority of thoughts are, it hovers round its creator and is always ready to react upon him whenever he is for a moment in a passive condition....

Each man travels through space enclosed within a case of his own building, surrounded by a mass of the forms created by his habitual thought. Through this medium he looks out upon the world, and naturally he sees everything tinged with its predominant colors, and all rates of vibration which reach him from without are more or less modified by its rate. Thus until the man learns complete control of thought and feeling, he sees nothing as it really is, since all his observations must be made through his medium, which distorts and colors everything like badly made glass.

If the thought-form be neither definitely personal nor specially aimed at someone else, it simply floats detached in the atmosphere, all the time radiating vibrations similar to those originally sent forth by its creator. If it does not come into contact with any other mental body, this radiation gradually exhausts its store of energy, and in that case, the form falls to pieces; but if it succeeds in awakening sympathetic vibration in any mental body near at hand, an attraction is set up, and the thought-form is usually absorbed by that mental body.[130]

To this picture of the mental body, Yogi Ramacharaka adds a further description of the mental world as such:

Places and localities are often permeated by the thought of persons who formerly lived there, who have moved away or died many years ago....The occultist knows that this thought-atmosphere of a village, town, city, or nation is the composite thought of those dwelling in it or whom have previously dwelt there. Strangers coming into the community feel the changed atmosphere about it, and, unless they find it in harmony with their own mental character, they feel uncomfortable and desire to leave the place.

Annie Besant — Fabian Socialist (and George Bernard Shaw's lover) who later became head of the Theosophical Society

If one, not understanding the laws operating in the thought world, remains long in a place, he is most likely to be influenced by the prevailing thought-atmosphere, and in spite of himself a change begins to be manifest in him and he sinks or rises to the level of the prevailing thought....

In the same way dwellings, business-places, buildings, etc., take on the predominant thought of those inhabiting them or who have dwelt in them.[131]

An example of the perception of thoughtforms is provided by the famous medium Eileen Garrett:

One sees lines and colors and symbols. These move, and one is wholly concentrated on them and their movement. I say "symbols" here for want of a better word. I frequently see curving lines of light and color that flow forward in strata, and in these strips or ribbons of movement there will appear other sharply angled lines that form and change and fade like arrow heads aimed and passing in various directions. And in this flow of energy that is full of form and color, these arrow heads will presently indicate the letter H. Each line of the H will be an independent curve, and their combination will not remain identifiable for very long. But I shall have caught it; and holding it suspended in awareness, I continue to watch the process develop and unfold. Soon a rapidly drifting A appears in the field of concentration, and then, let us suppose, an R; and presently I have gathered the word HARRY out of the void, either as a proper name or as a verb temporarily without either subject or object. Whether it is actually a noun or a verb will depend upon the context of the perception as a whole.

This process is infinitely rapid. But I have achieved an alertness of attention, of awareness, of being, which is equal to this rapid flow of immaterial line and color and symbol, and out of this alertness, poised above the flowing stream of differentiated energy, I gather a message with a meaning — a message which has come to my consciousness out of the objective world as factually as the reflected light from the distant Moon may reach my consciousness by way of my sense of sight.[132]

EILEEN GARRETT (1893-1970), FOUNDER OF THE PARAPSYCHOLOGY FOUNDATION, NEW YORK

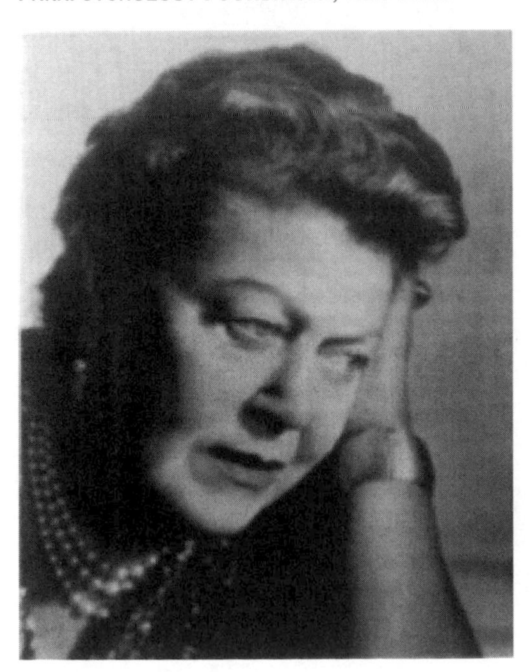

The existence of the mental world implies a view of nature incorporating meaning as well as mechanism. We are no longer dealing with blind forces bouncing aimlessly throughout the universe. The substance of the mental world is imbued with purpose. Minds, or monads, are constantly emitting radiation of an intelligent nature. Every thought may be thought of as an active spiritual force. Iconoclastic researcher Andrija Puharich, M.D., has coined the term *inergy*, meaning "intelligent energy" to refer to this realm of spirit or thought.

The Aura

Theosophical, psychic and mystical lore has it that the emotions and thoughts of an individual distinguish themselves by their form and color. This is thought of as the *aura* or *astral body* which is visualized as an egg-shaped envelope around the human being. Is the *aura* simply composed of our thoughts (or our thoughts about someone else's thoughts) or does it have an independent physical existence? The answer, of course, depends on what we mean by *aura* — which is hardly an operationally defined scientific term. There are many different meanings for the term. (For example, in medicine, an *aura* refers to sensations which develop prior to the onset of an epileptic seizure.)

C. W. Leadbeater, one of the Theosophists who was responsible for popularizing the term "astral" plane, claims that it was inherited from the medieval alchemists. The term means starry and was applied to the plane above the physical because of its luminous appearance. Furthermore, the emotional currents were thought to be influenced by the planetary positions. The meaning of the different colors that appear in the astral body is recorded by Yogi

Ramacharaka whose writings mimic those of Madame Blavatsky, founder of the Theosophical Society, on this topic:

> AURIC COLORS AND THEIR MEANINGS
>
> *Black* represents hatred, malice, revenge, and similar feelings. Gray, of a bright shade, represents selfishness.
>
> *Gray,* of a peculiar shade (almost that of a corpse), represents fear and terror.
>
> *Gray,* of a dark shade, represents depression and melancholy.
>
> *Green,* of a dirty shade, represents jealousy. If much anger is mingled with the jealousy, it will appear as red flashes on the green background.
>
> *Green,* of almost a slate-color shade, represents low deceit.
>
> *Green,* of a peculiar bright shade, represents tolerance to the opinions and beliefs of others, easy adjustment to changing conditions, adaptability, tact, politeness, worldly wisdom, etc., and qualities which some might possible consider "refined deceit."
>
> *Red,* of a shade resembling the dull flame when it bursts out of a burning building, mingled with the smoke, represents sensuality and the animal passions.
>
> *Red,* seen in the shape of bright-red flashes resembling the lightning flash in shape, indicates anger. These are usually shown on a black background in the case of anger arising from hatred or malice, but in cases of anger arising from jealousy they appear on a greenish background. Anger arising from indignation or defense of a supposed "right," lacks these backgrounds, and usually shows as red flashes independent of a background.
>
> *Crimson* represents love, varying in shade according to the character of the passion. A gross sensual love will be dull and heavy crimson, while one mixed with higher feelings will appear in lighter and more pleasing shades. A very high form of love shows a color almost approaching a beautiful rose color.
>
> *Brown,* of a reddish tinge, represents avarice and greed.
>
> *Orange,* of a bright shade, represents pride and ambition.
>
> *Yellow,* in its various shades, represents intellectual power. If the intellect contents itself with things of a low order, the shade is a dark, dull yellow; and as the field of the intellect rises to higher levels, the color grows brighter and clearer, a beautiful golden yellow betokening great intellectual attainment, broad and brilliant reasoning.
>
> *Blue,* of a dark shade, represents religious thought, emotion, and feeling. This color, however, varies in clearness according to the degree of unselfishness manifest in the religious conception. The shades and degrees of clarity vary from a dull indigo to a beautiful rich violet, the latter representing the highest religious feeling.
>
> *Light Blue,* of a peculiarly clear and luminous shade, represents spirituality. Some of the higher degrees of spirituality observed in ordinary mankind show themselves in this shade of blue filled with luminous bright points, sparkling and twinkling like stars on a clear winter night.

The student will remember that these colors form endless combinations and blends, and show themselves in greatly varying degrees of brightness and size, all of which have meanings to the developed occultist.

> In addition to the colors mentioned above, there are several others for which we have no names, as they are outside of the colors visible in the spectrum, and consequently science, not being able to perceive them, has not thought it necessary to bestow definite names upon them, although they exist theoretically. Science tells us that there are also what are known as "ultraviolet" rays and "ultra-red" rays, neither of which can be followed by the human eyes, even with the aid of mechanical appliances, the vibrations being beyond our senses. These two "ultra" colors (and several others unknown to science) are known to occultists and may be seen by the person with certain psychic powers. The significance of this

statement may be more fully grasped when we state that when seen in the Human Aura either of these "ultra" colors indicates psychic development, the degree of intensity depending upon the degree of development. Another remarkable fact, to those who have not thought of the matter, is that the "ultraviolet" color in the Aura indicates psychic development when used on a high and unselfish plane, while "the ultra-red" color, when seen in the Human Aura, indicates that the person has psychic development, but is using the same for selfish and unworthy purposes — "black magic," in fact. The ultraviolet rays lie just outside of an extreme of the visible spectrum known to science, while the "ultra-red" rays lie just beyond the other extreme. The vibrations of the first are too high for the ordinary human eye to sense, while the second comprises vibrations as excessively low as the first is excessively high. And the real difference between the two forms of psychic power is as great as is indicated by the respective positions of these two "ultra" colors. In addition to the two "ultra" colors just alluded to, there is another which is invisible to the ordinary sight — the *true* primary yellow, which indicates of the Spiritual Illumination and which is faintly seen around the heads of the spiritually great. The color which we are taught characterizes the seventh principle, Spirit, is said to be of pure white light, of a peculiar brilliancy, the like of which has never been seen by human eye — in fact, the very existence of *absolute* "white light" is denied by Western science.

The Aura emanating from the Instinctive Mind principally comprises heavier and duller shades. In sleep, when the mind is quiet, there appears chiefly a certain dull red, which indicates that the Instinctive Mind is merely performing the body's animal functions. This shade, of course, is always apparent, but during the waking

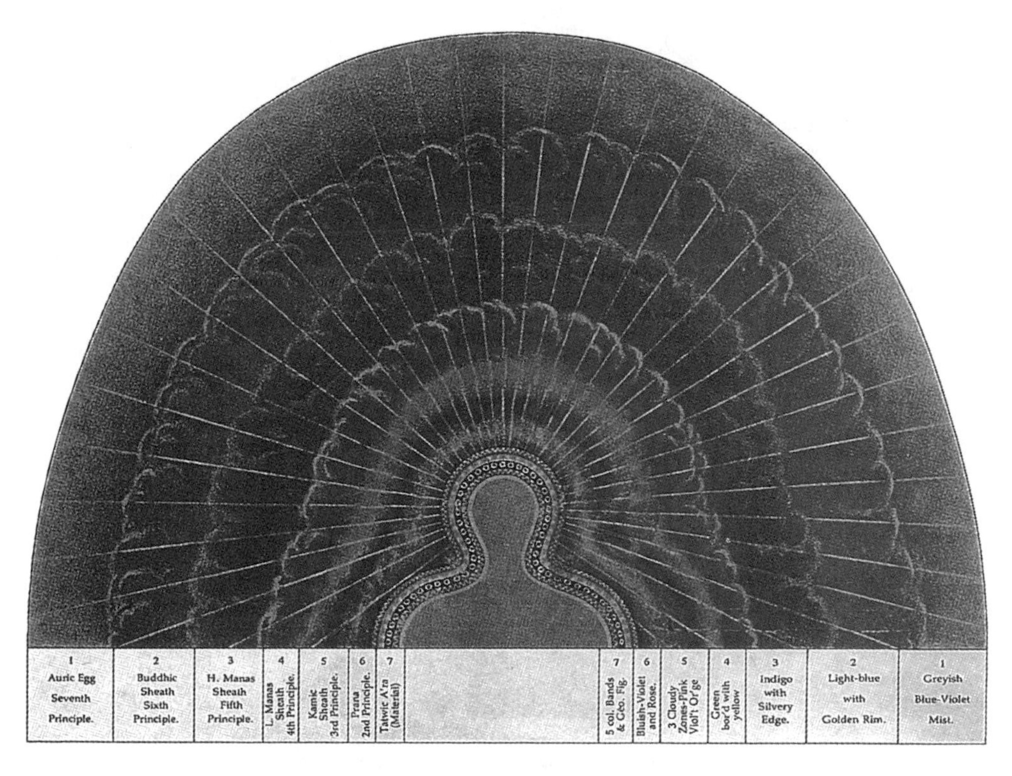

SEVEN LAYERS OF THE HUMAN AURA ACCORDING TO A THEOSOPHICAL MANUSCRIPT PUBLISHED IN 1896. EACH LAYER IS SAID TO REPRESENT FINER PARTICLES OF MATTER.

hours it is often obscured by the brighter shades of the passing thoughts, emotions or feelings.

Right here it would be well to state that even while the mind, emotions or feelings remains calm there hover in the Aura shades which indicate a man's predominant tendencies, so that his stage of advancement and development as well as his "tastes" and other features of his personality may be easily distinguished. When the mind is swept by a strong passion, feeling, or emotion, the entire Aura seems to be colored by the particular shade or shades representing it. For instance, a violent fit of anger causes the whole Aura to show bright red flashes upon a black background, almost eclipsing the other colors. This state lasts for a longer or shorter time, according to the strength of the passion. If people could but have a glimpse of the Human Aura when so colored, they would become so horrified at the dreadful sight that they would be far more hesitant about flying into rage — it resembles the flames and smoke of the devil's "pit" and, in fact, the human mind in such a condition becomes a veritable hell temporarily. A strong wave of love sweeping over the mind will cause the entire Aura to show crimson; the shade will depend upon the character of the passion. Likewise, a burst of religious feeling will bestow upon the entire Aura a blue tinge, as explained in the table of colors. In short, a strong emotion, feeling, or passion causes the entire Aura to take on its color while the feeling lasts. You will see from what we have said that there are two aspects to the color feature of the Aura; the first depending upon the predominant thoughts habitually manifesting in the mind of the person; the second depending upon the particular feeling, emotion, or passion (if any) is dominating him, at that particular time.

The student … will realize readily that as the man develops and unfolds he becomes less and less prey to passing passions, emotions, or feelings emanating from the Instinctive Mind, and that Intellect, and then Spiritual Mind, manifest themselves instead of lying dormant in a latent condition. Remembering this, he will readily see how great a difference there must be between the Aura of an undeveloped man and that of the developed man. The one is a mass of dull, heavy, gross colors, the entire mass being frequently flooded by the color of some passing emotion, feeling, or passion. The other shows the higher colors and is very much clearer, being but little disturbed by feelings, emotion, and passions, all of which have been brought largely under the control of the will.

The man who has Intellect well developed shows an Aura flooded with the beautiful golden yellow betokening intellectuality. This color in such cases is particularly apparent in the upper part of the Aura, surrounding the head and shoulders of the man, the more animal colors sinking to the lower part of the Aura. Read the remarks under the head of "Yellow" in the color table in this lesson. When the man's Intellect has absorbed the idea of spirituality and devotes itself to the acquirement of spiritual power, development, and unfoldment, this yellow will show around its edges a light blue of a peculiarly clear and luminous shade. This peculiar light blue is indicative of what is generally called "spirituality," but which is simply "intellectual-spirituality," if you will pardon the use of the somewhat paradoxical term — it is not the same thing as Spiritual Mind, but is merely Intellect impregnated by Spiritual Mind, to use another troublesome term. In some cases when this intellect is in a highly developed state, the luminous light blue shows as a broad fringe or border often being larger than the center itself, and in addition, in special cases, the light blue is filled with brilliant luminous points, sparkling and twinkling like stars on a clear winter night. These bright points indicate that the color of the Aura of the Spiritual Mind is asserting itself, and shows that Spiritual Consciousness has either become momentarily evident to the man or is about to become so in the near future. This is a point upon which much confusion has

> arisen in the minds of students and even teachers of occultism. The next paragraph will also shed further light upon the matter.
>
> The Aura emanating from the Spiritual Mind, or sixth principle, bears the color of the true primary yellow, which is invisible to ordinary sight and which cannot be reproduced artificially by man. It centers around the head of the spiritually illumined, and at times produces a peculiar glow which can even be seen by undeveloped people. This is particularly true when the spiritually developed person is engaged in earnest discourse or teaching, at these times his countenance seems fairly to glow and to possess a luminosity of a peculiar kind. The nimbus shown in pictures of mankind's great spiritual leaders results from a tradition based on a fact actually experienced by the early followers of such leaders. The "halo" or glory shown on pictures arises from the same fact.[133]

Because most of Ramacharaka's descriptions accord with the teachings of Blavatsky, Leadbeater and other Theosophists, it would be a mistake to assume that he is writing on the basis of either careful measurements or personal experience. From the perspective of a biocomputer model of mental functioning, one might view the perception of auric colors as a particular way some individuals program their minds to function, i.e., as the result of cultural conditioning or autoconditioning enhanced by the altered state of consciousness induced through meditation and yogic practices.

If we view the mind/brain system as a biocomputer, we could say that there are various sensory inputs (eyes, ears, nose, mouth, skin) and various internal perceptual display systems (sight, hearing, smell, taste, touch). It is entirely possible that the input from one sensory mode could be displayed internally using a modality normally reserved for a different sensory mode. Thus, under the influence of hypnotic suggestion or psychedelic drugs, individuals often report "seeing music." This well-known phenomena is referred to as *synesthesia.*

A very reasonable explanation of the human aura as reported by psychics is that this is also a form of *synesthesia* — a special way we can program ourselves to display information in the *sensorium* of our minds. The inputs for this display pattern could conceivably arrive from any sensory (or extrasensory) modality, could be derived from intuitive or logical processing, or could be generated from the biocomputer programming (i.e., cultural conditioning and autoconditioning) itself.

An amusing anecdote relating to the perception of the human aura on the "astral level" comes from the Texas psychic Ray Stanford. Ray, who seems to be very proficient at seeing auras, visited his twin brother Rex, a parapsychologist then at the University of Virginia; he gave a demonstration of his talents before a small group of researchers assembled by his brother. One of the guests was Dr. Robert Van de Castle, the director of the sleep and dream laboratory. Ray noticed a number of pink spots in the aura around Van de Castle's abdomen. This perception puzzled him since it is one he normally associated with pregnant women, and he remarked to Dr. Van de Castle, "If I didn't know better, I'd say you were pregnant." This drew some laughter from the audience. However, Van de Castle then reflected that he had been analyzing the dreams of pregnant women all morning and had even remarked earlier that day that he was beginning to feel like a pregnant woman himself.[134]

Experimental Tests

You might think it would be relatively simple for scientists to test the objectivity of the aura, by comparing the independent observations of a number of psychics. In fact, the problem is difficult and there has been very little systematic research. For one thing, if the observations are being made at different moments of time, it is possible the aura could change appearance. Also, a truly objective study would want to rule out any other sensory cues that could be confused with the aura. Charles Tart has suggested that the target person for such a study be hidden behind an opaque screen shaped so only the aura should be visible beyond its perimeter and not the physical body at all.[135] To my knowledge, twenty years after Tart's proposal, there have still been no satisfactory experiments of this type.

In a study conducted by Dr. A. R. G. Owen of Toronto, fourteen different psychics made independent observations of the aura of a single subject. The reported descriptions show wide variation that, according to Owen "seems to go beyond that degree of variability in the aura, that according to percipients of auras, is to be expected as a result of temporal variations in the physical, emotional or mental state of the possessor of the aura." However, the study took place over a one-year period. Going over the data, I myself was struck by the similarity of reports made by different observers on the same day. Owen maintained that there was no cogent evidence the subject was in different physical or emotional states during the different days of experimentation. It does not appear that he was looking for subtle emotional changes. The fact that lighting conditions were different on the different days of experimentation further confuses the data. Furthermore, some subjects saw the aura with their eyes open, while at least one subject viewed the aura with his eyes closed.[136]

The Vital Body

In addition to the *astral body*, which seems to correlate with thoughts and emotions, some occultists refer to the *vital body* or sometimes *etheric body* — more associated with life energy and health and more suggestive than the *astral body* of having a measurable physical basis. It is interesting to note that the term *etheric body* developed at a time in history prior to the Michelson-Morley experiments which disconfirmed the physical theory of the *ether* as a medium permeating the known universe. It is probable that the term *etheric body* (like *astral body*) developed from what was once legitimate scientific speculation. Today such terms belong clearly in the realm of occult folklore. Max Heindel, founder of the Rosicrucian Fellowship, describes the *etheric* or *vital body*:

> The vital body of plant, animal, and man, extends beyond the periphery of the dense body as the Etheric Region, which is the vital body of a planet, extends beyond its dense part, showing again the truth of the Hermetic axiom "As above, so below." The distance of this extension of the vital body of man is about an inch and a half. The part which is outside the dense body is very luminous and about the color of a new-blown peach-blossom. It is often seen by persons having very slight involuntary clairvoyance. The writer has found, when speaking with such persons, that they frequently are not aware they see anything unusual and do not know what they see.
>
> The dense body is built into the matrix of this vital body during ante-natal life, and with one exception, it is an exact copy, molecule for molecule, of the vital body. As the lines of force in freezing water are the avenues of formation for ice crystals, so the lines of force in the vital body determine the shape of the dense body. All through life the vital body is the builder and restorer of the dense form. Were it not for the etheric heart the dense heart would break quickly under the constant strain we put upon it. All the abuses to which we subject the dense body are counteracted, so far as lies in its power,

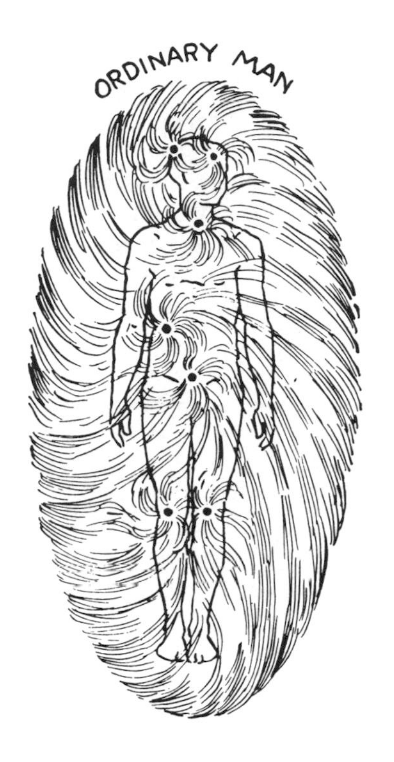

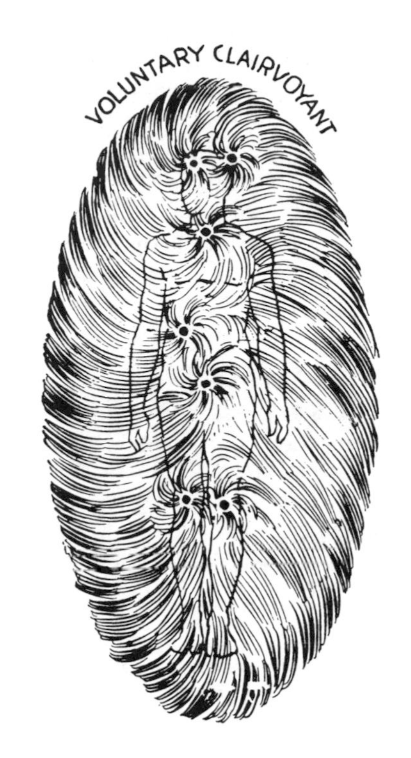

by the vital body, which is continually fighting against the death of the dense body.

The exception mentioned above is that the vital body of a man is female or negative, while that of a woman is male or positive. In that fact we have the key to numerous puzzling problems of life. That woman gives way to her emotions is due to the polarity noted, for her positive, vital body generates an excess of blood and causes her to labor under an enormous internal pressure that would break the physical casement were not a safety-valve provided in the periodical flow, and another in the tears which relieve the pressure on special occasions — for tears are "white bleeding."

Man may have and has as strong emotions as a woman, but he is usually able to suppress them without tears, because his negative vital body does not generate more blood than he can comfortably control.

Unlike the higher vehicles of humanity, the vital body (except under certain circumstances, to be explained when the subject of "Initiation" is dealt with) does not ordinarily leave the dense body until

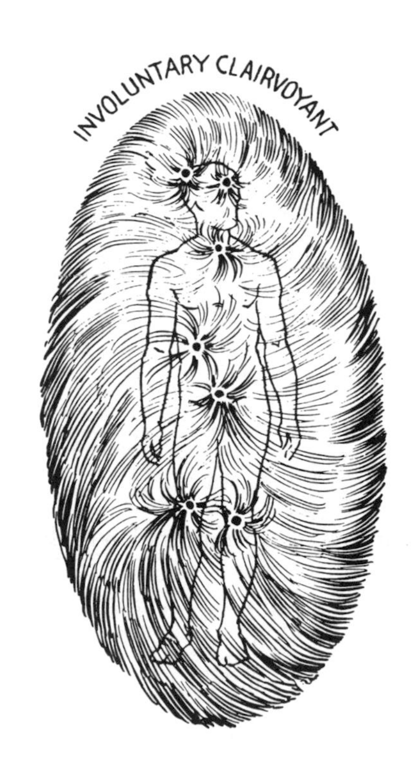

According to Rosicrucian Max Heindel, one's level of clairvoyant functioning is reflected in the aura as diagrammed above.

the death of the latter. Then the chemical forces of the dense body are no longer held in check by the evolving life. They proceed to restore the matter to its primordial condition by disintegration so that it may be available for the formation of other forms in the economy of nature. Disintegration is thus due to the activity of the planetary forces in the chemical ether.

There are certain cases where the vital body partly leaves the dense body, such as when a hand "goes to sleep." Then the etheric hand of the vital body may be seen hanging below the dense arm like a glove and the points cause the peculiar pricking sensation felt when the etheric hand re-enters the dense hand. Sometimes in hypnosis the head of the vital body divides and hangs outside the dense head, one half over each shoulder, or lies around the neck like the collar of a sweater. The absence of prickly sensation at awakening in cases like this is because during the hypnosis part of the hypnotist's vital body had been substituted for that of the victim.

When anesthetics are used the vital body is partially driven out, along with the higher vehicles, and if the application is too strong and the life ether is driven out, death ensues. This same phenomenon may also be observed in the case of materializing medium and an ordinary man or woman is just this: In the ordinary man or woman the vital body and the dense body are, at the present stage of evolution, quite firmly interlocked, while in the medium they are loosely connected. It has not always been so, and the time will again come when the vital body may normally leave the dense vehicle, but that is not normally accomplished at present. When a medium allows his or her vital body to be used by entities from the Desire World who wish to materialize, the vital body generally oozes from the left side — through the spleen, which is its particular "gate." When the vital forces cannot flow into the body as they do normally, the medium becomes greatly exhausted, and some of them resort to stimulants to counteract the effects, in time becoming incurable drunkards.

The vital force from the sun, which surrounds us as a colorless fluid, is absorbed by the vital body through the etheric counterpart of the spleen, wherein it undergoes a curious transformation of color. It becomes pale rose-hued and spreads along the nerves all over the dense body. It is to the nervous system what the force of electricity is to a telegraph system. Though there be wires, instruments, and telegraph operators all in order, if the electricity is lacking no message can be sent. The Ego, the brain, and the nervous system may be in seemingly perfect order, but if the vital force be lacking to carry the message of the Ego through the nerves to the muscles, the dense body will remain inert. This is exactly what happens when part of the dense body becomes paralyzed. The vital body has become diseased and the vital force can no longer flow. In such cases, as in most sickness, the trouble is with the finer invisible vehicles. In conscious or unconscious recognition of this fact, the most successful physicians use suggestion — which works upon the higher vehicles — as an aid to medicine. The more a physician can imbue his patient with faith and hope, the speedier disease will vanish and give place to perfect health.

During health the vital body specializes a superabundance of vital force, which after passing through a dense body, radiates in straight lines in every direction from the periphery thereof, as the radii of a circle do from the center; but during ill-health, when the vital body becomes attenuated, it is not able to draw to itself the same amount of force and in addition the dense body is feeding upon it. Then the lines of the vital fluid which pass out from the body are crumpled and bent, showing the lack of force behind them. In health the great force of these radiations carries with it germs and microbes which are inimical to the health of the dense body, but in sickness, when the vital force is weak, these emanations do not so readily eliminate disease germs. Therefore the danger of contracting disease is much greater when the vital forces are low than when one is in robust health.

> In cases where parts of the dense body are amputated, only the planetary ether accompanies the separated part. The separate vital body and the dense body disintegrate synchronously after death. So with the etheric counterpart of the amputated limb. It will gradually disintegrate as the dense member decays, but in the meantime the fact that the man still possesses the etheric limb accounts for his assertion that he can feel his fingers or suffers pain in them. There is also a connection with a buried member, irrespective of distance. A case is on record where a man felt a severe pain, as if a nail had been driven into the flesh of an amputated limb, and he persisted until the limb was exhumed, when it was found that a nail had been driven into it at the time it was boxed for burial.[137] The nail was removed and the pain instantly stopped. It is also in accordance with these facts that people complain of pain in a limb for perhaps two or three years after the amputation. The pain will then cease. This is because the disease remains in the still undetached etheric limb, but as the amputated part disintegrates, the etheric limb follows suit and thus the pain ceases.[138]

Heindel's description is typical of the type of writing found in the occult and mystical literature from many cultures and periods of time.

A word of caution here. There are a few effects of an optical or physiological nature that might easily be taken for an aura by a careless, or uninformed, observer. In a clever series of experiments, Canadian researcher A. R. G. Owen determined that many people will see such "rim" auras, glowing about an inch or two from the edge of inanimate objects even more distinctly than around living plants, animals, and humans. Many people were unable to distinguish between the aura that appeared around a piece of cardboard shaped as a hand and that observed around a real human hand. Other observers, particularly those who saw a much larger and more vivid aura, were quite able to make the distinction. In any case, almost all of the subjects were able to see some aura-like visual phenomena. These perceptions are attributed to the active role the retina and the visual cortex take in organizing and interpreting visual contours while the eye itself is constantly making tiny movements, scanning whatever is observed.[139]

You can easily experience this yourself simply by focusing on the contours of the word written in the margin. You can see a "rim aura" when you stare at the outline of the letters. See what you notice. The power of suggestion also is active here. This effect is highlighted by the sharp black and white contrast.

On the other hand, Dr. Owen was able to repeatedly demonstrate a most unusual and vivid aura-like appearance on the end of a rod while it was the focus of concentration from two gifted psychics. A number of observers were able to independently verify this perception, which was not normally seen around the rod.[140] However the exact conditions for replication of this effect are not known.

A. R. G. OWEN

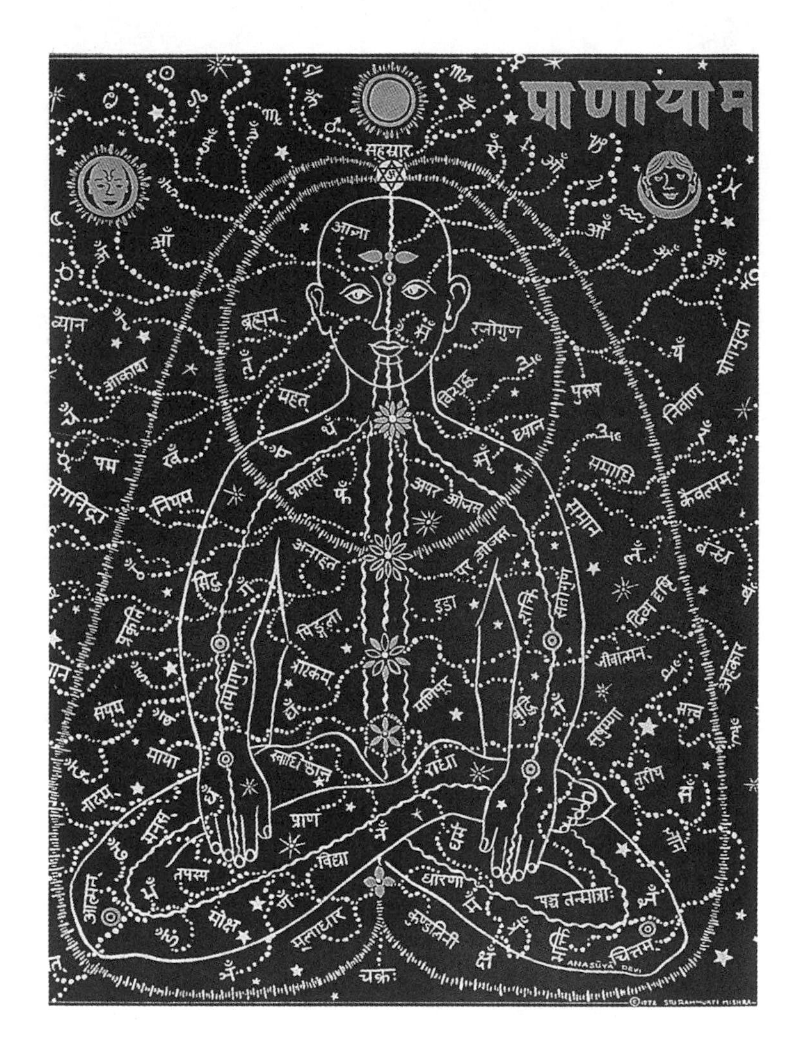

PRANAYAMA ENERGY CHART SHOWING CURRENTS OF LIFE ENERGY PASSING FROM THE COSMOS THROUGH THE CHAKRAS INTO THE BODY, AS UNDERSTOOD IN PRANAYAMA YOGA. THIS CHART WAS DESIGNED FOR USE AS A FOCUS DURING MEDITATION BY SRI RAMMURTI MISHRA, M.D.

The Chakras

The word *chakra* in Sanskrit means wheel; and according to the Theosophical tradition, the *chakras* are "a series of wheel-like vortices existing in the surface of the etheric body."[141] The etheric body[142] is part of the human aura closest in proximity to the skin. It is sometimes referred to as the health aura, and I think can be equated to the electromagnetic field of the body or the bioplasma without doing injustice to the Theosophical system. The *chakras* actually extend out beyond the etheric body to the more subtle parts of the aura — such as the astral body. Some individuals perceive the etheric body as a faintly luminous mist extending slightly beyond the body.

In 1927, the Reverend Charles Leadbeater wrote a book on the *chakras* based largely on his own psychic perceptions:

> When quite undeveloped they appear as small circles about two inches in diameter, glowing dully in the ordinary man; but when awakened and vivified they are seen as blazing, coruscating whirlpools, much increased in size, and resembling miniature suns....If we imagine ourselves to be

> looking straight down into the bell of a flower of the convolvulus type, we shall get some idea of the general appearance of a *chakra.* The stalk of the flower in each springs from a point in the spine.
>
> All these wheels are perpetually rotating, and into the hub or open mouth of each a force from the higher world is always flowing....Without this inrush of energy the physical body could not exist.[143]

Leadbeater also has uncovered descriptions of such vortices, similar to his own, in the works of the seventeenth-century German mystic Johann Georg Gichtel, a pupil of Jacob Boehme. Gichtel assigned an astrological planetary influence to each of the seven centers in his system. It is uncertain to me whether he was influenced by the Sanskrit tradition. However, on the title page of his book, *Theosophia Practica,* he claims to be presenting...

> A short exposition of the three principles of the three worlds in humanity, represented in clear pictures, showing how and where they have their respective Centres in the inner person; according to what the author has found in himself in divine contemplation, and what he has felt, tasted and perceived.[144]

In Los Angeles at the Higher Sense Perception Research Foundation, Dr. Shafica Karagula, a neuropsychiatrist, has for many years made clinical observations of individuals gifted with extraordinary perception. One of her subjects, whom she calls "Diane," reported the ability to visualize vortices of energy, like spiral cones, which seemed to be in remarkable agreement with Leadbeater's descriptions. She described the etheric body as a sparkling web of light beams in constant movement "like the lines of a television screen when the picture is not in focus." There were eight major vortices of force and many smaller vortices. Seven of the vortices seemed to be directly related to the different glands of the body. Diane was able to successfully diagnose various diseases by noticing disturbances in the

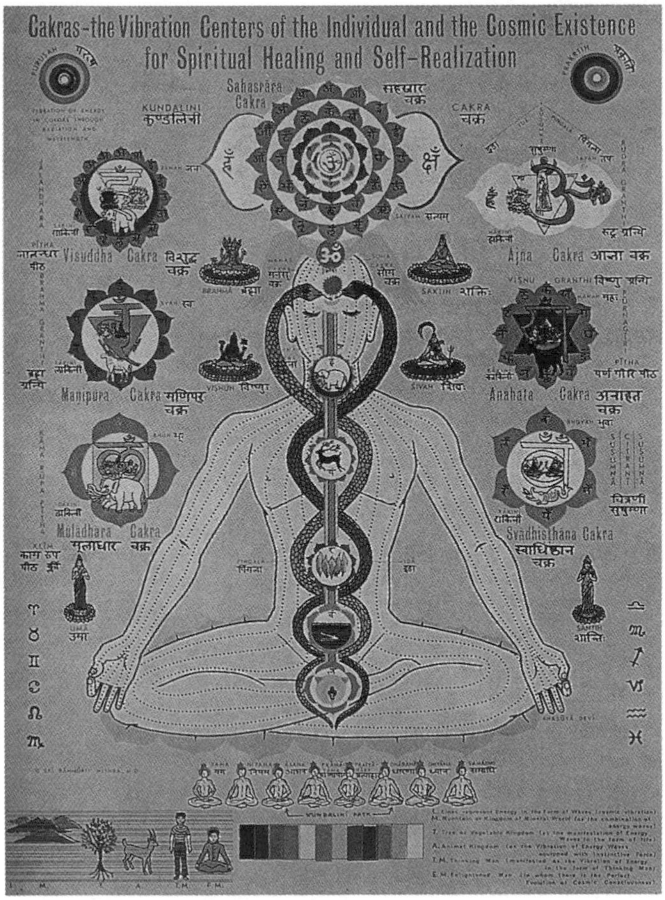

Chart showing the chakras of yoga, associated with the meridians of acupuncture. Designed by Sri Rammurti Mishra, M.D.

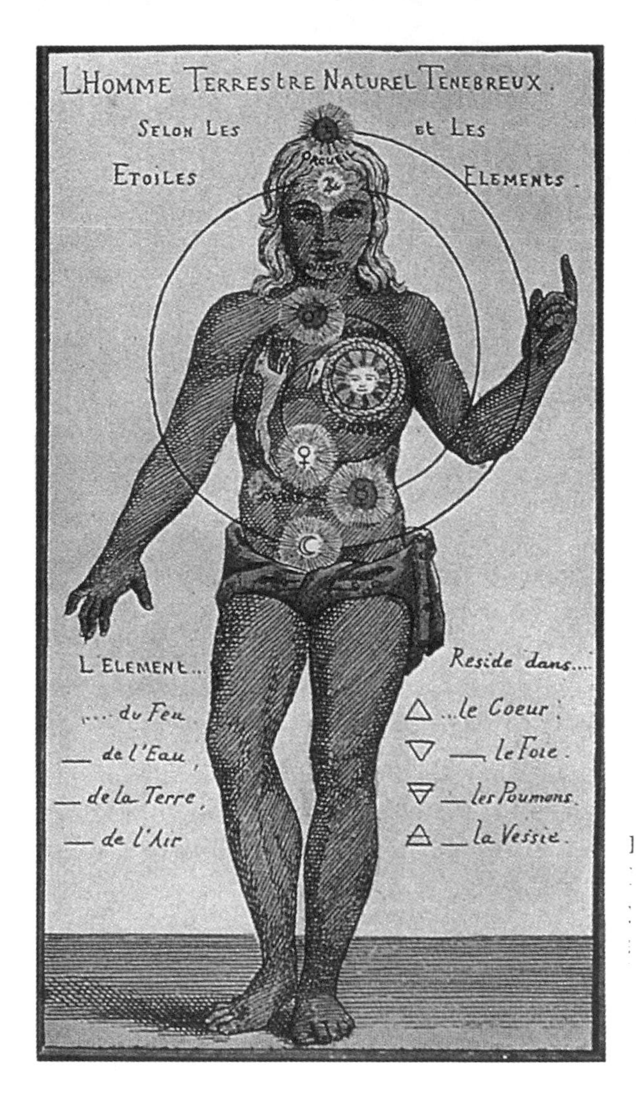

SEVEN CHAKRA-LIKE CENTERS WITH ASTROLOGICAL CORRELATIONS, ACCORDING TO THE SEVENTEENTH-CENTURY GERMAN MYSTIC, JOHANN GEORG GICHTEL.

vortices. Karagula tested this ability by taking Diane to an endocrine clinic of a large New York hospital and having her read the auras of patients selected at random in the waiting room. Then Diane's observations were checked against the medical case records.

Karagula claims that she was amazed at the accuracy of Diane's diagnoses over a large number of cases. However, she provides no exact figures in her book or in her published reports, and we are not informed if independent judges and experimental controls were used.[145,146] It is difficult to ascertain the extent to which Dr. Karagula or her subjects may have had their perceptions colored by the Theosophical tradition. Many other psychic individuals I have been acquainted with report an ability to visualize *chakras.* However, I know of no tested psychics who have indicated the ability to perceive *chakras* prior to any occult training.

When it comes to making any physiological sense out of the *chakras,* the whole matter is filled with confusion. One widely quoted approach equates the first *chakra* with the reproductive system. Others associate the second *chakra* with sexuality and reproduction. Sometimes the sixth *chakra* or third eye is associated with the pineal gland, sometimes with the pituitary. The third *chakra* is sometimes associated with the solar plexus, sometimes with the spleen, and sometimes with the digestive system. Sometimes the second *chakra* is associated with the spleen. Sometimes all of the *chakras* are associated with nerve plexus; sometimes they are all associated with the endocrine glands. In the Tibetan system, the sixth and seventh *chakra* — the third eye and the "thousand petalled lotus" — are thought of as one. The cabalistic system divides the body into ten centers. Ironically, all these systems will go into great detail in specifying the circuitry — often called *nadis* — connecting the *chakras* together. I find it easiest to confront all of these paradoxical interpretations with a certain curiosity and humility (although I

SHAFICA KARAGULA, M.D.

tend to think some writers masked their lack of understanding with dogmatic assertion). Paradoxes of a comparable sort are not uncommon in the physical and natural sciences, and generally exist on the frontiers of knowledge. Most researchers tend to ignore these uncomfortable, and poorly substantiated, reports.

One ingenious hypothesis was developed by Dr. William Tiller at Stanford University. Tiller was impressed with the apparent relationship of location and function between the *chakras* and the endocrine glands. He wondered how these so-called etheric organs might interact with the glands. Drawing from concepts used by electrical engineers, he suggested that this interaction could be analogous to a process of transduction. Imagine great energy streams flowing through space and passing through our bodies, unabsorbed and unnoticed. Tiller suggests that perhaps the *chakras* can be tuned in to couple with this power source and transduce some of its energy from the astral or etheric levels into the glands. One can think of the *chakras* and glands as electrical transformer loads that will deliver maximum power if they are balanced with respect to each other.[147] One might say that ideas are speculative in the extreme. While such ideas have little or no scientific merit, they serve the function of providing a modern metaphor for ancient teachings.

An interesting approach to the *chakras* has been taken by Lee Sannella, M.D. He noticed that the classic literature of yoga refers to a process of psychic awakening known as the *rising of kundalini.* This is pictured metaphorically as the rising of a coiled snake-like energy from the base of the spine to the top of the head. As the *kundalini* rises, it energizes or awakens each of the *chakra* centers.

Sanella encountered many cases of individuals who reported symptoms similar to the classic descriptions of *kundalini* rising. These include many strange bodily

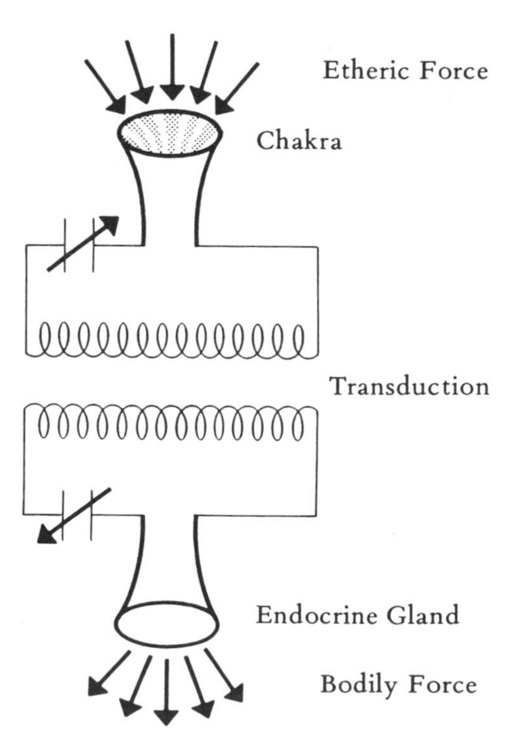

TRANSDUCTION OF ETHERIC FORCE THROUGH THE CHAKRA INTO THE BODILY FORCE WITHIN AN ENDOCRINE GLAND (COURTESY WILLIAM TILLER).

LEE SANNELLA (FROM *WHAT IS KUNDALINI?* VIDEOTAPE, COURTESY THINKING ALLOWED PRODUCTIONS)

sensations of vibration and heat, combined with visionary experiences and apparent psychic awareness. He suggests that the classic yoga descriptions may be more appropriate than the medical tendency to label such experiences as "psychotic."[148]

Do *chakras* have some objective existence, or are they are the creations of minds who claim to observe them? The same problem is actually encountered in all fields of human knowledge. Do atoms exist? Are quarks real? Where is humor? Such concepts serve as maps to guide us through our experience; or, to use another metaphor, they are menus. We would be foolish to confuse the map for the territory or the menu for the meal or the metaphor for that which is denoted by it. Sometimes, however, by a subtle consensus of agreement, this is exactly what we do.

Dr. Hiroshi Motoyama of Tokyo is a student of raja yoga who has attempted to give a literal interpretation to the *chakra* metaphor. In addition to wearing the hats of medical researcher and psychiatrist, Motoyama is also a Shinto priest. Using his intuitions, and those of several observers, Dr. Motoyama divided a yoga class of a hundred members into three groups: (A) the yogi group in which the *chakras* had been clearly awakened; (B) those in whom the *chakras* had been slightly awakened; and (C) those in whom the *chakras* had not yet been awakened. The *chakras* are often visualized as lotus blossoms that when fully awakened appear in full bloom. In this case, no controls seem to have distinguished between "awakened *chakras*" and skill in practicing yoga. A number of investigations were then made to determine if there were physiological differences between these three groups.

Examining the "disease tendency" of the different internal organs corresponding to *chakras*, such as the heart, the digestive system, the genitourinary system, and the nervous system, Motoyama found significantly greater instability of these systems in class A and B subjects. Acupuncture points associated with these organs were stimulated and measurement of skin current values were made on the palms of the hands before and after stimulation. Again the highest level of response was found in the A group. Motoyama also measured differences in the current of the fingertips and toes on right and left sides. This time greater imbalances were found in the A group of "yogis" with awakened *chakras*. From these studies, he concluded that the nervous system and the autonomic functioning of individuals with awakened *chakras* shows a much wider range and flexibility of response than with ordinary individuals.[149]

Certainly the study as reported could be criticized. One might easily suggest that Motoyama was drawing inferences from random data in order to fulfill his own expectations. Perhaps the findings seem cogent and consistent with other studies in which yoga and zen masters are able to dramatically vary heartbeat and brainwave measurements. A safer interpretation is simply to suggest that quasi-scientific work of this sort, while it contributes almost nothing to our scientific understanding, serves to perpetuate psychic folklore and polish it with the gleam of seeming scientific approval.

According to yogic tradition, the *chakras* themselves are not to be confused with any actual physical organs of the body. Dr. Rammurti S. Mishra — endocrinologist, Sanskrit scholar, and yogi — in his translation of the *Yoga Sutras of Patanjali* states that the seven *chakras* are purely psychological classifications adopted as *focuses of concentration* in yoga. He also added that through the *chakras* mindstuff is able to operate upon the anatomical parts and physiological activities.[150] We might say that *chakras* are important parts of the software programmed into our biocomputers. As one becomes deeply involved in yogic meditation, one is taught practices associating particular sounds or mantras, images, and mythological patterns to each *chakra*. Thus, to an extent the *chakras* are brought into awareness by a creative thought process, acting upon the unformed substance we can loosely call the human aura, bioplasm, consciousness, or imagination.

Lama Anagarika Govinda, an Indian national of European descent belonging to a Tibetan Buddhist order, describes this process quite succinctly:

> "Thinking is making," this is the fundamental principle of all magic, especially of all mantric science. By the rhythmic repetition of a creative thought or idea, of a concept, a perception or a mental image, its effect is augmentized and fixed (like the action of a steadily falling drop) until it seizes upon all organs of activity and becomes a mental and material reality: a deed in the fullest sense of the word.[151]

Chinese Acupuncture

Another theory dealing with subtle physiological systems of the human body is the Chinese healing art of acupuncture which unites ancient cosmology and astrology with a concept of life-energy, or *Qi*, flowing through channels in the body. One of the best ways to experience acupuncture is through a massage technique which focuses on the acupuncture points and meridians. This requires only a very gentle touch and is not difficult to learn or apply. Instructions can be found in several good books.[152,153] It has been my experience that such a massage, in addition to being healthful and sensual, provides an excellent way for a person to actually feel the flow of something (call it *Qi*, or *Chi*, if you like) inside and around the body. For about twenty-four continuous hours after I have had acupuncture massage, I have clearly felt the awareness of my body flow extend about a foot out from my skin. This is something you really should try. The experience is extraordinary, but not scientifically evidential.

There has been a lot of testimony regarding the successful use of acupuncture as a cure for all diseases and as an anesthetic. However, many Western doctors and researchers, unable to accept the "mystical" Chinese system, tend to ascribe these "miracles" to the power of suggestion.

Drs. Theodore Xenophone Barber and John Chaves of Medfield State Hospital in Massachusetts exemplify this view in an article published in *Psychoenergetic Systems.* They maintain that acupuncture can only be used successfully as an anesthetic when the patient is not fearful and has a strong belief in its efficacy. Furthermore, they add that additional sedatives, narcotics, and local anesthetics are generally used in combination with acupuncture. They also point out that the acupuncture needles can act as a counter-irritant, distracting the mind from the pain occasioned by surgery.[154]

This view is, in fact, consistent with the "gate control" theory of pain. You have probably had the experience yourself, when you were in pain, of being able to alleviate your suffering by softly stroking or scratching some other part of your body. The suggested explanation for this phenomena is the "spinal gate" in the *substantia gelatinosa* through which pain signals must pass to be received in the brain. Fewer pain signals can get through this gate if there are other non-painful stimuli activating the nerves which they must pass through. This theory is still problematic, but remains generally accepted among Western scientists.[155]

Essentially, explanations of the sort Barber and Chaves have proposed are based on the assumption there is simply no validity to the concepts of *chi* energy or acupuncture meridians. Dr. Felix Mann, a Western researcher who at one time accepted the traditional theory, now argues differently:

> The Chinese have so many connections in their acupuncture theory that one can explain everything just as politicians do....in reality I don't believe the meridians exist. I think that the meridians of acupuncture are not very much more real than the meridians of geography.[156]

Mann points out that the meridians for the large and small intestines are never used by the Chinese in treating intestinal problems. The only reason the twelve meridians are there, he claims, is in order for acupuncture theory to be consistent with Chinese astrology. This argument is questionable as the S.I. meridians are used for treating a number of other problems. Nevertheless, experienced healers pragmatically avoid using any unnecessary points. Mann proposes that the effectiveness of acupuncture is actually due to stimulation of neural pathways mediated by spinal and ganglionic reflexes. In spite of his rejection of the Chinese theory, Mann still follows the traditional methods in his therapeutic practice.[157]

Wilhelm Reich and Orgone Energy

A concept parallel to *chi* energy and *prana* is the notion of *orgone energy* developed by Wilhelm Reich, a Freudian psychiatrist noted for his analysis of character based on muscle tensions. The term *orgone* comes from "organism" and "orgasm" and refers to the orgasm reflex of repeated expansions and contractions as the basic formula of all living functioning. Reich made the bold assumption that he had discovered a new form of energy — underlying the pulsations of life — neither heat nor electricity, magnetism nor kinetic energy, chemical energy nor an amalgam of these. Most historians agree that in his early years Reich was an influential theorist. He is credited as being a father of psychotherapeutic systems, such as bioenergetics, which work primarily with the human body. However, many claim that Reich went insane in his later years. He was accused of medical quackery and died in a federal prison in 1955. Reich's story may be viewed as a sad example of the social, political and psychological dangers inherent in forcing a premature marriage of science and mysticism.

Researches in the late 1930s in Norway led Reich to assume that he had discovered *bions,* which he regarded as the basic units of *orgone.* Using high quality optical microscopes with magnification from 2000x to 4000x, Reich observed sterile solutions of organic compounds in water. He would, for example, take coal dust and heat it to incandescence in a gas flame and then, while it was aglow, put it into a sterile nutritive solution. Under the microscope, tiny vesicles were seen pulsating rhythmically in a soft, organic manner. Reich claimed to clearly distinguish this motion from the random, angular Brownian movements also observed at that magnification. Eventually these vesicles, or *bions*, seemed to take on a blue glimmer, unlike the black carbon from which they seemed to originate. In fact, at a certain stage in their development, according to Reich, they took on a positive blue stain reaction to a biological Gram stain, unlike the carbon particles. The bions were about one micron in diameter (or one millionth of a meter).

In the same series of experiments Reich also claimed to discover smaller elongated, red bodies, approximately 0.2 microns in length. He called these bodies T-bacilli and felt, through a series of experiments beyond the scope of the present book, that they were the cause of cancer. The essential point for now is that Reich felt that he had observed the creation of life within his test tubes.[158,159]

Later experiments led Reich to postulate that orgone energy permeated the entire universe and that it could be concentrated in a special device he called the *orgone accumulator.* Inside the accumulator he observed, in addition to the small blue ion dots, a diffuse bluish-grey light and rapid straight yellowish rays — all manifestations of orgone. Reich began to observe these forms in dark rooms, and outdoors throughout nature.

The accumulators themselves were simply boxes with walls made from alternating layers of an organic material, like wood, and an inorganic material, like iron. Sometimes as many as twenty layers have been used. The idea is something like a greenhouse effect such that orgone energy enters into the accumulator but cannot leave it. Most significant, from the standpoint of possible experimental proof, was Reich's claim that the temperature inside the orgone box and also outside the walls was generally slightly higher than the temperature in the room or outside air about it. The difference averaged about one degree centigrade. Furthermore, this temperature difference was greater on dry days than in humid weather. This experiment, if verified, provides concrete evidence of some new and unknown form of energy generating heat.[160]

Reich took his findings directly to the most famous scientist of his day, Dr. Albert Einstein. After some correspondence, Reich visited Einstein in Princeton on January 13, 1941. For nearly five hours that day, Reich discussed his theories with Einstein. He actually demonstrated the visible radiation within the accumulator and explained the temperature difference effect. Einstein, realizing the importance of this work, offered to test the orgone accumulator himself for the temperature difference effect. He did so and arrived at the results predicted by Reich. However, in a letter to Reich, he added that his assistant had come up with an alternative explanation — the temperature difference was due to air convection currents in the cellar of Einstein's home where the experiment took place. Reich retested the phenomena in the open air and with sufficient controls to rule out the possibility of air currents. His results were again positive; however, Einstein refused to answer any of his further correspondence. Reich's letters at this time show reasonable arguments and thorough research.[161] Nevertheless, Einstein's rejection led him to turn away from all establishment science.

Eventually Reich's work with cancer and his rental of orgone accumulators brought him into conflict with the U.S. Food and Drug Administration. In 1954, Reich was brought to trial but refused to testify, claiming that his researches were a matter for scientific, and not legal, jurisdiction. He was sentenced to prison for two years for contempt of court. His books were actually burned by the government and withheld from the market. Nine months after sentencing he died in a federal prison. Careful examination of his writings shows that while they often lacked a scientific precision, they revealed a scientific willingness to be led by the facts. For all his faults, Reich was a genius and by no means a cancer quack. His imprisonment and death were a great setback to those who were interested in pursuing his researches.[162]

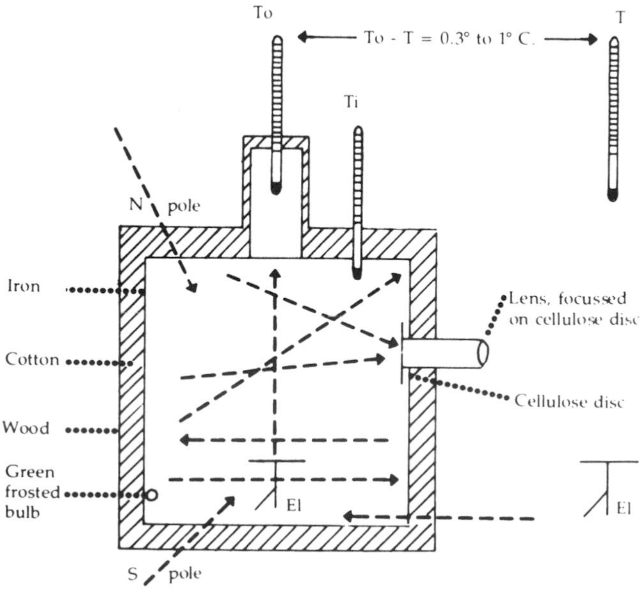

While I am aware of several scientists (such as Dr. Bernard Grad at McGill University in Montreal) who claim to have observed the formation of bions under the microscope, there are — to my knowledge — no published replications of this crucial finding from independent laboratories. Neither are there any published refutations. In *The Cancer Biopathy*, Reich does include a letter from Dr. Louis Lapique of the University of Paris who had observed the pulsating bions and was prepared to offer a physical-chemical interpretation of this effect. Reich also states that his findings had been experimentally confirmed in 1937 by Prof. Roger DuTeil in Nice.[163] However, there is no independent report. The temperature difference experiment has been replicated and the results published in the orthodox Reichian *Journal of Orgonomy* (Nov. 1971).

Reichian research continues only as a fringe study outside of the boundaries of the scientific community. It is unlikely that Reich's *orgone* theories will ever be taken seriously by most mainstream scientists. Eventually, science may progress to the point where it will be able to integrate the Reichian anomalies (if such there truly be). At the present time, there is only dwindling interest in this area.

Some studies have pointed toward the unusual properties of orgone accumulators. For example, at UCLA in the early 1970s, experiments were conducted with an orgone accumulator and an identical-looking control box (built by an

Basic design of orgone accumulator. T_o = temperature above accumulator; T_i = temperature within accumulator; T = control (temperature of air in room). EL = electroscope. Arrows indicate direction of ostensible radiation.

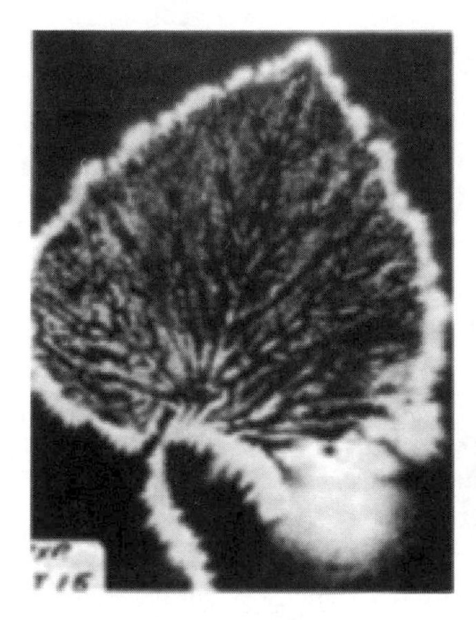

LEFT, HIGH-VOLTAGE PHOTOGRAPH SHOWING ELECTRICAL CORONA AROUND A LEAF KEPT IN AN ORGONE ACCUMULATOR AFTER FIFTEEN DAYS. (COURTESY THELMA MOSS).

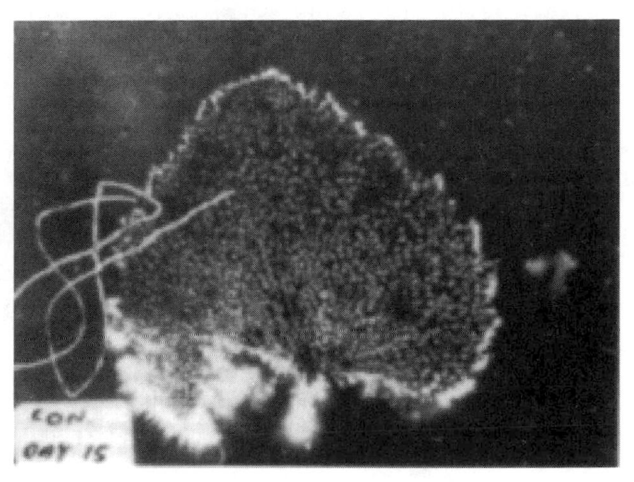

RIGHT, HIGH-VOLTAGE PHOTOGRAPH OF LEAF KEPT IN A WOODEN BOX FOR FIFTEEN DAYS (COURTESY THELMA MOSS).

undergraduate student, Roger MacDonald) made out of wood. Into each of these boxes was placed a tray containing ten leaves all plucked from the same plants. High-voltage photographs were then taken of the leaves every day for one week by an experimenter who did not know which leaves were in the orgone box and which were in the control. After seven days, eight of the ten experimental leaves were easily photographable and produced bright images, while only three of the control leaves produced pictures. Even after fifteen days, eight of the leaves placed in the orgone box were still producing high-voltage images, while all but two of the leaves in the control box were wilted and dying to the point they were not photographable.[164] This finding, like other research in high-voltage photography, has largely been dismissed by the scientific community because of inadequate experimental controls against possible extraneous influences.

Another series of experiments with orgone accumulators was conducted by Dr. Bernard Grad of McGill University. Using careful experimental controls, Dr. Grad tested the effects of treatment in an orgone accumulator upon cancerous rats. The results of Grad's studies are complex. While the orgone treatment alleviated the symptoms of cancer, it did not really prolong the animal's lifespan.[165]

Yet Reichian ideas are a fertile source of folklore. *Orgone blankets* are still sought as a cancer treatment. And there are those who claim that *cloudbusters* developed by Reich are capable of controlling weather patterns.

The Soviet Concept of Biological Plasma

The Soviet concept of *biological plasma* is the latest version of what is essentially Mesmer's old notion of *animal magnetism.* The term plasma in physics refers to a gaseous collection of positive and negative ion — sometimes regarded as a fourth state of matter as it is not quite the same as a molecular solid, liquid or gas. The atmospheres of stars, which extend out to interstellar space, are composed of such plasma. The idea that a coherent plasma body might surround and interact with biological organisms was first proposed in 1944 by V. S. Grischenko, a physicist and engineer. Dr. Victor Inyushin, a biophysicist at Kirov State University in Alma-Ata, Kazakhstan, has been the leading theoretical spokesman for the biological plasma body.

In contrast to inorganic plasma, biological plasma is said to be a coherent, organized system. The entropic, chaotic motion of particles is reduced to a minimum. Like the visible human body, the bioplasmic body is thought to be relatively stable in varying environmental conditions — although it is particularly susceptible to electrical and magnetic perturbations.

> All kinds of oscillations of bioplasma put together create the biological field of the organism. In the complex organism and its cerebral structures a complicated wave structure — a biologogram — is being created, characterized by its great stability as far as the maintenance of the wave characteristics is concerned.[166]

The euphonious term, *biologogram,* appears to be an application of the hologram idea — a three-dimensional image formed by wave interference patterns. The entire image can be reconstructed from any portion of the hologram. This model is very popular among consciousness researchers. Holographic analogies explain why brain functioning is not severely impaired when portions of that organ are removed.

The theory of the bioplasmic body was useful in a communist country where the official dogma was materialist — and researchers had to be careful to avoid heretical doctrines. However, the Soviets acknowledge that the biological plasma theory was originally conceived in the absence of any experimental proof.[167] The concept is now used as an umbrella explanation of all sorts of phenomena ranging from hypnosis to astrology, telepathy, psychokinesis, and high-voltage photography. The explanations I have seen in the translated literature seem like rather awkward efforts to fill in the gap in our knowledge left unfilled because of insufficient experimentation. Bioplasma is still, as far as I can tell, an entirely speculative concept. That plasmic phenomena occur in connection with biological organisms is not doubted, but if such fields are organized into coherent and stable patterns, a deeper explanation will be required.

The research finding that lends support to the concept of bioplasma is the preliminary report that changes in the corona discharge of humans and certain animals can be shown to vary with the emotional state of the organism, or state of consciousness, in a way independent of other physiological variables that might affect the discharge. If true, this finding is most unusual since we generally associate a number of physiological parameters with changes in emotional intensity. None of the reported experiments has been described in sufficient detail to be taken at face value.

High-Voltage Photography

The Soviets reported that high-voltage (as developed by Semyon and Valentine Kirlian) photographs are sensitive to changes in the emotions, thoughts and states of consciousness of human subjects. Additional apparent support for this theory came from data gathered by Dr. Thelma Moss and her colleagues at the UCLA Neuropsychiatric Institute.[168] Studies with subjects in relaxed states produced by meditation, hypnosis, alcohol and drugs generally showed a wider and more brilliant corona discharge on the fingertips. Preliminary research seemed to indicate that these photographic indicators were independent of such physiological measurements as galvanic skin response, skin temperature, sweat, or constriction and dilation of the blood vessels. This is a difficult finding to accept, and not thoroughly documented in published reports. Other studies showed a brighter and wider corona in subjects who were in the presence of a close friend or someone of the opposite sex.[169]

THELMA MOSS

In 1970, Lynn Schroeder and Sheila Ostrander published in *Psychic Discoveries Behind the Iron Curtain* a rumor regarding research with the Soviet healer Col. Alexei Krivorotov:

> At the moment when he seemed to be causing a sensation of intense heat in a patient, the general overall brightness in Krivorotov's hands decreased and in one small area of his hands a narrow focused channel of intense brilliance developed. It was almost as if the energy pouring from his hands could focus like a laser beam.[170]

These reports aroused the interest of Western researchers who were determined to investigate this phenomenon for themselves. E. Douglas Dean of the Newark College of Engineering in New Jersey, using Czechoslovakian-designed equipment, had the opportunity to conduct similar experiments with a psychic healer by the name of Ethel E. De Loach. Dean took several sets of her fingers when she was at rest and when she was thinking of healing. In every case, Dean reported that the flares and emanations were much larger in the pictures when she was thinking of healing.[171] Some of the effects with Mrs. De Loach were very striking:

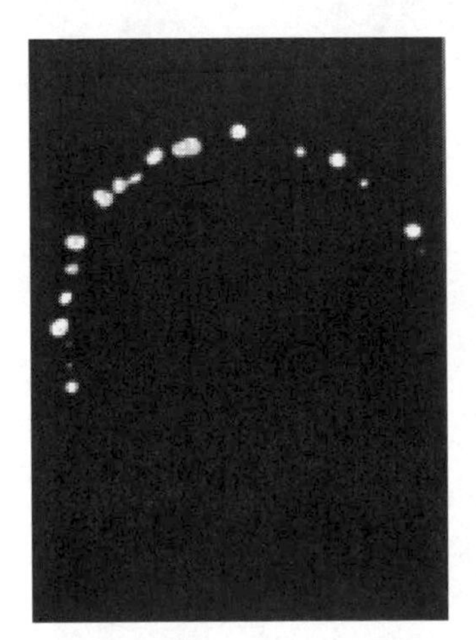

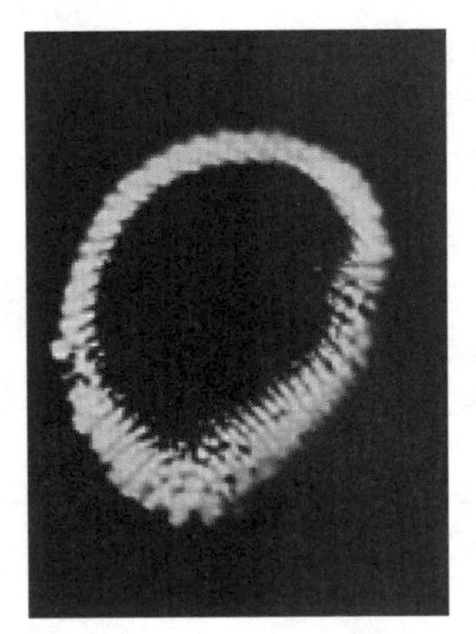

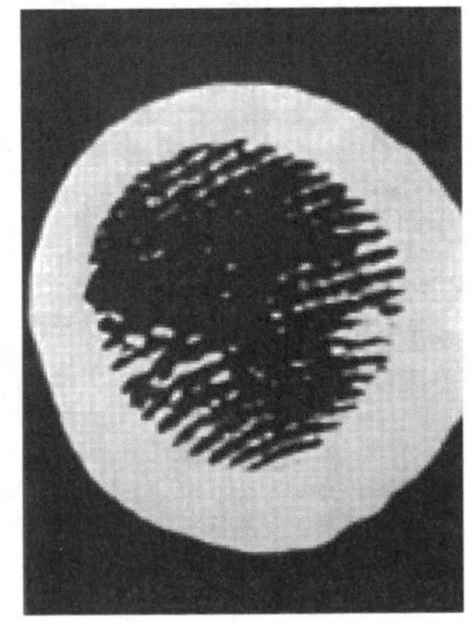

LEFT, HIGH-VOLTAGE PHOTOGRAPH SHOWING ELECTRICAL CORONA AROUND FINGERTIP OF INDIVIDUAL BEFORE ALCOHOL CONSUMPTION. *MIDDLE,* SAME FINGERTIP AFTER CONSUMPTION OF NINE OUNCES OF BOURBON. *RIGHT,* IN STATES OF AROUSAL, TENSION, OR EMOTIONAL EXCITEMENT, THE RESEARCHERS OBSERVED THE APPEARANCE OF BLOTCHES ON THE COLOR FILM.

One time Ethel was doing a healing and she knew I was so happy about getting this big orange flare on the photograph. She asked me if I would like a green one. Well I said, "My goodness, yes! You mean you can make a green one to order?" She said, "yes." So we set up the equipment and we got a green flare, a small one.[172]

Further research along these lines was conducted by Dr. Thelma Moss and her associates working at the UCLA Center for the Health Sciences. Using high-voltage photography, they observed an apparent energy transfer from healer to patient. After the healer had finished a treatment, the corona around his fingertip was diminished. On the other hand, an increase in the brilliance and width of the corona of the patient was observed after treatment. Volunteers with no experience in healing were unable to replicate the same effect.[173]

In another series of experiments, the UCLA group explored the healing interactions between people and plants. In this study, the "healers" were people who claimed to have a "green thumb"; in other words, people who had the ability to make plants flourish under their care. In each experiment there was both an experimental leaf and a control leaf. Both leaves were photographed after being freshly plucked from the same plant. Then each leaf was mutilated and photographed again. Typically this caused the leaf to become dimmer on film. Then the "healer" would hold his hand about an inch above the experimental leaf for as long as he felt was necessary, and the experimental leaf would be photographed again. Most of the twenty "green thumb" volunteers were able to cause an increased brightness in the leaves after treatment. These leaves also remained brighter for many weeks longer than the control leaves.[174]

Moss and her coworkers found a number of subjects who claimed to have a "brown thumb" with plants — plants always seemed to get sick and die under their care. When these subjects attempted the leaf experiment, they were able to cause the corona around the leaf to disappear.[175]

One of America's most well-known ostensible psychic healers, Olga Worrall,

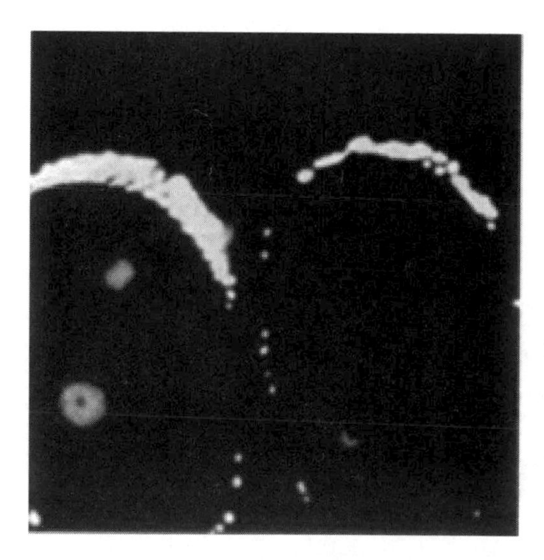

HIGH-VOLTAGE PHOTOGRAPH SHOWING ELECTRICAL CORONA AROUND TWO FINGERTIPS OF INDIVIDUAL BEFORE USING MARIJUANA.

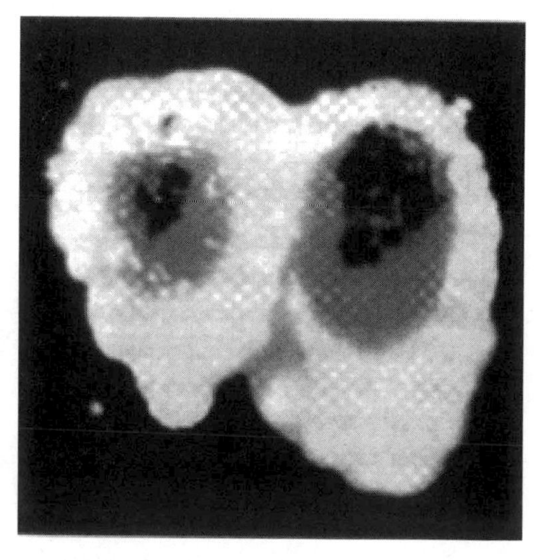

HIGH-VOLTAGE PHOTOGRAPH OF SAME FINGERTIPS DURING MARIJUANA INTOXICATION (COURTESY THELMA MOSS).

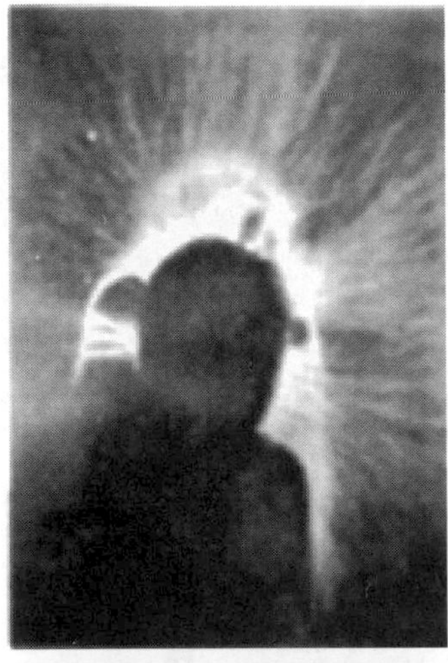

TOP, HIGH-VOLTAGE PHOTOGRAPH SHOWING ELECTRICAL CORONA AROUND FINGERTIP OF MRS. ETHEL DE LOACH PRIOR TO HEALING.

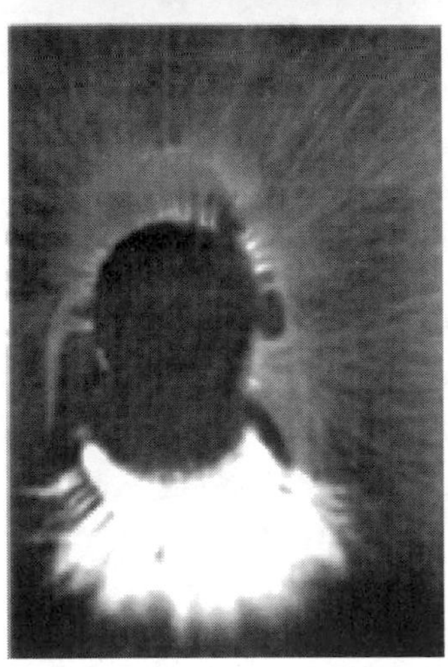

BOTTOM, HIGH-VOLTAGE PHOTOGRAPH OF SAME FINGERTIP WHILE MRS. DE LOACH IS THINKING ABOUT HEALING (COURTESY DOUGLAS DEAN).

OLGA WORRALL

exhibited apparent conscious control over the energy interactions being photographed. Oddly enough, the leaf had almost disappeared in the photograph of Mrs. Worrall's first test run. Thelma Moss commented:

> This was deeply disturbing to us: how could we tell Dr. Worrall, a lady for whom we had the deepest respect, what she had done to the leaf? But, obviously we had to tell her. She looked at the photographs with quiet dignity, and then asked if she might repeat the experiment. She believed she had given the leaf "too much power," and thought a more gentle treatment might have different results. The experiment was, of course, repeated....the second, mutilated leaf...after a more gentle treatment has become brilliant. This was the first time someone had been able, deliberately, to reverse the direction of the bioenergy. Since then, we have had another subject who was able to predict the direction of the energy flow.[176,177]

A report by skeptical researchers Arleen J. Watkins and William S. Bickel at the University of Arizona has identified six different physical factors that affect Kirlian photographs: photographic paper, pressure, voltage discharge, exposure time, moisture in the sample, and photographic developing time.[178] Undoubtedly, there are other factors as well. It is not clear that any of the published studies purporting that this method produces interesting results of a psychic or psychological nature, has sufficiently controlled for all of these factors.

High-Voltage Photography Anecdotes

Several related findings have been reported from the UCLA radiation field photography laboratory. One study attempted to observe the fingertips of pairs of

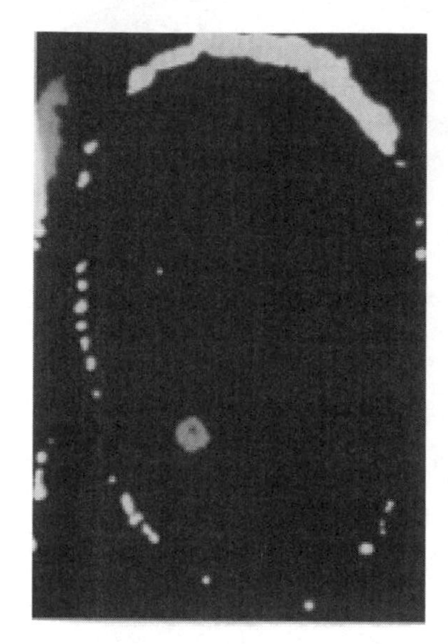

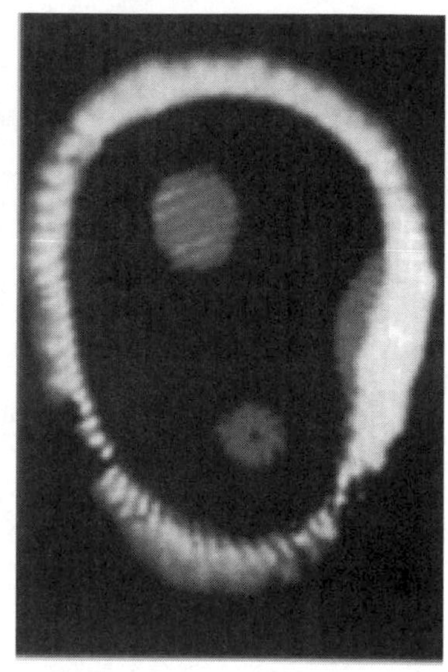

Left, High-voltage photograph showing electrical corona around fingertip of patient before psychic healing treatment

Right, High-voltage photograph of same fingertip after psychic healing treatment (courtesy Thelma Moss)

individuals, holding their fingers close together, but not touching, as they stared into each other's eyes. Frequently they found, for no apparent explanation, that one of the fingertips in each pair would practically disappear. One of the subjects was a professional hypnotist, and it was repeatedly discovered that he could blank out the fingertips of any one of a number of partners. In a rather striking experiment, one subject was asked to visualize sticking a needle into her partner, who was known to be afraid of needles. The high-voltage photograph of their fingertips shows a sharp red line darting out of the aggressor's finger toward her imagined victim whose emanations appear to be retreating. On the other hand, the photographed corona of two individuals taken while they have meditated together, according to Moss, has typically shown a merging and uniting of the two individual coronas.[179]

Sometimes when two persons were able to generate feelings of hostility towards each other, the corona between their fingers would abruptly cut off, leaving a gap so sharp and clear it became known as the "haircut effect." In some instances a bright bar, like a barrier, would appear between the two photographed fingerpads. Further studies with family groups engaged in family therapy were conducted. Group photographs were taken with the fingerpads of each member of the family. Typically one member of the group, generally the son, would not photograph at all. Other photographs in this study suggested to the researchers that high-voltage photography could provide insights into the emotional reactions between people.[180,181]

The Phantom Leaf Effect

The most startling finding of high-voltage photography research was called the "phantom leaf" effect. Ostrander and Schroeder in *Psychic Discoveries Behind the Iron Curtain* first reported that the Soviets were often able, after removing a portion of a plant leaf, to photograph a corona pattern around the leaf as if the whole leaf were still there. This suggested to researchers that radiation of energy around the leaf formed a holographic pattern acting as an organizing force field for physical matter.[182] The Soviets dubbed this hypothesized organizing field the *biological plasma body*.

For several years American experimenters tried unsuccessfully to duplicate this effect. While the relevant procedural variables were still unknown, scientists such as William Tiller maintained that this single observation was "of such vast importance to both physics and medical science that no stone should be left unturned in seeking the answer!"[183]

In 1973, Kendall Johnson, after more than five hundred trials, succeeded in producing a "phantom leaf" with clear internal details. Immediately researchers suggested that the results were due to an artifact — possibly from an electrostatic charge left on the electrode's surface before the leaf was cut.

John Hubacher, a graduate student working in Thelma Moss' laboratory then produced about a dozen phantom leaves that show an internal structure — presumably belonging to the cut-off section of the leaf. Experimenting in the spring months (which was suggested as a relevant variable), Hubacher came to expect clear phantom images in about 5% of his attempts and partial images in another 20%. He was unable to ascertain the variables that resulted in a perfect image. He claims that he was careful to cut the leaf before it was placed on the electrode in order to avoid the possibility of an electrostatic artifact. In fact, he went further and attempted to deliberately create a pseudo-phantom effect by pressing the leaf against the film emulsion before cutting a section off. The results of these efforts did not create any good-looking phantoms.[184]

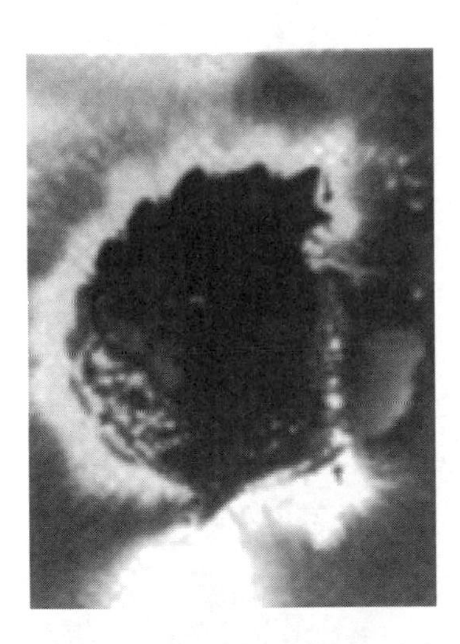

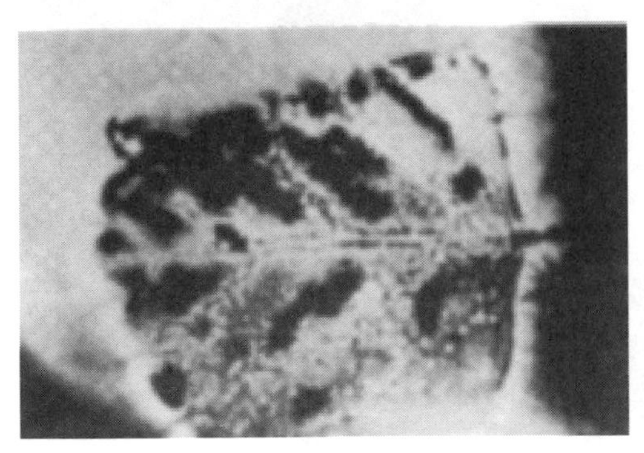

Above, Phantom leaf high-voltage photograph

Below, Phantom leaf high-voltage photograph

Perhaps the most encouraging efforts in this direction were the motion pictures taken of the fading phantom leaf through a special transparent electrode. The speed of the camera was slowed to about six frames a second. This work was in Dr. Moss' laboratory with the help of Clark Dugger, a graduate student in UCLA's noted cinema department. Both black-and-white and color high-voltage photographs showed the "phantom" sparkling brilliantly and pulsing for several seconds before it disappeared. In these experiments, the leaf was reportedly always cut before it was placed on the electrode; and the phantom leaves were obtained only during spring months.[185,186]

Working in Moss' laboratory, and also at the Washington Research Center in San Francisco, I was able to reproduce partial phantom effects with little difficulty. However, I am unable to make any claims for the phenomena as it would have taken many months, perhaps years, of intensive research to control all of the possible sources of artifact. The leaf being photographed, for instance, must be grounded with an electrode; and the placement of this electrode, a possible source of additional corona discharge, seems crucial. Sometimes unaccountable images appeared on high-voltage photographs of normal leaves, fingertips, and also inanimate objects. William Joines and his colleagues of the electrical engineering department at Duke University have been able to produce a "phantom leaf" effect, for example as a result of film buckling.[187]

The phantom leaf effect, if true, carries such significance for science that it is essential the experiments be replicated under tightly controlled conditions that can provide a secure foundation for theoretical models. While only further well-controlled studies can resolve these tenuous problems, the scientific community has turned away from Kirlian high-voltage photography as a productive research tool.

Kurt Lewin's Field Theory

In order to explain these uncanny photographic events, some researchers have drawn upon the efforts of psychologist Kurt Lewin (1890-1947) to apply the concepts of physical fields to the study of human personality. One of the unique characteristics of Lewin's theory was the use of diagrammatic representations of internal and external personality interactions. The following diagram is one Lewin used:

The individual is described graphically by the quality of psychological environment (or aura) around him. Person b, for example, is one with a thicker boundary. The outer world has little influence on the life-space and vice versa. The life-space of person a is more open and expansive.[188]

Lewin has often been criticized for the unjustified application of physical concepts and terminology to the realm of personality where they did not belong. It was claimed that his diagrams were an attempt to appear scientific without using the requisite controls and measurements of science. Furthermore, it was difficult for these critics to see what these diagrams had to do with the "real world." Proponents of Kirlian high-voltage photography suggest that Kirlian photographs can be read almost as if they were Lewin diagrams of personality fields.[189] This claim goes far beyond what Lewin himself ever actually suggested. However, in the

WILLIAM JOINES (COURTESY FOUNDATION FOR RESEARCH ON THE NATURE OF MAN)

analytical psychology of Carl Gustav Jung, it has been proposed many times by Jung himself that the archetypal world — although it exists within the mind — should be thought of as objective reality. It resembles Plato's realm where ideas themselves exist as visible thoughtforms.

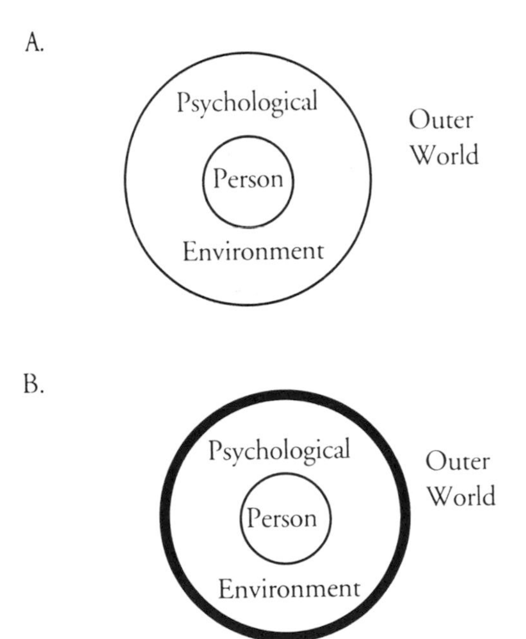

Communication with Higher Intelligence

The notion of higher intelligences that influence the affairs of the human race has an old and venerable tradition and can be found in all religions. Ancient Sumeria, for example, is generally regarded as the first major civilization. To Sumeria we owe the inventions of the wheel, writing, arithmetic and geometry, and money. The Sumerian's own legend, as recorded by the ancient historian Berosus around 400 B.C., is that the arts of civilization were taught to the savage inhabitants of the fertile crescent region by an unknown creature who possessed superhuman intelligence.

> There appeared, coming out of the sea where it touches Babylonia, an intelligent creature that men called *Oan[nes]* or *Oe[s]*, who had the face and limbs of a man and who used human speech, but was covered with what appeared to be the skin of a great fish, the head of which was lifted above his own like a strange headdress. Images are preserved of him to this day.
>
> This strange being, who took no human nourishment, would pass entire days in discussions, teaching men written language, the sciences, and the principles of arts and crafts, including city and temple construction, land survey and measurement, agriculture, and those arts which beautify life and constitute culture. But each night, beginning at sundown, this marvelous being would return to the sea and spend the night far beyond the shore. Finally he wrote a book on the origin of things and the principles of government which he left his students before his departure. The records add that during later reigns of the prediluvian kings other appearances of similar beings were witnessed.[190]

Angels and Guardian Spirits

St. Augustine (354-430 A.D.) described in his *City of God* a very vivid picture of the evolution of the soul through stages towards the heavenly kingdom. Like later Church fathers, he proclaimed that while magic was real, it was the work of the devil and therefore evil. On the other hand, while he repudiated pagan magic, Augustine fervently believed in the protection angels and guardian spirits would provide to Christians:

> They watch over and guard us with great care and diligence in all places and at all hours, assisting, providing for our necessities with solicitudes; they intervene between us and Thee, O Lord, Conveying to Thee our sighs and groans, and bringing down to us the dearest blessings of Thy grace. They walk with us in all our ways; they go in and out with us, attentively observing how we converse with piety in the midst of a

IRINA TWEEDIE (FROM *SPIRITUAL TRAINING* VIDEOTAPE, COURTESY THINKING ALLOWED PRODUCTIONS)

> perverse generation, with what ardour we seek Thy kingdom and its justice, and with what fear and awe we serve Thee. They assist us in our labours; they protect us in our rest; they encourage us in battle; they crown us in victories; they rejoice in us when we rejoice in Thee; and they compassionately attend us when we suffer or are afflicted for Thee.[191]

The Glance of the Master

The Sufi tradition, which originated in Persia, involved singing, dancing and storytelling as techniques for exploring the inner mind. Many of the wonders described in the *Tales of the Arabian Nights* are of Sufi origin. Snake-charming and fire-eating practices still exist as testimony to the faith self-control of certain Sufi mystics.

One well-known Sufi was the Sheikh Shahab-el-Din. Idries Shah relates the following story about him:

> It is related of him that he once asked the Sultan of Egypt to place his head in a vessel of water. Instantly the Sultan found himself transformed into a shipwrecked mariner, cast ashore in some totally unknown land.
>
> He was rescued by woodmen, entered the nearest town (vowing vengeance against the Sheikh whose magic had placed him in this plight) and started work there as a slave. After a number of years he gained his freedom, started a business, married and settled down. Eventually, becoming impoverished again, he became a free-lance porter, in an attempt to support his wife and seven children.
>
> One day, chancing to be by the seashore again, he dived into the water for a bath.
>
> Immediately he found himself back in the palace at Cairo, again the King, surrounded by courtiers, with the grave-faced Sheikh before him. The whole experience, though it had seemed like years, had taken only a few seconds.
>
> This application of the doctrine that "time has no meaning to the Sufi" is reflected in a famous instance of the life of Mohammed. It is related that the Prophet, when setting out on his miraculous "Night Journey," was taken by the angel Gabriel to Heaven, to Hell and to Jerusalem. After four-score and ten conferences with God, he returned to earth just in time to catch a pot of water that had been overturned when the angel took him away.[192]

The Sufis loved to tell such stories. Their traditions seem to be a mixture of teaching stories in the genre of Aesop's *Fables* and the *Tales of the Arabian Knights* with a very profound understand of methods for transforming human nature.

Irina Tweedie is author of *Daughter of Fire,* a diary of five intensive years of spiritual training in India with a Sufi master. In the following excerpt of my *Thinking Allowed* interview with her, she describes aspects of her spiritual training which suggest that some ancient techniques have survived to the present day. She refers to what the Sufis call *the glance* — the unanticipated look of the guru that affects one profoundly. At that moment, one realizes one's connection with that teacher:

> According to the Sufi tradition, the moment the teacher looks at you for the first time, you are born again.

I was struck by the many doubts which Mrs. Tweedie experienced during her period of Sufi training. She seemed to alternate between moments of great peace and moments of the most profound inner torture. She said:

> I remember at the end I was suicidal. I decided to throw myself from the bridge, at Kampur, the city where my teacher lived. It was on the Ganges, which is deep. So I thought, Well, it won't hurt very much.
>
> He seemed to know my thoughts, because suddenly he turned to me. He was sitting in the garden, and I was so disgusted I didn't want to look at him. He said, "Mrs. Tweedie, look at me." So I looked at him. I sort of — aahhh! He was full of blinding light. I sort of just looked, speechless. And he said, "Mrs. Tweedie, do you think I would waste my powers if you really were hopeless?" And perhaps half a day before he had told me that I was utterly hopeless.[193]

This particular teacher seemed to be acting in an irrational manner, apparently to help his student "stop her mind," to get her outside of her conditioned intellect. This is sometimes referred to as the tradition of crazy wisdom. The purpose of the teaching is not so much to instruct people about it, because there are many books, but to lead people to it.

Robert Frager, founder of the California Institute of Transpersonal Psychology, is on the faculty of the Institute for Creation Spirituality at Holy Names College. He is the coauthor of *Personality and Personal Growth.* A fifth-degree black belt of Aikido, he was a student of Osensei, the original founder of Aikido. He is also a Sufi sheik. In the following excerpt from my *Thinking Allowed* interview with him he describes his first encounter with a Sufi teacher which is very much reminiscent of Irina Tweedie's description of the glance:

> I was sitting in my office at the California Institute of Transpersonal Psychology, with my feet up on my desk, talking on the phone to someone about something administrative. Two of my colleagues had invited a Sufi group to come talk at the school for a few days; and a heavy-set Turkish man walked by my office and looked at me.
>
> The look must have taken a half a second at most. He did not break stride. He did not stop and stare. But the moment he looked at me, time stood still, absolutely. Time stopped for me, and I had this impression that all the data of my life was being read into a high-speed computer — that he somehow knew everything that led up to my being in that office with my feet on the desk.
>
> Many years later, someone else in another Sufi tradition, when I mentioned this story,

Top, Robert Frager (from *Common Threads in Mysticism* videotape, courtesy Thinking Allowed Productions)

Bottom, Joseph Chilton Pearce, renowned author of *Crack in the Cosmic Egg* and *Magical Child,* is a devotee of Gurumayi Chidvilasananda. (From *The Guru Principle* videotape, courtesy Thinking Allowed Productions)

> said, "My God, that's a perfect description of what we call *the look of the sheik*. That itself is an initiation at some level."
>
> I suddenly found myself sitting at mealtime with this man, and then he started telling stories, and it was as though no one had ever told spiritual stories before. The stories just knocked me over — the power, the wisdom. It felt as though there was no one else in the room, that he was literally telling the stories just to me, and that they were all designed for me. And then after the stories were over I glanced around and I said, "My goodness, there are other people here too."[194]

I asked Dr. Frager if a cynic might not say at this point that even a sophisticated professor like himself could become hypnotized. He responded that it was more like falling in love, in a Platonic sense, with the teacher. The essence of the practice is one of opening the heart.

The practice of devotion to a guru originated in the Hindu *Bhakti Yoga* tradition. The guru is said to communicate higher states of consciousness through his or her very presence or being. Certain gurus, with thousands of Western devotees, currently encourage their disciples to practice this method. They include Gurumayi Chidvilasananda and Da Love-Ananda, both of whom claim to have received confirmation of their spiritual attainment from Swami Muktinanda.

Space does not permit extensive comments on this modern manifestation of an ancient tradition. It is worth noting, however, that press scrutiny of virtually all contemporary claimants to various degrees of divine or holy stature strongly suggest, if one is willing to make conventional judgments, that such guru figures have failed to maintain an unwavering state of higher consciousness. Many claimants to a state of higher consciousness — including Maharishi Mahesh Yogi, Bhagwan Rajneesh and Swami Muktinanda — have been reported to have feet of clay.[195,196] Of course, such individuals can be effective teachers. However, if a guru encourages the worship of devotees, a prospective disciple would be wise to consider the effect that the human flaws of the master may have on such a practice.

Da Love-Ananda (also known as Da Free John), a Western seeker formerly named Franklin Jones, has become a prolific writer on the "great tradition" of spiritual seeking, as well an individual whose claim to having attained the highest state of awareness is viewed seriously by many scholars.[197] For several years, he has been compiling an annotated bibliography of spiritual literature, organized according to his own system which views the spiritual journey as a process of seven stages.[198] Several of Da Love-Ananda's other books attempt to explicate the telepathic processes by which a spiritual master interacts with devotees.[199,200,201]

Cabala[202]

Cabala is the word for the Jewish mystical tradition that acknowledges the personal experience of the absolute. The tree of life diagram shows the ten emanations of God that are the attributes of both humanity and the universe. In later occult systems the tree of life was used as a philosophical basis for integrating the tarot cards with astrology as well as a guide for meditation and reveries. In these ecstatic states one progressed through a hierarchy of visions that lead to ultimate mystical union. One might say that the tree of life served as a very sophisticated map of the inner spaces through which consciousness progresses.[203]

One of the most sophisticated interpretations of Cabala is that offered by Stan Tenan of the MERU Foundation in San Anselmo, California. Tenan maintains that the Cabalists discovered in the ancient languages a schematic for the unfolding of the universe from unity to multiplicity. This schematic, he maintains, is isomorphic to ideas that are currently being generated in contemporary cosmology.[204]

Emmanuel Swedenborg

Emmanuel Swedenborg (1688-1772) the single individual who combined within himself the most intense spiritualistic exploration with the most sophisticated scientific expertise was born the son of a devout Swedish bishop whose family was ennobled by the king when he was thirty-one. Being the eldest son, Baron Emmanuel Swedenborg took a position in the Swedish House of Nobles.

EMMANUEL SWEDENBORG

During his long life Swedenborg published scientific papers on a wide variety of topics. They include soils and muds, stereometry, echoes, algebra and calculus, blast furnaces, astronomy, economics, magnetism, and hydrostatics. He founded the science of crystallography and was the first to formulate the nebular hypothesis of the creation of the universe. He spent many years exploring human anatomy and physiology and was the first to discover the functions of the ductless glands and the cerebellum.

In addition to mastering nine languages, he was an inventor and a craftsman. He built his own telescope and microscope. He designed a submarine, air pumps, musical instruments, a glider and mining equipment. Throughout his life he worked as a mining assessor in Sweden. He participated in the engineering of the world's largest dry dock. He developed an ear trumpet, a fire extinguisher, and a steel rolling mill. He learned bookbinding, watchmaking, engraving, marble inlay, and other trades. At one point he engineered a military project for the king of Sweden which transported small battleships fourteen miles over mountains and through valleys. At the age of fifty-six, Swedenborg had mastered the known natural science of his day and stood at the brink of his great exploration of the inner worlds.

He began by surveying all that was understood by scholars in the area of psychology and published this information in several volumes along with some observations of his own. Then he started writing down and interpreting his own dreams. He developed yoga-like practices of suspending his breathing and drawing his attention inward, thus enabling him to observe the subtle symbol-making processes of his mind. He carefully probed the hypnogogic state, the borderland between sleep and waking in which the mind forms its most fantastic imagery.

As he intensified this process he gradually began to sense the presence of other beings within his own inner states. Such a sensation is common to the hypnogogic state. However for Swedenborg, these occasional glimpses into another world came to full fruition quite suddenly in April of 1744. From that time until his death, twenty-seven years later, he claimed to be in constant touch with the world of spirits. He regularly probed the vast regions of heaven and hell and engaged in long and detailed conversations with angels and spirits.

The following passages provide us with a typical example of Swedenborg's later thought:

> Of the Speech of Spirits and Angels
>
> The discourse or speech of spirits conversing with me, was heard and perceived as distinctly by me as the discourse or speech of men; nay, when I have discoursed with them whilst I was also in company with men, I also observed, that as I heard the sound of man's voice in discourse, so I heard also the sound of the voice of spirits, each alike sonorous; insomuch that the spirits sometimes wondered that their discourse with me was not heard by others; for, in respect to hearing there was no difference at all between the voices of men and spirits. But as the influx into the internal organs of hearing is different from the influx of man's voice into the external organs, the discourse of the spirits was heard by none but myself, whose internal organs, by the divine mercy of the Lord, were open. Human speech or discourse is conveyed through the ear, by an external way, by the medium of the air; whereas the speech or discourse of spirits does not enter through the ear, nor by the medium of the air, but by an internal way, yet into the same organs of the head or brain. Hence the hearing in both cases is alike....
>
> The words which spirits utter, that is, which they excite or call forth out of a man's memory, and imagine to be their own, are well chosen and clear, full of meaning, distinctly pronounced, and applicable to the subject spoken of; and, what is surprising, they know how to choose expressions much better and more readily than the man himself; nay, as was shown above, they are acquainted with the various significations of words, which they apply instantaneously, without any premeditation; by reason, as just observed, that the ideas of their language flow only into those expressions which are best adapted to signify their meaning. The case, in this respect, is like that of a man who speaks without thinking at all about his words, but is intent only on their sense; when his thought falls readily, and spontaneously, into the proper expressions. It is the sense inwardly intended that calls forth the words. In such inward sense, but of a still more subtle and excellent nature, consists the speech of spirits, and by which man, although he is ignorant of it, has communication with them.
>
> The speech of words, as just intimated, is the speech proper to man; and indeed, to his corporeal memory: but a speech consisting of ideas of thought is the speech proper to spirits; and, indeed, to the interior memory, which is the memory of spirits. It is not known to men that they possess this interior memory, because the memory of particular or material things, which is corporeal, is accounted every thing, and darkens that which is interior: when, nevertheless, without interior memory, which is proper to the spirit, man would not be able to think at all. From this interior memory I have frequently discoursed with spirits, thus in their proper tongue, that is, by ideas of thought. How universal and copious this language is may appear from this consideration, that every single expression contains an idea of great extent: for it is well known, that one idea of a word, may require many words to explain it, much more the idea of one thing; and still more the idea of several things which may be collected into one compound idea, appearing still as a simple idea. From these considerations may appear what is the natural speech of spirits amongst each other, and by what speech man is conjoined with spirits.[205]

It is tempting to think that Swedenborg went insane at this point. However, he otherwise showed no signs of mental weak-

Melancholy,
Albrech Durer,
1514

ness. He continued to serve as a mining assessor, for instance, throughout his life. Yet, during this twenty-seven year period he wrote some 282 works in the above manner describing his inner explorations.

When asked how he could write so much, he casually answered that it was because an angel dictated to him. Numbers of people witnessed him speaking with invisible figures, yet he could always be interrupted in the midst of these states to deal with a visitor or a business matter.[206]

He described the world to which we all go after death like a number of different spheres representing various shades of light and happiness, each soul going to that for which his spiritual evolution has fitted him. The light of higher states seems painful and blinding to one who is not yet ready. These spheres resembled the earthly society that Swedenborg knew. His descriptions of life in the spheres are written with the careful mind of a scientist. He speaks of the architecture, the flowers and fruits, the science, the schools, the museums, the libraries, and the sports.

The great German philosopher Emmanuel Kant set about to examine the Swedenborg phenomena with an aim toward discrediting them. However, Kant was at a loss to explain the well-reported incident in 1756, when Swedenborg, then in Gottenburg, clairvoyantly saw a fire raging three hundred miles away in Stockholm. This incident occurred in front of fifteen very distinguished observers.[207]

Due to the voluminous quantity of his erudite writings, Swedenborg's popularity has not been large among the general population. Often, his spiritual visions do seem to degenerate into arbitrary theological interpretations of scripture. After his death, the Church of the New Jerusalem was founded to preserve his teachings, which can be found in the encyclopedic *Heaven and Hell*, *The New Jerusalem* and the *Arcana Coelestia* as well as in several excellent biographies.

Swedenborg's thought was to exert a particular influence on two of Europe's great artistic geniuses, William Blake (1757-1827) and Johann Wolfgang von Goethe (1749-1832).

Gustav Theodor Fechner and Psychophysics

Until the middle of the nineteenth century, there was no recognized branch of experimental science whose domain of exploration was man's psyche. While there was generally a strong public interest in the researches of Mesmer and Reichenbach, there were no trained academicians or established professionals who were competent to research and judge extraordinary claims. Psychology was thought of as a branch of philosophy, until the pioneering research of Gustav Theodor Fechner (1801-1887) established psychology as an independent branch of science.

Fechner's formal training lay in medicine and physics. Like the ancient shamans he showed a natural sensitivity to the subtle levels of his own inner world that he could not suppress. Writing under the pseudonym of Dr. Mises, he published a number of works both satirical and symbolic. His biographer, Dr. G. Stanley Hall, describes one of these books, written in 1825, entitled *Comparative Anatomy of the Angels*:

> These are not symbolic, but real, living angels, which stand in the organic world a little higher than man, who is not the highest nor the most beautiful. Even the ass thinks his own type ideal. The human form is a strange aggregate of surfaces and curves, hollows and elevations. There are no flat surfaces and therefore curves and specifically the sphere are the ideal forms and these change (as, indeed, Plato had said). The parts of man's body are beautiful as they approach it, but the eyeball is most complete. It is the organ of light and in light angels live. Earth is not their fitting residence. They belong to higher bodies like the sun, the stars, or light. Just as the air is the element of the angels, who are simply free and independent eyes, all eye, or the eye-type in its highest and most beautiful development. Thus, what in man is a subordinate organ, in the angels is of independent worth. In

animals the eyes look backward or sideward, whereas in man they look forward. But angels are single eyes. Their language is light and their tones are colors. The eye-language of love hints at the speech of angels, these creatures of the sun with their ethereal bodies. Their skin is merely connected vapors, like soap bubbles. Their transparent nature can take on colors. They change their form and expand and contract according to their feelings. They are attraction or repulsion, and with this goes the wonderful color play. They are organisms. They move by hovering and sweeping along. General gravitation, which relates all bodies, is their sense. They feel the farthest thing in the universe and the slightest change in it. They are, in short, living planets and, in fine, the planets are angels.[208]

In his search for the archetypal form of angels, Fechner's work can be seen as in the scientific tradition of Goethe, his countryman, who attempted to reconcile science and poetry. In his perception of the earth and planets as living organisms, he is bearing witness to the ancient esoteric teachings.

One of the defining characteristics of Fechner's life is that he suffered a disease very much akin to the initiatory sickness known among shamans. In 1840 his eyesight began to fail him. Soon he could neither read nor write. He found he could not eat or drink and he was unable to endure society. He lost all control over his thoughts or his attention. His dreams tormented him. His own state seemed to him like the condition of a puppet. By the end of 1843, people believed him to be incurably blind and completely insane. He spent months in solitude in a dark room; and at a level deep within himself he never lost hope. It was in this state, literally the dark night of his soul, he felt that he was called upon by God to do extraordinary things for which his sufferings had prepared him. He recovered after this self-perception and soon discovered within himself even greater physical strength and psychic sensitivity. The whole world now revealed itself to him in a splendor and detail exceeding his earlier visions. He resumed his academic work, no longer in physics but as a philosopher.

This work led to some very cogent philosophical explorations into the nature of consciousness itself and from there to his pioneering experimental work in *psychophysics*:

As to the origin of consciousness, we have a series of thresholds, upper and lower waves. The highest consciousness is God, who planned vaguely at first and is realizing his purpose in all the world processes, so that his plan progresses and becomes more definite and conscious. Thus, as Paracelsus and Jacob Boehme thought, God is growing in our experience, which, as it gives him character, also contributes to his consciousness and adds to his achievements. God comes to consciousness in us....

Life and consciousness never arose, he said, but are original activities of the universe; they are two expressions of the same thing and differ only as a circle seen from within differs from one seen from without. From without all is manifold, from within all is unity, and both together constitute all there is. The soul is not punctual but is pervasive throughout all the body. Those processes immediately bound up with consciousness are *psycho-physic movements* and they are primordial and cosmogonic.[209]

The physical world operates under one law and we must assume that the spiritual world is no less so. There must be then, a priori, some exact mathematical relationship between the physical and the psychical, some law of concomitant variations, for all that is psychic is but the self-appearance of the physical; a material process runs parallel to every conscious process.[210]

This Pythagorean insight as to the mathematical relationship between the physical and psychic worlds led directly to the development of modern, scientific psychology. In all fairness, however, to contemporary psychologists, most are ignorant of Fechner's mystical background — and would probably be shocked by it.

The Theosophical Society

A most intriguing chapter in consciousness history involves the Theosophical Society, founded in 1875 by Madame Helena Petrovna Blavatsky (H. P. B. for short), a most notorious character.

Madame Blavatsky declared herself to be a *chela* or disciple of a brotherhood of spiritual adepts in Tibet whose members had acquired psychic powers beyond the reach of ordinary men. She asserted that they took a special interest in the Theosophical Society and all initiates of occult lore, being able to communicate intelligently with individuals by visiting them in a phantom or astral form. These beings were called the *Mahatmas* and are described in Blavatsky's book *Isis Unveiled*:

> Travelers have met these adepts on the shores of the sacred Ganges, brushed against them on the silent ruins of Thebes, and in the mysterious deserted chambers of Luxor. Within the halls upon whose blue and golden vaults the weird signs attract attention, but whose secret meaning is never penetrated by idle gazers, they have been seen, but seldom recognized. Historical memoirs have recorded their presence in the brilliantly illuminated salons of European aristocracy. They have been encountered again on the arid and desolate plains of the Great Sahara, or in the caves of Elephanta. They may be found everywhere, but they make themselves known only to those who have devoted their lives to unselfish study and are not likely to turn back.

According to William Q. Judge, a New York lawyer who was one of the co-founders of the Theosophical Society, such a Mahatma appeared to the first Theosophists when they held a meeting to frame their constitution. A "strangely foreign Hindoo," came before them, left a package and vanished. On opening the package they found the necessary forms of organization, rules, etc., that were adopted. The early history of the society was based largely on such miracles. Blavatsky's wonderworking

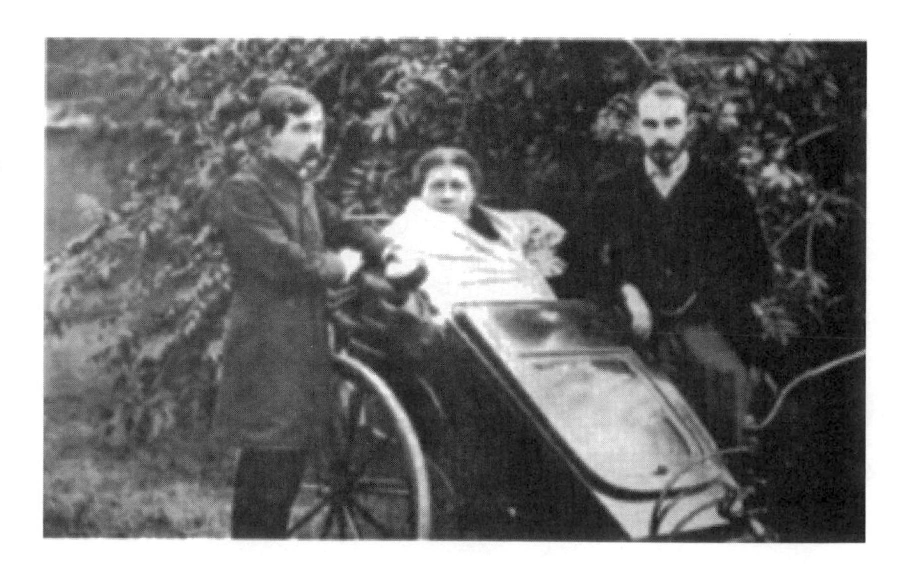

MADAME BLAVATSKY IN PERAMBULATOR, ATTENDED BY TWO STUDENTS JAMES M. PRYSE (LEFT) AND G. R. S. MEAD (RIGHT)

and teaching attracted such notable students as Thomas Edison, Sir William Crookes, Alfred Russell Wallace, British prime minister William Gladstone, Alfred Tennyson, and later U. S. vice-president Henry Wallace, and Annie Besant (the former mistress of George Bernard Shaw, who succeeded Blavatsky as head of the movement).

After seeing the society well established in New York, Madame Blavatsky moved to India. Marvelous phenomena of an occult nature were alleged to have taken place there at the Adyar headquarters. Mysterious, ghostly appearances of Mahatmas were seen, and messages were constantly received by supernatural means. One of the apartments, named the Occult Room in the headquarters, contained a sort of cupboard against the wall, known as the Shrine. Ghostly letters from the Mahatmas were received in this shrine, as well as sent. Skeptics were convinced, and occult lodges spread rapidly. Madame Blavatsky and other Theosophists were interviewed in England by members of the Society for Psychical Research (SPR), who were favorably impressed.

At this point in 1884, a scandal broke out. Two members of Blavatsky's staff claimed that they had conspired with Madame, forging Mahatma letters and placing them in the shrine through a trap door. To back up their claim, they submitted private correspondences from H. P. B. Madame Blavatsky countered with charges of her own. Leaders of the SPR considered the matter significant enough to send Richard Hodgson to India in order to investigate the matter personally. What followed was perhaps the most complicated and confused investigation in the history of psychical research.

Hodgson concluded Madame Blavatsky was a phony — "one of the most accomplished, ingenious, and interesting impostors

Cartoon, "A Mahatma at home."

of history." His two-hundred-page report attempted to reconstruct in detail all of the mechanisms by which she impersonated every sort of phenomena. He hired handwriting experts, for example, who determined that the Mahatma letters were really written in Madame's handwriting. Most of the evidence was of a circumstantial nature as the original shrine had been destroyed by the time Hodgson had arrived at Adyar.[211]

More recently, Theosophical apologist Victor Endersby has written a book challenging the Hodgson report point for point. Endersby cites independent testimony from handwriting experts who clearly disagree with those hired by Hodgson.[212]

The Theosophical Society is still active. The teachings of the Theosophists continue to have an enormous impact on the esoteric folklore of Western culture and for that reason are quoted several times in this book.

A Course in Miracles

Since the original publication of *The Roots of Consciousness*, there have emerged numerous examples of ostensible contact with higher intelligence. One of the foremost among these is *A Course in Miracles*, which is a system of spiritual transformation. Hundreds of thousands of individuals have used this material; and its emphasis on love, forgiveness and freedom from guilt have had an influence on a new generation of spiritual seekers comparable to that of the Theosophical Society in previous years. *A Course in Miracles* suggests that a miracle is really a shift in perception — to see the spirit that lies behind all forms.

Judith Skutch Whitson is the president of the Foundation for Inner Peace, the organization which published *A Course in Miracles.* In the following excerpt from a *Thinking Allowed* interview, she describes the origins of this material:

> In 1975, I met William Thetford and Helen Suchman, two medical psychologists at Columbia Presbyterian School of Physicians and Surgeons, who served as scribes for the material. The way it came to them was through their relationship. They had had a very long period together, teaching, researching, writing grant proposals; yet their life together and among their faculty was not very harmonious. They described it as one of the most stress-filled domains in the world — academia, medical academia.
>
> One day the quieter of the two of them, Bill Thetford, who was a very gentle man, a very thorough scientist, a very solid person — he just blew up. He said in a very meaningful way to her, so that she heard him, that he was sick and tired of the attitudes that that stress seemed to have promoted between the two of them, and that they just were not getting along, that there had to be a better way to live in the world.
>
> There had to be a better way, and he was determined to find it. Instead of laughing at him — because she was quite an acerbic woman, very sharp, the older of the two of them by fourteen years — she actually took his hand, and she said, "You know, Bill, I think you are right. I do not know what the better way could be, but I will help you find it."
>
> Two people joined to find a better way of being in the world — in other words, to heal their relationship. Not too long after, Helen started to experience what she called heightened visual imagery which gave her the feeling that there was something within her catching her attention and very gently taking her along the way, through experience, to an opening up. After many of these visions, she started to become very familiar with an inner voice which spoke with a gentleness and yet an authority she could not avoid listening to.
>
> One day, she was at home, unable to sleep, and she was actually feeling, hearing the words: "This is a course in miracles. Please take notes."
>
> She did not know what to do. That was quite startling. She called up Bill on the phone, and she said, "You know that voice I told you about? It will not go away, and it is saying something very peculiar."
>
> He said, "What is it saying?" And she told him: "This is a course in miracles. Please take notes."
>
> He is a very pragmatic fellow. He said, "Well, you've been having interesting experiences which I've been taking down so we have a record of them. Why don't you just do what it says? You take very fast shorthand. Why don't you just do it?"
>
> So she did, and what she took down startled her a great deal, but the next morning she brought it into the office. Before the staff came in, they locked the door and pulled down the shades so no one should catch them at this. She actually read from her notebook to him what she had

JUDITH SKUTCH WHITSON (FROM *A COURSE IN MIRACLES* VIDEOTAPE, COURTESY THINKING ALLOWED PRODUCTIONS)

taken down, and he typed it up. It was an introduction to *A Course in Miracles.*

It said: "This is a required course; only the time you take it is voluntary. Free will does not mean you can establish the curriculum, only the time in which you need to take it."

It said the opposite of love is fear, but what is all-encompassing can have no opposite. It also said the course could be summed up very simply this way: "Nothing real can be threatened, and nothing unreal exists. Herein lies the peace of God."

Well, she was threatened. At that time she called herself a militant atheist, and he was an agnostic, and here was something that mentioned G-O-D, and it just was not in her vocabulary.

So he convinced her that it was beautifully written, and whatever it was, if it should happen again, to keep on doing it. That began a seven-year collaboration. Any time she wished, when she was ready, she could pick up her shorthand notebook and her pen, and literally start from where she left off before, without even checking what the book said. With Bill typing what she had taken down every day, *A Course in Miracles* came into being.

I asked Helen, the first day I met her, "Did the voice have a name? Did it identify itself, such as the Seth material and others?" She said, "I was afraid you were going to ask that," and Bill Thetford said, "Why don't you tell her, dear? She is going to read it." And she said, "It says it is Jesus."

I said, "Well, is it?" And she said, "Of course," which was interesting, because on the one hand, as a Jewish lady, she did not believe in it, but on a metaphorical level she knew it was true. I think we are all in that position in our lives. There is something that we know is true, but we do not agree to believe in it.

I could not have predicted, in 1975, that there would be three hundred thousand copies of *A Course in Miracles* in circulation. I have no idea how many people study one copy, so I am guessing over a half a million folk are students of *A Course in Miracles.* I think we are probably laying the foundation, along with many others, who share basically the same point of view but go about it in different ways — a foundation for a tremendous change of mind, which I call the great transformation.[213]

The Invisible College

Earlier we discussed the image of the *Invisible College* used by the early Rosicrucians to symbolize their internal contact with higher intelligence. (This same term has been used by scientists, such as Jacques Vallee and J. Allen Hynek, to refer to the loosely connected network of scientists investigating the UFO data.[214])

An interesting perspective on the *Invisible College* comes from a report from Dr. Shafica Karagula, director of the Higher Sense Perception Research Foundation in Los Angeles. Karagula specialized in clinical studies of individuals who are gifted with unusual perceptive talents. One of her subjects, whom she called Vicky, described a series of experiences she had in her sleep where she seemed to be visiting a college and attending classes in many different subjects. Her vision was quite lucid, recalling the architecture of the buildings, and the subject matter of her lectures. The lectures follow an orderly sequence and Karagula claimed to have carefully recorded a number of them from Vicky.

On one occasion, Vicky remembered that a friend of hers, who lived across the United States, was in the classroom with her. After some cautious questioning on the telephone, this person verified that he also remembered being present although he did not recall the details of the lecture as clearly as she did.[215]

Although similar experiences have been reported by many people, and are known to dream researchers, they have yet to be more systematically probed.

The notion of the *Invisible College*, of

course, stems from the Rosicrucian writings of Francis Bacon. Peter Dawkins, a Francis Bacon scholar, tells a story of his own involvement with this work that falls very much in the *Invisible College* tradition. The following excerpt is from my *Thinking Allowed* interview with him:

> One day we went on a retreat, my wife and I, and on that retreat was a lady who was the secretary of the Francis Bacon Society in England. She introduced me to Francis Bacon, who I had not really studied before. I knew of him, but hadn't really bothered much about him. A great sort of gap in my education, that was. I went away not really knowing what to do about this.
>
> And I was woken up with a vivid dream. Now, I don't often remember dreams, but this one I was woken up to remember. It was quite vivid. There was a certain gentleman making a certain gesture that was important in the dream, with a very short message to send to this lady. So I had to do this at four o'clock in the morning. There was no way I could not do it. It was a request that could not be refused. I wrote, and I got a letter back from her by return post saying, "Thank you very much. I've waited twenty years for this. Now we can begin our work."
>
> She asked me a series of questions, and as I read them, suddenly I could answer them, whereas a few days before it probably would have been alien to me. Something changed in my consciousness at that time, and it opened the doorway to another level of consciousness, which I've been working with ever since.[216]

To me, Dawkin's story is plausible because it echoes a dream experience that has been relevant to the development of my own career. That experience took place in 1972 at a time when I was still a graduate student in criminology at the University of California. My interest in criminology reflected my fascination with human deviance. However, I was feeling very uncomfortable studying only negative forms of deviance. I deeply wanted to reorient my career focus.

One evening I felt inspired to tell myself, and to accept without doubt, that I would have a dream which would provide an answer to my career dilemma. Then I did have such a dream.

I dreamed that I was visiting some friends in Berkeley, who were not at home. Knowing where they hid their housekey, I took the key and let myself into their apartment. I walked into the living room where I found a magazine sitting in the middle of the floor. In the dream it was called *Eye* (a popular magazine at that time). I picked it up and began paging through it. While I was dreaming, I had a distinct feeling of elation. I knew that somehow the answer I was seeking existed in that magazine.

I awoke early in the morning and, like Dawkins, felt drawn to act on the inspiration of the dream. Immediately I dressed and ran four miles across Berkeley to the apartment I had dreamed of. My friends were not home,

PETER DAWKINS (FROM *FRANCIS BACON AND WESTERN MYSTICISM* VIDEOTAPE, COURTESY THINKING ALLOWED PRODUCTIONS)

but I did know where they kept their key. Breaking the bounds of conventionality, I let myself into their home. To my delight, there was a single magazine in the middle of their living room floor. It was not called *Eye*; it was called *Focus*. And this magazine literally brought focus to my life. It was the magazine of listener-sponsored television and radio in San Francisco.

As I sat there paging through *Focus* magazine, I was struck with the idea that I would redirect my career through involvement with public broadcast media. I applied to volunteer at KPFA-FM, Berkeley's listener-sponsored radio station — and within three weeks I was asked to host and produce a program twice a week called "The Mind's Ear."

Suddenly, I found that my life was transformed. Every Tuesday and Thursday I had the opportunity to hold intimate, hour-long, uninterrupted discussions with leaders of the human potential movement, yogis, scientists, psychics, psychologists, visionaries, humorists, etc. I felt as though I had found my home in the universe.

It was this experience that gave me the confidence to pursue a unique doctoral diploma in parapsychology at the University of California, Berkeley, and to write the first edition of *The Roots of Consciousness.* The inspiration of that dream still motivates my life twenty years later as I produce the *Thinking Allowed* television series and prepare this revised edition.

My own dream experience certainly does not reflect a contact with higher intelligence in the romanticized or stereotyped manner characteristic of theosophical and rosicrucian legend. It does suggests a synchronistic connection (which implies some higher intelligence) that has been integrated into the movements and actions of my life pattern.

Other Worlds

Scholars, such as F. W. H. Myers (author of *Human Personality and Its Survival of Bodily Death*), who have attempted to catalog the range of psychic experiences, maintain that there is a continuous spectrum of experiences leading from common dream states to heightened creativity to ostensibly spiritualistic and supernatural manifestations of a bizarre nature. Such a spectrum inevitably suggests a relationship between ostensible psychic contact with higher forms of intelligence and ostensible contact with intelligences from other planets or other dimensions of time and space.

Much of this exploration overlaps with UFOlogy — the controversial study of unidentified flying objects — and SETI — the search for extra-terrestrial intelligence. In fact, it is extremely difficult, if not impossible, to determine if any of the purported contactee claims in these areas are anything other than folklore. Even in those rare cases which seem to produce some independent physical suggestion of extra-terrestrial visitations via spacecraft the mythological and archetypal process is also very active.

Swedenborg

For example, the Swedish seer, Emmanuel Swedenborg, whose descriptions of angels have been previously quoted, also claimed to have engaged in extensive psychic communication with inhabitants of other planets. His descriptions certainly imply some sort of direct experience:

> The inhabitants of the Moon are small, like children of six or seven years old; at the same time, they have the strength of men like ourselves. Their voices roll like thunder, and the sound proceeds from the belly, because the Moon is in quite a different atmosphere from the other planets.

The role of imagination in producing such descriptions cannot be denied. Swedenborg himself was a very convincing individual — and a great scientist in his own day. In spite of its obvious inconsistencies with known physical facts (i.e., the moon has *no* atmosphere), I sometimes find myself *wanting* to believe in the validity, on some level, of claims such as these. Self-discipline demands that such yearnings be subjected to and augmented by rational analysis. And I am not alone. It is a human dilemma — caught as we are between the alienation of existential separation and the promise of spiritual unity.

From India to the Planet Mars

In 1899, Prof. Theodore Flournoy, a psychologist at the University of Geneva, published a book called *From India to the Planet Mars* which detailed the trance communications of Mlle Catherine Muller (known under the pseudonym of Helen Smith). In this book he documented striking incidents of telepathy and even a good deal of evidence for a past reincarnation in India — complete with historical accuracy and a knowledge of Sanskrit. He further claimed to have observed incidences of psychokinesis. In spite of such accomplishments, Mademoiselle Muller's greatest claim was to have established mediumistic contact with people from the planet Mars. The medium actually furnished the investigators with a Martian language, complete with its own unique written characters. A number of other investigators defended the extra-terrestrial origin of the language.

Flournoy, however, challenged this claim and produced evidence that the Martian language was actually a sub-conscious elaboration based on French grammar, inflections, and construction. Had it not been for Flournoy's careful analysis, Mademoiselle Muller might have been credited as the first human being to have established intelligent communication with Mars.

A more recent analysis of the speech produced by "channeled entities" was conducted by Sara Grey Thomason, professor of linguistics at the University of Pittsburgh. The typical patterns are very inconsistent, suggesting that these verbalizations are, at least partially, the result of conscious or unconscious personification.[217]

The Fatima Appearances

The 1917 appearances of "the Virgin Mary" at Fatima, Portugal, seem to bear characteristics of a psychically triggered UFO manifestation.

Three young children, Lucia dos Santos, Francisco Marto and his sister Jacinta, figure in this series of extraordinary events. According to their testimony, they were first visited by an angel who asked them to pray. In May 1917, they were visited by the figure of a lady who spoke to them and told

MARTIAN LANDSCAPE DRAWN BY CATHERINE MULLER WHILE IN TRANCE

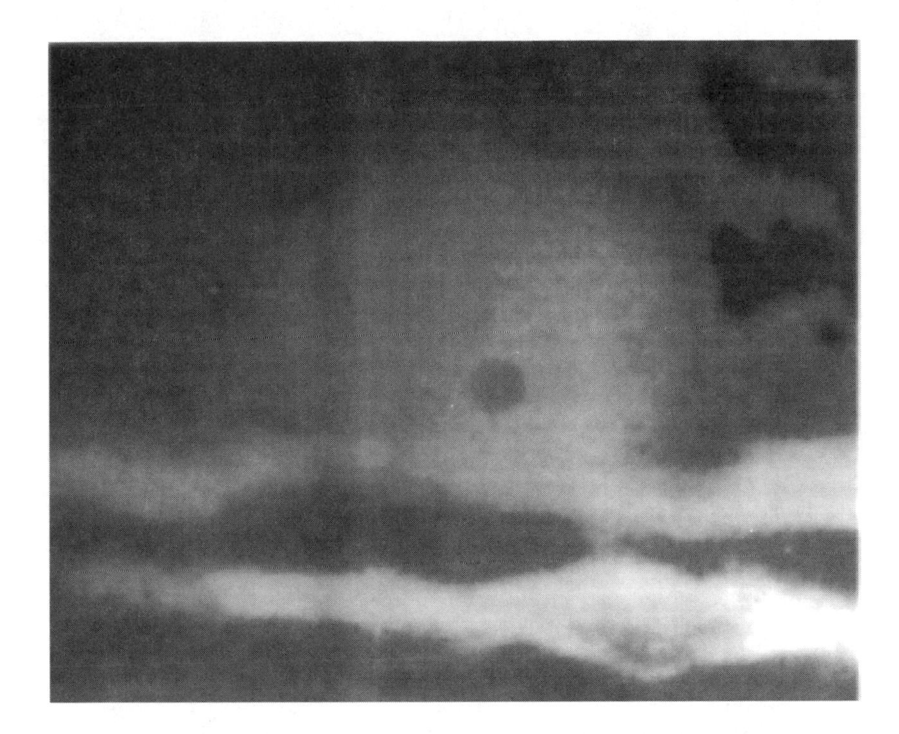

RARE PHOTO OF DISK WHICH APPEARED AT FATIMA, PORTUGAL, ON OCT. 13, 1917. THE DISK IS SHOWN AFTER IT HAD DESCENDED, NEAR THE HORIZON (COURTESY ASSOCIATION FOR THE UNDERSTANDING OF MAN. TAKEN FROM THE BOOK *FATIMA PROPHECY: DAYS OF DARKNESS PROMISE OF LIGHT*)

them to return to the same field on the thirteenth of each month for more messages. Each month more and more people gathered with them to behold the appearances. Many witnesses noted strange lights and sounds, but only the children reported actual contact with the radiant lady. The children claimed to receive much information that was passed on to Roman Catholic Church officials, with instructions that it be released to the public in 1960.

In October, so many people were aware of the phenomena that over fifty thousand had gathered at the Cova da Iria to witness the event. Much to the delight of freethinking skeptics, it was raining that day. The sky was completely overcast. Some spectators saw a column of blue smoke in the vicinity of the children that appeared and disappeared three times. Then suddenly the rain ceased and through the clouds was seen a radiant disk, not the sun, spinning, and throwing off fantastic streamers of light — a constantly changing montage of red, violet, blue, yellow, and white. This continued for about four minutes.

Then the disk advanced toward the earth until it was just over the crowd. The heat was enormous, and many were terrified that the end of the world had come. When it finally retreated into the sky, the shaken masses realized that their clothes and the ground were completely dry — although they had been soaked to the skin a few minutes before.

On October 1930, after eight years of investigation, the Catholic Church announced that the apparitions seen had been genuine visitations of the Virgin Mary. However, for its own reasons, the Church has decided not to publish the prophecies given to the children.[218]

More recently, in 1968, a series of apparitions, seemingly of the Virgin Mary, appeared above the roof of a Coptic Orthodox Church in Zeitoun, Egypt, a suburb north of Cairo. For a period of several months, thousands of people observed and photographed these images. To this day, no rational explanations have been offered for the phenomena. Hundreds of spontaneous healings were reported at the site which were investigated by a commission of medical doctors headed by Dr. Shafik Abd El Malik of Ain Shams University. The apparitions ceased to appear only after the Egyptian government cordoned off the area and began selling tickets to the throngs who had come to observe the phenomena.[219]

UFOs as Apparitions

Ghostly apparitions not infrequently reveal themselves to be similar to some UFO reports. The philosopher C. J. Ducasse mentions a case in which the Reverend Abraham Cummings, holder of a Master of Arts degree from Brown University, attempted to expose the apparition of a deceased woman he had assumed was a hoax:

> Some time in July 1806, in the evening, I was informed by two persons that they had just seen the Spectre in the field. About ten minutes after, I went out, not to see a miracle for I believed they had been mistaken. Looking toward an eminence twelve rods distance from the house, I saw there as I supposed, one of the white rocks. This confirmed my opinion of their spectre, and I paid no attention to it. Three minutes' after, I accidentally looked in the same direction, and the white rock was in the air; its form a complete globe, with a tincture of red...and its diameter about two feet. Fully satisfied that this was nothing ordinary I went toward it for more accurate examination. While my eye was constantly upon it, I went on for four or five steps, when it came to me from the distance of eleven rods, as quick as lightning, and instantly assumed a personal form with a female dress, but did not appear taller than a girl seven years old. While I looked upon her, I said in my mind "you are not tall enough for the woman who has so frequently appeared among us!" Immediately she grew up as large and tall as I considered that woman to be. Now she appeared glorious. On her head was the representation of the sun diffusing the luminous, rectilinear rays every way to the ground. Through the rays I saw the personal form and the woman's dress.[220]

According to Cummings this apparition appeared many times, speaking and delivering discourses sometimes over an hour long. In his pamphlet on the subject, he produces some thirty affidavits from persons who had witnessed the specter. Each time the manifestation began as a small luminous cloud that grew until it took the form of the deceased woman. Witnesses also observed the form vanish in a similar manner.

Carl Jung's Interpretation of UFOs

Carl Jung, the great Swiss psychiatrist, saw UFO descriptions as an archetypal myth-making process within the collective unconscious, or subliminal mind of mankind.[221] He pointed out that, faced with decaying religious values and mythological structures, men attempt to create a new sense of cosmic unity and belonging. The round saucer shape itself has historically symbolized wholeness. When the traditional conceptions of religion are no longer potent, the mind's image-forming processes invest a great deal of psychic energy in forming new images and new unconscious links with the creator. Jung saw this process reflected in his patient's dreams as well as in modern art and fiction. This unconscious activity he felt, could account for many UFO experiences with religious or occult overtones.

However, it was not his intention to deny the reality of such experiences. After ten year's research into the UFO literature written by such respected scientists as Edward Ruppelt, former head of the Air Force Project Blue Book,[222] and Maj. Donald Keyhoe, director of National Investigations Committee on Aerial Phenomena (NICAP).[223]

Jung felt there was no room for doubting that many UFOs sighted were physically real. He suggested that some religiously oriented UFO experiences were simply occasioned by or projected upon actual sightings. (He also hypothesized that certain psychic projections could throw back an echo upon a radar screen or result in other physical manifestations.) Claims of extra-terrestrial contact, within a religious or spiritual context, can be found throughout history.

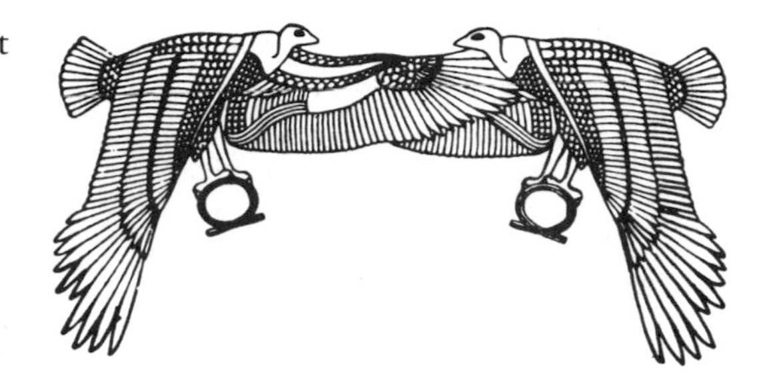

EGYPTIAN BUZZARDS, GUIDING THE SOUL BETWEEN WORLDS

Uri Geller and UFOs

Andrija Puharich, M.D., the colorful and controversial scientist-inventor who has done the most to bring Uri Geller's ostensible psychokinetic feats to the educated public's attention, maintained that Geller was being used as an agency of other dimensional intelligences in order to prepare mankind for psychic adulthood within a larger cosmic community. The evidence to support this claim, which has been presented in his book *Uri*, is actually rather scant.[224] Alleged tape-recorded messages from the UFO occupants have disappeared. Essentially all that is left is the testimony of individuals who have seen UFOs in Geller's presence or in ways connected with him, as well as a number of photos taken by Geller or Puharich.

Puharich implies that we may soon experience a major contact between our civilization and the distant intelligent powers who work through Geller. In his account Puharich noted that at stressful times a white, hawklike bird often appeared and seemed to renew his faith in the intelligent powers working through Geller. The following description is typical:

> At times one of the birds would glide in from the sea right up to within a few meters of the balcony; it would flutter there in one spot and stare at me directly in the eyes. It was a unique experience to look into the piercing "intelligent" eyes of a hawk. It was then that I knew I was not looking into the eyes of an earthly hawk. This was confirmed about 2 p.m. when Uri's eyes followed a feather, loosened from the hawk, that floated on an updraft toward the top of the Sharon Tower. As his eyes followed the feather to the sky, he was startled to see a dark spacecraft parked directly over the hotel. We all looked where he pointed, but we did not see what he saw. But I believed that he saw what he said he saw.[225]

Such a statement might well be taken as face-value evidence that Puharich has lost any claim to objectivity. Geller himself proposed a rather subtle interpretation of the hawk and spacecraft phenomena.

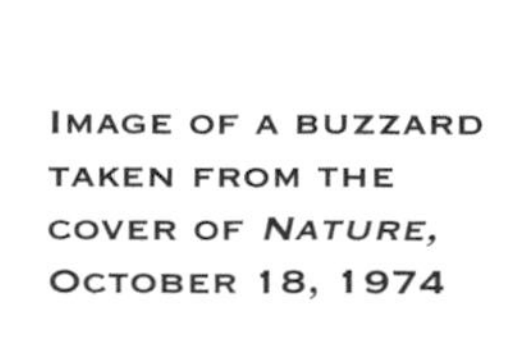

IMAGE OF A BUZZARD TAKEN FROM THE COVER OF *NATURE*, OCTOBER 18, 1974

He felt that this was simply a form taken by IS [an abbreviated term for Intelligence in the Sky], just as they took the form of a spacecraft, because it suited their purposes.[226]

Others besides Puharich have also noted the appearance of a white hawk in situations favorably connected with Geller. Ila Ziebel, a psychologist from Madison, Wisconsin, was with Geller and Puharich during some of the most dramatic sightings in Israel.

Ray Stanford, the psychic from Austin, Texas, who works with the Association for the Understanding of Man, has seen the hawk as a symbolic, yet real, form of the intelligence that works through Geller. Stanford also associates the Hawk with UFO phenomena.[227] Appropriately enough, Puharich used the term "Horus" when referring to this hawk.

Another Hawk coincidence occurs on the cover of the October 18, 1974, issue of *Nature* magazine, which carried an article on the research with Geller conducted at Stanford Research Institute by physicists Harold Puthoff and Russell Targ. This illustration came from a century-old issue of *Nature* in which is mentioned the view of Alfred Russell Wallace, a spiritualist and co-author with Darwin of the theory of evolution, that man is more than a highly complex machine devoid of free will.

The Stella Lansing Case

Some of the purported data suggests patterns lending support to psychic interpretations. For example, there is the case of Mrs. Stella Lansing of Palmer, Massachusetts, which has been studied for many years by the eminent psychiatrist, Dr. Berthold E. Schwarz.[228,229]

Since 1967, Mrs. Lansing has produced over fifty rolls of 8mm movie film depicting saucer-shaped UFOs and other unidentified objects, using a variety of cameras under many different conditions. It has become

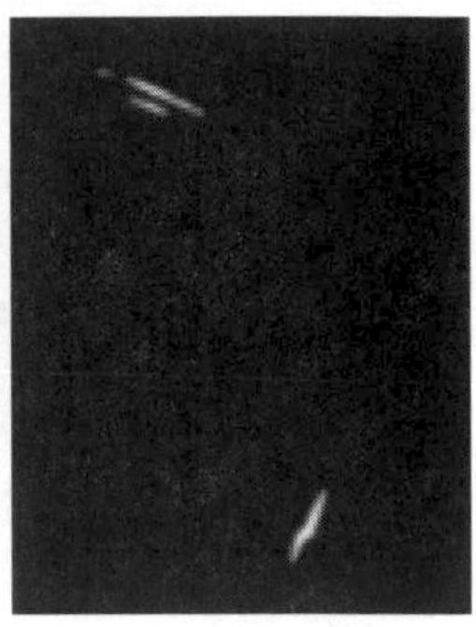

LEFT, CLOCKLIKE UFO PATTERNS PHOTOGRAPHED BY STELLA LANSING (COURTESY BERTHOLD SCHWARZ). RIGHT, ENLARGEMENT OF UFO IMAGE

something of a hobby for her — almost a compulsion. Following her intuitive impressions, at all hours of the day or night, in all sorts of weather, often accompanied by researchers, she roams about the countryside with her camera. Many times the strange lights in the sky she manages to photograph have been witnessed by friends, neighbors and independent observers. Some of her sightings have been verified by officials at the local airport. Sometimes other people accompanying Mrs. Lansing are also able to photograph the strange lights.[230]

On the other hand, many of her pictures are quite unlike the lights that have been witnessed. Objects and faces appear on the film very reminiscent of Ted Serios' psychic photographs. In fact, investigation suggests that Mrs. Lansing can exert a psychokinetic effect upon film, videotape, and even large static objects. Strange sounds and voices also appear on audio tapes in her presence. Some of her apparent psychic photographs show images of flying saucers in a clocklike formation. Often Mrs. Lansing claims to see these strange images that show up on the film, even though they are invisible to observers. Other individuals standing next to Mrs. Lansing are unable to photograph these odd images.

She has always shown complete openness

and willingness to cooperate with scientific investigators. Years of psychiatric observations have convinced Dr. Schwarz of her honesty. The famous mentalist, Joseph Dunninger, a friend of Dr. Schwarz, also has examined Mrs. Lansing, finding no signs of fraud nor any logical explanation for her phenomena. According to Schwarz, her photographs point to the physical existence of possible UFOs and associated entities. However, the ostensible psychic events that accompany this evidence suggest other dimensions to this phenomena.[231]

Automobile Teleportation

One of the most unusual claims made by Puharich regarding the talents of Uri Geller involves Geller's alleged ability to teleport automobiles. (The topic is so bizarre that when I first interviewed Geller on the radio, in 1973, he asked me not to mention it for fear of ridicule.) Since Puharich's book, several other reports on auto teleportation have been filed.

Ray Stanford of Austin, Texas, discusses two instances, connected with Geller, in which his automobile was apparently teleported. One of these involved an automobile accident. Testimony from witnesses is recorded in the traffic-court transcripts. They observed Stanford's car suddenly appear in front of them, "like a light that had been switched on." The distance of this teleportation was about fifty feet. On the second occasion, the teleportation was even more dramatic. Driving with his wife along Interstate 10 in Texas, Ray suddenly noticed a silvery-metallic, blue glow around the car. Stuck in heavy traffic at the time, he actually mentioned to his wife he hoped that "Uri's intelligences would teleport us!" Then, according to his testimony he felt a strange sensation in his brain and the scene instantly changed. They had traveled thirty-seven miles in no time and using no gas. Later, as the car was not functioning well, it was hauled to a garage. The alternator and voltage regulator were completely burned out and all the wiring was completely charred.[232]

In South Africa, a couple reported that after they had observed a UFO from within their car, they lost control of the vehicle which continued at a high speed across the African plain. Except for a few short

MENTALIST DUNNINGER WITH STELLA LANSING AND HER PET DOG

intervals, they were unable to steer or brake the car for five hours. They also felt unusual coldness within the car. Under hypnosis, one of the couple stated that a humanoid from the UFO had entered their car and was with them, in the back seat, during the entire journey.

UFO investigators have files on at least five other cases of alleged automobile teleportation in South America. In three instances, individuals found themselves suddenly transported thousands of miles from Argentina or Brazil to Mexico. These cases are very difficult to carefully investigate. Many of the published reports simply lack significant details. Nevertheless, the reports do suggest a pattern of events.[233]

The Strange Case of Dr. X

Oddly enough, many reported UFO sightings and contacts involve incidences of what could be called paraconceptual healing. A prime example is "The Strange Case of Dr. X" which was reported by the French scientist Aime Michel in *Flying Saucer Review* in 1969.[234] "Dr. X" is the pseudonym for a well-known and respected physician who holds an important official position in southeast France.

Early one morning in November 1968, the doctor was awakened by the cries of his fourteen-month-old son. He got up painfully, due to an injury he had received a few days earlier while chopping wood, and found the baby pointing toward light flashes coming in through the shuttered window. Opening a large window, Dr. X observed two horizontal, disk-shaped objects that were silvery white on top and bright red underneath. The flashes were caused by a sudden burst of light between the two disks with a periodicity close to one second. As the doctor watched these disks they approached him and actually seemed to *merge* so there was only one disk from which emanated a single beam of white light. Then the disk began to flip from a horizontal to a vertical position, so it was seen as a circle standing on its edge. The beam of light came to illuminate the front of Dr. X's house and shone directly into his face. At that moment a loud sound was heard and the object vanished.

The doctor immediately woke up his wife to tell her what had occurred. It was then that she observed that the swelling and wound on his leg had disappeared. Furthermore, Dr. X had suffered from a partial paralysis of his right side from a war wound received ten years previously in Algeria. In the days following the sighting, these symptoms also disappeared.

There are many unusual aspects of this case which are still being investigated by a competent team of researchers. One study, reported several years after the original sighting, noted that an odd red pigmentation has periodically appeared in a triangle shape around the naval of both Dr. X and his young child. It would stay visible for two or three days at a time. This had happened even when the child was staying with his grandmother who knew nothing of the UFO sighting. Other incidents of a psychokinetic nature have been noted such as levitation and poltergeist-type phenomena. The sighting seems to have been a landmark event in Dr. X's life as he now faces life with a rather mystical acceptance, showing no fear of death or tragedy. This new attitude has been recognized by friends and relatives who also knew nothing of Dr. X's experience.[235]

Biological Effects of UFO Contact

Several other instances are known of unusual healings triggered by a flash of light coming from an unidentified flying object. A case was reported in Damon, Texas, in 1965, in which the wound received by a deputy sheriff from his pet alligator healed after being exposed to a brilliant flash of light and heat coming from a UFO. This incident was witnessed by a fellow officer.[236]

On other occasions, however, the effects of such contact more clearly resemble the symptoms of disease. Sometimes the effects are mixed. For example, in December 1972, a seventy-three-year-old Argentine watchman, Ventura Maceiras, observed a glowing craft hovering over a nearby grove of eucalyptus trees. The object was near enough so he was able to see figures staring at him through the windows in a round cabin. Then, as in the Dr. X case, the craft tilted toward him and he was momentarily blinded by a flash of light. Seconds later he had recovered and was able to watch the object move slowly away and disappear behind the trees on a low hill.

After his experience, Maceiras developed swollen pustules on the back of his neck. He suffered from nausea, headaches, and diarrhea. His eyes watered constantly and he experienced difficulty in speaking. At the site where the object had been, the tops of the eucalyptus trees were scorched and burned. On the positive side, since his experience, Maceiras has been growing a third set of new teeth.[237]

Another unusual healing associated with a UFO was reported in October 1957 in the mountains west of Rio de Janeiro, Brazil, where the daughter of a well-to-do family was sick with stomach cancer. Seven members of the family were present, as the girl was in agony and a strange glow shone outside the bedroom window. What followed could have been part of the intercosmic Red Cross emergency squad!

As the astonished family watched, two beings, just under four feet tall, with orange hair and slanting green eyes, emerged from a landed "saucer" and entered the sickroom, laying out their instruments as if in preparation for surgery! One of them placed his hand on the forehead of the father who began to communicate telepathically, he felt, the details of his daughter's illness. The rest of the operation is described in comparatively normal terms. By shining a light on the girl's stomach, the small surgeons lit up the inside of her abdomen so the cancerous growth became visible. The surgical removal took about half an hour. Before leaving, the beneficent visitors left some medicine for the girl with telepathic instructions for its use. Several weeks later, the girl's doctor verified that she had been cured of cancer.

If taken out of context, the Brazilian case seems completely unbelievable. However, it does contain elements similar to other UFO contact cases that researchers are inclined to take seriously. The following case, the most prominent of UFO contact stories, cannot easily be dismissed as fraud or delusion.

The Betty and Barney Hill Case

On September 19, 1961, returning late at night from a Canadian vacation along a lonely New Hampshire road, an interracial couple, Betty and Barney Hill, noticed a large glowing object in the sky above their car. It approached so close that Barney stopped the car, got out, and, taking his binoculars, actually saw humanoid occupants through what appeared to be portholes in a spacecraft. Terrified, he ran back to the car and stepped on the gas. As the Hills sped down the highway, they heard a strange beeping sound.

Although Barney very much wanted to forget this incident, Betty persisted in discussing the matter. She initiated reports to both the Air Force and the National Investigations Committee on Aerial Phenomena, a civilian organization in Washington, D.C. Barney participated reluctantly in the interviews that followed. NICAP investigators were particularly interested in their experience; after six hours of intense questioning a report was issued favoring the case's authenticity. Subsequent to this report, further questioning turned up the fact that there was a two-hour period of time between the sighting of the UFO and their arrival home that the couple could not account for. Both Barney and Betty seemed to be in amnesia's grip regarding some part of their UFO contact. Furthermore, strange markings were found on the car trunk which Betty associated with the beeping noises.

Eventually the Hills sought psychiatric aid in removing the memory blocks regarding their UFO experience. They were referred to Dr. Benjamin Simon, a highly regarded Boston practitioner particularly known for his skill in treating amnesia. Under Dr. Simon's skillful hypnotic induction, Betty and Barney each separately revealed memories from the forgotten two hours. What emerged under hypnosis was a most engaging tale of their abduction and physical examination by alien beings aboard a space craft, after which they were safely released. Betty actually engaged in some pleasant conversation with one of her captors before she and Barney were allowed to return. She was shown a star map marked with principle trade and exploration routes. Later she was able to produce a sketch of this map from memory.

Dr. Simon himself was not particularly interested in whether or not their reported experiences actually occurred. The purpose of his therapy, which lasted over six months, was to alleviate the stress that had resulted from the experience and the amnesia. It was his theory that the UFO sighting was real enough, but that the abduction story was a dream of Betty's which was somehow communicated to Barney. This seemed probable since nearly all of the material revealed by Barney under hypnosis also appeared in Betty's account. There were, however, some notable exceptions. For example, Barney described how the UFO occupants placed a cold cup over his groin during the examination. Later he developed a circular ring of warts around the groin. Barney, who would rather have believed it was a dream, was never able to accept Dr. Simon's theory simply because his own memories, after hypnosis, were too vivid.

Neither the Hills nor Dr. Simon ever initially sought any publicity for this case. Against their wishes, however, the story eventually received press coverage based on the incomplete information uncovered by one reporter. After this happened, the Hills, taking the advice of a lawyer, decided to allow John Fuller, a professional writer, to publish the complete story in a book. Fuller was given full access to the audio tapes from the hypnotic sessions. His book, *The Interrupted Journey*, implies that the abduction actually did occur. He puts the case within the context of other UFO sightings in the New Hampshire area, as well as other possible UFO abduction cases. While Dr. Simon would not go so far as to accept Fuller's conclusions, he participated in the book to the extent of insuring that accurate psychiatric information was provided.[238]

Betty Hill's sketch of the star map was reproduced in Fuller's book.

This map inspired Marjorie Fish, an

STAR MAP REPRODUCED BY BETTY HILL.

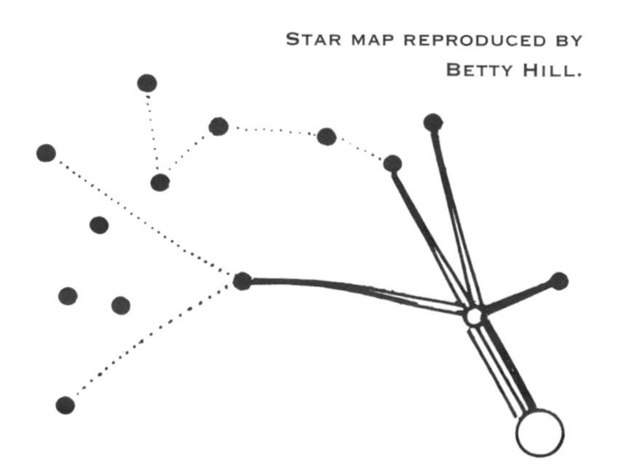

Ohio schoolteacher and amateur astronomer, to inquire whether Betty had been shown an actual pattern of celestial objects. Using beads dangling on threads suspended from the ceiling of her home, Ms. Fish constructed three-dimensional models of the stars in the vicinity of our own solar system. According to Fish, the view in Betty's map closely corresponded to the perspective one would see from the vicinity of a double-star system known as Zeta Reticuli, visible in the southern hemisphere, thirty-seven light years away.

A number of professional astronomers have carefully scrutinized Fish's data using sophisticated computer systems. Her hypothesis has been examined by scientists at Ohio State University, the University of Pittsburgh, Northwestern University, and the University of Utah. In each case, the likelihood of Fish's hypothesis has been confirmed. David R. Saunders, a statistics expert at the Industrial Relations Center of the University of Chicago, concluded that the odds of such a perfect match happening on a chance basis are about a thousand to one.[239] More recently this claim has come under severe scrutiny and has been highly questioned. Astronomer Carl Sagan has claimed, for example, that a computer analysis revealed little similarity between Betty Hill's map and the stellar perspective from the Zeti Reticuli system.[240]

Because the Zeta Reticuli system is one of sixteen nearby stars, similar enough to our own sun to support life-bearing planets, we would be most likely to visit there if we were looking for life as we know it in the vicinity of our solar system.

Another intriguing aspect of this case is the implication of telepathic interaction between the Hills and the UFO occupants. From the different accounts, it appears that the "extra-terrestrials" used a combination of speech, gestures, and ESP in their communication. Betty reports, for example, that although they spoke English, they asked her the meaning of several simple words, like "vegetable."

In another abduction case in 1967, policeman Herbert Schirmer of Ashland, Nebraska, also reported alien beings communicating to him with both ESP and speech. Schirmer also suffered from amnesia which could only be reversed through hypnosis. His case was evaluated by psychologist Leo Sprinkle of the University of Wyoming, who felt that the testimony was authentic.[241,242]

Robert Monroe UFO Encounter

Connections between out-of-body experience and ostensible UFO encounters stem from an experience reported by Robert Monroe, author of *Journeys Out of the Body.* Monroe's out-of-body experiences, which have been previously mentioned, actually began with an encounter strangely reminiscent of some of the more believable UFO abduction cases.

Monroe himself was not aware of the connection until many years after the following event when he first read Fuller's book on the Betty and Barney Hill case, *The Interrupted Journey.* There were three facts presented in the book that astonished him. The first of these was the physical examination that was described as taking place within the saucer vehicle. Second, were the beeping noises that the Hills heard. Last were the blemishes that appeared on the Hill's automobile. Fortunately, Monroe had kept careful records of his experiences and was able to check his notes for 1958 when this experience occurred.

He was alone one evening, in a ten-by-ten office he had built for himself in the Virginia Blue Ridge mountains. The first strange thing he noticed was a repeating tone about eight hundred cycles per second. Then it felt to him as if a beam of *heat* penetrated the walls of his office and hit his body, from a position about thirty degrees above the horizon. From that point on he was aware of the presence of three or four entities. He was unable to see their forms in detail, although he had a sense he was seeing some forms. He also had the experience of telepathic communication with them lasting for about two and a half hours. Particularly striking for Monroe was the sensation of having a cup-like object placed over his groin precisely as Barney Hill had described!

The examination and communication that followed was rather detailed. Monroe felt that the most important sensation was as if his brain were being probed — as if everything he knew or thought was examined. Of course, he was rather awestruck by the situation and did not make all of the observations which he later thought of. Today, he maintains the examination took place while he was outside of his body, but still in the physical environment of his office. However, being inexperienced at the time, he did not know how to interpret the event. Now he even hypothesizes that the incident may have changed his brain in some way that led to his subsequent out-of-body journeying.

Generally speaking, Monroe does not associate the out-of-body state with other UFO encounters. He acknowledges the presence of many strange creatures and beings in the out-of-body state, but feels that there is no evidence they are related to UFOs. Only on one other out-of-body trip did he experience anything like a UFO:

> This happened in the early days of my out-of-body journeys, from the same office. I rolled out of the physical and thought I would go play for a while. It was a beautiful cumulus cloud day. So I went up and was doing power rolls through clouds and things like that. As I came out and circled round this one large cumulus, there was this disk shaped object sitting between two cloud towers. The cockpit was, I would assume, maybe ten or eleven feet in diameter. There were two snake-like heads coming out of it. They were long heads without any shoulders. They could have been mechanical devices. They appeared to be rotating, like rods with helmets on top. I perceived a peculiar sort of frequency coming out of them. I was within two hundred feet. Then I got nervous and just turned around and went home before I was discovered.[243]

These are the only UFO experiences Monroe associates with the out-of-body state. However, he also describes peculiar visions preceded by a hissing sound and a sensation like a valve opening in his brain. Monroe claims that these visions have produced accurate scenes of the future in so far as he has been able to verify them. One such vision, which has not yet manifested, describes what might sound like mass UFO landings:

> Behind the first wave is row after row of the strange aircraft, literally hundreds of them. They are not like any airplanes I have seen before. No wings are visible, and each machine is gigantic, some three-thousand feet across. Each is shaped like the head of an arrow, V-shaped, but with no fuselage as in our swept-wing airplanes. The V shape is not a lifting surface, but houses the occupants in two or three decks. They sail majestically overhead, and I feel a tingle of awe at the mighty power they represent. I also feel fear, because I somehow know that these are not man-made.[244]

UFO Contactee Cults

A classic contactee is George Adamski, a trance medium, who received telepathic messages that led to his alleged encounters with numerous fair-haired and radiant Venusians, Martians and Saturnians. On one occasion several individuals signed sworn affidavits saying that they had witnessed the

landing of a small saucer-shaped craft as well as Adamski's conversation with the tall, radiant being, who stepped out and left footprints in the middle of the desert. This landing had been predicted by Adamski and was said to be an effort to warn earthlings about nuclear misuse. The credibility of this particular incident has been overshadowed by the unmistakably deceptive character of Adamski's other reports. He claimed, for example, to have been taken to the far side of the moon aboard a spacecraft. There he discovered UFO bases along with cities, lakes, rivers, and even snow-capped mountains. Of course, subsequent space probes have proved this impossible.

We will probably never know the complete story behind Adamski's claims. All of the varied evidence tends to mitigate against any simplistic interpretations.

An interesting sequel to this case took place in England on April 24, 1965. A Mr. E. A. Bryant of Scoriton, South Devon, was out walking in the country when he was confronted by a large aerial object that appeared out of thin air and landed forty yards in front of him. An opening appeared in the side of the "saucer" and three figures dressed in "diving gear" stepped forward. One of them stated in English that his name was "Yamski" or something similar. He went on to say that it was a pity that "Des" or "Les" was not there, as he would understand the visitation. Metallic fragments were left at the spot and later analyzed by members of the Exeter Astronomical Society. The odd aspect of this story is that George Adamski, who collaborated with Desmond Leslie in the book *Flying Saucers Have Landed*, died on April 23, 1965.[245]

Evidentially this is not a strong case, as there was only one witness. However, such testimony does illustrate the complexity of the phenomena we lump together in the UFO category. Many UFO cases are intertwined with parapsychological phenomena. However, alleged psychic phenomena that imply extra-terrestrial contact can not be taken at face value.

Ray Stanford UFO Research

The work of Ray Stanford in Austin, Texas, exemplifies a situation in which UFO phenomena of the type associated with Geller overlap with communications from the masters of the spiritual hierarchy. *Psychic Magazine* provides the following capsule description of Stanford's many activities:

> During his teens, Ray Stanford had a fascination with rockets and their principles of operation. In 1955, he was awarded the Texas Academy of Science's top award for his report on "Experiments with Multiple-Stage Principles of Rocketry." Some of his rockets weighed several hundred pounds each and challenged, if they did not break, amateur rocket altitude records.
>
> From 1954 through 1959, he had a series of witnessed unidentified flying object (UFO) observations, including a landing on October 21, 1956, which he had predicted. One sighting in 1959 was simultaneously filmed by 8mm and 16mm movie cameras. The case is listed as "unidentified" in the U.S. Air Force files.
>
> The turning point in his career, however, occurred in early 1960 during a group meditation, when he spontaneously passed into an ecstatic state which involved a sense of floating above his body experiencing "...a unity with light and love from beyond my normal state." Lapsing into an unconscious state, he awoke some two hours later to learn that strange voices and personalities had manifested through him calling themselves "Brothers." One of the voices announced that "Stanford's unconscious being could, when given appropriate suggestion, give medical clairvoyance and describe former life personalities and activities for individuals."...
>
> Final proof for him occurred in April of 1961 when he says that he along with four others in a lighted room in Austin, Texas, saw one of the Brothers, who had frequently spoken through him, suddenly materialize as a visible, glowing form over seven feet tall, complete with robe and metallic-like headpiece. That experience convinced him once and for all that the Brothers were something other than unconscious masquerades.[246]

One *Brother* who speaks through Ray is known as Kuthumi and appears to be — somewhat to Ray's embarrassment — a Mahatma as referred to by Madame Blavatsky. This entity bears all of the earmarks of an angelic, radiant, spiritual being. However, Ray himself is quite careful to distinguish his own trance personalities from those of the UFO occupants — with whom he claims to have had no direct contact.[247] Ray does suggest that the UFO occupants are very skilled in psychic abilities as well as in forms of travel that alter space-time relationships as we understand them. In terms of our own relationship with the universe of UFO occupants, Stanford's readings suggest:

> Planets are to civilizations what rivers are to salmon-spawning places. When we become adult we will have to go out into the sea of space and accept our place in...the cosmological community.[248]

Earth's Ambassador

I was personally involved for over ten years in the investigation of an individual who claimed to be in telepathic contact with other dimensional beings which he called Space Intelligences (SIs). For many years, Mr. Ted Owens supported his claim by making written predictions regarding a variety of unlikely events which he believed the SIs would be instrumental in causing. These events have included UFO sightings, hurricanes and storms, lightning striking preselected targets, mishaps on NASA spaceflights, power blackouts, earthquakes, and anomalous radar sightings. Some of these

RAY STANFORD (COURTESY ASSOCIATION FOR THE UNDERSTANDING OF MAN)

are described in the upcoming discussion on ostensible *macro-psychokinetic* phenomena.

A typical UFO-related claim was reported by Max L. Fogel, Ph.D., director of Science and Education for MENSA — an organization for individuals of superior intelligence. On November 9, 1973, Dr. Fogel wrote the following letter for Owens which was addressed "To Whom It May Concern":

> Ted Owens, who is known as "PK Man — The UFO Prophet," and is a member of Mensa, informed me by letter on Tuesday, October 23, 1973, that it was his intention to telepathically communicate with UFOs and ask them to appear within a 100 mile area of Cape Charles, Virginia, and show themselves to the police within that area. On October 25, 1973, two days later, a UFO appeared over the head of a policeman in Chase City, Virginia (within the specified 100 mile area) for 15 minutes, as described in the Richmond *Times-Dispatch* dated October 26, 1973.
>
> Thus, an example of the type of occurrence predicted in Mr. Owens' letter to me, written in advance of the occurrence, did take place.

The story published in the Richmond *Times-Dispatch* verified the police sighting in Chase City and recalled a remarkable UFO sighting that had occurred in that area in 1967:

> UFO SEEN, CHASE CITY POLICE SAY
>
> BY JOHN CLEMENT
>
> CHASE CITY — Sam Huff, a policeman for the past seven years, said he watched an unidentified flying object for 15 minutes around 2 A.M. Thursday.
>
> Huff said Marion Owen, the Chase City police radio dispatcher, first sighted the object after noticing an unusual light reflection on the window of the police station.
>
> Owen radioed Huff, who drove to the west side of town and parked on a railroad bridge, almost under the object, Huff said.

TED OWENS

Huff said the soundless object, fairly large, with a color that resembled a very bright star, hovered motionless over the town for about five minutes then moved rapidly to the north, reversed direction and returned to a position over his head.

It stayed motionless for another five minutes before heading rapidly in a westerly direction, he added.

UFOs are probably taken a little more matter-of-factly in this area since two well-publicized and still unexplained sightings occurred in 1967 in nearby South Hill and Lunenburg County.

In the South Hill incident, a warehouseman said, in April 1967 that he rounded a curve and encountered an object 12 feet in diameter, standing on three legs, that resembled an aluminum storage tank.

The object suddenly left the ground in a burst of brilliant light, he reported.

The blacktop street caught fire and when police arrived on the scene the tar was still hot and smoking.

Two months later, a rural Lunenberg County storekeeper said she was startled one night by a thundering roar.

She looked out her bedroom window and saw a bright light which she described as so bright "you could see every leaf on the tree." That object also allegedly left burn marks on the highway.

In reviewing about 140 claims by Owens, I arrived at a preliminary opinion that half of these cases involved uncanny events that could be construed to support his unusual claims. On November 7, 1976, Owens agreed to conduct a demonstration of his supposed powers for me. He predicted that within a ninety day period he would produce three major UFO sightings within a hundred miles of San Francisco, as well as other minor events such as power blackouts. Many strange events did occur during this period. One, of particular note, which I regard as perhaps the best-documented UFO sighting on record, occurred on December 8, 1976.

The craft was seen simultaneously from the air and the ground by hundreds of witnesses on the campus of Sonoma State University, just fifty miles north of San Francisco. It was photographed and videotaped by Bill Morehouse, professor and chairman of the art department at Sonoma State University. He lent his videotape to KQED-TV 9 in San Francisco which broadcast the event on its evening news program. A photo was published on the front page of the Berkeley *Gazette.* The object remained in public view for about ten minutes. It appeared while Steven Poleskie, a visiting professor of art, was performing an aerial artwork demonstration over the campus. Poleskie himself, as an artist, would be an obvious person to have faked such an event for the sake of an artistic effect. However, quite shaken and puzzled by the event himself, he has denied that it was either a hoax or an artistic effect.

The object appeared in his airzone at an

altitude of from five hundred to three thousand feet above ground. Witnesses were interviewed both by me and by Prof. James Harder of the University of California, Berkeley, a specialist in interviewing UFO witnesses and contactees.

In February 1977, an actual UFO abduction case was reported in Concord, California, about thirty miles east of San Francisco. The February 2, 1977, issue of the Concord *Transcript* read:

> FLYING SAUCER REPORT TO CONCORD PD
>
> A 24 year old Concord man told police early today he was whisked away and examined by five-foot grey beings from a flying saucer.
>
> According to the report he left a Willow Pass Road restaurant about 4:10 A.M. and was confronted by two short, grey-skinned men with enlarged skulls, no hair and black pupils.
>
> The next thing he knew, he said, was that he had been transported to a field at Willow Creek Elementary School. There he said he was facing a circular craft with a ladder extending toward him.
>
> Suddenly, he was inside the ship. While there, he said he stuck his left hand in a chamber, and "all sorts of lights went off."
>
> He asked what was happening, and telepathically he was told the aliens were on a "mission to study life habits" on Earth. The beings also noted that their craft was from a larger ship located outside the planet's atmosphere.
>
> The next thing he knew, according to the report, he was outside an apartment complex on Mohr Lane.
>
> For about 15 minutes, he said he was unable to move.
>
> He called the Concord Police Department at 5:33 A.M.
>
> The Oakland center of the Federal Aviation Administration noted it had no reports this morning of unidentified flying objects.

Professor James Harder and I both interviewed the contactee in this case. This individual, who had no desire for publicity, gave us the impression of being a legitimate witness. The man was married and employed as a salesman in the Concord area. He had had no previous interest in or experience with UFOs. Nor had he had any interest in or significant experience with psychic phenomena. He seemed to exhibit anxiety when referring to the apparently traumatic experience — about which he did not even want to tell his wife. He was not willing to subject himself to hypnotic regression.

Owens, who is now deceased, developed an extensive training program to enable other individuals to engage in the type of contact which he claimed. I have interviewed some individuals who have taken this training, who claim that it has resulted in UFO appearances. While I myself have taken the two-day training, I have had no desire in the subsequent years to use it for the purposes of developing UFO phenomena.

Throughout his life, Owens attempted with little success to convince scientists and others of the legitimacy of his claims. Although impressed with the results he produced, I never felt comfortable giving them any particular interpretation. It was my hope that I could convince others to conduct more rigorous investigations. This was not to be. The results produced by Owens were so striking that most individuals exposed to this material either denied it in a sarcastic and insulting manner or became frightened and confused. Constantly confronted with such reactions, Owens himself developed protective personality mechanisms that were occasionally unpleasant to deal with. The files are filled with stories of uncanny, yet unpleasant, events. An entire book could be written about this case. Owens died in 1987.

UFO Research Today

Research today on unidentified flying objects is largely in what might be called a prescientific phase. Spontaneous observations have been recorded. Documentary photographs have been taken. Researchers have certainly learned a great deal about what is not a UFO: weather balloons, sundogs, birds, aircraft, searchlights, reflections, satellites, and astronomical bodies. Numerous frauds have been also detected. At least five percent of the UFO sightings reported to the United States Air Force, during its period of investigations from 1948 until 1969, still remain unexplained in spite of reliable witnesses and adequate information.[249,250] Nevertheless, there has been no conclusive physical proof that UFOs are actually intelligently designed flying machines from a non-human source. No UFOs have been captured. None have even crashed and left parts. There have been no regularly acknowledged communication channels between scientists on earth and non-human intelligence. Nor have the UFOlogists provided us with any repeatable experiments.

It is hard for many individuals to accept the possibility that UFO occupants are interacting with the general public, when "officially" they have not been shown to even exist. The alleged parapsychological aspects of UFO reports add even more fuel to the skeptic's fire. In 1969, a government committee, headed by Dr. E. U. Condon at the University of Colorado, declared that UFO phenomena should no longer be considered of any scientific value.[251] The Air Force discontinued its Project Blue Book investigations the same year.

Nevertheless a number of dedicated scientists have continued to investigate UFO reports under the sponsorship of various civilian organizations. Dr. J. Allen Hynek, chairman of the department of astronomy at Northwestern University, and for twenty years a consultant to the Air Force study, has recently published a book, *The UFO Experience*, which takes issue with the Condon report.[252] Dr. Hynek's book was favorably reviewed in *Science*, the weekly journal published by the American Academy for the Advancement of Science.[253] While there has been no great clamor for another government investigation, it is quite clear the case against UFOs is no longer closed.

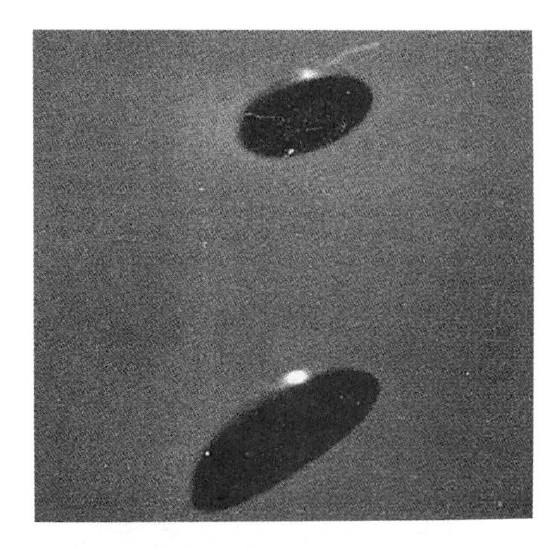

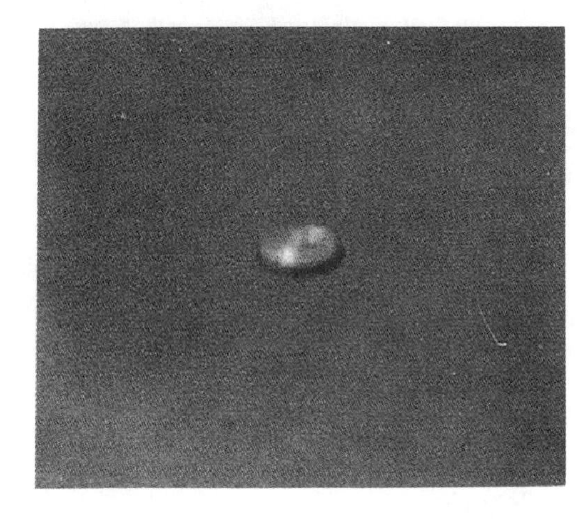

UNIDENTIFIED OBJECTS PHOTOGRAPHED IN YUNGAY, PERU, MARCH 1967 (COURTESY AERIAL PHENOMENA RESEARCH ORGANIZATION BULLETIN, JANUARY 1969)

Close Encounters

Since the original publication of *The Roots of Consciousness,* there has been considerable attention paid to cases of apparent close encounters with beings from extraterrestrial civilizations. James Harder, professor of hydraulic engineering at the University of California at Berkeley, is one of the world's foremost experts in hypnotic interviews with ostensible UFO contactees, having investigated over one hundred cases of this type. In a *Thinking Allowed* interview, he summarizes his findings from this work:

> Some alien visitors have been around a long time. They know what humans are like a lot better than we do in many ways. They certainly seem to have better ability to deal with certain kinds of illnesses, and better surgical techniques than we have. They are also the ones who induce a feeling of friendliness on the part of the people they pick up.
>
> People think of themselves as being picked up by old friends. They often have a kind of religious conversation, although not of a formal sort. They are much interested in helping their fellow man, perhaps even strongly interested in the peace movement. They are good people and they generally have been influenced in good ways.
>
> There is a lot of intelligence out there. It is like the Tibetan legend of that place somewhere high in the Himalayas, sometimes called Shambhala, where there are great spirits who are waiting patiently for the human race to get to the point where they are ready to understand the wisdom that is available to them. And it will be a while before they even get to the point where they are able to start to comprehend what is possible.
>
> It is there. There is extra-terrestrial technology. There is extra-terrestrial fantastic medicine, sociology and psychology.
>
> How do you get along if you have got a population of twenty-five thousand? The

average extraterrestrial civilization is not that big. A hundred thousand people is a lot. And yet, they get along with each other without a police system. They do not have any tyrannical dictatorships. (They do not have democracy either.)

Many of them have a remarkable feeling of absolute love for each other. The feeling that they have for each other could only be compared with the rapture of a human being newly fallen in love. But, of course, it is not sexual.[254]

JAMES HARDER (FROM *EXTRA-TERRESTRIAL INTELLIGENCE* VIDEOTAPE, COURTESY THINKING ALLOWED PRODUCTIONS)

Jacques Vallee's Analysis

Jacques Vallee, a computer scientist, is one of the world's foremost, and most thoughtful, investigators of UFO phenomena. Vallee, who served as a model for the French scientist in the film *Close Encounters of the Third Kind,* is author of *Passport to Magonia, The Edge of Reality, Anatomy of a Phenomena, The Invisible College, Messengers of Deception,* and *Dimensions.* In a *Thinking Allowed* interview, he mused:

> I am going to be very disappointed if UFOs turn out to be nothing more than visitors from another planet. Because I think they could be something much more interesting. I think the UFO phenomena is teaching us that we do not understand time and space. There are objects — I think we have to call them objects — that leave traces on the ground and that cause effects on witnesses, both psychological and physiological. Yet, they appear to be capable of manipulating time and space in ways beyond our current scientific understanding.
>
> What we know today about the UFO phenomenon is considerably more than we knew twenty years ago or ten years ago. We have to understand it at three different levels.
>
> The first level is the physical level. We can say that UFOs at the physical level contain a lot of energy in a small space — emitted through light and pulsed microwave energy.
>
> The second level is what happens to witnesses. What would happen to you and me if we were close to that source of energy? Again, we are beginning to understand the physiological and the psychological correlations of a close encounter. Those have to do with a loss of a sense of space, loss of orientation, and loss of a sense of time. People think that only ten minutes went by when three hours transpired. Very often effects on the skin. Sunburns (such as in [the movie] *Close Encounters*). Effects on the eyes — ranging from conjunctivitis to temporary blindness.
>
> There have been cases where, after a UFO close encounter, the healing process has been sped up. There are known techniques now using electromagnetic radiation that will heal fractures, for example, or will heal superficial wounds on the skin. But nothing that would heal as fast as the reported effects. So we are beginning to understand the symptoms of the exposure to UFOs.

JACQUES VALLEE (FROM *IMPLICATIONS OF UFO PHENOMENA* VIDEOTAPE, COURTESY THINKING ALLOWED PRODUCTIONS)

Then there is the third level, the level I was addressing in *The Messengers of Deception*, which is the social level, the impact on belief systems. This is very difficult to convey to the believers in UFOs, believers in little green men from space. At this level, it does not matter whether or not UFOs are real. If people believe that something is real, then it is real in its effects.

People act according to their beliefs. Could the UFO phenomenon be manipulating us? Could it be a teaching system of some sort? Perhaps something that we are creating ourselves. Perhaps a series of energies that we are projecting? I think Carl Jung came very close to expressing that idea.

Or, could it be manipulated purposely by people who have the technology to simulate UFO sightings? People say, "Of course not, who would do a thing like that?" Well, I would remind you that during the Watergate investigation, it was discovered that there was a plan, originated in the White House to surface a submarine off the coast of Cuba, and paint the second coming of Christ over the island of Cuba using holograms. This is well within our technology today. The idea was that, since there is a large Catholic population in Cuba, they would be so upset by this vision that they would saturate the communication channels, the telephone system in Cuba, long enough for an invasion to take place.

In psychological warfare, that kind of manipulation is well understood. I have personally investigated several apparently genuine UFO cases where there was, in fact, a manipulation taking place. It was not a hoax on the part of the witnesses. But a hoax on the part of somebody much better organized.

There is another way of thinking about this. We are at a time of crisis on earth. We have the means of destroying the planet, which we have never had before in human history. It may be that there is a collective unconscious. Perhaps we are creating the visions we need to survive, in order to transcend this crisis. Perhaps there are no UFOs in a manufactured sense.[255]

Life within Death — Death within Life

Survival of Consciousness after Death

I have no final conclusions to offer about ghosts, spirits, reincarnation, or any other manifestation suggestive of survival after death. As in most areas of consciousness exploration, a final opinion is of less value than an appreciation of and tolerance for ambiguity — as well as a willingness to carefully explore the evidence and claims, cultures and contexts. For the great by-product of the search for the life beyond is an extraordinary enrichment of our understanding of the life within.

In the first edition of this book, as I was writing about this topic, I found it difficult in my excitement to keep from floating off into romantic speculation about the great beyond. The feeling was something like looking at a mountain peak in the distance, being stimulated by the excitement of its beauty and majesty, just by knowing it is there. What *The Roots of Consciousness* is about (I had to remind myself) is the continuum of existence between birth and death. It is something like the space between sleeping and waking. Moving from one to the other is really a gradual process. Each step takes us to the next. Living and dying, breathing, sleeping, dreaming, being, communicating — our consciousness touches all of these worlds. Yet, in the spirit of the early Wittgenstein (the twentieth century's great positivist philosopher), there is truly nothing meaningful I can say about existence beyond the veil of mortal life (if such there be). What I shall expound upon is the consciousness that we the living develop in relationship to the mystery of death. For we are all touched by its silence.

Ancient Egypt

In Egypt, life after death was thought to be a natural continuation of life on earth, and one senses another inner reality merging with their technology.

The shamanic practices of earlier tribes became incorporated into the organized Egyptian national priesthood. However, the Egyptian religion is no longer practiced, and the ancient language is largely lost. The evidence remaining regarding the different cults and their practices and theologies is still shrouded in impenetrable mystery. Yet, these people devoted themselves so intensely to their cult of the dead, it is easy to imagine that many of their seer-priests actually did see the spirits of their departed.

The following testament comes from the pyramid of Unas, a Fifth Dynasty king:

> The heavens drop water, the stars throb, the archers go round about, the bones of Akeru tremble, and those who are in bondage to them take to flight when they see Unas rise up as a soul, in the form of the god who liveth upon his fathers and who maketh food of his mothers. Unas is the lord of wisdom and his mother knoweth not his name....The *kas* of Unas are behind him, the sole of his foot is beneath his feet, his gods are over him, his uraei are upon his brow, the serpent guides of Unas are in front of him and the spirit of the flame looketh upon his soul. The powers of Unas protect him: Unas is a bull in heaven, he directeth his steps where he will, he liveth upon the form which each god taketh upon himself, and he eateth the flesh of those who come to fill their bellies with the magical charms in the Lake of Fire.[256]

As one might infer from the above passage, the Egyptian version of existence in the afterworld is somewhat obscure. In certain instances it seems that the Egyptians actually believed in a physical existence after death for which the departed required worldly riches and sustenance. Other descriptions are embodied in the realm of mythology so that the nature of the afterlife is deeply symbolized in the godforms themselves.

The Egyptian concept of spiritual resurrection after death has a mythological basis in the story of *Osiris* — the lord of creation who was also a king of Egypt. At the height of his reign, *Osiris* is murdered by his jealous enemy *Set*. His body is enclosed in a chest that is placed at the mouth of the Nile, but eventually recovered by *Osiris'* wife, *Isis*. *Set*, however, finds it once more and dismembers it into fourteen pieces he scatters throughout the land.

Isis searches for the pieces of *Osiris'* body and finds all of them except the phallus. At this point, *Horus*, the sun, the hawk god, appears on the scene. As the god of the sun, he always existed. In fact, the hawk is probably the first living thing worshipped by the Egyptians, yet he is conceived by *Isis* from the dismembered body of *Osiris* lacking the progenitive organ! The appearance of *Horus* represents the resurrection of *Osiris*. Filled with his father's spirit, he defeats *Set* in battle.

The battle between *Horus* and *Set* is reenacted each day. You can watch this epic drama as it unfolds in the sky by getting up several hours before dawn and silently observing the sunrise.

The ancient Egyptians

OSIRIS, KING OF THE UNDERWORLD

believed that as the hawk arose from the dismembered body of *Osiris*, so would their awareness survive the bodily death.[257] However, they also seemed to believe that proper funeral rites were necessary to attain afterlife in heaven. It seems that we are dealing with popular superstitions mixed with higher philosophy and occult practice. In many ways the Egyptians seem to describe the afterlife as being quite physical and sensory — but perhaps this is a description of how real it was for them. This physical imagery is, perhaps, a poetical metaphor for the images the ancient seers had of the afterlife.

The Tibetan Book of the Dead

Buddhism, which originated in India, reached a height of consciousness exploration in Tibet. All states of existence for the Tibetan Buddhists other than pure Nirvana are reflections of the limited illusion of self-consciousness. *The Tibetan Book of the Dead* is a major document within this tradition; it describes the passage of consciousness from death to rebirth. Existence is divided into six *bardos*, three of which are experienced from birth to death and three of which occur from death to rebirth. Yet, Buddhist philosophy teaches that birth and death are not phenomena that occur only once in a human life; they are part of an uninterrupted process. Every instant something within us dies and something is reborn. The different bardos represent the different aspects of this process in our own lives. All of these states are in flux.[258]

chikhai bardo: The experience of the primary clear light and the secondary clear light at the moment of death.

chonyd bardo: The state of psychic consciousness. Experiencing lights, sounds and rays. Seeing the peaceful deities and then the wrathful deities.

sidpa bardo: Visions of the world into which one's karma leads one to be born. Visions of males and females in sexual union. Feelings of attachment and repulsion. Choosing and entering the womb.

bsam-gtan bardo: The dream state.

skye-gnas bardo: The everyday waking consciousness of being born into the human world.

These bardo states refer to the mental processes of the soul during the periods of life and rebirth. Outside of one's own consciousness there still remains another reality to be explored. Entrance to this reality is attained by recognizing at any point that the images and apparitions of the bardo state are merely the projections of one's own consciousness.

With every thought of fear or terror or awe
for all apparitional appearances set aside
May I recognize whatever visions appear as
reflections of mine own consciousness
May I not fear the bands of peaceful and
wrathful deities, mine own thoughtforms.[259]

In the *sidpa bardo*, before rebirth, there occurs a judgment of the good and bad deeds of the soul of the dead.

> If thou neither prayest nor knowest how to meditate upon the Great Symbol nor upon any tutelary deity, the Good Genius, who was born simultaneously with thee, will come now and count out thy good deeds with white pebbles, and the Evil Genius, who was born simultaneously with thee, will come and count out thy evil deeds with black pebbles. Thereupon, thou wilt be greatly frightened, awed, and terrified, and wilt tremble; and thou wilt, attempt to tell lies, saying "I have not committed any evil deed."
>
> Then the Lord of Death will say, "I will consult the Mirror of Karma."
>
> So saying, he will look in the Mirror, wherein every good and evil act is vividly reflected. Lying will be of no avail.[260]

This situation very closely parallels the weighing of the heart described in *The Egyptian Book of the Dead* as well as the judgments appearing in later Greek and Christian traditions. It seems to reflect an

Judgment scene from *The Egyptian Book of the Dead*. The heart of the deceased is being weighed against a feather

TIBETAN JUDGMENT OF THE SOUL AFTER DEATH. GOOD AND BAD DEEDS ARE WEIGHED AGAINST EACH OTHER

archetypal reality that permeates the deep consciousness of many cultures, perhaps the ultimate symbol of meaning and order in the universe. Modern research is still attempting to investigate this reality, though perhaps with less sophistication than the Tibetans who claim to have developed ways to communicate with the departed spirit after death in order to aid in its passing through the *bardo* states.

Interestingly enough, *The Tibetan Book of the Dead* has been used as a "constant companion" by Carl Jung whose psychological theories tend to unite scientific logic with mysticism. It was also used by Timothy Leary, Richard Alpert and Ralph Metzner as the basis for understanding the nature of the *psychedelic* experience.[261]

The *bardo* states are accessible to everyone. Yet there is a subtle process involved in attaining all higher states of consciousness — a kind of trying not to try — which is embodied most clearly in the life of Tibet's great saint, Milarepa, who never uttered a word that did not come out in song and poetry.[262] This delicacy of living is manifest also in Chinese Taoism.

Every culture has a set of beliefs regarding the afterlife. Some of the most interesting are those developed by scientists themselves.

The Visions of Gustav Theodore Fechner

The great psychophysicist Gustav Theodore Fechner — one of the foremost instigators of modern, scientific psychology — wrote extensive speculations in a work called the *Book of Life After Death* which G. Stanley Hall describes thus:

> *How* now do the dead live on? First and chiefly in us. Fechner takes his leading concept from the mystic way in which Christ lives in his followers, who are members of his body and branches of his vine. To this larger life of his in the Church, his earthly career is only a grain of mustard seed. Gloriously his soul has gone marching on. Just so the dead press in upon us, yearning to add their strength to ours, for thus they not merely live, but grow. New impulsions and sudden insights in us are inspirations from them. Not only do the great and good dead influence and pervade us all the time, but we are exposed also to the bad. Many of them are always bad, and so if our will is weak and our personality unorganized, they may dominate us. Their visitation is insistent. They do not crave incarnation in the flesh, like Plato's spirits, but in our moral life, that therein they may be made perfect. We all have in us sparks from the lives of Luther, Goethe, Napoleon, etc., who think and act in us "no longer restrained by the limitations of the body, but poured forth upon the world which in their lifetime they moulded, gladdened, swayed, and by their personality they now supply us with influences which we never discern as coming from them." Each great dead soul extends itself into many and unites them in a spiritual organism. Thus, the dead converse with each other in us. They also fight the good and bad in each other in us, causing strife in our souls....
>
> There is, however, a higher soul in which we and all things live, move, and have our being, and in which and only in which spirits are real. We are, in fact, what we have become. The brain is a kind of seed which decays that the soul may live. The individual soul may mount on the collective souls of the dead as a sparrow is carried up on an eagle's back to heights it never could attain, but, when there, can fly off and even a little higher. At death the soul seems to drop below a threshold and the spark of consciousness might be conceived to go out but for the fact that the soul is not projected into an empty world but into one where it incessantly meets varying resistances that keep personality above the point of submergence or any other extinction without appeal to the conservation of energy. Just as attention moves about from point to point within the body, so after death the soul moves around the world.[263]

When the phenomena of spiritualism became popular in mid-nineteenth century Europe and America, Fechner sat with zeal at a number of séances. He was one of the few men of his age who, while not detecting trickery, had the depth of wisdom with which to incorporate but also transcend the sensationalism and trivia of the popular spiritualist impulse.

Spiritualism

Spiritualism as a social movement apparently began in the small New York town of Hydesville in March of 1848, where several months earlier, the Fox family had taken over an old farmhouse about which the previous tenants had complained of strange noises. The Foxes themselves soon noticed unusual rapping sounds that occurred in the night frightening the two younger daughters, Margaret and Kate, who then insisted on sleeping with their parents. On the fateful evening of March 31, the youngest daughter Kate playfully challenged the raps to repeat the snapping of her fingers. Her challenge was answered. Within hours many of the neighbors were brought over to the house to witness the uncanny demonstration.

By asking that the sounds be repeated twice for a negative answer and only once for an affirmative, the people assembled were soon able to carry on a dialogue with the rapping — which had revealed itself to apparently be coming from a spirit source. One of the neighbors, Deusler, suggested naming the letters of the alphabet and having the spirit rap when they reached certain letters in order to spell out letters and sentences. In this way the spirit revealed himself to have been a traveling peddler who was murdered in the house by a previous owner and buried in the cellar. Digging commenced immediately; however, a high water level prevented any immediate discoveries.

Meanwhile, hundreds of neighbors continued visiting the Fox house, day and night, listening to the spirit's rapping. They also formed an investigation committee to take testimony. The case was even studied by the Hon. Robert Dale Owen, a member of the U.S. Congress, and a founder of the Smithsonian Institution. In the summer of 1848, more digging unearthed human teeth, some fragments of bone and some human hair.

While the testimony is ambiguous, some neighbors reported that the raps continued in the Fox house even when family members were not present. However, it became apparent that this form of mediumship centered on the Fox sisters, though it soon spread to many other people as well.

Fifty-six years later in 1904, the gradual disintegration of one of the cellar walls of the Fox house exposed to view an entire human skeleton.

Other mediums, using the alphabet method, also claimed to be in contact with the spirits of the deceased. Their messages were generally not reliable, however. It seems that for every apparently genuine medium there were many deluded or phony imitators.

THE HOUSE IN WHICH AMERICAN SPIRITUALISM WAS BORN, HYDESVILLE, NEW YORK

MARGARETTA FOX

While receiving the sympathetic attention of Horace Greeley, the editor of the New York *Tribune*, who later became a candidate for the U.S. presidency, the sisters remained a center of controversy. In 1871, Charles F. Livermore, a prominent New York banker, sent Kate Fox to England in gratitude for the consolation he had received through her powers.[104] At that time she was examined by the physicist Sir William Crookes, who later received the Nobel prize for his discovery of thalium:

> For several months I have enjoyed the almost unlimited opportunity of testing the various phenomena occurring in the presence of this lady, and I especially examined the phenomena of these sounds. With mediums, generally, it is necessary to sit for a formal séance before anything is heard; but in the case of Miss Fox it seems only necessary for her to place her hand on any substance for loud thuds to be heard in it, like a triple pulsation, sometimes loud enough to be heard several rooms off. In this manner I have heard them in a living tree — on a sheet of glass — on a stretched iron wire — on a stretched membrane — a tambourine — on the roof of a cab — and

LEAH FOX

During November 1849, the Spiritualists held their first public meeting in the largest hall available in Rochester, New York. Three different citizens' committees in Rochester were invited to investigate the Fox sisters. All three made favorable reports indicating that the sounds heard were not produced by ventriloquism or machinery. The public was outraged at these reports. A riot resulted and the girls had to be smuggled away from an angry crowd.

The Fox sisters made a career of their mediumship. They toured the country under the auspices of the showman, P. T. Barnum.

KATE FOX

on the floor of a theatre. Moreover, actual contact is not always necessary; I have heard these sounds proceeding from the floor, walls, etc., when the medium's hands and feet were held — when she was standing on a chair — when she was suspended in a swing from the ceiling — when she was enclosed in a wire cage — and when she had fallen fainting on a sofa. I have heard them on a glass hermonicon — I have felt them on my own shoulder and under my own hands. I have heard them on a sheet of paper, held between the fingers by a piece of thread passed through one corner. With a full knowledge of the numerous theories which have been started, chiefly in America, to explain these sounds, I have tested them in every way that I could devise, until there has been no escape from the conviction that they were true objective occurrences not produced by trickery or mechanical means.[264]

The main argument used by skeptics to discredit the Fox sisters was that they created the rapping sounds themselves by cracking the bones in their toes and knuckles. This hypothesis, however, does not seem sufficient to explain the different kinds of sounds that appeared, their loudness, the fact that they often occurred in arpeggios and cadenzas, and the fact that they seemed to emanate from different places.

Nevertheless, in 1888, Margaret Fox made a public statement denouncing the spiritualists, claiming that she had made the noises by cracking her toes. Kate, who was with her at the time remained silent, as if in agreement. The following year, however, Margaret recanted, saying that she had fallen under the influence of people who were inimical to spiritualism and who had offered her money. Both sisters were alcoholics at this time. At no time in their careers were they actually detected in a fraudulent act.

Crookes also recorded an experience of *direct writing* with Ms. Fox:

A luminous hand came down from the upper part of the room, and after hovering near me for a few seconds, took the pencil from my hand, rapidly wrote on a sheet of paper, threw the pencil down, and then rose up over our heads, gradually fading into darkness.[265]

The Spiritism of Allan Kardec

From the ranks of the spiritualists, investigations were also conducted, although along somewhat different lines than that of the scientists. The former attempted to describe the world according to the teachings of the spirits themselves. This theorization of spiritualism was mainly due to L. H. D. Rivail (1803-1869), a doctor of medicine who became celebrated under the pseudonym Allan Kardec.

Kardec's theories were simple enough: After death the soul becomes a spirit and seeks reincarnation, which, as Pythagoras taught, is the destiny of all human souls; spirits know the past, present, and future; sometimes they can materialize and act on matter. We should let ourselves be guided by good spirits, Kardec maintained, and refuse to listen to bad spirits.[266,267]

Kardec wrote many books which achieved enormous popularity in his own lifetime. His works also spread to Brazil, where he still has a huge following, and where postage stamps were recently issued in his honor.[268] His intellectual energy certainly deserves admiration. However, he built his theory on the untenable hypothesis that mediums, embodying a so-called spirit, are never mistaken, unless their utterances are prompted by evil spirits. This notion does not of course, take into account the possibilities of suggestion, multiple personality, or unconscious influences which were quickly developed as alternative hypotheses to outright fraud by skeptical scientific investigators such as Michael Faraday.

ALLAN KARDEC

Founding of the Society for Psychical Research

Sir William F. Barrett, a professor of physics at the Royal College of Science in Dublin, had been conducting experiments in the 1880s testing the notion of *thought-transference.* Barrett conceived of the idea of forming an organization of spiritualists, scientists, and scholars who would join forces in a dispassionate investigation of psychical phenomena. F. W. H. Myers, Edmund Gurney, and Henry Sidgewick attended a conference in London that Barrett convened, and the Society for Psychical Research (SPR) was created with Sidgewick, who had a reputation as an impartial scholar, accepting the first presidency.

The society set up six working committees, each with a specific domain for exploration:

> 1. An examination of the nature and extent of any influence which may be exerted by one mind upon another, apart from any generally recognized mode of perception.
>
> 2. The study of hypnotism, and the forms of so-called mesmeric trance, with its alleged insensibility to pain; clairvoyance and other allied phenomena.
>
> 3. A critical revision of Reichenbach's researches with certain organizations called "sensitive," and an inquiry whether such organizations possess any power of perception beyond a highly exalted sensibility of the recognized sensory organs.
>
> 4. A careful investigation of any reports, resting on strong testimony, regarding apparitions at the moment of death, or otherwise, or regarding disturbances in houses reputed to be haunted.
>
> 5. An inquiry into the various physical phenomena commonly called spiritualistic; with an attempt to discover their causes and general laws.
>
> 6. The collection and collation of existing materials bearing on the history of these subjects.[269]

The great American psychologist, William James, met Gurney in England in 1882 and immediately they struck up a close friendship. Later James also became a close friend of Myers. In 1884, Barrett toured the United States and succeeded in arousing the interest of American scholars in forming a similar society, which was established in 1885, and in which William James took an active role. The American Society for Psychical Research constituted the first organized effort for experimental psychological research in the United States. For a period of many years, before the ascendency of the German experimental approach of Wilhelm Wundt, psychology in the United States was equated with the efforts of psychical research.

The evidence for life beyond death comes from several sources. There are cases of hauntings and apparitions, mediumistic communications and automatic writings, possessions, incidences of child prodigies, and ostensible reincarnation data. A large amount of this evidence was gathered during the heyday of spiritualism in the nineteenth century, and is recorded in *The Human Personality and Its Survival of Bodily Death* by F. W. H. Myers. Modern research places a much greater emphasis on laboratory studies and is predicated upon a much different approach to the evidence.

Many levels of the human personality clearly exist and are still generally unexplored and untapped. Yet a number of cases can be cited where even such explanations do not account for all the observed phenomena.

Human Personality and Its Survival of Bodily Death

Many other phenomena were explored by the SPR during its early years. The major attempt to synthesize the great mass of data which had been gathered was undertaken by Frederick Myers and published in 1903 after his death in a work called *The Human Personality and Its Survival of Bodily Death.* Myers was widely read in all the fields of knowledge of his day. His work is a testimony to his wide-reaching and poetical mind and his deep interest in the work of the psychoanalysts. He was, in fact, the first writer to introduce the works of Freud to the British public, in 1893. His book is still regarded by many as the most important single work in the history of psychical research. Even those who do not accept his hypothesis of the survival of the soul are indebted to his explorations of the unconscious or *subliminal* regions of personality.

Myers maintained that the human personality was composed of two active coherent streams of thoughts and feelings. Those lying above the ordinary threshold of consciousness were considered *supraliminal,* while those that remain submerged beneath consciousness are *subliminal.* The evidence for the existence of this *subliminal* self derives from such phenomena as automatic writing, multiple personalities, dreams, and hypnosis. These phenomena all expose deeper layers of the personality that normally remain unseen. In many cases these layers seem autonomous and independent of the *supraliminal* self. For example, certain memories uncovered through hypnosis and dreams are normally inaccessible to the conscious mind. Or in the case of certain people of genius, complete works of art will emerge from dreams. Automatic writers can sometimes maintain two conversations at once, each unaware of the other one.

Myers examined all of these phenomena carefully and felt that they were part of a continuum ranging from unusual personality manifestations to telepathic communications, traveling clairvoyance, possession by spirits, and actual survival of the subliminal layers of personality after the death of the body. He felt that each experience in this spectrum was integrally related to the other states of being. This insight was his deepest theoretical penetration into the roots of consciousness.

Myers began his analysis by looking at the ways in which the personality was known to disintegrate. Insistent ideas, obsessing thoughts and forgotten terrors lead up to hysterical neuroses in which the subliminal mind takes over certain body functions from the supraliminal. Gradually these maladies merge with cases of multiple personalities. He noted that the subliminal personalities often represented an improvement over the normal conscious self, and suggested that,

> As the hysteric stands in relation to ordinary men, so do we ordinary men stand in relation to a not impossible ideal of sanity and integration.[270]

Thus from the disintegrated personality which reveals some of the negative aspects of the subliminal self, Myers moved naturally to look at people of genius, within whom,

F. W. H. MYERS

according to Myers, "some rivulet is drawn into supraliminal life from the undercurrent stream." He discussed mathematical prodigies and musicians whose works spring fully formed into their consciousness. Of particular interest was Robert Louis Stevenson who deliberately used his dream life in order to experiment with different dramatizations of his stories. Not mentioned by Myers, but certainly applicable here, would be the incredible inventions which entered the minds of Thomas Edison (himself a spiritualist) and Nicola Tesla, whose genius led him to develop alternating current and many modern electrical appliances. Myers did cite the poet Wordsworth as being particularly sensitive to this aspect of the creative process, which he described in "The Prelude, or Growth of a Poet's Mind":

> That awful power rose from the mind's abyss,
> Like an unfathomed vapor that enwraps,
> At once, some lonely traveller. I was lost;
> Halted without an effort to break through;
> But to my conscious soul I now can say —
> "I recognize thy glory;" in such strength
> Of usurpation, when the light of sense
> Goes out, but with a flash that has revealed
> The invisible world, doth greatness make
> abode.[271]

In addition to people of genius, Myers included saintly men and women whose lives have absorbed "strength and grace from an accessible and inexhaustible source."

From neurosis, genius and sainthood Myers moved to a state of being all individuals experience — sleep, which he describes as the abeyance of the supraliminal life and the liberation of the subliminal. The powers of visualization, for instance, are heightened during the *hypnogogic* state as one passes into sleep and in the *hypnopompic* state as the dream lingers into waking consciousness. Myers also discerned the heightened powers of memory and reason that occur in some dreams, and further cases of clairvoyance and telepathy in dreams. And he cited cases of what seem to be "psychical invasions" in dreams by spirits of both living and departed persons. He concluded by suggesting that sleep is every person's gate to the "spiritual world."

Hypnosis was described as the experimental exploration of the sleep phase of human personality. The unusual phenomena that occur in hypnosis were ascribed to the power of the *subliminal* self that is appealed to in such states. The *subliminal* self appears to enjoy greater control over the body than the *supraliminal*. Myers also pointed out the relationship of hypnosis to other phenomena such as faith healing, the miraculous cures at Lourdes, and the use of magical charms. He emphasized the experimental work done in telepathic hypnotic induction at a distance as well as telepathy, clairvoyance and precognition observed in the hypnotized subject.

From hypnosis Myers moved to visual and auditory hallucinations psychical researchers have labeled sensory automatisms. When hearing a sound or seeing a color or form carries with it an association of images from another sense, this process is within the brain and is termed *entencephalic.* The stages leading from such percepts to ordinary vision include entoptic impressions due to stimuli from the optic nerve or eye and after-images which are formed in the retina. Stages leading further inward from *entencephalic* vision include memory images, dreams, images of the imagination and hallucinations. Many hallucinations cited were shown to contain information that was later verified. Other hallucinations clearly seemed to hold positive benefits for the personality and were not associated in any way with disease. Crystal gazing is a possible positive use of the mind's ability to hallucinate. Other hallucinations include the phantasms of the living and the dead which we have already discussed.

From *sensory automatisms* Myers moved to *motor automatisms* — including automatic writing and speaking in tongues. Most of these phenomena can be attributed to the subliminal mind within the automatist's own brain. Other cases lead one to suspect telepathy and possible communication from deceased spirits. There are cases of automatic writing, for example, in which the handwriting of a deceased person is alleged. A further development of this would be possession by another personality other than the subliminal self. However, it is very difficult to distinguish cases of spirit possession from cases of multiple personality. The personal identity of such a spirit must be clearly distinguished by its memory and its character. Yet this is a phenomena common to all religious traditions that has also been observed at least once, Myers felt, by SPR researchers. He noted that such possession did not appear to have an injurious effect on the medium.

It is on the basis of this continuum of experiences that Myers asserted that the *subliminal* self is able to operate free from the brain in ways that modify both space and time as they appear to the *supraliminal* self. Just as the *subliminal* self is able to control physiological functions of the brain and body, as best exemplified through hypnotic experiments, so is it able to exert force on other physical objects accounting for levitations, materializations, spirit rapping, etc.

The Watseka Wonder

The case of the *Watseka Wonder* is listed by Myers as an incidence of multiple personality strongly suggesting the spiritualist hypothesis. It was originally published in the *Religio-Philosophical Journal* in 1879 and later in pamphlet form with the title "The Watseka Wonder," by E. W. Stevens. The editor of the journal, highly regarded as a skillful and honest investigator by Myers, spoke highly of Dr. Stevens and claimed to have taken great pains to "obtain full corroboration of the astounding facts" from competent witnesses. The case briefly is the alleged possession of thirteen-year-old Lurancy Vennum by the spirit of Mary Roff, a neighbor's daughter who had died at the age of eighteen when Lurancy was a child of about fifteen months.

Myers quotes Dr. Stevens, with his abridgements in square brackets:

> [Mary Lurancy Vennum, the "Watseka Wonder," was born April 16th, 1864, in Milford township, about seven miles from Watseka, Illinois. The family moved to Iowa in July 1864 (when Lurancy was about three months old), and returned to within eight miles from Watseka in October 1865 (three months after the death of Mary Roff). Lurancy was then about a year and a half old. After two other moves in the neighbourhood, the family moved into

Watseka on April 1st, 1871], locating about forty rods from the residence of A. B. Roff. They remained at this place during the summer. The only acquaintance ever had between the two families during the season was simply one brief call of Mrs. Roff, for a few minutes, on Mrs. Vennum, which call was never returned, and a formal speaking acquaintance between the two gentlemen. Since 1871 the Vennum family have lived entirely away from the vicinity of Mr. Roff's, and never nearer than now, on extreme opposite limits of the city.

"Rancy," as she is familiarly called, had never been sick, save a light run of measles in 1873.

[On July 11th, 1877, she had a sort of fit, and was unconscious for five hours. Next day the fit recurred, but while lying as if dead she described her sensations to her family, declaring that she could see heaven and the angels, and a little brother and sister and others who had died. The fits or trances, occasionally passing into ecstasy, when she claimed to be in heaven, occurred several times a day up to the end of January 1878; she was generally believed to be insane, and most friends of the family urged that she should be sent to an insane asylum.

At this stage Mr. and Mrs. Asa B. Roff, whose daughter, Mary Roff, as we shall see, had had periods of insanity, persuaded Mr. Vennum to allow him to bring Dr. E. W. Stevens of Janesville, Wisconsin, to investigate the case.]

On the afternoon of January 31st, 1878, the two gentlemen repaired to Mr. Vennum's residence, a little out of the city. Dr. Stevens, an entire stranger to the family, was introduced by Mr. Roff at four o'clock P.M.; no other persons present but the family. The girl sat near the stove, in a common chair, her elbows on her knees, her hands under her chin, feet curled up on the chair, eyes staring, looking every like an "old hag."...She refuses to be touched, even to shake hands, and was reticent and sullen with all save the doctor, with whom she entered freely into conversation giving her reasons for doing so; she said he was a spiritual doctor, and would understand her.

[She described herself first as an old woman named Katrina Hogan, and then as a young man named Willie Canning, and after some insane conversation had another fit, which Dr. Stevens relieved by hypnotizing her. She then became calm, and said that she had been controlled by evil spirits. Dr. Stevens suggested that she should try to have a better control, and encouraged her to try and find one. She then mentioned the names of several deceased persons, saying there was one who wanted to come, named Mary Roff.

Mr. Roff being present, said: "That is my daughter; Mary Roff is my girl. Why, she has been in heaven twelve years. Yes, let her come, we'll be glad to have her come." Mr. Roff assured Lurancy that Mary was good and intelligent, and would help her all she could; stating further that Mary used to be subject to conditions like herself. Lurancy, after due deliberation and counsel with spirits, said that Mary would take the place of the former wild and unreasonable influence. Mr. Roff said to her, "Have your mother bring you to my house, and Mary will be likely to come along, and a mutual benefit may be derived from our former experience with Mary."

[On the following morning, Friday, February 1st, Mr. Vennum called at the office of Mr. Roff and informed him that the girl claimed to be Mary Roff, and wanted to go home. He said, "She seems like a child real homesick, wanting to see her pa and ma and her brothers."

Mary Roff was born in Indiana in October 1846. . . . Mary had had fits frequently from the age of six months, which gradually increased in violence. She had also had periods of despondency, in one of which, in July 1864, she cut her arm with a knife until she fainted. Five days of raving mania followed, after which she recognized no one, and seemed to lose all her natural senses, but when blindfolded could read and do everything as if she saw. After a few days she returned to her normal condition, but the fits became still worse, and she died in one of them in July 1865. Her mysterious illness had made her notorious in the neighbourhood during her life-time, and her alleged clairvoyant powers are said to have been carefully investigated "by all the prominent citizens of Watseka," including newspaper editors and clergymen.

It was in February 1878 that her supposed "control" of Lurancy began. The girl then became "mild, docile, polite, and timid, knowing none of the family, but constantly pleading to go home," and "only found contentment in going back to heaven, as she said, for short visits."]

About a week after she took control of the body, Mrs. A. B. Roff and her daughter, Mrs. Minerva Alter, Mary's sister, hearing of the remarkable change, went to see the girl. As they came in sight, far down the street, Mary, looking out of the window, exclaimed exultantly, "There comes my ma and sister Nervie!" — the name by which Mary used to call Mrs. Alter in girlhood. As they came into the house she caught them around their necks, wept and cried for joy, and seemed so happy to meet them. From this time on she seemed more homesick than before. At times she seemed almost frantic to go home.

On the 11th day of February, 1878, they sent the girl to Mr. Roff's, where she met her "pa and ma," and each member of the family, with the most gratifying expressions of love and affection, by words and embraces. On being asked how long she would stay, she said, "The angels will let me stay till some time in May;". . . .

The girl now in her new home seemed perfectly happy and content, knowing every person and everything that Mary knew when in her original body, twelve to twenty-five years ago, recognizing and calling by name those who were friends and neighbours of the family from 1852 to 1865, when Mary died, calling attention to scores, yes, hundreds of incidents that transpired during her natural life. During all the period of her sojourn at Mr. Roff's she had no knowledge of, and did not recognize any of Mr. Vennum's family, their friends or neighbours, yet Mr. and Mrs. Vennum and their children visited her and Mr. Roff's people, she being introduced to them as to any strangers. After frequent visits, and hearing them often and favourably spoken of, she learned to love them as acquaintances, and visited them with Mrs. Roff three times.

One day she met an old friend and neighbour of Mr. Roff's, who was a widow when Mary was a girl at home. Some years since the lady married a Mr. Wagoner, with whom she yet lives. But when she met Mrs. Wagoner she clasped her around the neck and said, "O Mary Lord, you look so very natural, and have changed the least of any one I have seen since I came back." Mrs. Lord was in some way related to the Vennum family, and lived close by them, but Mary could only call her by the name by which she knew her fifteen years ago, and could not seem to realize that she was married. Mrs. Lord lived just across the street from Mr. Roff's for several years, prior and up to within a few months of Mary's death; both being members of the same Methodist church, they were very intimate. . . .

One evening, in the latter part of March, Mr. Roff was sitting in the room waiting for tea, and reading the paper, Mary being out in the yard. He asked Mrs. Roff if she could find a certain velvet head-dress that Mary used to wear the last year before she died. If so, to lay it on the stand and say nothing about it, to see if Mary would recognize it. Mrs. Roff readily found and laid it on the stand. The girl soon came in, and immediately exclaimed as she approached the stand, "Oh, there is my head-dress I wore when my hair was short!" She then asked, "Ma, where is my box of letters? Have you got them yet?" Mrs. Roff replied, "Yes, Mary, I have some of them." She at once got the box with many letters in it. As Mary began to examine them she said, "Oh, ma, here is a collar I tatted! Ma, why did you not show to me my letters and things before?" The collar had been preserved among the relics of the lamented child as one of the beautiful things her fingers had wrought before Lurancy was born, and so Mary continually recognized every little thing and remembered every little incident of her girlhood. . . .

In conversation with the writer about her former life, she spoke of cutting her arm as hereinbefore stated, and asked if he ever saw where she did it. On receiving a negative answer, she proceeded to slip up her sleeve as if to exhibit the scar, but suddenly arrested the movement, as if by a sudden thought, and quickly said, "Oh, this is not the arm; that one is in the ground," and proceeded to tell where it was buried, and how she saw it done, and who stood around, how they felt, &c., but she did not feel bad. I heard her tell Mr. Roff and the friends present, how she wrote to him a message some years ago through the hand of a medium, giving name, time, and place. Also of rapping and of spelling out a message by another medium, giving time, name, place, &c., &c. which the parents admitted to be all true. . . .

During her stay at Mr. Roff's her physical condition continually improved, being under the care and treatment of her supposed parents and the advice and help of her physician. She was ever obedient to the government and rules of the family, like a careful and wise child, always keeping in the company of some of the family, unless to go in to the nearest neighbours across the street. She was often invited and went with Mrs. Roff to visit the first families of the city, who soon became satisfied that the girl was not crazy, but a fine, well-mannered child.

As the time drew near for the restoration of Lurancy to her parents and home, Mary would sometimes seem to recede into the memory and manner of Lurancy for a little time, yet not enough to lose her identity or permit the manifestation of Lurancy's mind, but enough to show she was impressing her presence upon her own body.

[On May 19th, in the presence of Henry Vennum, Lurancy's brother, Mary left control for a time, and "Lurancy took full possession of her own body recognizing Henry as her brother. The change of control occurred again when Mrs. Vennum came to see her the same day.]

On the morning of May 21st Mr. Roff writes as follows: —

"Mary is to leave the body of Rancy to-day, about eleven o'clock, so she says. She is bidding neighbours and friends good-bye. Rancy to return home all right to-day. Mary came from her room upstairs, where she was sleeping with Lottie, at ten o'clock last night, lay down by us, hugged and kissed us, and cried because she must bid us good-bye, telling us to give all her pictures, marbles, and cards, and twenty-five cents Mrs. Vennum had given her to Rancy, and had us promise to visit Rancy often."

[Mary arranged that her sister, Mrs. Alter, should come to the house to say good-bye to her, and that when Lurancy came at eleven o'clock she should take her to Mr. Roff's office, and he would go to Mr. Vennum's with her. There was some alternation of the control on the way, but the final return of the normal Lurancy Vennum took place before they reached Mr. Roff's office, and on arriving at her own home she recognized all the members of her own family as such, and was perfectly well and happy in her own surroundings. A few days later, on meeting Dr. Stevens, under whose care she had been at Mr. Roff's house, she had to be introduced to him as an entire stranger, and treated him as such. The next day she came to him spontaneously, saying Mary Roff had told her to come and meet him, and had made her feel he had been a very kind friend to her, and she gave him a long message purporting to be from Mary.[272]

In 1890, Richard Hodgson visited Watseka and interviewed many of the principal witnesses of this case. Their testimony was in agreement with Dr. Stevens' presentation. However, Hodgson was unable to get in touch with Lurancy Vennum herself. He draws the following conclusions to the case:

I have no doubt that the incidents occurred substantially as described in the narrative by Dr. Stevens, and in my view

> the only other interpretation of the case — besides the spiritistic — that seems at all plausible is that which has been put forward as the alternative to the spiritistic theory to account for the trance-communications of Mrs. Piper and similar cases, viz., secondary personality with supernormal powers. It would be difficult to disprove this hypothesis in the case of the Watseka Wonder, owing to the comparative meagreness of the record and the probable abundance of "suggestion" in the environment, and any conclusion that we may reach would probably be determined largely by our convictions concerning other cases. My personal opinion is that the "Watseka Wonder" case belongs in the main manifestations to the spiritistic category.[273]

Apparitions and Hauntings

Working as honorary secretary of the SPR and active on the literary committee, Edmund Gurney soon discovered that the largest single class of occurrences reported were what came to be labelled *crisis apparitions.* These occur when the figure or the voice of a living person who is experiencing a crisis — such as an accident or a death — is seen or heard.

Probably you or your friends have had such experiences, that are strangely confirmed by the news, later on, of the actual crisis.

Within one year of its organization, the SPR had collected more than four hundred reports of such cases and in 1886, Gurney published a thirteen-hundred-page document entitled *Phantasms of the Living* in which 702 different apparition cases were analyzed. All of the evidence was obtained firsthand from the percipients and was generally backed by corroboratory testimony. Witnesses were also interviewed by SPR members who appraised the value of all testimony.

Gurney described several categories of apparition cases. These are cases of spontaneous telepathy, which occur when the sender is undergoing some shock or strong emotion. For example, a lady lying in bed may feel a pain in her mouth at the exact moment when her husband is accidently struck in the jaw. Then come cases where the percipient's experience is not an exact reproduction of the agent's experience, but is only founded upon it, the receiver building a detailed picture from his or her own mind. There are many cases of this type where a person about to arrive at a location is actually seen there by someone not expecting him before his arrival. It is very unlikely that the agent will have in his mind the image of himself as others see him. Finally, Gurney refers to the cases in which the agent may be dead or dying while the phantom appears in quite normal behavior and clothing.

Gurney felt that these cases could be explained as hallucinations induced in the mind of the percipient by means of a telepathic message from the agent. What was harder to explain were collective apparitions in which several people independently perceive the identical phantom. There were also reciprocal cases whereby a person imagining himself to be at a distant scene is actually seen at that location by others.

Phantasms of the Living was soon criticized by the eminent American philosopher C. S. Pierce and several others on the grounds that the cases reported did not meet sufficient conditions to be acceptable as evidence. Most of these critical individuals simply did not read the entire book. Their criticisms focused on the weakest cases and overlooked certain cases that were very well documented in all regards. However, Gurney felt that if only a few single cases were strongly evidential, the conclusions for crisis telepathy were inescapable. He stressed the extent to which the skeptical arguments would have to be pushed in order to dismiss the entire bundle of data:

> Not only have we to assume such an extent of forgetfulness and inaccuracy, about simple and striking facts of the immediate past, as is totally unexampled in any other range of experience. Not only have we to assume that distressing or exciting news about another person produces a havoc in the memory that has never been noted in connection with stress or excitement in any other form. We must leave this merely general ground, and make suppositions as detailed as the evidence itself. We must suppose that some people have a way of dating their letters in indifference to the calendar, or making entries in their diaries on the wrong page and never discovering the error; and that whole families have been struck by the collective hallucination that one of their members had made a particular remark, the substance of which had never entered that members head; and that it is a recognized custom to write mournful letters about bereavements which have never occurred;...and that when a wife interrupts her husband's slumber with words of distress or alarm, it is only for fun, or a sudden morbid craving for underserved sympathy; and that when people assert that they were in sound health, in good spirits, and wide-awake, at a particular time which they had occasion to note, it is a safe conclusion that they were having a nightmare, or were the prostrate victims of nervous hypochondria. Every one of these improbabilities is perhaps, in itself a possibility; but as the narratives drive us from one desperate expedient to another, when time after time we are compelled to own that deliberate falsification is less unlikely than the assumptions we are making, and then again when we submit the theory of deliberate falsification to the cumulative test, and see what is involved in the supposition that hundreds of persons of established character, known to us for the most part and unknown to one another, have simultaneously formed a plot to deceive us — there comes a point where reason rebels.[274]

Phantasms of the Living did not deal with apparitions of persons who had been dead for more than twelve hours. However, according to an article published by Mrs. Eleanor Sidgewick, the society had some 370 cases in its files "which believers of ghosts would be apt to attribute to agency of deceased human beings." While the majority of these cases might be dismissed as hallucinations, four were types of cases that did seem to support the notion that some aspect of personality survives death.

1. Cases in which the apparition conveyed to the percipient accurate information that was previously unknown to him.

2. Cases in which the "ghost" seemed to be pursuing some well-defined objective. The spirit of Hamlet's father who makes Hamlet swear to seek revenge for his murder is a famous literary example of this type.

3. Cases in which the phantom bears a strong resemblance to a deceased person who is unknown to the percipient at the time of the manifestation. A case of this sort, incidentally, recently made headlines in the Berkeley *Gazette*, as the phantom was observed in the Faculty Club of the University of California.

4. Cases in which two or more people had independently seen similar apparitions: Into this category falls your typical haunting ghost or apparitions associated with a particular location. Often such phantoms are seen by individuals who are ignorant of previous sightings. These phantoms rarely seem to speak or take notice of humans, although voices and noises may be associated with them, and they are generally not seen for more than a minute before they vanish.[275]

Apparitions and personal experiences of seeing the dead still occur and there is a great need for people to feel comfortable discussing them openly. The following article with a front-page headline appeared in the Berkeley *Gazette* on March 19, 1974. The reason for the headline was not that this experience with a phantom was unusual, but rather that it was uncommon — and commendable — for a person of professional standing in the community to speak so directly about his experiences.

A HAUNTING AT THE FACULTY CLUB

BY RICHARD RAMELLA

Dr. Noriyuki Tokuda did not believe in ghosts until he encountered some recently in his room at the Faculty Club on the University of California campus here.

The visiting Japanese scholar, described by a local friend as "an intelligent, rational man," had no pat explanation to give for what he saw the evening of March 9.

In a half-somnolent state, he recalls, he saw a "very gentlemanly" looking Caucasian man, sitting on a chair and peering at him. As Dr. Tokuda shook out of his sleep, he next saw "something like two heads, floating, flying high across the room."

A moment later the apparitions had vanished.

Later, when Dr. Tokuda told club personnel about his unsettling experience, they told him the room in which he was staying had been occupied for 36 years by a University of California professor who died two years and a week before Tokuda checked into the room.

Officials described the professor. To Tokuda, there seemed to be a resemblance.

Tokuda says he, his wife and children lived in Berkeley from 1967 to 1969. "I love Berkeley very much." Given this feeling, he discounts any possibility of his vision being engendered by being in a strange place.

He recalls what happened:

"On March 9 I flew to Berkeley from Boston. Prof. Chalmers Johnson of the political science department took me from the airport to the faculty club. He advised me to take a short nap because of the three hour gap in time. I was tired, so I took a nap.

"At 7 p.m., while sleeping, I had a funny impression — felt some kind of psychological pressure. I was almost awake. I saw something in my dream. I felt some old gentleman — Western, white — sitting on the chair by the bed, watching quietly. It was quite strange.

"I opened my eyes then and saw a funny picture — two heads with a body passing out of my sight and disappearing."

With that Dr. Tokuda opened his eyes very wide and saw nothing strange. "I was surprised. Maybe it was a dream or a fantasy. I went out to dinner, came back and slept. I wasn't visited again."

After a trip to Stanford, Dr. Tokuda returned to the faculty club. By then he had told his strange story to his friends, Prof. Chalmers Johnson and his wife Sheila. They insured that he did not get Room 19 during the second stay.

When Tokuda checked out of the club yesterday, an official there told him his former room had for 36 years been the home of a solitary professor who died (not in the room) in March 1971.

"You're kidding me. Don't scare me," Tokuda responded.

"I cannot believe in ghosts," says Tokuda, a political scientist specializing in modern China.

But still, he smiled and said: "I think perhaps Prof. X still likes to live there and doesn't want to leave."

Tokuda, 42, is senior researcher and chairman of East Asian studies at the Institute of Developing Countries in Tokyo. He teaches at Keio University in Japan's capital.

"We have lots of ghost stories in Japan, too," the scholar says, "but I have never been so serious about it except when I was a child."

Professor Tokuda was quite clear about the fact that he was not in a normal state of consciousness during his experience. Yet one senses from his statements and the fact that he was motivated to mention the incident publicly that whatever he perceived was much more real to him than the hypnopompic imagery which typically precedes full awakening. That the apparition seemed to resemble the deceased former resident is also interesting. Tokuda's apparition, however, is not typical in many respects. Most apparition sightings reported to psychical researchers are, in fact, much more vivid. Culling over hundreds of case studies, G. N. M. Tyrell provides us with a picture of the "perfect apparition."[276] If the "perfect apparition" were standing next to a normal individual, we would find points of resemblance:

> (1) Both figures would stand out in space and be equally solid.
> (2) We could walk around the apparition and view it from any perspective as vividly as the normal individual.
> (3) The two figures would appear the same in any sort of lighting conditions, whether good or bad.
> (4) On approaching the apparition, one could hear it breathing and making other normal noises, such as the rustling of its clothes.
> (5) The apparition would behave as if aware of our presence. It might even touch us, in which case it would feel like an ordinary human touch.
> (6) The apparition could be seen reflected in a mirror just as a real person.
> (7) The apparition might speak and even answer a question, but we would not be able to engage it in any long conversation.
> (8) If we closed our eyes, the apparition would disappear from view just as the ordinary person would.
> (9) The apparition would appear clothed and with other normal accessories such as a stick or a package, perhaps even accompanied by a dog.
> (10) Many times, when close to, or touched by, the apparition, we would feel an unusual sensation of coldness.
> (11) If we tried to grab hold of the apparition, our hand would go through it without encountering any resistance. The apparition might disappear when cornered in this fashion.
> (12) The apparition would generally not remain more than half an hour. It might vanish through the walls or floor. Or it might simply open the door and walk out.
> (13) Apparitions differ in the extent to which they are able to actually affect physical objects, open doors, cast a shadow, be photographed. The "perfect apparition" cannot really cause any objectively measurable effects, although it may cause the subjective appearance of doing so.

When several individuals, independently or simultaneously, observe the same phantom under conditions that make deception or suggestion unlikely, the event can no longer be interpreted as a totally subjective experience. Some kind of parapsychological explanation is probably required.

Collective cases of this sort account for approximately eight percent of the total number of reported apparitions. Naturally, many times only one person is present to see the phantom. Collective cases are more common when several potential observers are present. In a group situation, if one person sees an apparition, there is about a forty percent likelihood that others will share his perception.[277]

However, even collective cases of apparitions of a person known to be dead do not provide certain evidence for survival. The research on out-of-body experiences suggests that it may be possible for one to cause one's own apparition to appear to others. Likewise, there is evidence that an individual, through concentration, can create the apparitional appearance of a different person as well. Such experiments were documented in 1822 by H. M. Wesermann, who was the government assessor and chief inspector of roads at Dusseldorf. The account of the appearance is recorded by one of the percipients, a Lieutenant S. He says Herr n had come to spend the night at his lodgings.

> After supper and when we had undressed, I was sitting on my bed and Herr n was standing by the door of the next room on the point also of going to bed. This about half-past ten. We were speaking partly about indifferent subjects and partly about the events of the French campaign. Suddenly the door out of the kitchen opened without a sound, and a lady entered, very pale, taller than Herr n, about five foot four inches in height, strong and broad in figure, dressed in white, but with a large black kerchief which reached down to the waist. She entered with bare head, greeted me with the hand three times in a complimentary fashion, turned around to the left toward Herr n, and waved her hand to him three times; after which the figure quietly, and again without creaking the door, went out. We followed at once in order to discover whether there were any deception, but found nothing.[278]

Even though the woman had been dead for five years, Wesermann claimed that he was the agent of her appearance. Furthermore, it appears as if he had expected Herr n to be asleep alone in his bedroom at the time instead of in the anteroom with Lieutenant S. Thus the observation of the phantom by two awake individuals went beyond even the intentions of the agent. By appearing in the anteroom, the phantom seemed to show a will independent of the agent, Wesermann; it was perhaps a product of the percipient's mind as well.

Wesermann's experiments were conducted at a time when many researchers, including the French government commission, were reporting unusual effects with mesmerism. Wesermann himself acknowledged that the successful production of such a phantom was, indeed, a rare event.

The point of such evidence is that the human mind seems capable of generating apparitional appearances identical to those often attributed to "departed spirits." However, in nearly all apparitions, there is no living individual attempting to create the appearance. It is unlikely that most apparitions were consciously created although they may have been unconscious human productions.

Furthermore, even if an apparition results from efforts by a once-living person, the apparition itself is not yet evidence of a fully conscious disembodied spirit. Such phenomena may well be mere images or thoughtforms hanging around in the psychic space. Very rarely do they show the characteristics of a well-developed personality, even though they exhibit some independent consciousness. The apparition evidence suggests that we are continually swimming in a sea of thoughts and images that exist independently of our own minds and that occasionally intrude dramatically into our conscious awareness.

Near-Death Experiences

Some interesting data relating apparitions to survival appears in a survey of physicians' and nurses' observations of dying patients. The 640 respondents to the survey, conducted by Karlis Osis of the American Society for Psychical Research, had witnessed over thirty-five thousand incidents of human death. Of these, only about 10% of the patients were conscious in their last hours. These dying individuals often experienced states of exultation or hallucination that could not be attributed to the nature of their illness, or to drug usage. While many patients had visions of spiritual worlds opening up to them in accordance with their particular religious beliefs, most of the hallucinations were of individuals already dead. Half of the percipients stated that the apparitions were coming to help them "enter the other world." The education, age or sex of the patients seemed to have nothing to do with the manifestation of such apparitions. However, the apparitions seemed most likely when the dying patient was in a state of physiological and psychological peace and equilibrium.[279] Oddly enough, a number of these apparitions were of individuals whose death was unknown to the dying patient.

Dr. Raymond Moody is a psychiatrist whose classic book, *Life After Life*, published in 1976, was a major impetus for a new wave of research on the phenomenon of near-death experience. A founding member of the International Association for Near-Death Studies, he is a professor of psychology at West Georgia College. He is also author of *Reflections on Life After Life* and *The Light Beyond*. In a *Thinking Allowed InnerWork* interview he summarizes his years of investigation into near-death experience:

> When patients have a cardiac arrest — they have no heartbeat, no detectible respiration — very often their physicians will say something such as, "Oh my God, he's dead, we've lost him." The patients tell us that from their perspective they feel more alive than ever. They say they float up out of their bodies, and they watch the resuscitation going on from a point of view immediately below the ceiling of the operating room. From this perspective they can see exactly what is going on down below. They can understand the remarks and the thoughts of the medical personnel who are around. It does not seem to make any sense to them.
>
> After a while they realize that although they can see clearly and understand perfectly what is going on, no one seems to be able to see or to hear them. So they undergo an experience in which they realize that this is something to do with what we call death, and at this point they experience what we might characterize as a turning inward of the sense of identity.
>
> One woman described this to me by saying, "In this experience, at this point, you are not the wife of your husband, you are not the mother of your children, you are not the child of your parents. You are totally and completely you."
>
> And at this point, at this moment of isolation and realizing that this is what we call death, then very unusual, transcendental experiences begin to unfold. I call them transcendental partly because the patients say that these later steps of the near-death experience are absolutely ineffable. Try as they may, they can not find any words to describe the amazing feelings and experiences.
>
> They say they become aware of what is described as a tunnel, a passageway, a portal, and they go into this tunnel, and when they come out, they come out into a very brilliant, warm, loving and accepting light. People at this point describe just amazing feelings of peace and comfort.
>
> In this light they say that relatives or friends of theirs who have already died seem to be there to help them through this transition.
>
> Another thing they will often tell us is that at this point they are met by some religious

KARLIS OSIS

figure. Christians say Christ, Jews say God or an angel. This being, in effect, asks them a question. Communication does not take place through words as you and I are now using, but rather in the form of an immediate awareness: "What have you done with your life? How have you learned to love?"

At this point, they say, they undergo a detailed review of everything they have done in their lives. This is displayed around them in the form of a full-color, three-dimensional panorama, and it involves every detail of their life, they say, from the point of their birth right up to the point of this close call with death.

The people who go through this say that clock time is not a factor in these experiences. Interestingly enough, patients often report that they review the events of their lives from a third-person perspective, displaced out to the side or above. They can also empathetically relate to the people with whom they have interacted.

They take the perspective of the person that they have been unkind to. Accordingly, if they see an action where they have been loving to someone, they can feel the warmth and good feelings that have resulted. The patients who go through this tell us that no one asks them about their financial well-being or how much power they have had. Rather they were faced with the question of how they had learned to love, and whether they had put this love into practice in their lives.

It certainly does not give us scientific evidence in a rigorous sense, or proof, that we live after we die. But I do not mind saying that after talking with over a thousand people who have had these experiences, it has given me great confidence that there is a life after death. As a matter of fact, I must confess to you in all honesty, I have absoutely no doubt, on the basis of what my patients have told me, that they did get a glimpse of the beyond.

It gives them a great sense of peace. Never again do they fear death — not that any of them would want to die in a painful or unpleasant way, or that they would actively

RAYMOND MOODY, M.D. (FROM *LIFE AFTER LIFE* VIDEOTAPE, COURTESY THINKING ALLOWED PRODUCTIONS)

seek this out. As a matter of fact, they all say that life is a great blessing and a wonderful opportunity to learn, and that they do not want to die anytime soon. But what they mean is that when death comes in its natural course of events, they are not going to be afraid. They do not fear it in the least anymore as being a cessation of consciousness.

We see this again and again — no more fear of death, renewed commitment to loving others, living in the present and not worrying about the future and a great sense of contentment.[280]

Mediumship

The classic line of argumentation for human survival beyond death is based upon cases of spiritualistic mediumship — an area fraught with fraud and chicanery. Another problem with this type of mediumship was that the information coming spontaneously from the medium might have already been stored in the unconscious memory. What was needed to counter this objection was a medium who could consistently produce accurate information on demand and without advance notice.

Mrs. Piper

Perhaps the best claimant to successful mediumship was Mrs. Leonora E. Piper of Boston, Massachusetts. Her mediumship began spontaneously in 1884, on the occasion of going into a trance during the séance of another medium. At first her controlling spirits rather pretentiously claimed to be Bach and Longfellow. Then appeared a self-styled French doctor who gave the name of Phenuit and spoke in a gruff male voice full of Frenchisms, Negro patois, and vulgar Yankee slang, nevertheless offering successful diagnoses and prescriptions. Often the deceased relatives of the sitters would speak through Mrs. Piper at her séances.

In 1886 William James, the great American psychologist, anonymously attended one of her séances. He was sufficiently impressed with the information that she revealed to him. He sent some twenty-five other people, using pseudonyms, to her. Fifteen of these people reported back to James that they had received from Mrs. Piper names and facts it was improbable she should know. In 1886, James issued a report to the SPR:

> My own conviction is not evidence, but it seems fitting to record it. I am persuaded of the medium's honesty, and of the genuineness of her trance; and although at first disposed to think that the "hits" she made were either lucky coincidences, or the result of knowledge on her part of who the sitter was and of his or her family affairs, I now believe her to be in the possession of a power as yet unexplained.[281]

Despite his interest in Mrs. Piper, William James gave up the inquiry at this point. Having convinced himself of her validity, he chose to give his other work higher priority at that time. The following year, however, Richard Hodgson, who had gained a reputation as a skeptical researcher for his debunking of Madame Blavatsky, arrived in Boston to head the American branch of the SPR. He was astounded when Mrs. Piper was able to offer many details about his family in Australia. To check on her honesty, he even had her and her family shadowed for some weeks by detectives. James and Hodgson decided that it would be wise to test Mrs. Piper in another environment, where she would have neither friends nor accomplices to aid her. Accordingly, she was invited to England by the SPR organization there and set off in November of 1889.

Leonora E. Piper

The results with Mrs. Piper in England were mixed. On a good day she was able to produce a mass of detailed information about the sitters which generally left them dumbfounded. On a bad day, her control, Phenuit, would behave in a most obnoxious manner, keeping up a constant babble of false assertions and inane conversation, blatantly fishing for information, and generally provoking the sitters. On no occasions was it concluded that Phenuit was anything more than a secondary personality of Mrs. Piper's.

During one séance Mrs. Piper revealed to Sir Oliver Lodge a great deal of information regarding an uncle of his who had been dead for twenty years. Lodge sent an agent to inquire in the neighborhood where the uncle had lived. In three days he was unable to unearth as much information as Mrs. Piper had provided. All of her remarks were eventually verified by surviving relatives.

In 1890 Mrs. Piper returned to the United States where she worked very closely with Richard Hodgson who spent the next fifteen years investigating her mediumship.

Nandor Fodor gives us a picture of Hodgson's research with Mrs. Piper:

> His first report on the Piper phenomena was published in 1892....In it no definite conclusions are announced. Yet, at this time Hodgson had obtained conclusive evidence. But it was of a private character and as he did not include the incident in question in his report, he did not consider it fair to point out its import. As told by Hereward Carrington in *The Story of Psychic Science*, Hodgson when still a young man in Australia had fallen in love with a girl and wished to marry her. Her parents objected on religious grounds. Hodgson left for England and never married. One day, in a sitting with Mrs. Piper, the girl suddenly communicated, informing Dr. Hodgson that she had died shortly before. This incident, the truth of which was verified, made a deep impression on his mind.[282]

At first Hodgson felt that Mrs. Piper's knowledge came to her telepathically. However, during a sitting in March of 1892 a new controlling spirit came who identified himself as a George Pellew, a prominent young man who had been killed a few weeks earlier and who was casually known to Hodgson. Five years previously he had had one anonymous sitting with Mrs. Piper. Pellew eventually replaced Phenuit as the main control and as the intermediary between the sitters and the spirits of their deceased friends. This particular control was very realistic and seemed to Hodgson to be more than a mere secondary personality. He showed an intimate knowledge of the affairs of the actual George Pellew, by recognizing and commenting on objects that had belonged to him. Out of 150 sitters who had been introduced to him, he recognized exactly those thirty people with whom the living Pellew had been acquainted. He even modified the topics and style of conversation with each of these friends, and showed a remarkable knowledge of their concerns. Very rarely did the Pellew personality slip up.

Mrs. Piper had never once in her career as a medium been detected in a dishonest action. Frank Podmore, the severest critic in the SPR, became convinced of the genuineness of her telepathic phenomena and, based on the Pellew material, the skeptical Richard Hodgson was inclined toward a spiritualistic position.

To credit spiritualism, he based his arguments for this position largely on the

RICHARD HODGSON

fact that a good amount of verified evidence produced by Pellew was unknown to anyone in the room at that time, and therefore could not have been picked up telepathically by any of the sitters.

In 1897, Hodgson published a report on Mrs. Piper in the *Proceedings of the Society for Psychical Research*:

> At the present time I cannot profess to have any doubt that the chief communicators to whom I have referred in the foregoing pages are veritably the personages that they claim to be, and that they have survived the change that we call death, and that they have directly communicated with us whom we call living, through Mrs. Piper's entranced organism.

Mrs. Eleanor Sidgewick argued against this position, emphasizing the occasions that the personality of the control did seem to degenerate.

Eventually several other spirits seemed to take control over Mrs. Piper's mediumship including that of the departed Reverend Stainton Moses.

In the presence of William James, America's foremost psychologist, the "Hodgson control" was able to describe incidents that Hodgson and James had intimately experienced together in life and that were unknown to other individuals. The personality was quite clear and distinct. At other times this was not the case and the "spirit" seemed like an obvious personation from Mrs. Piper's mind. In analyzing this data, James suggested that several factors were at play:

> Extraneous "wills to communicate" may contribute to the results as well as a "will to personate," and the two kinds of will may be distinct in entity, though capable of helping each other out. The will to communicate, in our present instance, would be, on a *prima facie* view of it, the will of Hodgson's surviving spirit; and a natural way of representing the process would be to suppose the spirit to have found that by pressing, so to speak, against "the light," it can make fragmentary gleams and flashes of what it wishes to say mix with the rubbish of the trance-talk on this side. The two wills might thus strike up a sort of partnership and reinforce each other. It might even be that the "will to personate" would be comparatively inert unless it were aroused to activity by the other will. We might imagine the relationship to be analogous to that of two physical bodies, from neither of which, when alone, mechanical, electrical or thermal activity can proceed. But, if the other body be present, and show a difference of "potential," action starts up and goes on apace.
>
> *I myself feel as if an external will to communicate were probably there* — that is, I find myself doubting, in consequence of my whole acquaintance with that sphere of phenomena, that Mrs. Piper's dream-life, even if equipped with telepathic powers, accounts for all the results found. But if asked whether the will to communicate be Hodgson's, or be some mere spirit counterfeit of Hodgson, I remain uncertain and await more facts, facts which may not point clearly to a conclusion for fifty or a hundred years.[283]

While James affirmed his belief in the reality of the Hodgson spirit, based on his sense of dramatic probabilities, he acknowledged that the case was not a good one because Hodgson had known Mrs. Piper so well in life. There was no way of proving that any of the evidential material did not simply come from her unconscious mind.

Cross-Correspondences

One of the more significant cases of mediumistic communication concerns the many messages received by Sir Oliver Lodge from his deceased son, Raymond. An eminent physicist, Lodge pioneered in the development of radio technology, which actually was as much his brainchild as it was

Marconi's, although he did not pursue its commercial development to the same extent as his Italian colleague. Lodge (an SPR founder) was already satisfied with the evidence for survival which had been gathered by Myers and others before his son's death during a mortar attack on September 14, 1915.

Actually, the story of Raymond's "spirit communications" begins a few weeks before his death, on August 8, when a message allegedly came from the spirit of Myers through Mrs. Piper in America. Hodgson's "spirit" claimed to be in control of the medium at the time he delivered a message for Lodge, which he claims to have received from Myers. The enigmatic message stated:

> Now Lodge, while we are not here as of old, i.e. not quite, we are here enough to take and give messages. Myers says you take the part of the poet and he will act as Faunus. Yes. Myers. *Protect.* He will understand. What have you to say, Lodge? Good work. Ask Verrall, she will also understand.[284]

In order to interpret this message, Lodge wrote to Mrs. Verrall — a medium, psychic researcher, and wife of a deceased Cambridge classical scholar — asking her to interpret the message. She replied at once referring to Horace (*Carm.* II. xvii. 27- 30), saying that the reference was to an account of the poet's narrow escape from death, from a falling tree. Faunus, the guardian of poets, lightened the blow and saved him.

On September 25, Raymond's mother, Mrs. Lodge, attended a sitting with a reputable medium, Mrs. Osborne Leonard. The visit was anonymous, and there was no intention of contacting Raymond; the purpose being, rather, to accompany a grieving friend whose two sons had also been killed in the war. In fact, it seemed as if the spirits of those sons did communicate through Mrs. Leonard. However, on that occasion, a message also came through purporting to be from Raymond:

> R: Tell father I have met some friends of his.
> ML: Can you give any name?
> R: Yes. Myers.

Two days later, Sir Oliver himself attended anonymously a sitting with Mrs. Leonard. The voice speaking through Mrs. Leonard was her childlike "spirit control," Feda, who described Raymond's condition, saying he was being taught by an old friend, M., and others. Feda also made an allusion to the Faunus message:

> ...Feda sees something which is only symbolic; she sees a cross falling back on you; very dark, falling on you; dark and heavy looking; and as it falls it gets twisted round and the other side seems all light, and the light is shining all over you....The cross looked dark and then it suddenly twisted around and became a beautiful light....Your son is the cross of light.[285]

This message seemed to be perceived symbolically as a thoughtform. One might

SIR OLIVER LODGE

complain that the allusion was too vague to be evidential, although it cannot be denied that it is remarkably consistent with the original Faunus message. In many respects this is typical of the complex series of more than three thousand cross-correspondence messages to develop between a number of mediums over the next several decades. Taken as a whole, they seem to weave a pattern indicative of a unifying intelligence.

That afternoon, after seeing Mrs. Leonard, Lady Lodge visited another medium, separately and strictly anonymously. The following is a transcript of Mrs. Lodge's sitting, with Sir Oliver's own annotations in brackets:

> Was he not associated with chemistry? If not, someone associated with him was, because I see all the things in a chemical laboratory. That chemistry thing takes me away from him to a man in the flesh [O. J. L., presumably as my laboratory has been rather specially chemical of late]; and connected with him, a man, a writer of poetry, on our side, closely connected with spiritualism. He was very clever — he too passed away out of England.
>
> [This is clearly meant for Myers, who died in Rome.]
>
> He has communicated several times. This gentleman who wrote poetry — I see the letter M — he is helping your son to communicate....If your son didn't know the man he knew of him.
>
> [Yes, he could hardly have known him, as he was only about twelve at the time of Myers' death.]
>
> At the back of the gentleman beginning with M, and who wrote poetry, is a whole group of people. [The SPR group, doubtless.] They are very interested. And don't be surprised if you get messages from them even if you don't know them.[286]

At this sitting the "spirit control" also made particular reference to a photograph of Raymond with a group of other men in which his walking stick could be seen. This puzzled Lady Lodge as she and her husband knew of no such photograph. However, several months later they received a letter from the mother of one of Raymond's fellow officers with an offer to send a copy of a group photo which she had.

Two days later, Sir Oliver also had a sitting, anonymously, with the same medium and received material from the "spirit control":

> Your common-sense method of approaching the subject in the family has been the means of helping him to come back as he has been able to do; and had he not known what you had told him, then it would have been far more difficult for him to come back. He is very deliberate in what he says. He is a young man that knows what he is saying. Do you know F W M?
>
> O. J. L. — Yes I do.
>
> Because I see those three letters. Now, after them, do you know S T; yes, I get S T, then a dot, and then P? These are shown to me; I see them in light; your boy has shown these things to me.
>
> O. J. L. — Yes, I understand. [Meaning that I recognised the allusion to F. W. H. Myers' poem *St. Paul*.]
>
> Well he says to me: "He has helped me so much more than you think. That is F W M."
>
> O. J. L. — Bless him!
>
> No, your boy laughs, he has got an ulterior motive for it; don't think it was only for charity's sake, he has got an ulterior motive, and thinks that you will be able by the strength of your personality to do what you want to do now, to ride over the quibbles of the fools, and to make the Society, *the* Society, he says, of some use to the world.[287]

About five weeks later, Lodge again sat with Mrs. Leonard, who by this time grasped his identity. He asked Raymond, through the control Feda, to describe further the group photograph, which had not yet arrived.

Additional details were given in terms of the position Raymond took relative to the man behind him who was leaning on his shoulder. These details were confirmed when the picture finally arrived.

Communications from Raymond, filled with evidential material, continued for many years through Mrs. Leonard and also through other mediums. Lodge's entire family participated in these sittings and all became convinced of the reality of Raymond's departed spirit. On one occasion sittings were held simultaneously at two different locations and Raymond successfully managed to convey information from one group to the other. The complete account of these many sittings is recorded in Lodge's book *Raymond*, published in 1916, which was written to further the cause of spiritualism.

Raymond clearly conveys the excitement which Lodge and his family felt at the time. However, Raymond himself never actually seems to control a medium. He speaks either through a "spirit control" or through automatic writing or table-rapping. There is a consistency to his personality, but not with the vividness of the Hodgson or the Pellew controls experienced through Mrs. Piper.

Mrs. Leonard's integrity has never been called into question. For over forty years her mediumship was the subject of exhaustive study by members of the SPR. Throughout this time, Feda was her only "control," although with a few sitters she would sometimes allow other spirits to speak directly through the medium. In these cases, the characterizations were brilliant and seemed to go much beyond mere reproduction of mannerisms. For years, one occasional communicator, a person Mrs. Leonard had never met in life, gave message after message to former loved ones without ever speaking out of character or using inappropriate emotional inferences.[288] If one refuses to accept the survival hypothesis to explain such cases, one must at least acknowledge extraordinary ESP capabilities on the part of the medium.

The super-telepathy theory is strained somewhat in dealing with the phenomena of cross-correspondences. The Faunus message which was received by Sir Oliver Lodge and then alluded to by another psychic is a minor example. The idea is creating a kind of jigsaw puzzle in the messages coming through different mediums. Any individual piece, when taken alone, seems to have no meaning. But when the separate pieces are put together, they form a coherent whole, and provide evidence for a constructive mind behind the entire design.

The major messages seem to have been directed by the spirit of F. W. H. Myers who died in 1901. Records show that the notion never occurred to him while he was alive. Other deceased members of the SPR also seem to have originated cross-correspondence messages. The mediums received these messages about the same time in places as distant from each other as London, New York, and India. Often the messages were filled with Greek and Latin allusions which were beyond the understanding of the different mediums. In fact, the messages seemed to contain the type of humor, style, and scholarship which was characteristic of the deceased researchers. The messages were often so complex that the puzzles could only be understood by classical language scholars.[289]

The "Margery" Mediumship

A baffling cross-correspondence series occurred in the United States, supposedly originating through "Walter," the deceased brother of Mina Crandon. Mrs. Crandon, the wife of a professor of surgery at Harvard University, has been one of the most controversial mediums of the twentieth century. Her psychokinetic manifestations were verified by many researchers throughout the world under apparently strict conditions. Yet at other times she was accused of fraud.

The Crandons lived in Boston; however, Walter also appeared through other mediums in New York, Niagara Falls, and Maine. On one occasion he announced a cross-correspondence in which "Margery" (as Mrs. Crandon was called) would make up a problem and two other mediums would each provide half of the answer. The problem written automatically by the medium was: "11 X 2 — to kick a dead." The mediumistic circle in New York was rung by telephone and told by Judge Cannon that Walter had given a message: "2 — no one stops." The next morning a telegram was received from Niagara Falls announcing this fragment: "2 horse." When the fragments are put together, one can see that the problem which Walter worked out and communicated — assuming there was no conspiracy to cheat — was this: "11 X 2 = 22. No one stops to kick a dead horse."[290]

One might argue that this case is an instance of group telepathy as the medium obviously knew the entire puzzle. Other evidence, however, also strengthens the case for Walter's autonomous existence. At times he was able to speak with a "direct voice — without using the vocal cords of either the medium or the sitters. His voice just appeared in the room. Furthermore, some cross-correspondences devised by Walter were in Chinese — a language which Margery did not know. Walter claimed he was getting help from some Chinese spirits."[291] Even if telepathy were at play in the transmission of information, it is hard to explain the actual design of the puzzles with that hypothesis.

Reincarnation [292]

Another approach to the question of survival focuses on reincarnation. The popular view of reincarnation is that after a person dies, the spirit of that person is reborn in another body. This process, of entering the womb, is vividly described in *The Tibetan Book of the Dead.* The concept is much more complex, however, when considered in the light of mystical philosophies which see one underlying reality beyond and within all time and individuality.

A belief in reincarnation has been held throughout the world. Lyall Watson, noted biologist and author of *Supernature* and *The Living World of Animals*, has commented extensively on reincarnation in his book, *The Romeo Error: A Matter of Life and Death*:

> No other single notion has ever received so widespread a cultural endorsement. It could be argued that this in itself might have kept a meaningless concept alive for a long time, but the belief stems from so many diverse and culturally unconnected origins that I cannot believe it has not basic biological validity.

The *Manavadharmasastra* or *Laws of Manu* constitute a classic of Hindu juridical

CRUDE "ECTO-PLASMIC" HAND EXUDING FROM NAVEL, PHOTO-GRAPHED AT SÉANCE WITH "MARGERY," IN BOSTON, 1925

theory. This work, which may perhaps be dated one or two centuries B.C., condenses in the form of diversified maxims all of the content of *dharma*. One passage dealing with the rules for ascetics — or those seeking enlightenment — elaborates the Hindu concept of reincarnation:

> Let him reflect on the transmigrations of men, caused by their sinful deeds, on their falling into hell, and on the torments in the world of *Yama* [hell], on the separation from their dear ones, on their union with hated men, on their being overpowered by age and being tormented by diseases, on the departure of the individual soul from this body and its new birth in another womb, and on its wanderings through ten thousand millions of existences, on the infliction of pain on embodied spirits, which is caused by demerit, and the gain of eternal bliss, which is caused by the attainment of their highest aim, gained through spiritual merit.

Remembrance of former existences (*pubbenivasanussati*) is considered one of the six higher spiritual powers of Buddhism. Regarding this power, the Buddhist scripture states:

> He remembers manifold former existences (*pubbenivasanussati*), such as one birth, two, three, four and five births...hundred thousand births, remembers many formations and dissolutions of worlds: "There I was, such name I had...and vanishing from there I reappeared here." Thus he remembers, always together with the marks and peculiarities, many a former existence.

The Jewish historian Josephus (died A.D.101) wrote, "All pure and holy spirits live on in heavenly places, and in course of time they are again sent down to inhabit righteous bodies." Josephus refers to reincarnation as being commonly accepted among the Jews of his time. The Zohar (Vol. II, *fol.* 99, *et seq.*), the basic text of Jewish Kabbalistic mysticism is also quite explicit:

> All souls are subject to the trials of transmigration; and men do not know the designs of the Most High with regard to them; they know not how they are being at all times judged, both before coming into this world and when they leave it. They do not know how many transmigrations and mysterious trials they must undergo; how many souls and spirits come to this world without returning to the palace of the divine king.
>
> The souls must reenter the absolute substance whence they have emerged. But to accomplish this end they must develop all the perfections, the germ of which is planted in them; and if they have not fulfilled this condition during one life, they must commence another, a third, and so forth, until they have acquired the condition which fits them for reunion with God.

The Universal Jewish Encyclopedia also states that reincarnaton became a "universal belief" in the mystical Jewish Hassidic tradition. The beloved Jewish writer Sholem Asch (1880-1957) provides a vivid description of reincarnation in his novel, *The Nazarene*:

> Not the power to remember, but its very opposite, the power to forget, is a necessary condition of our existence. If the lore of the transmigration of souls is a true one, then these, between their exchange of bodies, must pass through the sea of forgetfulness. According to the Jewish view we make the transition under the overlordship of the Angel of Forgetfulness. But it sometimes happens that the Angel of Forgetfulness himself forgets to remove from our memories the records of the former world; and then our senses are haunted by framentary recollections of another life. They drift like torn clouds above the hills and valleys of the mind, and weave themselves into the incidents of our current existence....then the effect is exactly the same as when, listening to a concert broadcast through the air, we suddenly hear a strange voice break in, carried from afar on another ether-wave and charged with another melody.

The mystic and mathematician, Pythagoras, was one of the earliest known advocates of the doctrine of reincarnation. Pythagoras claimed that he had lived as a prophet named Hermotimus, who was burned to death by his rivals about two hundred years earlier. In one of Plato's dialogues, Socrates indicates that teaching is not a matter of something being placed in one person by another, but of eliciting something already present. He was not interested in drawing out the petty things like names and dates that we retrieve under hypnosis, but "traces of knowledge garnered by the soul in its timeless journey."

The concept of reincarnation existed in Christianity until it was attacked in 543 A.D. by the Byzantine emperor Justinian and finally condemned by the Second Council in Constantinople in 553 A.D. Recent evidence advanced by Catholic scholars throws a new light on the whole matter.

The Catholic Encyclopedia (under "Councils of Constantinople") gives some information permitting the conclusion on technical grounds, that there is no barrier to belief in reincarnation for Catholic Christians. For example, Pope Vigilius, although he was in Constantinople at the time, refused to attend, in protest against the way in which the Emperor Justinian exerted absolute control over the Church patriarchy.

Catholic scholars are beginning to disclaim that the Roman Church actually took any official part in the anathemas against the doctrine of the pre-existence of the soul, suggesting that during the many centuries when the Church believed it had condemned reincarnation, it was mistaken.

While reincarnation is not emphasized in contemporary Islamic faith, the Koran (2.28) explicitly asks, "How can you make denial of Allah, who made you live again when you died, will make you dead again, and then alive again, until you finally return to him?"

One of the most eloquent testimonials to the beauty of reincarnation comes from the writings of the great Sufi poet, Jalaluddin Rumi (died 1273) who wrote:

> Like grass I have grown over and over again. I passed out of mineral form and lived as a plant. From plant I was lifted up to be an animal. Then I put away the animal form and took on a human shape. Why should I fear that if I died I shall be lost? For passing human form I shall attain the flowing locks and shining wings of angels. And then I shall become what no mind has ever conceived. O let me cease to exist! For non-existence only means that I shall return to Him.

The Swiss psychiatrist, Carl Gustav Jung, expresses my own thoughts:

> Rebirth is an affirmation that must be counted among the primordial affirmations of mankind. These primordial affirmations are based on what I call archetypes....There must be psychic events underlying these affirmations which it is the business of psychology to discuss — without entering into all the metaphysical and philosophical assumptions regarding their significance.

Some evidence for reincarnation comes from cases where individuals under hypnosis produce memories from what might be taken as a prior lifetime. These memories come through with a vividness of emotion and detail very much like early childhood memories. Often the reincarnation dramas seem to explain important characteristics in the subject's psychological makeup. This type of testimony is very interesting from a psychodynamic point of view. However, it cannot constitute acceptable evidence for reincarnation until it is shown that the descriptions match actual life-histories which are unknown to the subject — even then it could be merely postcognition.

Hypnotic regressions to ostensible "past lives" are fascinating psychological events that have attracted attention in professional journals and in the popular press. A large number of case reports have been published, but few of these cases were researched exhaustively, few were based on extensive hypnotic interviews, and few authors reported negative as well as positive findings.

One recently reported case developed a methodology for research in this area: Sixty hypnotic interviews were conducted with a single subject, all relevant archives were exhausted, and negative as well as positive data was reported. The reincarnation hypothesis was juxtaposed with a more conventional hypothesis emphasizing normal factors such as suggestion, role-playing, loss of inhibition, dissociation (including cryptomnesia), and desire to please the hypnotist. A method was introduced in which information found in public libraries and popular texts was contrasted with information found in foreign archives. The results of this rigorous procedure led to the acceptance of conventional explanations for the reported.[293]

Still another explanation for so-called reincarnation evidence would be simple spirit-possession.

Even if we were to conclude from other evidence that reincarnation was real, we would not need to assume that it always occurs. Some cultures maintain the belief, for example, that one reincarnates only if one dies prematurely by accident.

The cases which are coming under serious scientific scrutiny are typically those in which a small child, two to four years old, begins talking to the parents about another lifetime. Generally, the parents will dismiss such talk as nonsense — even in cultures where reincarnation is believed to occur. However, the child may persist and even insist upon visiting the community of his former residence. If the child supplies many details the parents may initiate an inquiry. Ideally at this stage, a scientific investigator is introduced to the scene. Careful records are made of all the child's statements. Verification can then begin by visiting the indicated community. If a family exists meeting the child's descriptions of his former household, the investigator can arrange for the two families to visit. Tests are then arranged to determine if the child can recognize places, objects, and people. Often it seems that these memories are lost as the child grows older.

Unfortunately, most investigations do not proceed so smoothly. Nevertheless, more than a thousand such cases have now been investigated, and a very convincing body of evidence is accumulating. As an example of an actual study, the case of Swarnlata Mishra is instructive.

On March 2, 1948, Swarnlata was born the daughter of the district school inspector in Chhatarpur, Madya Pradesh, India. At the age of three and a half, while on a trip with her father passing through the town of Katni, she made a number of strange remarks about her house in this village. The Mishra family had never lived closer than a hundred miles from this town. Later she described to friends and family further details of a previous life. Her family name, she claimed, had been Pathak. She also performed unusual dances and songs which she had had no opportunity to learn.

At the age of ten Swarnlata recognized a new family acquaintance, the wife of a college professor, as a friend in her former lifetime. Several months later, this case was brought to the attention of Sri H. N. Banerjee, of the Department of Parapsychology, University of Rajasthan, Jaipur. He interviewed the Mishra family; then, guided by Swarnlata's statements, he located the house of the Pathak family in Katni.

Banerjee found that Swarnlata's statements seemed to fit the life history of Biya, a daughter of the Pathak family and deceased wife of Sri Chintamini Panday. She had died in 1939.

In the summer of 1959, the Pathak family and Biya's married relatives visited the Mishra family in Chhatarpur. Swarnlata was able to recognize and identify them. She refused to identify strangers who had been brought along to confuse her. Later Swarnlata was taken to Katni and the neighboring towns. There she recognized additional people and places, commenting on changes which had been made since Biya's death. Unfortunately, Sri Banerjee was not present during these reunions.

It was not until the summer of 1961 that Dr. Ian Stevenson, an eminent psychiatrist and psychical researcher from the University of Virginia, visited the two families and attempted to verify the authenticity of the case.

Stevenson determined that of forty-nine statements made by Swarnlata only two were found to be incorrect. She accurately described the details of Biya's house and neighborhood as they were before 1939. She described the details of Biya's disease and death as well as the doctor who treated her. She was able to recall intimate incidents known only to a few individuals. For example, she knew Sri Chintamini Panday had taken twelve hundred ruples from a box in which Biya had kept money. He admitted this act, when questioned, and stated that no one but Biya could have known of the incident. She accurately identified former friends, relatives, and servants in spite of the efforts of the witnesses to deny her statements or mislead her. Most of the recognitions were given in a way which obliged Swarnlata to provide a name or state a relationship. It was not a case of asking, "Am I your son?" but rather, "Tell me who I am."

Perhaps because of her family's tolerance, Swarnlata's impressions of Biya's life have not faded. In fact, Swarnlata continues to visit Biya's brothers and children and shows great affection for them. Remarkably, she continues to act as an older sister to the Pathak brothers — men forty years older than she. Furthermore, the Pathak family was rather Westernized and did not believe in reincarnation before their encounter with Swarnlata.

Swarnlata also talked about another intermediate life as a child named Kamlesh in Sylket, Bengal, where she died at the age of nine. While this claim has not been verified in detail, many of her statements were found to correlate with the local geography. Her songs and dances were also verified as Bengali, although she had lived all her life only among Hindi speaking people.[294]

If one rules out the possibility of fraud in such cases — and there are many which are as evidential as this one — one might assume that a child like Swarnlata was recalling the memories of stories which she had overheard during her very early childhood or infancy. The other explanation — as with mediumship — involves ESP along with a remarkable skill for impersonation.

Xenoglossy

Perhaps the most extraordinary cases which challenge the super-ESP hypothesis are those which involve *xenoglossy* or the ability to speak a language one has never learned. The Bengali songs and dances which Swarnlata was able to recite offer a minor example of xenoglossy. Other cases are far more intriguing.

Dr. Ian Stevenson documents the case of a Russian-Jewish woman living in Philadelphia who, under hypnosis, claimed to be a Swedish peasant named Jensen

Jacoby. Furthermore, she was able to carry on rather involved conversations in this state using a mixture of Swedish and Norwegian with proper grammar and inflectional intonations. Speaking in a gruff male voice, she vividly portrayed the personality of the illiterate peasant and was also able to accurately identify objects borrowed from the American Swedish historical museum in Philadelphia. Most of these hypnotic sessions were taperecorded.

Stevenson spent more than six years researching this case, interviewing witnesses and family members in order to determine if there was any possibility that the subject had been exposed to the Swedish language at any time in her life. The case was not merely a question of reciting memorized or remembered passages — but rather one of carrying on an active dialogue. After extensive and thorough research, Stevenson felt that there was no period in the subject's life when she would have been able to acquire the languages spoken in trance.

The lady and her husband, the medical doctor who hypnotized her, were both subjected to a battery of personality, language, aptitude, and lie detector tests. The indications from these tests further added to the authenticity of the case. Stevenson feels that while ESP might account for the informational aspects of a foreign language, it does not necessarily explain the skill of using the language conversationally in a meaningful way. Thus, the case strongly points toward the survival hypothesis — even though the historical existence of the Swedish peasant has not been fully documented.[295]

It is impossible to base an airtight argument for survival after death upon cultural traditions, apparitions, near death experience, mediumship, possession, cross-correspondences, reincarnation, or xenoglossy — either individually or in combination. Nevertheless as one investigates the extraordinary depths of the human personality which are illustrated in the range of well-documented survival material, it does become apparent that events and processes do occur which seem to challenge all of our conceptions.

Unusual Powers of Mind over Matter

Among the most marvelous, most frightening and certainly most unbelievable possibilities suggested by psychic folklore is that human beings may be able to exert an observable influence upon the physical world — simply through the power of conscious intention; or unconscious intention or; by some accounts, through the assistance of spiritual intelligences; or as a result of a mysterious principle known as *synchronicity.* Some scholars — such as Stephen Braude, professor of philosophy at the University of Maryland — take such reports very seriously, claiming that no honest person can examine the case study reports and easily dismiss them:

> I have now spent more than five years carefully studying the non-experimental evidence of parapsychology — in fact, just that portion of it which is most contemptuously and adamantly dismissed by those academics....I started with the expectation that the received wisdom would be supported, and that my belief in the relative worthlessness of the material would merely be better-informed. But the evidence bowled me over. The more I learned about it, the weaker the traditional skeptical counter-hypotheses seemed, and the more clearly I realized to what extent skepticism may be fueled by ignorance. I was forced to confront the fact that I could find no decent reasons for doubting a great deal of strange testimony. It became clear to me

that the primary source of my reluctance to embrace the evidence was my discomfort with it. I knew that I had to accept the evidence, or else admit that my avowed philosophical commitment to the truth was a sham.

I am hardly comfortable about announcing to my academic colleagues that I believe, for example, that accordions can float in mid-air playing melodies, or that hands may materialize, move objects, and then dissolve or disappear....But I have reached my recent conclusions only after satisfying myself that no reasonable options remain.[296]

Skeptics (as well as most psi researchers) adamantly insist that it is absurd to give any credence to such reports until they meet the highest scientific standards. (Ironically, why would anyone bother to expend the large amounts of time and money required for meticulous scientific testing of such claims unless they were to give some credence to the non-scientific accounts?)

An interesting insight into the psychological dynamics of such events is provided by the great Swiss psychiatrist Carl G. Jung — who developed the concept of *synchronicity* as an acausal explanatory principle. In 1909, Jung visited his mentor Sigmund Freud in Vienna, and at one point asked him his opinion of psychic phenomena. Although Freud later changed his mind on the subject, at that time he dismissed the likelihood that such events could occur. Jung narrates an uncanny incident that occurred in the course of this conversation:

While Freud was going on in this way, I had a curious sensation. It was as if my diaphragm was made of iron and becoming red-hot — a glowing vault. And at that moment there was such a loud report in the bookcase, which stood right next to us, that we both started up in alarm, fearing the thing was going to topple over us. I said to Freud: "There, that is an example of a so-called catalytic exteriorisation phenomenon."

STEPHEN BRAUDE

"Oh come," he explained. "That is sheer bosh."

"It is not," I replied. "You are mistaken, Herr Professor. And to prove my point I now predict that in a moment there will be another loud report!" Sure enough, no sooner had I said the words than the same detonation went off in the bookcase.

To this day I do not know what gave me this certainty. But I knew beyond a doubt that the report would come again. Freud only stared aghast at me. I do not know what was in his mind, or what his look meant. In any case, this incident aroused his mistrust of me, and I had the feeling that I had done something against him. I never afterwards discussed the incident with him.[297]

The theme of mistrust characterizes the entire history of macro-psychokinetic claims. It is probably fair to state that no one, since Jesus Christ, has ever made such claims and been trusted (and there are many who distrust the supposed miracles of Christ). Furthermore, although mistrust may well blind us against considering vital possibilities, it is clearly warranted by the simple fact that numerous cases of fraud have been exposed in this area.

Perhaps, at a deeper level, both the fraud and the mistrust which it justifiably produces are part of a deeper protective mechanism developed within the *collective unconscious* (to use a Jungian term) of humanity. For, given our present level of ethical development, what awesome horrors might we wreak upon ourselves if we were able to harness psychokinesis in a disciplined manner? There are reasons to think that, if psychokinesis is real, it is a Pandora's box that is best left unopened by humankind — even if the price for this is our ignorance.

I personally feel comfortable with our lack of progress in this area. As a result of personal experiences which I shall recount, I accept the possibility of large-scale psychokinesis. I am also convinced that our planetary culture must demonstrate a willingness to solve the obvious problems of hunger, pollution, political inequality and war before we will be capable of responsibly wielding the full power of our own minds. The following examples provide some hints as to what that full power might possibly entail.

D. D. Home — The Greatest Medium Who Ever Lived

Perhaps the greatest ostensible physical medium who has ever lived was Daniel Dunglas Home. He was born in 1833 near Edinburgh, in Scotland. However, at an early age he went to New England to live with his aunt who adopted him. At the age of seventeen he had a vision of the death of his mother, which was soon verified. After that time the household was frequently disturbed with loud raps and moving furniture. Declaring that he had introduced the devil to the household, his aunt threw him out. He began living with his friends and giving séances for them.

Among those who were convinced of his abilities in this early period were Judge John Edmunds of the New York State Supreme Court and Robert Hare, an emeritus professor of chemistry at the University of Pennsylvania.

Home never accepted any payments for his séances. He exhibited religious reverence for the powers and knowledge that manifested through him along with a scientific curiosity to seek rational explanations. He did, however, accept presents from his wealthy patrons. Napoleon III of France provided for his only sister. Czar Alexander of Russia sponsored his marriage. He conducted séances with the kings of Bavaria and Wurtemburg as well as William I of Germany and assorted nobility throughout Europe. Noted literati also consulted with him.

To Lord Bulwer Lytton's satisfaction, Home called up the spirit that influenced him to write his famous occult novel, *Zanoni.* He conducted a séance for Elizabeth Barrett Browning and her husband Robert. Although his wife protested, Robert Browning insisted that Home was a fraud and wrote a long poem called "Mr. Sludge, The Medium," describing an exposure which never took place. In fact, throughout his long career, Home was never caught in any verifiable deceptions — although there were some apparent close calls.

In 1868, Home conducted experiments with Cromwell Varley, chief engineer of the Atlantic Cable Company and afterwards before members of the London Dialectical Society, who held fifty séances with him at which thirty persons were present. Their report, published in 1871, attested to the observation of sounds and vibrations, the movements of heavy objects not touched by any person, and well-executed pieces of music coming from instruments not manipulated by any visible agency, as well as the appearance of hands and faces that did not belong to any tangible human beings, but that nevertheless seemed alive and mobile.[298] This report inspired Sir William Crookes to investigate Home for himself.

Crookes conducted two very ingenious experiments with Home in which he tested alterations in the weight of objects and the playing of tunes upon musical instruments under conditions rendering human contact with the keys impossible. For the first experiment, Crookes developed a simple apparatus measuring the changes in weight of a mahogany board.

> One end of the board rested on a firm table, whilst the other end was supported by a spring balance hanging from a substantial tripod stand. The balance was fitted with a self-registering index, in such a manner that it would record the maximum weight indicated by the pointer. The apparatus was adjusted so that the mahogany board was horizontal, its foot resting flat on the support. In this position its weight was three pounds, as marked by the pointer of the balance.[299]

DANIEL DUNGLAS HOME

Crookes and eight other observers, including Sir William Huggins, a physicist and member of the Royal Society, observed Home lightly place his fingertips on the end of the board and watched the register descend as low as nine pounds. Crooks noted that since Homes' fingers did not cross the fulcrum, any tactile pressure he might have exerted would have been in opposition to the force that caused the other end of the board to move down. This experiment was conducted many times. On some occasions, Home never even touched the board: he merely placed his hands three inches over it. In other experiments, Crookes used a recording device to make a permanent record of the fluctuations in the weight. This was done to confute the argument that he himself was a victim of hallucinations.

In order to test the stories about music being played on an instrument, Crookes designed a cage in which to place an accordion he purchased specifically for these experiments (see illustration). The cage would just slip under a table, allowing Home to grasp the instrument on the end opposite to the keys, between the thumb and the middle finger. Again many witnesses were present:

> Mr. Home, still holding the accordion in the usual manner in the cage, his feet being held by those next him, and his other hand resting on the table, we heard distinct and separate notes sounded in succession, and then a simple air was played. As such a result could only have been produced by the various keys of the instrument being acted upon in harmonious succession, this was

CAGE USED BY CROOKES IN THE HOME ACCORDION EXPERIMENT

> considered by those present to be a crucial experiment. But the sequel was still more striking, for Mr. Home then removed his hand altogether from the accordion, taking it quite out of the cage, and placed it in the hand of the person next to him. The instrument then continued to play, no person touching it and no hand being near it.[300]

Crookes submitted his experimental papers to the Royal Society in order to encourage a large-scale investigation of the phenomena, which he felt were caused by a psychic force. However, the secretary of the society rejected his papers and refused to witness his experiments.

Crookes also testified to having seen many other phenomena with Home, including levitation of Home's body, levitation of objects, handling of hot coals, luminous lights, and apparitions.

Home himself bitterly resented any fraud or deception. In his book, *Lights and Shadows of Spiritualism*, written in 1878, he took an aggressive stance against phony mediums or even those who were unwilling to cooperate with scientists. Unlike most mediums, Home was always willing to be tested under well-lit and closely supervised conditions.[301]

Sir William Crookes' Researches

Despite the rejection of his psychical research by the scientific establishment, Crookes asserted the validity of his work throughout his life. In 1913, he was elected president of the Royal Society, but unfortunately he had by then long since abandoned his experimental work with mediums and found it wise not to discuss his work often in public. The phenomena that Crookes reported have been beyond the experience of all researchers before or since his time. Often his experimental reports were inadequate by contemporary standards since he simply assumed that his own word was sufficient to establish general acceptance of a phenomenon. We cannot hastily conclude that Crookes was deluded or duped, for he was at the height of his intellectual creativity at the time he conducted this research. In the words of his friend, Sir Oliver Lodge, "It is almost as difficult to resist the testimony as it is to accept the things testified." His most amazing experiments were conducted with a medium named Florence Cook.

Cook's ostensible ability to materialize the forms of various spirits had caused a stir among spiritualists. The most notable spirit to appear identified herself as Katie King, the daughter, in a former life, of the buccaneer Henry Morgan.

The phenomena of spirit materialization had actually attracted public attention a few years earlier through a Mrs. Samuel Guppy, the protégée of Alfred Russell Wallace, a prominent spiritualist who was also noted as one of the discoverers with Darwin of the theory of evolution. Mrs. Guppy introduced into her work the use of a tightly sealed cabinet in which she was placed in order to build up sufficient "power" for the construction of a spirit form which could then stand the scrutiny of the light outside the cabinet. The cabinet also provided, of course, an ideal opportunity for subterfuge on the part of the medium, which was undoubtedly taken advantage of on many occasions, for rarely were any medium and her spirit seen together at the same time.

Crookes attended séances with Florence Cook for a period of over three years and studied her intensively for several months in a laboratory in his own home. He also made numerous observations of Katie King and took more than forty photographs of her. On several occasions he had the opportunity of seeing both Florence and her spirit, Miss King, at the same time and even of photographing them together. Katie appeared quite solidly before the guests at the séance, sometimes staying and conversing with them for as long as two hours. Crookes even reports having embraced and kissed her. At other times she seems to have vanished instantaneously and soundlessly. It is difficult to believe that an accomplice could have continued such an intimate masquerade, in Crookes' own home, for several months without detection. He gives several reasons why he feels Florence Cook could not have committed fraud:

SIR WILLIAM CROOKES

> During the last six months Miss Cook has been a frequent visitor at my house, remaining sometimes a week at a time. She brings nothing with her but a little handbag, not locked; during the day she is constantly in the presence of Mrs. Crookes, myself, or some other member of my family, and, not sleeping by herself, there is absolutely no opportunity for any preparation....I prepare and arrange my

library myself as the dark cabinet, and usually, after Miss Cook has been dining and conversing with us, and scarcely out of our sight for a minute, she walks direct into the cabinet, and I, at her request, lock its second door, and keep possession of the key all through the séance.[302]

Katie's height varies; in my house I have seen her six inches taller than Miss Cook. Last night, with bare feet and not "tip-toing," she was four and a half inches taller than Miss Cook. Katie's neck was bare last night; the skin was perfectly smooth to touch and sight, whilst on Miss Cook's neck is a large blister, which under similar circumstances is distinctly visible and rough to the touch. Katie's complexion is very fair, while that of Miss Cook is very dark. Katie's fingers are much longer than Miss Cook's, and her face is also larger.[303]

Crookes also indicates that Miss Cook was willing to submit to any test he wished to impose. Ironically enough, on two occasions, in 1872 and in 1880, individuals claimed to have exposed Florence Cook fraudulently masquerading as her spirit.[304]

It is not unreasonable to suggest any of several contradictory hypotheses: (1) that Crookes himself may have been deluded or enchanted by Florence Cook, (2) that while Crookes himself did observe genuine phenomena, Cook sometimes lost her abilities and resorted to fraud, (3) that the alleged exposures were not genuine, or (4) that Crooke's accounts were fraudulent. Psychical phenomena have always had an ironic and paradoxical nature, and Crookes' experimental methodology was certainly not sufficient to answer all of the questions one might like to ask.

KATIE KING

It is so difficult to maintain that a man of Crookes' scientific caliber could have been taken in by cheap tricks, some of his critics have assumed that he himself was in on the fraud. They have claimed that Crookes had been involved in a love affair with Florence Cook, and that he testified to her phenomena in order to shield her reputation and hide his emotional entanglements with her. However, even if it were so, other matters would remain quite unsolved. If Crookes was involved with Miss Cook, who was only fifteen years old at the time, this hypothesis cannot account for the phenomena he reported with both Home and Miss Fox. Nor does it begin to explain the research on the same phenomena reported by a number of other eminent scientists. Nevertheless, the accusation of experimenter fraud still continues to haunt psychical researchers, and will continue to do so as long as people are reinforced in their expectation of fraud by periodic publicly exposed episodes.

Marthe Béraud

Another extraordinary physical medium whose ectoplasmic materializations were observed and photographed by many investigators was Marthe Béraud. Nobel laureate physiologist Charles Richet described the production of a phantom, called Bien Boa, under experimental conditions that he felt negated the possibility of theatrical props or accomplices:

BIEN BOA

He seemed so much alive that, as we could hear his breathing, I took a flask of baryta water to see if his breath would show carbon dioxide. The experiment succeeded. I did not lose sight of the flask from the moment I put it into the hands of Bien Boa who seemed to float in the air on the left of the curtain at a height greater than Marthe could have been even if standing up....

A comical incident occurred at this point. When we saw the baryta show white (which incidentally shows that the light was good), we cried "Bravo." Bien Boa then vanished, but reappeared three times, opening and closing the curtain and bowing like an actor who receives applause.

However striking this was, another experiment seems to me even more evidential: Everything being arranged as usual....after a long wait I saw close to me, in front of the curtain which had not been moved, a white vapour, hardly sixteen inches distant. It was like a white veil or handkerchief on the floor; it rose up still more, enlarged, and grew into a human form, a short bearded man dressed in a turban and white mantle, who moved, limping slightly, from right to left before the curtain. On coming close to General Noel, he sank down abruptly to the floor with a clicking noise like a falling skeleton, flattening out in front of the curtain. Three or four minutes later...he reappeared rising in a straight line from the floor, born from the floor, so to say, and falling back on it with the same clicking noise.

The only un-metapsychic explanation possible seemed to be a trap-door opening and shutting: but there was no trap-door, as I verified the next morning and as attested by the architect.

Several photographs were taken....The softness and vaporous outline of the hands are curious; likewise the veil surrounding the phantom has indeterminate outlines....A thick, black, artificial-looking beard covers the mouth and chin....Bien Boa would seem to be a bust only floating in space in front of Marthe, whose bodice can be seen. Low down, between the curtain and Marthe's black skirt, there seem to be two small whitish rod-like supports to the phantom form.[305]

Paraffin Hands

The most impressive evidence for ectoplasmic materializations comes from molds of "spirit hands" made in paraffin. Richet reports his careful studies:

> [Gustav] Geley and I took the precaution of introducing, unknown to any other person, a small quantity of cholesterin in the bath of melted paraffin wax placed before the medium during the séance. This substance is insoluble in paraffin without discolouring it, but on adding sulphuric acid it takes a deep violet-red tint; so that we could be absolutely certain that any moulds obtained should be by the paraffin provided by ourselves....
>
> During the séance the medium's hands were firmly held by Geley and myself on the right and on the left, so that he could not liberate either hand. A first mould was obtained of a child's hand, then a second of both hands, right and left; a third time of a child's foot. The creases in the skin and veins were visible on the plaster casts made from the moulds.

By reason of the narrowness of the wrist these moulds could not be obtained from living hands, for the whole hand would have to be withdrawn through the narrow opening at the wrist. Professional modellers secure their results by threads attached to the hand, which are pulled through the plaster. In the moulds here considered there was nothing of the sort; they were produced by a materialization followed by dematerialization, for this latter was necessary to disengage the hand from the paraffin "glove."[306]

The plaster casts from these molds — including a cast of intertwining hands are still available for inspection at the Metapsychic Institute in Paris. A physiologist of the first order, Richet sums up his research on ectoplasmic materialization:

> There is ample proof that experimental materialization (ectoplasmic) should take definite rank as a scientific fact. Assuredly we do not understand it. It is very absurd, if a truth can be absurd.
>
> Spiritualists have blamed me for using this word "absurd"; and have not been able to understand that to admit the reality of these phenomena was to me an actual pain; but to ask a physiologist, a physicist, or a chemist to admit that a form that has circulation of the blood, warmth, and muscles, that exhales carbonic acid, has weight, speaks, and thinks, can issue from a human body is to ask of him an intellectual effort that is really painful.
>
> Yes, it is absurd; but no matter — it is true.[307]

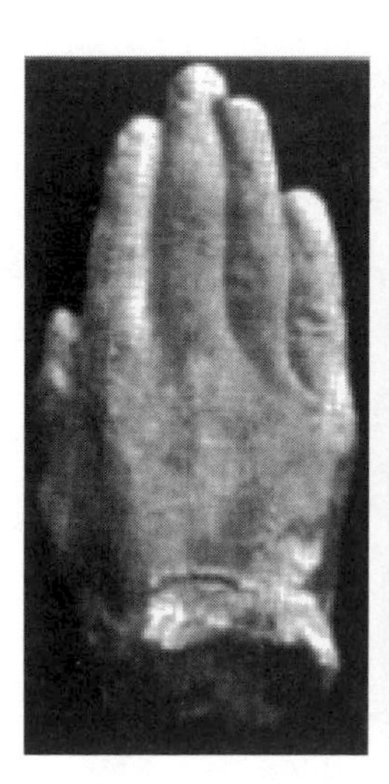

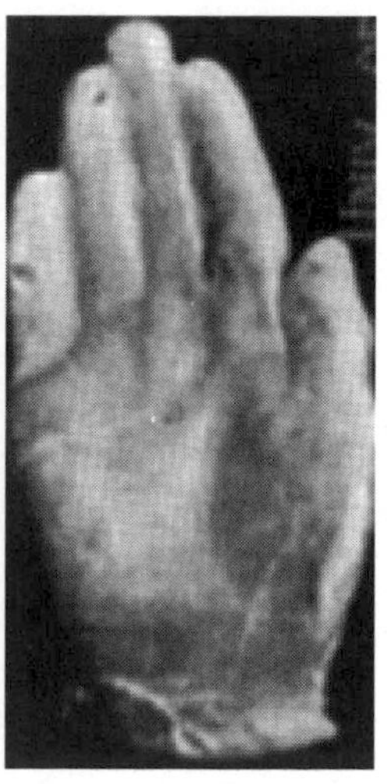

Eusapia Palladino

One of the most extraordinary physical mediums in the history of psychical research was Eusapia Palladino, a rough peasant woman from Naples. She came to the attention of the learned world through séances held with the eminent Italian sociologist Cesare Lombroso. These séances continued to be held in Italy until 1894 when the French physiologist Charles Richet invited her to his private island to attend séances with Frederick Myers and Sir Oliver Lodge as well as J. Ochorowicz, a Polish researcher. It was Richet's belief he would be able to prevent Eusapia from using props or accomplices while she was on the island. The group witnessed most of the phenomena that had been previously reported: levitations, grasps, touches, lights, materializations, raps, curtains billowing, scents, and music. At all times the researchers were holding Eusapia's hands and feet.

The following excerpts are from the published account of one of these sessions:

> Richet held both arms and one hand of E., while M. held both feet and her other arm. R. then felt a hand move over his head and rest on his mouth for some seconds, during which he spoke to us with his voice muffled. The round table now approached. R.'s head was stroked behind....The round table continued to approach in violent jerks....A small cigar box fell on our table, and a sound was heard in the air as of something rattling....A covered wire of the electric battery came on to the table and wrapped itself around R.'s and E.'s heads, and was pulled till E. called out....The accordion which was on the round table got on the floor somehow, and began to play single notes. Bellier [Richet's secretary] counted 26 of them and then ceased

counting. While the accordion played, E.'s fingers made movements in the hands of both M. and L. in accord with the notes as if she was playing them with difficulty....Eusapia being well held, Myers heard a noise on the round table at his side, and turning to look saw a white object detach itself from the table and move slowly through the clear space between his own and Eusapia's head....Lodge now saw the object coming past Myers' head and settling on the table. It was the lamp-shade coming white side first....The "chalet" [music box] which was on the round table now began to play, and then visibly approached, being seen by both Myers and Lodge coming through the air, and settled on our table against Myers' chest....During the latter half of the sitting, Eusapia had taken one of Myers' fingers and drawn some scrawls with it outside Richet's flannel jacket, which was buttoned up to his neck. Myers said, "She is using me to write on you," and it was thought no more of. But after the séance, when undressing, Richet found on his white shirt front, underneath both flannel jacket and high white waistcoat, a clear blue scrawl: and he came at once to bedrooms to show it.[125]

EUSAPIA PALLADINO

Myers, Lodge, and Richet were convinced of the genuineness of the phenomena that they reported and soon arranged for Eusapia to repeat her performance before SPR members in Cambridge. Again a number of phenomena were noted. Protuberances observed coming out of Eusapia's body and the billowing of curtains were particularly hard to explain away. However, at Hodgson's insistence the Cambridge group relaxed their controls over Eusapia's hands and feet to see if she would cheat if given an opportunity. Under these conditions, Eusapia conducted several séances producing nothing but fraudulent phenomena, whereupon Hodgson insisted that none of her other phenomena could be trusted. Other investigators acknowledged that she would cheat if given a chance, but that nevertheless, under controlled conditions she did produce authentic phenomena.

The SPR maintained a firm policy of rebuffing the phenomena of any mediums who have ever been found guilty of systematic fraud. Members were urged to ignore any future reports of experiments with Eusapia.

Reports concerning Eusapia, however, continued to flow in. In 1897, the noted French astronomer Camille Flammarion reported on a series of séances in which "spirit" impressions were made in wet putty. Flammarion gives us a description of the event:

> I sit at the right hand of Eusapia, *who rests her head upon my left shoulder*, and whose right hand I am holding. M. de Fontanay is at her left, and has taken great care not to let go of the other hand. The tray of putty, weighing nine pounds, has been placed upon a chair, twenty inches behind the curtain, consequently behind Eusapia. She cannot touch it without turning around, and we have her entirely in our power, our feet on hers. Now the chair upon which was

the tray of putty has drawn aside the hangings, or portieres, and moved forward to a point above the head of the medium, who remained seated and held down by us; moved itself also over our heads, — the chair to rest upon the head of my neighbor Mme. Blech, and the tray to rest softly in the hands of M. Blech, who is sitting at the end of the table. At this moment Eusapia rises, declaring that she sees upon the table another table and a bust, and cries out, "*E Fatto*" ("It is done"). It was not at this time, surely, that she would have been able to place her face upon the cake, for it was at the other end of the table. Nor was it before this, for it would have been necessary to take the chair in one hand and the cake with the other, and she did not stir from her place. The explanation, as can be seen, is very difficult indeed.

Let us admit, however, that the fact is so extraordinary that a doubt remains in our mind, because the medium rose from her chair almost at the critical moment. And yet her face was immediately kissed by Mme. Blech, who perceived no odor of the putty.[126]

Finally in 1909, the SPR did publish a report of another series of séances with Eusapia conducted by a group of experimenters known for their exposure of other fraudulent mediums — the Hon. Everard Fielding, Hereward Carrington, and W. W. Baggally. They observed a number of levitations and materializations under good lighting conditions. These séances occurred in the middle room of a three-room hotel suite they had rented for the purpose in order to rule out the possibility of confederates. Their account is quite detailed and thorough, having been dictated minute by minute to a professional stenographer. They were favorably impressed with what they had observed. However, the following year Eusapia's abilities, whatever they were, seem to have faded and it was simply too late to conduct further research with her.

Psychic Photography

One interesting technique for measuring psychokinesis is thought-photography. Claims of spirit photographs, where extra faces appear on developed film, go back as far as the history of photography itself. Some have even claimed to photograph actual human thoughtforms. Photography of this sort almost inevitably provoked accusations

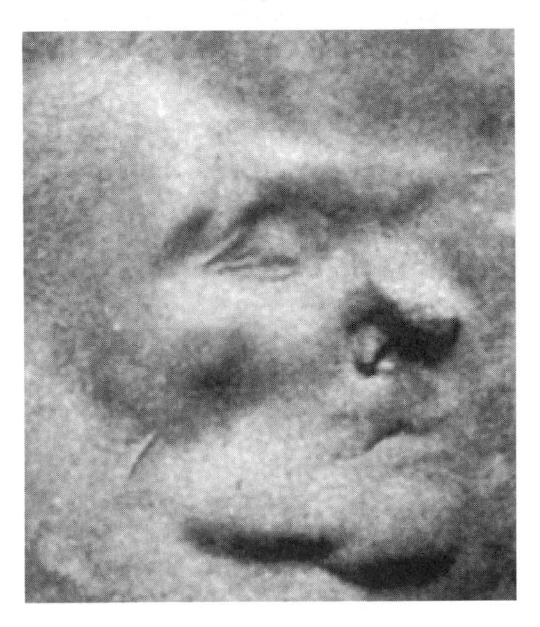

of fraud that were difficult to disprove. In 1910, Dr. Tomokichi Fukurai, a professor of literature at the Imperial University of Tokyo, conducted a series of experiments in thoughtography. The publication of his findings aroused such hostility among Japanese scientists that he was forced to resign his position. He then continued his work at a Buddhist university associated with a temple of the esoteric Shingon sect of Buddhism on top of Mt. Koya.[308] His works were translated into English in 1931 in a book titled *Spirit and Mysterious World.* Although it showed a carefully planned scientific investigation, even the psychical researchers of the time were not ready to deal with this type of data, embedded as it was in Buddhist philosophy.[309]

It was not until the late 1950s that a claim for psychic photography was taken seriously by researchers. The special gift for creating these photographs was discovered in Ted Serios, a Chicago bellhop who had little formal education. The phenomena began when Serios allowed a friend to hypnotize him just to pass away the time. Serios claimed to be able to describe the locations of buried treasure. The friend then suggested that he concentrate on making photographs of the locations when he pointed a camera at a blank wall and triggered the shutter. They did not find buried treasure, but to their amazement, actual images appeared on the Polaroid prints of things that were not visible in the room.

The phenomena came to the attention of members of the Illinois Society for Psychic Research who eventually persuaded a Denver psychiatrist, Dr. Jule Eisenbud, to observe one of Ted's demonstrations. After a long string of failures, Serios managed to produce a striking success for Eisenbud, who, although he had engaged in previous psychical exploration within the context of psychoanalysis, was unprepared for phenomena of this sort. After a sleepless night, he invited Serios to Denver for further study. Eisenbud spent two years conducting well-controlled studies with Serios. He was quite aware of the history of fraud and gullibility in research of this sort and claims that he took every precaution to guard against it. His book, *The World of Ted Serios,* published in 1966, contains the results obtained from his examinations.

The way in which Ted's mind ostensibly shaped the pictures was sometimes quite remarkable. In one session, in front of several witnesses, Ted first tried to reproduce images of the medieval town of Rothenburg. Then the experimenters asked him to try to reproduce an image of the old Opera House in Central City, Colorado. Serios agreed, and then asked the experimenters if they would like a composite of both images. The results are extraordinary. The photograph shows a striking resemblance to the livery stable across from the old Opera House. However, instead of the brick masonry, the image shows a kind of imbedded rock characteristic of the buildings in the medieval town.

The large photograph shown is an enlargement of a Polaroid "thoughtograph" of the Denver Hilton Hotel. Eisenbud held the camera, which was pointed at Serios' forehead. Ted, at the time, was trying to produce an image of the Chicago Hilton. ("I missed, damn it.") Eisenbud claimed that this image could only have been made with a lens different from that of the Polaroid 100, from an angle well up in the air, between the tree tops. This suggests that the thoughtographs are associated with out-of-body or traveling clairvoyance states.

Eisenbud's book is noted for detailed observation, but even more remarkable is the penetrating study of this anomalous phenomena and the reaction to it of scientists and educators. To Eisenbud, the photo-

LEFT: ACTUAL PHOTOGRAPH OF LIVERY STABLE IN CENTRAL CITY, COLORADO. INSET IN UPPER LEFT SHOWS BRICK PATTERN. *RIGHT:* SERIOS "THOUGHTOGRAPH" OF LIVERY STABLE (COURTESY OF JULE EISENBUD)

graphic manifestations seemed to follow a pattern pointing to the active operation of the animistic powers known to ancient people.

> As to building blocks for a theoretical structure that might bridge the gulf on other fronts between the mental and physical,...I can't think of a better place to begin than right where Ted is (and hopefully where others like him will be). For in a study of images and imagery of this sort — and in phenomenon like dreams, hallucinations, and apparitions, which prove no less remarkable and even more familiar than Ted's image — we are confronted by various organized entities with one leg in the world of reality and one leg in that extraordinary world we ordinarily term appearance.[310]

Adequate understanding of the Serios phenomena can only be obtained through detailed study of the experimental reports. During the following years, studies were also conducted by researchers at the Division of Parapsychology of the University of Virginia Medical School. These researchers failed to detect any signs of fraud in their cooperative subject, and they successfully obtained numerous striking photographs. While they were calling for further study of this puzzling phenomena, Serios' abilities began to fade and he has remained less active for the past twenty years.[311]

Skeptics claim that Ted Serios was definitely exposed by Charlie Reynolds and David Eisendrath, both amateur magicians and professional photographers. They presented their account in a *Popular Photography* piece (October 1967) based on one weekend with Serios and the psychiatrist Jule Eisenbud, whose book, *The World of Ted Serios,* had sparked their ire.

However, the November 1967 issue of *Popular Photography* published Eisenbud's response letter:

> I hereby state that if, before any competent jury of scientific investigators, photographers and conjurors, any chosen by them can in any normal way or combination of ways duplicate, under similar conditions,

Serios "thoughtograph" of the Denver Hilton Hotel. (Courtesy of Jule Eisenbud)

> the range of phenomena produced by Ted, I shall (1) abjure all further work with Ted, (2) buy up and publicly burn all available copies of *The World of Ted Serios,* (3) take a full-page ad in *Popular Photography* in order to be represented photographically wearing a dunce cap, and (4) spend my spare time for the rest of my life selling door-to-door subscriptions to this amazing magazine. No time limit is stipulated.

An article in *Fate,* August 1974, revealed that only one magician had responded to this delectable invitation. The Amazing James Randi couldn't resist the bait, but on learning of the conditions he backed out. According to Randi, one of the conditions was that he perform in a state of alcoholic intoxication, as Serios had typically done. As a non-drinker, Randi found this condition unacceptable.

Nina Kulagina

Meanwhile, in the Soviet Union, researchers claimed to have discovered a woman, Nina Kulagina, who could exert a psychokinetic influence upon static objects. In 1968, Western researchers attending a conference in Moscow were shown a film of her in action. This film, which has since been seen many times in the United States, shows Kulagina apparently moving small objects, without touching them, across a table top. The Soviets claimed that this woman also known as Nelya Mikhailova had been studied by some forty scientists, including two Nobel laureates. They also reported that, like Serios, Madame Kulagina was able to cause images to appear on photographic film.[312] The communist scientists, who were by no means inclined to take a spiritualistic world view, felt that they had encountered a new force in nature.[313] Very thorough studies of the electrical fields around her body as well as the electrical potentials in her brain were conducted by Dr. Genady Sergeyev, a well-known physiologist working in a Leningrad military laboratory. Exceptionally strong voltages and other unusual effects were observed:

> There is a large gradient between the electrical characteristics in the forward part of Mikhailova's brain versus the back part of the brain (fifty to one), whereas in the average person the gradient is four to one. The usual force field around Mikhailova's body is ten times weaker than the magnetic field of the earth.
>
> During PK, her pulse rises to 240 per minute. There is activation of deeper levels of the occipital lobe and reticular formation. This enhances polarization in the brain between front and back, says Sergeyev. When the gradient between front and back of the brain reaches a certain level, and there is most intense activity in the occipital lobe, radiation of electrostatic and electromagnetic fields are detected by the force field detectors four yards from the body....Heartbeat, brain waves, and force field fluctuations are in ratio. The fields around the PK medium are stronger further away than close to the head. Mikhailova appears to focus these force field waves in a specific area.[314]

Detailed physiological studies of this sort with outstanding psychics are so rare they raise more questions than they answer. Kulagina has received a certain amount of adverse publicity. However, since 1968, several groups of Western researchers have had opportunities to test her under differing circumstances. In most cases, their reports attested to the authenticity of her psychokinetic abilities.[315,316,317,318]

Her mediumship has led to a strain on her health leading to a heart attack, and her doctors have suggested that she limit this type of activity. The former Soviets, however, are reported to have found others who have developed talents for psychokinesis, and are also researching ways to train this ability in normal individuals. The training begins with long hours practicing to move the needle of a compass.[319]

Uri Geller

The most unusual psychokinetic effects currently being reported by scientists are associated with the Israeli psychic Uri Geller. Dr. Andrija Puharich, a physician known for his theoretical efforts to grasp the physics and physiology of psychic phenomena,[320] as well as for his previously mentioned researches into psychic healing, in August of 1971,

NINA KULAGINA

encountered Geller in Israel, where he arranged to conduct an extensive series of experiments with him. Eventually he brought Uri to the United States where his research continued and where he negotiated for further testing at the Stanford Research Institute in Menlo Park, California. It was at a symposium in Berkeley, sponsored by KPFA-FM at the University of California, that Andrija Puharich made the first public presentation of experimental research with Uri Geller.

URI GELLER

Puharich carefully went over his investigations with Geller, indicating the conditions under which he had observed Geller bend and break metal objects, erase magnetic tape, make things disappear and reappear elsewhere, and cause the hands of a clock to change time. He also discussed how his sessions with Geller led him to believe that there was some other intelligent form of energy working through Geller, possibly from an extra-terrestrial or extra-dimensional source.

The following week, the controversy over Geller deepened as *Time* magazine published a story claiming that Geller was a fake. Physicists Harold Puthoff and Russell Targ of Stanford Research Institute also presented a paper about their research with Geller at a physics colloquium at Columbia University.

The SRI scientists primarily emphasized the telepathic studies they had done with Geller. However, they did report on two significant psychokinetic experiments with Uri:

> A precision laboratory balance was placed under a Bell jar. The balance had a one-gram mass placed on its pan before it was covered. A chart recorder then continuously monitored the weight applied to the pan of the balance. On several occasions Uri caused the balance to respond as though a force were applied to the pan. The displacement represented forces from 1.0 to 1.5 grams. These effects were ten to 100 times larger than could he produced by striking the Bell jar or the table or jumping on the floor. In tests following the experimental run, attempts were made to replicate Geller's results using magnets and static electricity. Controlled runs of day-long operation were obtained. In no case did the researchers obtain artifacts which resembled the signals Geller had produced.[321]

Subsequent to the presentation of the above report, the SRI researchers backed away from the Bell jar study claim, having been convinced that the result could have resulted from artifacts. The lesson of this incident is that time is indeed necessary to sift through and evaluate experimental claims in the area of psychokinesis. Simply because a claim is presented in a scientific format, one cannot assume that it will ultimately withstand the test of scrutiny.

On several occasions, a group of nearly

eighteen scientists, organized by me and Dr. Joel Friedman of the philosophy department at the University of California, Davis, met with Geller and observed a wide variety of unusual phenomena in his presence. However, none of them occurred under conditions of sufficient control for us to feel confident about publishing the results. One of our researchers, Saul-Paul Sirag (author of the material in the Appendix), conducted an experiment with Geller in which Saul-Paul unexpectedly handed Geller a bean sprout and asked him to "make the movie run backwards." Uri closed his fist over the sprout and when he opened his hand some thirty seconds later there was no longer a sprout, but a whole solid mung bean. This effect, if verified by further replication, seems to indicate a psychokinetic influence involving time.

Another study the Berkeley research group conducted was a follow-up survey of the reactions of individuals who witnessed Geller's performances. Many people reported experiencing unusual visual or telepathic phenomena and several reported that, after watching Geller's demonstrations, they also were able to produce various psychokinetic effects. On occasions when I have broadcast radio interviews with Uri, dozens of listeners have reported psychokinetic phenomena in their own homes.

Perhaps even more remarkable, thousands of individuals in England, France, Germany, Switzerland, Norway, Denmark, Holland and Japan are now reporting that they can also use PK to bend spoons after having only seen Geller on television.[322,323] Ironically, the same social phenomena seems to occur when skeptics, masquerading as psychics make similar radio and television appearances.

In a letter published in the April 10, 1975, issue of *Nature*, J. B. Hasted, D. J. Bohm, E. W. Bastin, and B. O'Regan report on the apparent partial dematerialization of a single crystal of vanadium carbide, encapsulated in plastic. The authors claimed that "there is no known way of producing this effect within the closed capsule and no possibility of substitution." The letter stressed the need for scientists to remain open-minded toward such extraordinary phenomena and to pay attention to psychological variables that can affect experiments. The crystal disappearance was not regarded as conclusive evidence as the authors did not actually observe or measure the change as it occurred. Nevertheless, they claimed to have "significant work in progress."

At a conference on The Physics of Paranormal Phenomena held in Tarrytown, New York, it was estimated that psychokinetic metal-bending has been witnessed in at least sixty different people.[324]

Metallurgical analyses have been made of several objects bent or fractured by Geller. In many instances, the results were no different from those of similar objects broken by the scientists as controls. In some instances, fatigue fractures were observed, even though the metal was new (i.e., key blanks) and was bent without the application of known physical stress.

Perhaps the most interesting finding related to a platinum ring that spontaneously developed a fissure in Geller's presence — although he was not touching it. This ring was analyzed by physicist Wilbur Franklin with a scanning electron microscope. He claimed that adjacent areas of the ring indicated totally different conditions resembling (1) fracture at a very low temperature, such as with liquid nitrogen, (2) distortion as if by a mechanical shear, and (3) melting at a very high temperature. Although the ring was fractured at room temperature, conditions (1) and (3) were observed at locations only one hundredth of an inch apart. Franklin pointed out that there was no known method to duplicate such findings at room temperature — and that such findings were extremely difficult to fabricate even by known laboratory techniques.[325]

Poltergeist Cases

An altogether different line of PK investigation has been poltergeist research. The word *poltergeist* is German and means a noisy and rattling spirit. Modern investigators, however, view the poltergeist as a spontaneous, unconscious, recurring psychokinetic phenomena centering around a person, usually an adolescent simmering with repressed feelings of anger. Unable to vent these feelings in a normal fashion, he manifests them through psychic means.

William G. Roll, of the Psychical Research Foundation (affiliated with West Georgia College in Carrollton, Georgia), is one of the foremost American researchers of poltergeist phenomena. One typical case occurred in a Miami warehouse full of glasses, ashtrays, plates and novelties. The disturbance, which involved more than two hundred incidences, took place in January 1967. Police officers, insurance agents, a magician and others were unable to explain it. Roll describes his approach:

> It soon became clear that the incidents were concentrated around one employee, Julio, a nineteen-year-old shipping clerk. Certain areas of the large warehouse room where the disturbances took place were more frequently affected than others and these became the focus of the investigation. The investigators designated certain parts as target areas and placed objects in them hoping that the objects would be affected while Julio and the other employees were under observation.[326]

In several cases this is precisely what did happen. Julio was brought to the Psychical Research Foundation (then located in Durham, North Carolina) for further testing which revealed his strong feelings of hostility, especially towards parental figures, which he could not express openly and from which he felt personally detached. PK tests with a dice-throwing machine produced suggestive results with Julio. In addition there was a poltergeist disturbance of a vase in a hallway in the laboratory while Julio was standing with the researchers several feet away. Within recent decades there have been about thirty well-documented poltergeist cases.[327]

Matthew Manning

Perhaps the most intriguing "poltergeist person" to be studied so far is Matthew Manning, who since 1966, at the age of eleven, has been the center of various psychokinetic outbreaks. Dr. A. R. G. Owen, former Cambridge mathematician and geneticist, who authored perhaps the most comprehensive book on poltergeists, claims that Manning "is probably the most gifted psychic in the western world."[328]

William G. Roll

In addition to typical psychokinetic outbreaks, Matthew has shown an apparent ability to communicate with spirits via automatic writing and drawing. Although his schoolmaster claims that he has never shown any particular drawing talent, he is able to reproduce — without any apparent effort or concentration — detailed and precise works of art in the style of deceased masters such as Durer, Picasso, Beardsley, and Matisse. Automatic writing has been produced in languages with which Manning was unfamiliar. Often verified information, and even psychic diagnoses, come through in this way. Thus the phenomena contain the kinds of evidence we might really associate with spirit phenomena.[329,330]

Particularly since the public demonstrations of Uri Geller, Manning has exhibited intentional psychokinetic effects amenable to scientific testing. When tested by Nobel laureate physicist Brian Josephson in Cavandish Laboratory at Cambridge University, Matthew demonstrated an unusual spinning effect over a compass needle. Ironically, when further instrumentation was used to record magnetic changes in the vicinity of the compass, the needle of the compass would only remain stationary. Nevertheless, the instruments did detect magnetic changes. Josephson maintains that until further data is collected, his results will still have to be labelled "inconclusive."[331,332]

In other tests, conducted at the New Horizons Research Foundation in Toronto, Manning was able to demonstrate metal-bending, on demand, which was actually recorded on motion picture film. Several tests were conducted that recorded physiological measures such as muscle tension and brain waves during psychokinetic activity.[333]

No unusual muscular activity was noted. However, rather profound changes were seen in the electrical activity of the brain which have been described by Dr. Joel Whitton as a *ramp function* (actually a rather pictorial description of the chart printout). The ramp functions appeared similar to the EEG patterns in a patient suffering from an overdose of a hallucinatory drug and is suspected to stem from the older and deeper areas of the brain.[334]

These findings led the Toronto scientists to speculate on neurophysiological psi interrelationships. Dr. Whitton conducted a small-scale investigation with a number of known psychics to determine if they had any common childhood experiences. The answer was quite fascinating — for the one experience that all of the psychics had suffered in common was a severe electric shock before the age of ten. Although Matthew Manning

Julio being tested for psychokinesis (courtesy Psychical Research Foundation)

Broken glass from Miami poltergeist (courtesy Psychical Research Foundation)

did not recall such an incident, his mother informed the scientists that she had been so severely shocked three weeks before Matthew was born she was afraid she would lose him.

This line of research seems to have implications for psychical research. Perhaps the increasing number of children who can now ostensibly demonstrate PK is associated with the greater number of electronic gadgets in modern homes — with the correspondingly increased probability of electric shocks. However, even if further inquiry in this direction proves revealing, it will still fail to account for another type of poltergeist case also documented by the Toronto group.

Philip the Ghost

One most exciting PK case of the poltergeist variety actually did not involve a real ghost, or an individual, but rather an imaginary spirit named Philip. This unusual situation developed in Toronto as a group of members of the local Society for Psychical Research decided to meet regularly in an effort to conjure an apparition they created. They invented the character of Philip, an aristocratic Englishman who died of a tragic remorse during the seventeenth century. Every week for an entire year the group met for meditation, concentrating on Philip's story, in an attempt to manifest an apparition.

There was no success, but in the summer of 1973 they learned about similar efforts made in England since 1964 by Batcheldor, Brookes-Smith, and Hunt. The British approach had been directed toward producing the physical phenomena of the old type séances of the Victorian era. Instead of quiet meditation, they created an atmosphere of jollity, together with singing songs, telling jokes, and exhortations to the table to obey the sitter's commands. Consequently, the Toronto group decided to take this approach.

Extraordinary things began to happen: The table began to produce raps which became louder and more obvious as time went on. Using one rap for yes and two for no, the table was actually able to answer questions and recreate the personality of Philip. Occasionally, however, the answers were out of character for Philip.

These raps occurred in a fair amount of light, with all the participants' hands in view on the table. The thickly carpeted floor generally prevented foot-tapping. At least four members, of the original group of eight, were necessary to produce this phenomena. However, no single person was found to be essential. Eventually the table began to move around the room at great speed with no one touching it. On one occasion, the table completely flipped over.

These phenomena are still continuing and are now being duplicated by other groups who are learning how to unlock their own hidden PK abilities. All efforts at investigation have so far been unable to detect fraud and a two-hour film has even been made documenting these occurrences.[335,336,337,338,339]

This imaginary communicator, created by a group consciousness, seems to suggest that other alleged spirits, ghosts, entities, and perhaps even flying saucers also originate from within us.

On several occasions the Philip group has been able to produce psychokinetic phenomena for live television audiences in Toronto. Indications were, in fact, that the large audience aided in the production of more dramatic phenomena. Reports state that there were two other groups within the Toronto Society for Physical Research also able to produce spirit-like psychokinetic phenomena.

One of these, the "Lillith group" has concocted a fictitious ghost story as the focus of their concentration. Like the Philip story,

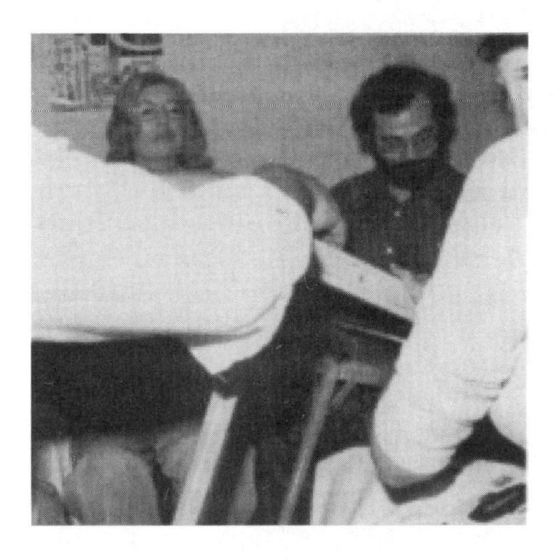

LEFT, THE "PHILIP" TABLE TILTING WHILE EVERYONE'S HANDS ARE SAID TO BE IN LIGHT CONTACT WITH THE TOP (COURTESY ROBIN E. OWEN, NEW HORIZONS FOUNDATION).

BOTTOM, THE "PHILIP GROUP" FROM LEFT TO RIGHT: IRIS M. OWEN, SID KORMAN, BERNICE MANDRYK, AL PEACOCK, MARGARET SPARROW, DOROTHY O'DONNELL, ANDY HENWOOD, LORNE HENWOOD (COURTESY ROBIN E. OWEN, NEW HORIZONS FOUNDATION).

it has all the proper dramatic elements of romance and tragedy. Learning from the Philip group, the Lillith group was able to enter into the jovial atmosphere conducive to phenomena without spending time on meditations or visualizations. The phenomena they produced have been quite striking, including table levitations said to be more impressive than those caused by the original group. The Lillith group is now attempting to produce voices on magnetic recording tape. Preliminary results are encouraging.

During the annual Christmas party of the Toronto SPR, a large group of individuals were able to spontaneously develop psychokinetic table-rapping. Somebody asked the "spirit" if it were Santa Claus and from then on the responses continued as if it were old Saint Nick himself rapping. Since then a third Toronto group has developed psychokinetic table-rapping, this time ostensibly coming from Dickens' Artful Dodger.[340]

Since the metal-bending demonstrations of Uri Geller and Matthew Manning in Toronto, the Philip group has also shown some success in this direction. In one instance, a metal medallion, which was partially bent during the group session, continued to bend after the group departed until it completely crumpled.

Perhaps the most significant development in the Philip story is the qualitative acoustic measurement of psychokinetic table-rapping. Normal raps on the table used in the Philip session produced a sound that typically lasted for about half a second. On the other hand, many of the raps produced by Philip were shown to last only 0.16 sec. This was true in spite of the similarities in loudness and frequency of the raps.[341]

Further research along these lines may provide a clearer notion of how the sounds are produced. Although, it would seem likely that once a clear understanding of the phenomena is gained the quality of the raps themselves will change.

Ted Owens — The "PK Man"

Earlier in the discussion of UFOs, I presented some material suggesting that Ted Owens, now deceased, had an ability to create various large-scale effects through telepathic communication with "space intelligences." Owens, himself, vacillated as to whether these effects were due to his own PK abilities or to the intervention of beings from another dimension.

Owens learned about psychokinesis in the late 1940s, when, as a Duke University student (after having served in the Navy during the war) he was a research assistant in the Parapsychology Laboratory under the direction of J. B. and Louisa Rhine. He claims that he discovered that it was just as easy, in terms of mental effort, to produce large-scale psychokinetic events as it was to produce small-scale events such as with Rhine's dice experiments (which will be discussed in Section III). Before he died, he expressed his hope that this "discovery" would, one day, be termed the "Owens Effect." Here are some examples:

Lightning Striking

A letter from the Ted Owens files dated August 12, 1967, addressed "To Whom It May Concern" and signed by Charles Jay of Morton, Pennsylvania, reads:

> Several weeks [ago] I took my friend, Kenneth Batch, over to Philadelphia to visit Ted Owens. It was a rainy day, and we had heard of Ted Owens' ability to make lightning strike...so we asked Ted Owens to give us a demonstration of his so-called power...by having...lightning strike in given areas we would designate.
>
> The three of us went out onto a balcony outside of Ted Owens' apartment...and my friend and I asked Ted Owens to have lightning strike at or near the top of the City Hall. In the ensuing period of time there were three massive strokes of lightning in that exact direction. And those were the only three bolts that struck in the entire sky...just where Ted Owens had pointed his hand.
>
> To test this, we then asked Ted Owens to make lightning strike in an entirely different portion of the sky. He pointed his hand...and the lightning appeared in that different area, exactly where we had asked it to appear. No other bolts appeared anywhere in the sky at any time during our experiments, except exactly where Ted Owens pointed his hand.
>
> My friend and I were in complete agreement that the experiment was a complete success.

The above testimonial is not an isolated incident. I interviewed an attorney who wishes to remain anonymous, a partner in a Philadelphia law firm where Owens worked as a typist in 1967. Although my interview was ten years after the event, this lawyer vividly remembered the afternoon *he* challenged Owens to influence lightning.

It was an overcast day in May of 1967. There was neither rain nor lightning. The law firm was located in an office tower overlooking the Camden bridge. The attorney, Mr. M., challenged Owens to make lightning strike the bridge — on the spot. Owens pointed his hand at the bridge and seemed to concentrate. Within minutes a bolt of lightning struck the bridge. According to Mr. M., it was the only lightning in the sky. His signed affidavit is in my files.

Weather Control

On February 12, 1974, Owens wrote a letter to Ed Busch, of radio station WFAA in Dallas, Texas. Owens, who had appeared on Busch's radio program the week before, made a claim:

> If you recall, on the program itself you requested that I make it snow instantly, and your colleague wanted heat. All right....[I] will cause freakish weather and, of course, heat. Normal summer heat, coming up, should be amplified tremendously, perhaps to break a record. You will have great storms, lightning attacks, etc. But into this will be the intelligence not to cause death or injury to Texas people, but to show how I...can control the weather anyplace in the world.

On February 16, 1974, newspaper clippings record than an earthquake centered in the Texas panhandle shook parts of Texas, Oklahoma, and Kansas. The tremor registered between 4 and 4.5 on the Richter scale. On March 20, 1974, a storm developed over Texas and moved rapidly to the northeast. By the time it arrived in Georgia, winds reached up to a hundred miles per hour. A meteorologist with the National Weather Service said, "It's the strongest wind I've ever seen in the continental United States."

Ed Busch wrote a statement testifying to these and other events, dated May 7, 1974. He stated:

> Owens sent me a letter, stating that he...would produce a "major demonstration" of weather control over Texas. Following Owens' letter Texas was struck by an earthquake, 4.5 on the Richter scale. Then Texas was struck by high winds and tornados. Then Texas had the coldest weather ever in its history. Then Texas was struck with hot winds that destroyed half of the Texas wheat crop.
>
> I am submitting this statement of fact to Owens at his request. It is true and accurate to the best of my knowledge. Whether Owens had any connection with the above weather phenomena, I do not know; perhaps it was mere coincidence.

Dozens of similar "demonstrations" appear in my records. One of them occurred in the San Francisco Bay area in early 1976 and was the cause of my learning about Owens' remarkable claims. On January 30, 1976, Owens sent the following letter to Harold Puthoff and Russell Targ at SRI International, a giant research organization located in Menlo Park, California, just south of San Francisco:

> Last night over TV the evening news showed a stricken California. No water. "The worst drought in 72 years." "Only three times in the entire history of the State of California...has such a drought appeared." Crops are dead and dying...and the animals are in pitiful condition.
>
> Now I, Ted Owens, PK Man...will change all of that. Within the next 90 days from the time of this letter...I will pour and pour and pour rains onto the State of California...until it is swimming in water, and the dangerous drought is completely over. There will be storm after storm, lightning attack after lightning attack, and high winds....

A UPI clipping from February 1, 1976,

confirms Owens' statement about the drought:

> The cost of a California winter-drought has mounted to about $310.5 million....Ten more days of drought could precipitate an emergency in the livestock industry. But there is little moisture in sight.

However, by February 6, 1976, the headlines changed:

> SAN FRANCISCO SNOWED BY A RECORD SNOWFALL
>
> The biggest snowfall in exactly 89 years hit the city and surrounding areas....The storm also featured lightning and sleet. A giant television tower on Mt. San Bruno, south of San Francisco, was hit by lightning about 8:30 p.m. Wednesday knocking several TV stations off the air.

The following news clip was sent to Owens by Puthoff and Targ, who had received Owens' prediction only one day earlier. From the Palo Alto *Times*, Thursday, February 5, 1976:

> RARE SNOWFALL ENDS DROUGHT ON PENINSULA
>
> The unexpected and unfamiliar weather was at odds with a forecast Wednesday that the dry spell would continue in the Bay Area....Not since the morning of Jan. 21, 1962, have Mid-peninsulans awakened to find their homes blanketed with snow.

The Oakland *Tribune* of February 5, 1976, stated that the storm brought with it:

> ...nearly every phenomenon in the weatherman's book throughout the Bay Area....Snow, hail, sleet, light rain, thunder and lightning hit the Bay Area after weeks of dry, balmy weather....Varying amounts of rain fell upon the lower two-thirds of the state....In northwestern California, there are gale warnings.

On February 10, a UPI story stated:

> The rainy season continued in California for the sixth consecutive day. Some mountainous regions of the state have received 6 to 8 inches of rain and coastal areas have measured 3 to 4 inches.

UFO sightings, power blackouts and fireballs were also reported during this period.

The Owens case is extremely complex, involving more than a hundred ostensible macro-PK events, synchronicities, UFO appearances, poltergeist-type phenomena, as well as apparitions and appearances of monster-like creatures. It was further complicated by Owen' own colorful personality which was far from saintly and far from conducive to thoughtful scholarly exploration. In addition, many of his seeming demonstrations involved deaths and accidents. If Owens' supposed powers were real, they were sometimes very dangerous. This situation alone led several researchers to reject any possibility of seriously studying or testing Owens' claims.

My years of involvement with the Owens case suggest to me that humanity is far from ready to confront the possibility of large-scale PK phenomena of this sort. On the other hand, if such abilities are possible, it is not wise to neglect their study.

Firewalking

Firewalking has been reported all over the world. Giovanni Ianuzzo reviews thirty-three scholarly accounts of firewalking, worldwide, dating back to 1894.[342] He touches upon various possible explanations for successful firewalking including fraud, calloused feet, skin moisture, physiological explanations related to altered states of consciousness, and psychic ability. His somewhat circular conclusion is "that the phenomena of fire immunity is related to a modification of the human organism to intensive thermal stimuli in altered states of consciousness."

A particularly dramatic episode of firewalking in Singapore has been described by anthropologist Ruth Inge Heinze. Eight hundred devotees participated in this ceremony, spending several days and nights in ritual purification, involving praying and fasting, within a Hindu temple. More than a thousand logs were placed into a twenty-foot pit.

Heinze, who observed the entire ceremony, noted that two people fell into the pit and had to be rescued by temple attendants. One man died an hour after taking part in the ceremony. Approximately forty people, or 5% of those participating, received burns. Heinze observed, "I could predict whether people would survive the ordeal by looking at their faces before they entered the pit: in particular, doubt appeared on the faces of some white-collar workers, and they got burned, sinking into the coals up to their ankles."[343]

Writing in a mainstream professional psychological journal, chemist Meyn Reid Coe, Jr. chronicles his own successful attempts with firewalking and a variety of related behaviors:

> Touched red-hot iron with my fingers.
>
> Touched red-hot iron with my tongue.
>
> Touched molten iron with my tongue. (No sensation! Can't feel it.)
>
> Bent red-hot steel bars by stamping them with my bare feet.
>
> Ran barefoot on red-hot iron.
>
> Walked on red-hot rocks.
>
> Plunged my fingers into molten lead, brass and iron.
>
> Took a small quantity of molten lead in my mouth and spat it out immediately. (Once I allowed it to solidify in my mouth and almost was burned. Never try this.)
>
> Carried red-hot coals around in my mouth.
>
> Popped red-hot coals into my mouth.
>
> Chewed charcoal off burning sticks. (This is easy if done fast enough.)
>
> Walked on beds of red-hot coals, taking eight steps to cross a fourteen-foot pit.
>
> Placed my fingers, hands and feet in candle flames until covered with carbon black. (No burns! Not hot! Only warm!)
>
> Held my face, hands and feet in fire for a short time.[344]

Psi researcher Larissa Vilenskaya, a Soviet emigree, studied firewalking procedures with American guru, Tolly Burkan. Burken, and one of his students, Anthony Robbins, claim to have taught firewalking to over ten thousand individuals — as of mid-1984. Vilenskaya, herself, has been among those trained by Burkan to instruct seminars in the art of firewalking and has generously written several accounts of her experiences.[345,346]

Doherty reports an experiment by noted physicist Friedbert Karger in the Fiji Islands. Karger, using temperature sensitive paints, determined that a native firewalker stood on a specific rock for seven seconds which had a temperature of 600 degrees Fahrenheit (315 degrees Centigrade). The paint on the man's feet revealed that they had not been hotter than 150 degrees Fahrenheit.[347]

Without any preparation, I myself had an opportunity to participate in a firewalking ritual with a group of Kailas Shugendo Buddhists in San Francisco under the direction of Dr. Ajari Warwick. The religious practices of these individuals include daily fire rituals of several kinds, maintaining an ambulance rescue service (pulling people out of plane wrecks and fires), as well as mountain climbing — and country-western music. Unlike many "spiritual groups," the Kailas Shugendo people make no effort to proselytize. In fact, they actually discourage would-be converts. They are extremely disciplined, yet they possess an overflowing humor. It was in one such peak of gaiety that Ajari invited me to come to a ritual with my

Larissa Vilenskaya

camera and tape recorder. I regarded the invitation as an honor because I knew the group was very cautious about allowing the public to treat the practices as a circus sideshow. I did not expect to attend, as I was without transportation at the time and the ceremony took place on a remote beach. I put the idea out of my head. However, by a coincidence, a friend with a car appeared at 7:30 a.m. on the appointed day — and off we went.

The ceremony was modest — simply a six-foot pit of flaming logs that we walked over dozens of times, quite briskly, generally stepping once with each foot. The flames rose up and singed the hair on my legs, although I felt no pain and suffered no burns. I had complete confidence in Ajari who asked that I follow him across the pit. Microphone in hand, I recorded my impressions on tape as we went over the flames. I must admit that I actually felt protected in some way. It was a totally uplifting experience. Later on, some psychic readers mentioned that I was surrounded by a white light. Perhaps they noticed my silly smile.

Actually the phenomena of handling or footing hot coals provides a very tricky problem for logical analysis. The first experimental tests of firewalking were conducted by the University of London Council for Psychical Investigation in 1935 under the direction of Harry Price. In his initial report Price discussed several sessions held with the Indian fakir, Kuda Bux, who also performed acts of blindfolded clairvoyance of questionable authenticity. According to Price, the blindfolds always allowed a line of vision along the side of the nose. His firewalking was more impressive. In nearly a year of advertising for firewalkers with which to conduct experiments, Kuda Bux was the only individual to step forward.

Before a large audience of newsmen and scientists, he walked barefoot across a twelve-foot pit of burning coals. During one demonstration it was windy and the surface temperature of the fire was measured at 806 degrees F., while the body of the fire was 2552 degrees Fahrenheit — hot enough to melt steel. Kuda Bux took four steps across the pit and suffered no burns. His feet were carefully inspected both before and after his performance to eliminate the possibility that he could have used chemicals of any sort to protect himself. The entire event was also recorded on film.

Kuda Bux claimed that he could convey an immunity to other individuals who followed him across the coals. Unfortunately, this was not the case during the first set of experiments. All other individuals who followed him over the coals suffered minor burns.

Human flesh scorches more easily than cotton fabric, and experiments with a wooden shoe covered with calico indicated scorching in less than a second when placed on the hot embers. However, the scientists noticed that no portion of the skin was in contact with the hot embers for as long as half a second. Perhaps, they thought, the art of firewalking merely involved the skill of stepping quickly and properly.[348]

A second series of experiments seemed to

ABOVE, MICROPHONE IN HAND, THE AUTHOR FOLLOWS AJARI OVER THE FLAMES. "IT DOESN'T HURT."

LEFT, AJARI WARWICK FIREWALKING; *RIGHT,* KUDA BUX FIREWALKING.

confirm this opinion. This time the firewalker was another fakir from India, Ahmed Hussain. He showed approximately the same ability as did Kuda Bux. Interestingly enough, the temperature of his feet was found to be 10 degrees Fahrenheit lower after the firewalk than before, indicating a certain amount of autonomic physiological regulation. However, when the length of the trench was increased to twenty feet, Hussain also suffered burns. Furthermore, several amateurs now found that they could walk across the twelve-foot fire trench without suffering burns. These tests led Price to conclude:

> ...any person with the requisite determination, confidence, and steadiness, can walk unharmed over a fire as hot as 800 degrees Centigrade. The experiments proved once

and for all that no occult or psychic power, or a specially induced mental state, is necessary in a firewalker.[349]

Price's conclusion is supported by the measurements of feet movement. In normal walking, it was found that the time from the contact of the heel with the floor until the big toe left the floor was 0.65 second. For only 0.05 second was the entire sole of the foot in contact with the floor. During the brisk firewalk contact was even less. Price's argument entirely depends upon this brief contact time.

The literature on fire-handling is much more difficult to deal with. Careful measurements such as Price's have not been made, but the observations seem to mitigate against a simple physical interpretation. A description of a fire-test with the nineteenth-century medium, D. D. Home, was written by Lord Adare who later became the Earl of Dunraven:

> He went to the fire, poked up the coals, and putting his hand in, drew out a hot burning ember, about twice the size of an orange; this he carried about the room, as if to show it to the spirits, and then brought it back to us; we all examined it. He then put it back in the fire and showed us his hands; they were not in the least blackened or scorched, neither did they smell of fire, but on the contrary of a sweet scent which he threw off from his fingers at us across the table. Having apparently spoken to some spirit, he went back to the fire, and with his hand stirred the embers into a flame; then kneeling down, he placed his face right among the burning coals, moving it about as though bathing it in water. Then, getting up, he held his finger for some time in the flame of a candle. Presently, he took the same lump of coal he had previously handled and came over to us, blowing upon it to make it brighter. He then walked slowly around the table, and said, "I want to see which of you will be the best subject. Ah! Adare will be the easiest..." Mr. Jencken held out his hand saying, "Put it in mine." Home said, "No, no, touch it and see." He touched it with the tip of his finger and burnt himself. Home then held it within four or five inches of Mr. Saal's and Mr. Hurt's hands, and they could not endure the heat. He came to me and said, "Now if you are not afraid, hold out your hand;" I did so, and having made two rapid passes over my hand, he placed the coal in it. I must have held it for half a minute, long enough to have burned my hands fearfully; the coal felt scarcely warm. Home then took it away, laughed, and seemed much pleased. As he was going back to the fireplace he suddenly turned round and said, "Why, just fancy, some of them think that only one side of the ember was hot." He told me to make a hollow of both of my hands; I did so, and he placed the coal in them, and then put both his on the top of the coal, so that it was completely covered by our four hands, and we held it there for some time. Upon this occasion scarcely any heat at all could be perceived.[350]

Sir William Crookes also describes a fire-handling incident with Home. Crookes stated that he tested, in his laboratory, a fine cambric handkerchief the medium had folded around a piece of red charcoal, then fanned to white heat with his breath without damaging the handkerchief. Crookes concluded that the cloth "had not undergone the slightest chemical preparation which could have rendered it fire-proof."[351]

A similar fire-test was performed by Jack Schwarz, of Selma, Oregon, before physicians of the Los Angeles County medical and hypnosis associations. After having been examined by the doctors, Schwarz put his hands into a large brazier of burning coals, picked some up, and carried them around the room. Subsequent examination showed no burns or other signs of heat on his hands.[352]

A number of observations of similar fire-handling among the "saints" of the Free Pentecostal Holiness Church are reported by Dr. Berthold E. Schwarz. Members of this church, in states of religious ecstasy are well known for handling poisonous snakes, swallowing strychnine, and handling fire. Schwarz witnessed such an incident:

> Once this saint, when in a relatively calm mood, turned to a coal fire of an hour's duration, picked up a flaming "stone-coal" the size of a hen's egg and held it in the palms of his hands for 65 seconds while he walked among the congregation. As a control, the author could not touch a piece of burning charcoal for less than one second without developing a painful blister.[353]

Apparently the saints' immunity is related to trance. Schwarz describes an incident in which a "brother" applied a coal oil torch to the palm of his hand for several seconds with complete immunity. However, when he noticed that a piece of wick was breaking off, he woke from his trance and suffered a burn.

In so far as these reports cannot be explained on the basis of Price's theory of deft and speedy handling, science has yet to arrive at an adequate explanation of the fire-tests. One hint of a theory comes from the notion of Prof. Clark Maxwell's Sorting Demons — tiny little beings who can stop, strike, push or pull atoms and molecules in such a fashion as to insure there would always be a layer of cool, fresh molecules between the skin and the frenzied, spinning, energetic molecules at red-heat. If the answer is to be found on the molecular of atomic levels, further investigation of the phenomena of fire-handling will certainly expand our knowledge of bio-physics.

Dr. George Egely of Hungary has developed a mathematical model for calculating the temperature distribution on the human sole during firewalking. His model suggests that one can walk or run, with relative safety on a surface as hot as 400 degrees Centigrade. However, he admits that this model is incomplete as he lacks data on perspiration and other possible cooling mechanisms of the skin that might allow for the possibility of firewalking over materials of even hotter temperatures.[354]

No good explanation has yet been offered for this heat resistance. However, if one assume that psi abilities are of the type that respond to real human needs, the firewalking experience provides a repeatable context for the observation of a response to genuine survival needs at work. It is highly unlikely that such a context could ever be duplicated in a scientific laboratory. Even if motivational and need-related factors could be simulated, such experiments would never (quite properly) withstand the scrutiny of "human-subjects research" and "ethics" committees.

One study did conduct psychological tests, with ninety-eight individuals, participating in firewalking "workshops" led by Tolly Burkan and Anthony Robbins. Of this group, fifty-two subjects walked over the coals for the first time. Sixty-four percent of these completed the walk successfully, while 36% received some indication of skin exposures to high temperatures, ranging from slight discoloration of the skin to blisters.

Several pretest factors seemed to distinguish between those who received blisters from the experience and those who did not. Ironically, the non-blisterers were more generally disposed to experience anxiety and more willing to attribute great power to others in their lives (i.e., the seminar leaders). Those who received blisters were less prone toward anxiety and more willing to assume control for their own lives. However, a greater shift in attitude was noticed by those who successfully completed the firewalking experience. They achieved the same willingness to assume power in their own lives as those who were blistered had before the experience.[355] One can only assume from this puzzling data that firewalking is not for everybody.

Psionics — Practical Application of Psychic Awareness

Can psychic powers be used for detrimental purposes? What are the limits of psychic ability? Certainly some inferences can be obtained by drawing upon the history, the literature, and the folk-wisdom of psi. I sometimes use the term "psionics," taken from the science fiction literature,[356] to describe the applied branch of psi exploration.

Psionics is not particularly concerned with the truth value of scientific, pseudoscientific or religious theories about the nature of psychic functioning. It is concerned with the practical utility of such theories for individual practitioners. It is concerned with reliability, consistency and magnitude of psi effects, not in the laboratory, but in the world of business and professional affairs.

Harmful Purposes

Many traditions teach that psychic abilities can only be used for good purposes, for instance, healing. Other harmful applications are said either not to work or to rebound back upon the evil-wisher. Dr. Louisa E. Rhine, who had made a lifelong study of spontaneous psychic experiences, took such a stance in responding to the question of a seventh-grade inquirer:

> No Nancy, ESP could not possibly be used to hurt anyone physically or mentally. It is true that sometimes people get the false idea that someone is influencing them by ESP. They think it is by telepathy, but this is very unlikely. Telepathy seldom, if ever, works that way, for no one can send his thought to another and make him take it....
>
> The only way a person could be hurt would be by his belief that he could be so affected. It is possible sometimes for a person to "think himself sick" for other reasons and in the same way he could think himself sick by believing that someone was affecting him by telepathy. But, if so, his sickness would be caused by mistaken suggestion, not by telepathy.[357]

Mrs. Rhine's reassuring answer reflects an understanding of the psychological mechanisms involved in mediating psi. It is reasonable to think that individuals can reject telepathic suggestions as easily as you, the reader, might reject any statement in this book. For an aware and enlightened individual this would certainly be the case. It is also the case that much of what we think of as psychic phenomena is merely due to suggestion.

The anthropological literature regarding tribal cultures indicates that the violation of a taboo and the placement of a hex can result in death within a few days. This has been attributed to an extreme operation of the stress-response syndrome by modern researchers. We might consider the reported instances of death, illness, and accidents from hexes, voodoo, spells, and curses to be the result of suggestion. But we might just as easily ask ourselves whether, if psi could heal people independently of suggestion, it could not also harm them. Perhaps psi — like electricity — is a neutral force from a moral perspective? A number of apparent hexes seem to have occurred without the knowledge of the victim.[358,359]

The research of the Soviet physiologist Leonid Vasiliev suggests that telepathic hypnotic induction may be occasionally instrumental in effective behavior manipulation over distances.[360] Similar telepathic experiments have been used to awaken sleeping subjects, with slightly less success. However, few subjects are so susceptible, and we have yet to understand the mechanisms that differentiate good and poor subjects.

National Security Applications

Ancient History and Folklore

The Bible. In the Bible as related in II Kings 6, the prophet Elisha used clairvoyant abilities to inform the king of Israel about the battle plans which the king of Syria had formed against him. Informed of Elisha's abilities, the Syrian king sent a host of chariots and horsemen to capture the prophet. However, according to the biblical account, Elisha used his abilities to blind and confuse the Syrians so that they would be captured by the Israelites.

Similar and even more dramatic tales are told of the exodus of the Jewish nation from Egypt; of the original Hebrew conquest of Canaan; and of the subsequent military conquests of Saul, David, and Solomon.

Asian Martial Arts. The earliest treatise on warfare, *The Art of War*, written in 500 B.C. by the Chinese general Sun Tzu, details the intimate link between success in battle and the skilled management of a "force" which was called *ch'i*. The basic principles are elucidated in this amazing document.[361]

Sun Tzu argues that wars are won through a combination of conventional military tactics and extraordinary methods which involve the knowledge and control of *ch'i*, which flows through the body of the warrior and can be used to influence the mind of the enemy to produce illusions, deception and weakness.

The warrior cultivated *ch'i* through self-knowledge gained by following the mystical Taoist traditions. Mental stillness and other psi-conducive states enabled the warrior to obtain a poise and concentration so intense that it was effortless in its deadly spontaneity. Such training emphasized the ability to maintain the meditative state in the midst of intense physical activity. The martial arts trained the warrior to take advantage of the slightest break in the enemy's concentration.

Joan of Arc. A peasant girl with no military training followed her visions and voices to lead the bedraggled armies of France to victory against the English. Many ostensibly miraculous events — the subject of continuing historical debate — led to the French Dauphin's appointment of Joan as titular head of his army. Joan was burned at the stake in 1431 as a witch. In 1456, an ecclesiastical court proclaimed the iniquity of her first trial and annulled its judgment. In 1920, she was canonized as a saint.

The World Wars

In June 1919, in an action against the Hungarian Republic, Czech soldiers were put into a hypnotic state and asked to clairvoyantly scan the landscape to determine the enemy's strength and position. A pamphlet titled *Clairvoyance, Hypnosis and Magnetic Healing at the Service of the Military*, written in 1925 by Karl Hejbalik, reports that the information obtained through these means always proved correct when later checked through normal means. The contemporary Czech psychotronic researcher Zdenek Rejdak interviewed the individuals involved in the Czech psi maneuvers. According to Rejdak, they confirmed Hejbalik's account "in all details."[362]

The Nazis are said to have assembled many powerful occult adepts from Tibet and Japan to train and advise them. One of the most important and powerful groups in Germany was the Nazi Occult Bureau, which attempted to use occult forces for espionage and the magical control of events including a conscious, pseudo-Nietzschean attempt to replace Christianity with the ancient Teutonic myth of the war god, Wotan. Coincidentally, in the early 1930s, the great Swiss psychiatrist Carl Jung noticed a marked pattern of imagery of the war god Wotan in the dreams of his German patients. The situation was so dramatic that it prompted him to write, in an essay in 1933, that the German people were subconsciously preparing themselves for war.

Hitler obtained extensive occult training from the German nationalistic Vril Society and the adept circle known as the Thule Group. Teachings of these groups account for many aspects of Nazi culture which are inexplicable in terms of ordinary historical scholarship.[363]

Hitler's personal psychic abilities, especially clairvoyant and precognitive visions, were by some accounts instrumental in many of the dramatic tactical victories during the early period of the war. Eventually, his intoxication with power and the use of drugs so poisoned his mind that he compulsively followed instructions received through visions, and these led to disastrous strategic errors.

British and American intelligence employed astrologers and clairvoyants to anticipate the occult advice being given to Hitler and his forces. According to Linedecker's *The Psychic Spy*, the Allies used out-of-body practitioners to scout key locations inside enemy territory from an island in the Atlantic.[364]

Lord Hugh Dowding, head of the Royal Air Force during World War II, and often called "the man who won the Battle of Britain," had experiences during the war which led him later to become a major figure in the spiritualist movement. Released secret documents of the British Army reveal that Dowding's wife was a sensitive. Using methods now known among psi researchers as "remote viewing," she was able to detect enemy air bases that the army had not discovered through conventional surveillance. These abilities so impressed Lord Dowding that he believed himself to be in contact with spirits of the British airmen who had been downed in battle.

Another of the great Allied commanders during World War II, U.S. Army General George S. Patton, reputedly possessed rare psi abilities. Patton believed himself to be the reincarnation of an ancient Roman general. General Omar N. Bradley, Patton's commanding officer during the war, has confirmed Patton's clairvoyant and precognitive abilities, referring to them as his "sixth sense." Bradley details a wartime example: After crossing the Moselle River near Coblenz with some three divisions moving south, Patton suddenly stopped his advance and collected his forces for no known reason. Questioned by subordinates about this strange behavior, Bradley expressed confidence that Patton had "felt" something that "was not apparent from the information we had at the time," which justified his action. The following day, Patton's forces were hit by a strong and otherwise unexpected counterattack which the general was able to repel only because he had earlier stopped to regroup.

Soviet interest in psi was kindled during World War II by a series of unusual events which transpired between Joseph Stalin and the well-known Polish psychic, Wolf Messing. By using telepathic hypnosis to suggest to Stalin's guards and servants that he was Lavantri Beria, the head of Soviet secret police, Messing was reputedly able to walk past them unchecked into Stalin's personal *dacha* and into the very room where Stalin was working. Stalin's subsequent tests of

Messing's abilities were published in the Soviet Journal, *Science and Religion.*

THE JEWELS OF WOLF MESSING

BY JEFFREY MISHLOVE

I had a dream, and I know that it's true
About the ancient jewelry of a great Polish Jew.
Wolf Messing, the mentalist, an entertainer by trade
Changed the fate of the world, on the day of his raid.
Alone he entered the dacha of Stalin.
The guards couldn't stop him; they had to allow him!

Wolf Messing the mentalist hypnotized them with ease.
They thought he was Avrenti Beria, Chief of Secret Police.
He used no costumes, nor any disguises.
He had only his jewelry, his psychic surprises.
The rumor now has it that the communists were wary.
They wanted such talents for the Soviet military.

Messing became the Russian's greatest entertainer.
He travelled the continent; he served as a trainer
For small groups of scientists, trained in physiology,
Bent on discovering the secrets of parapsychology.
So that the Soviet Union would one day become
Master of the world, through its psychic "A-bomb."

It has been forty years since Wolf Messing's project.
So, have the Russians achieved their object?
Are they the masters of space, time and beyond?
Can they call forth hidden powers, with the wave of a wand?
Oh, they've taken Kirlian photos, and psychotronics devised.
They've measured the "aura" and wood magnetized!

They hypnotize over distances, through mental control.
You couldn't fit all their psychics in the Hollywood bowl.
They've measured the language of plants talking to plants.
They've studied the voodoo and juju magic in African dance.
They've used psychic powers against the chess master Korchnoi
Could anything stop them against our American boys?

In the past forty years, they've done many a wonder.
Made a dog with two heads; they control rain and thunder.
They can manipulate brainwaves over thousands of miles
So that millions of people will nod when they smile.
But they've missed the greatest secret, and it's been just one thing:
They've lost the magic jewelry of the great Wolf Messing.

These jewels are very ancient, garish and tipsy.
They're the kind you might find on a wandering gypsy.
It was from wandering gypsies that Wolf Messing acquired them.
They've been charged with gypsy magic for over a millennium;
These jewels were the source of Messing's magic powers.
And when he died they returned to their owners.

But the gypsies can travel all over the place
From the heart of Romania, to the Bay to Breakers race.
There are gypsies in boxcars, there are gypsies on horses.
There are gypsies doing situps in the local par-courses.
The jewelry of Wolf Messing might be anywhere.
On a ship in the ocean or up in the air!

The jewels of Wolf Messing include pendents and rings.
Whoever can find them can live like ten kings.
They have all of the powers of Aladin's magic lamp.
Are they resting in the knapsack of some grizzled old tramp?
I dreamt of this jewelry and what it can do.
I dreamt of the people who are seeking them too.

There are fashion designers, and quaint demonologists,
Growth group leaders, movie producers and scientologists.
There are Mafia dons and KGB spies.

There are South American dictators, with greed in their eyes.
There are seekers of power and seekers of truth.
There are seekers of sex and seekers of youth.

The rings of Wolf Messing are like the holy grail.
Many will seek them and many will fail.
Like the Maltese Falcon, like the ancient Hebrew Ark,
These jewels can ignite a most passionate spark.
From the steppes of Siberia to the Argentine pampas,
From the Mongolian desert to the cornfields of Kansas.

I dreamt that I had them. I dreamt they were mine.
I could share with my friends all the pleasures divine.
My heart started pounding, my eyes rolled in their sockets.
All power could be mine if I put them in my pockets.
But such magic is an illusion from Capetown to Nome.
So I left the jewels in my dream, and brought you this poem.

How can I tell you that magic is not real?
What of psi research and the power to heal?
Why would I leave all those jewels in my dream?
Am I a skeptic after all, am I what I seem?
No, the jewels in my dream are where they belong.
Like the power of poetry and the power of song.

Magic is an illusion and magic is real.
This is the paradox that the jewels reveal.
Wolf Messing the mentalist lived an unusual life
Full of great glory that overcame horrible strife.
Wolf Messing was a master, and more than he seems:
He had the power to live from the jewels in his dreams.

Eastern Europe

In the 1920s, Prof. Lionid Vasiliev, director of Leningrad University's Department of Physiology, initiated a series of experiments into the effects of mental suggestion at a distance.[365] Vasiliev was motivated in part by reports of the French physiologist Pierre Janet, and perhaps also by the extraordinary power which the monk Rasputin once held over the entire Russian ruling family. Vasiliev began by attempting to influence a hypnotized subject to move his arm, leg, or even a specified muscle on cue without verbal instructions. Eventually, the hypnotist achieved success in the experiments with subjects separated by distances as great as seventeen hundred kilometers (i.e., from Leningrad to Sebastopol).

Contemporary Soviet interest in remote hypnotic manipulation has advanced considerably since this early research. Research and development now continues at the Institute of Cybernetics of the Ukrainian Academy of Science and the Institute of Psychology of the Moscow Institute of Control Problems. Experiments are no longer limited to influencing only trained subjects, but now also focus on hypnotic influence over untrained and unsuspecting persons, and occasionally even large groups.

In one Soviet study, reportedly conducted at Kharkov University, a telepathist is claimed to have been able to stimulate the brain of a dead rat for three minutes after clinical death. In another Soviet study, the psychokineticist Nina Kulagina is reported to have influenced a frog's heart to stop beating. In another Soviet experiment conducted by Prof. Veniamin Pushkin, at the Research Institute of General and Pedagogical Psychology, the same psi practitioner was reportedly able to influence the blood volume in the brain of other individuals. The subjects became so dizzy that they could no longer stand and had to sit or lie down.

The Soviets have also practiced the strategic application of telepathic manipulation. Engineer Larissa Vilenskaya, a Soviet emigre engaged in various forms of psi practice and investigation, reported on an NBC Brinkley Magazine television special that researchers recruited gifted subjects for the purpose of negatively influencing foreign political leaders while watching them on television.

The Czechoslovakian who pioneered the hypnotic method of training ESP, Dr. Milan Ryzl, defected to the United States in 1967 when he was made to understand that the Czech government wished to support his research for military and espionage purposes. Ryzl wrote that secret psi research associated with state security and defense was going on in the former USSR. One such project was for the purpose of using telepathic hypnosis to indoctrinate and "reeducate" antisocial elements.

United States

As early as 1952, the U.S. Department of State used visualization exercises to train its operatives in the use of intuitive psi faculties. A number of CIA-funded secret reports are now available through the Freedom of Information Act on projects incorporating psi research, including Projects Bluebird, Artichoke and MK-ULTRA.[366] One of the goals of each of these operations was to achieve reliable psi capability in laboratory subjects.

It was during the Eisenhower administration, according to knowledgeable sources, that the CIA set up an interagency committee to follow psi research. This committee has been active for three decades, and has sponsored a number of international scientific conferences to which Soviet neurophysiologists and cyberneticists were invited. Counter-intelligence cases during this period led the CIA to infer that the Chinese military had achieved significantly superior mind-control abilities — presumably thanks to training by the Soviet Union.

There were some attempted applications of psi in the U.S. military during the Vietnam War. U.S. marines were trained to use dowsing rods to locate land mines during the war. The first report of such was by the weekly, *The Observer*, published for the U.S. forces in Vietnam in 1967. The report summarized the situation:

> Introduced to the Marines of the 2nd Battalion, 5th Marine Regiment, the divining rods were greeted with skepticism, but did locate a few Viet Cong tunnels.[367]

Many reports have emerged from Vietnam of spontaneous ESP experiences — often saving the lives of American troops under jungle guerilla war conditions. One marine sergeant has reported that entire platoons learned how to sensitize themselves to such intuitive signals, as a basic survival mechanism.

A significant acceleration of government-sponsored research in psychic research and related areas occurred during the Nixon administration. During this period, physicists at Stanford Research Institute, now SRI International, received increased funding from a number of government sources, including NASA, for psychic studies. They made the claim that select psychics, including scientologist "clear" Ingo Swann, Israeli psychic Uri Geller, and ex-police chief Pat Price (now deceased) produced clairvoyantly obtained evidence of remote physical sites (they called it "remote viewing") with such accuracy that the most secret reaches of any military installation of the surface of the earth — or Mars for that matter — were no longer safe from view.

These experiments persuaded the Office of Naval Research and the intelligence community to continue supporting the effort. In 1973, according to knowledgeable sources, the CIA and the National Security Agency — responsible for the codebreaking and the codemaking efforts of this country — arranged a top-secret demonstration of clairvoyance, or "remote viewing," at SRI. Swann and Price, given only geographic coordinates, sketched the target site accurately — an island in the Indian Ocean. The SRI research apparently demonstrated that secret military targets, in the United States and overseas, can be described in great detail. Objects as small as the head of a pin have been described by remote-viewers over distances of many kilometers.[368] Other experiments have successfully described military targets, such as airports, from distances of several thousand kilometers.[369]

From the military's point of view, such capabilities have clear application for obtaining otherwise unavailable information about enemy locations and operations. From the point of view of the intelligence community, a trained, accurate psi practitioner would be an ideal agent. He or she could use psi skill to break secret codes, penetrate guarded military installations and reveal strategic plans. Another important use of remote viewing could be for safety inspection of military equipment.

In 1972, according to John Wilhelm writing for the *New York Times* (1977), it sent a team of scientists, under the auspices of DARPA (Defense Advanced Projects Research Agency) to SRI to "objectively evaluate" the claims of researchers Russell Targ and Harold Puthoff. DARPA was particularly concerned over the interesting coincidence, if that is what it was, that its multimillion-dollar computer at SRI went inexplicably haywire while Uri Geller was attempting psychokinesis in a nearby lab. DARPA sent Ray Hyman, a noted psychologist, magician and skeptic; Robert Van de Castle, a psychologist and expert in sleep and dream research from the University of Virginia and also past-president of the Parapsychological Association; and George Lawrence, a second psychologist/skeptic. The report of the investigation was negative.[370] Nevertheless, reports that "remote viewing" replication was underway at Fort Mead, an important center of the National Security Agency, suggest that the military and intelligence communities do not take the certainties of either proponents or skeptics at face value.

Before becoming president, Jimmy Carter reported sighting an Unidentified Flying Object near his home in Georgia, and, as is well known, requested a full report of the phenomena upon taking office.

According to Uri Geller's report, while living in Mexico he developed a relationship with the wife of the president of Mexico, Mrs. Lopez-Portillo, who had a fascination with psychic phenomena. Researchers at an institute under the direction of the Mexican president's sister, Margarita Lopez Portillo, conducted some investigations of Geller 1977. President Carter, during a visit to Mexico, heard of the Mexican interest in psychokinesis and immediately ordered an extensive Defense Intelligence Agency investigation. A report resulted, titled *Paraphysics Research and Development — Warsaw Pact*, which was the third major report on psi research released by the Defense Intelligence Agency.[371]

In 1978, a survey of fourteen active psi research laboratories by Dr. Charles Tart revealed that five of those laboratories had been officially approached by officials or agents of the U.S. government who were gathering information on psi.[372] The total known figure, at the time, for funding to mainstream psi researchers amounted to several hundred thousand dollars a year. Almost all the researchers surveyed maintained that using psi for espionage or military purposes was a very real possibility, and several were certain it was being done.

Probably more than in other areas of

scientific investigation, it is information about who funds psi research that is classified, not only what is being done. Former White House staffer Barbara Honegger reported, for instance, that the very word "parapsychology" was classified at the CIA — that is, a directive existed that it is not to be used in telephone conversations except over secure lines; and that any report with the word in it is automatically classified.[373]

Of all the services, the Navy has historically been the most open-minded about taking psi research seriously and funding it. In 1975, the Navy reportedly funded SRI to see if psychics could detect sources of electromagnetic radiation at a distance, and, in 1976, to see whether they could influence a magnetometer at Stanford University. The SRI scientists reported that they could. Critics of this set of results, however, argue that the project was guilty of "optional stopping" to achieve its results. The Navy was interested because magnetometers, which measure magnetic fields, are important in detecting submarines.

The Navy, according to knowledgeable sources, also tested self-professed psychics to see whether they could accurately describe maneuvers of a foreign navy. A New York self-professed psychic, Shawn Robbins, has reported working with the intelligence community to track the movements of foreign nuclear submarines. Robbins was originally tested at the Maimonides Hospital Psychophysics Laboratory in Brooklyn, New York. (However, her psi abilities were not determined to be significant at that time.)

Columnist Jack Anderson claims that the Navy also funded the controversial research by polygraph expert, Cleve Backster, on the alleged ability of plants to detect and respond to unspoken thoughts and feelings of living organisms, ranging from humans to brine shrimp. However, according to other sources, it was the Army that funded Backster's research, in the hopes of "training" plants to cost-effectively detect intruders in dangerous, security-sensitive areas.

According to information revealed to Barbara Honegger, during the Reagan administration for the first time the CIA officer in charge of keeping abreast of psi research noted in his periodic report to the National Security Council that there is growing reason to take the field more seriously.

The fundamental reason for this increased interest is initial results coming out of laboratories in the United States and Canada that certain amplitude and frequency combinations of external electromagnetic radiation in the brainwave frequency range are capable of bypassing the external sensory mechanisms of organisms, including humans, and directly stimulating higher level neuronal structures in the brain. This electronic stimulation is known to produce mental changes at a distance, including hallucinations in various sensory modalities, particularly auditory.[374] The analogy of these results to some spontaneous case reports in the psychical research literature has not escaped notice by the CIA, which is following the research.

A development during the first months of the Reagan administration was the release by the House Science and Technology Subcommittee, chaired by Rep. Donald Fuqua, containing a chapter and supporting appendix on the "Physics of Consciousness" (misspelled "conscience" in its table of contents). The report recommends that psi research deserves serious attention by Congress for potential future funding. It states that "general recognition of the degree of interconnectedness of minds could have far-reaching social and political implications for this nation and the world."

The primary sources cited in the report are the research studies of Harold Puthoff and Russell Targ at SRI International, and a report prepared by William Gough, technical director of the Office of Program Assessment and Integration of the U.S. Department of Energy. Gough published the cited report under the auspices of the Foundation for Mind-Being Research.

A statement of U.S. government interest in psi-war scenarios appeared in the private publication, *Military Review,* by Lt. Col. John B. Alexander.[375] Alexander asserted that psychotronic weapons already exist and that their lethal capacity has been demonstrated. He was referring here predominantly to the claims of Lt. Col. Thomas E. Beardon that third-generation psychotronic weapons, including what he called a "photonic barrier modulator: which induces physiological changes at distances, near or far; the 'hyperspatial howitzer,'" which allegedly can transmit nuclear explosions to distant locations; and a radionics-type device which, Bearden contended, sank the U.S. nuclear submarine Thresher in 1963.[376,377]

Alexander's claims were unnecessarily alarmist in nature and a number of them are known to be either exaggerated or erroneous. He stated, for instance, that research in the Transcendental Meditation Sidhis Program has produced evidence that individuals are taught in the program to levitate. Investigations into this claim have found it unsupported by any valid observations or measurements.

In 1988, the U.S. Army commissioned a study on a variety of techniques purporting to enhance human performance.

Accident Prevention

The claim is made with the Transcendental Meditation (TM) program that the practice of TM by many individuals creates an unexplained effect on the "social field" which results in reduced accidents, foul weather, sickness and crime — as measured by statistical social indicators.[378,379,380,381,382] Although these claims are presented in a quasi-empirical fashion — with statistics and control groups — they have yet to be seriously presented or evaluated in the academic literature of either sociology or psi research. The methodological problems inherent in any study conducted by an organization for the purpose of promoting training offered by that organization are sufficient to merit skepticism in the absence of independent replication.[383]

Dowsing

Dowsing, a poorly understood technique for finding underground water and minerals, seems to be gaining popularity as a result of increasing need for efficient development of resources. Some researchers maintain that dowsing involves an extraordinary sensitivity to anomalies in weak magnetic fields. This is probably true, but still does not represent the entire picture.

Henry Gross, perhaps America's best known dowser, has in many instances been able to locate oil, water, minerals and even lost people by using only a map. Gross's abilities were confirmed somewhat in tests conducted by J. B. Rhine and published in 1950 — however, these laboratory studies were admittedly not related to dowsing as practiced by Gross in his everyday life.[384] Journalistic documentation of Gross's actual work including the prevailing conditions, the people and areas involved have been published in a series of books by Kenneth Roberts.

In many instances Gross was apparently successful in pinpointing wells when conventional geological techniques had failed or had indicated there was no water.

He dowsed water for many industrial concerns including RCA Victor and Bristol-Myers pharmaceutical company. He reportedly map-dowsed from Maine fresh water in Bermuda where none had been found in three hundred years. In Kansas he dowsed thirty-six wildcat oil wells and of the seventeen that were drilled, it is claimed he was correct in fifteen instances. Seismic predictions were wrong in nine out of seventeen cases. Although a strong advocate of Gross's abilities, Roberts also discusses a number of his failures.[385,386,387]

Henry Gross's talents were investigated by the New Jersey psychiatrist and psychical researcher, Dr. Berthold E. Schwarz. His investigations included psychiatric interviews, physiological studies and field trips.[388] Schwarz found Gross — a modest, friendly game warden — to be a man of complete honesty. The physiological data, as well as direct observation, indicated that Gross expended a great deal of energy in the dowsing process. In the field studies, Schwarz claims that he observed Gross successfully dowse seven oil wells in an area where oil was not geologically expected. Gross also apparently ascertained depth, flow and other quantitative measures that were presumably beyond the ability of normal sense perception.

At the Laboratory of Physiological Cybernetics at the University of Leningrad under the directorship of Prof. P. I. Gulyaev, research has been focused on the human ability to perceive faint electrostatic fields. This research has led to renewed Soviet interest in a phenomenon known in this country as dowsing, which the Soviets call the "bio-physical effect." Studies in this area were initiated by the Soviet geologist N. N. Sochevanov, who has reportedly documented several dozen cases in which dowsing has been successfully employed in mining and drilling projects. Dowsing is also currently taught to professional mineralogists and geologists in the former Soviet Union.[389]

MAHARISHI MAHESH YOGI

Dowsing has reportedly been successful in locating commercial-grade gold ore near Krasnoyarsk, tin deposits in Kirghizia and Tadzhikstan, iron in the southern Urals, copper-nickel ores near Krasnoyarsk, lead and zinc ores in the Tadzhik SSR, and gypsum in the Ukrainian SSR. Other reports describe finds in unspecified locations of molybdenum, bismuth, tungsten, bauxite and other economically and militarily valuable metals.

These findings may, if valid, be of strategic importance, given that the future security of a nation depends on continued access to mineral resources. Because of this importance, it would be reckless to overvalue the anecdotal evidence suggestive of dowsing or other psionic claims. Section III summarizes a body of psychological research demonstrating many types of cognitive errors to which all humans — skeptics and proponents alike — are susceptible. A clear perspective on dowsing (or any other folklore claims) can only be attained when skeptical arguments are carefully weighed against claims of proponents.[390]

Treasure Hunting

One of the most dramatic uses of psychic talent to recover treasure reportedly occurred here in the United States. The *National Enquirer* commissioned the Chicago psychic Olaf Jonsson[391] to assist treasure hunters in the search for the sunken ruins of Spanish galleons loaded with gold and silver bullion. Jonsson seemed to sense the spot as the search vessel approached it and he asked the crew members to form a circle and concentrate with him. Going into trance, he actually relived the sinking of the ships. Under his directions the divers reportedly recovered part of the fortune, valued at $300,000.[392]

Some psychics have a difficult time, probably for psychological reasons, using

their abilities for their own direct financial gain — although they perform satisfactorily when they charge others for "life readings," etc. Even Uri Geller fared very poorly in Las Vegas. (Although it would be interesting to test habitually successful gamblers for ESP.) The inabilities seem to be more a reflection of a person's personality, rather than a limitation upon psi itself.

Accuracy of Information Transmission

A number of case histories also testify to this possibility. For example, Dr. Georgi Lozanov, director of the Institute of Suggestology and Parapsychology in Sophia, Bulgaria, is said to have demonstrated a very impressive communications technique using the majority-vote technique. The telepathic receiver sits in front of two telegraph keys, one for each hand. Some distance away, the sender telepathically suggests that the receiver press either the right or left key, according to the beats of a metronome. Each telepathic suggestion is repeated ten times. The receiver must get six of these correct for the message to be considered received. Lozanov reported at the 1966 Parapsychology Conference in Moscow that phrases and entire sentences have been sent this way with about 70% accuracy. Thousands of such tests are said to have now been demonstrated before many scientists.[393]

As the name of his institute implies, Lozanov is concerned with many of the psychological factors effecting ESP scores.

Using techniques derived from yoga, Lozanov combines suggestion and relaxation in a way that is different from hypnosis in that his subjects remain in the waking state. Used in education, these techniques show phenomenal promise to increase language learning, memory, and artistic and musical ability.[394] Lozanov also is applying his techniques towards the development of mental healing and dermal vision.

One of Lozanov's many research activities involves the evaluation of the predictions made by the blind, peasant woman, Vanga Dimitrova, who may be the modern world's first government-supported prophetess. (In fact, the Institute of Suggestology and Parapsychology, with over thirty staff members, is supported by the Bulgarian government.) Studies are reported to have shown that Dimitrova's predictive abilities — particularly strong in terms of finding lost relatives and friends—are about 80% accurate.

In Prague, Czechoslovakia, things were somewhat different. Dr. Milan Ryzl, a biochemist at the Czech Institute of Biology, had spent years trying to interest the government in supporting psychic research — all with very little success. Undaunted, Ryzl continued his own studies which involved hypnotic techniques for developing ESP subjects. After practicing on some five hundred individuals, Ryzl claimed to have found fifty with very strong, testable psi abilities.

Ryzl used his psychic subjects to predict the winning numbers in the Czech public lottery. He was successful for weeks in a row, winning the equivalent of several thousand dollars. However, Ryzl's psychical research successes also proved to be detrimental to his safety. The Czechoslovakian regime became very interested in his work. He found himself constantly followed by secret agents. His manuscripts were stolen. Eventually he was asked, in rather forceful terms, to spy on his scientific colleagues in other countries. The authorities made it very clear that they were interested in the development of psi techniques for espionage purposes. The government exercised such control over his life that Ryzl had no choice but to comply or defect. His escape from Czechoslovakia was a masterpiece in precise timing. He actually contrived to leave the country with his entire family in three automobiles and many valuable possessions including his prized library. Ironically, Ryzl recalls that the details of his defection had been predicted for him fifteen years earlier by a psychic who had been a friend of the family.[395]

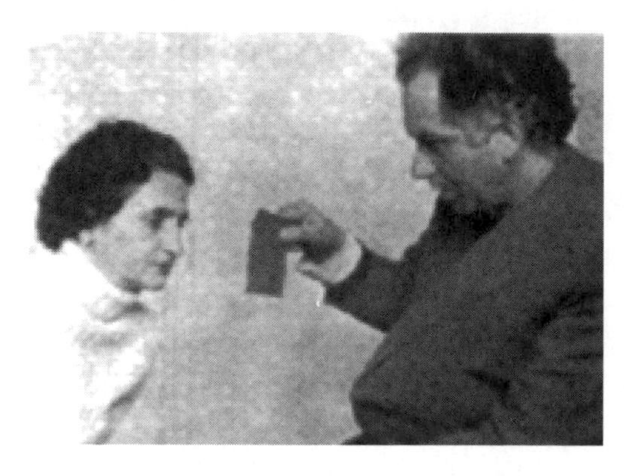

DR. GEORGI LOZANOV WITH A SUBJECT (COURTESY MILAN RYZL)

Researchers in socialist countries have continued the emphasis on the practical applications of ESP initiated by scientists such as Ryzl. Actually, since Ryzl's defection, Western psychical research has become somewhat more oriented toward practical uses.

Psychic Archeology

The use of psychics for archeological exploration has probably been the most extensively explored area of potential psi application. Its beginnings include the investigation of Glastonbury Abbey, perhaps England's oldest Christian ruin, by Frederick Bligh Bond.[396]

At the University of Toronto, Prof. J. N. Emerson of the Department of Anthropology has reported on his use of psychic assistance in doing archeological field work. His friend, a psychic, George McMullen, has shown extraordinary ability to *psychometrize* artifacts and relate accurate details about the history and circumstances surrounding the object. George has also proven his usefulness in examining archeological sites before the digging begins. Just by walking over a site, he has been able to describe its age, the people who lived there, their dress, dwellings, economy and general behavior. He has also provided specific excavation guidance. Emerson estimated that George's clairvoyance is 80% accurate. Furthermore, Emerson has been able to achieve even greater degrees of accuracy by using teams of several psychics and evaluating their reports using a majority-vote technique.[397,398,399]

MILAN RYZL AND THE AUTHOR ATTENDING THE PARAPSYCHOLOGICAL ASSOCIATION CONFERENCE IN 1973

In the former Soviet Union, techniques of dowsing are applied to archeology. Chris Bird reports that the Russian anthropologist Pushnikov has successfully used psychic dowsers to probe the remains of the Borodino battlefield, seventy miles from Moscow, where the Russians battled Napoleon in 1812. Other Russian excavations utilized the talents of dowsers in probing the estate of the legendary Czar, Boris Gudenov.[400]

The work of the Mobius Society is well known to the general public and the psychical research community. Stephan Schwartz began by looking at the role of psi in archeology. In his first book, *The Secret Vaults of Time*,[401] he described a dozen cases in which archaeologists have been successful in uncovering difficult-to-find locations and artifacts using psi methods. He then synthesized for himself a methodology, similar to the intuitive consensus method of William Kautz, which relied on the overlapping judgments of a number of independent practitioners. He has successfully used this method in a number of explorations. One of these, off the California coast on Santa Catalina Island, was broadcast on television. His explorations in Egypt have been the subject of several publications and scientific presentations.[402]

Following guidance obtained from interviews with psychic respondents, researchers from the Mobius Group, in Los Angeles, initiated an underwater archeological project in the Caribbean Sea. In September 1987, two respondents, Hella Hammid and Alan Vaughan were taken out in a small boat and within an hour had agreed on a site and dropped a buoy. The next morning divers noticed that a sequence of fire coral when viewed from one angle seemed unnaturally symmetrical. When one of the fire corals was chipped, it revealed what was later determined to be a bronze keel bolt. The buoy dropped by Vaughan and

Hammid was approximately ten feet from the site. Four weeks of excavation revealed an unusually intact wreck buried three to five feet beneath the eel grass and sand. Nothing was visible except the fire coral-covered keel bolts and some ballast mixed with natural rock. It required substantial excavation to uncover the remains of a collapsed American armed merchant brig that sank in the early decades of the nineteenth century.

What are the odds of finding the wreck described by luck? There is no completely satisfactory statistical answer to this question. The Mobius researchers justify their approach:

> Unlike a laboratory experiment with a known baseline, no absolute probability can be given in an archeological experiment; fieldwork applications of psi are inherently different from in-lab experiments. However, vigorous concurrent utilization of non-psi electronic location technologies can serve as field controls in a double or triple blind setting, producing results as significant, in the author's view, as low p values. For example, 85,000 shipwrecks identical to the one reported here could be fitted into just the Northern Consensus Zone, and under optimal conditions, it could take several months of magnetometer survey work to locate one such wreck. This wreck was psychically located, and the location verified, in less than five hours.[403]

Critics respond that the Mobius reports do not account for buoys dropped at other sites where shipwrecks were not located.[404]

Psychic Police Work

The use of psychics by police for solving crimes goes back many decades. As early as 1914, the Frenchman W. de Kerler, calling himself a psycho-criminologist, demonstrated on many occasions, without any reward or publicity, his ability to solve crimes that baffled police. Some of his many alleged exploits have been recorded.[405] In 1925, another case of clairvoyant detective work came to the attention of the German public. In this case, the psychic, August Drost, was on trial for fraud. The case resulted from an incident in which he had attempted, with little success, to help officials solve a burglary. During the trial, which lasted for several weeks, much of the testimony pointed toward Drost's successful ESP crime solving in other cases. He was acquitted and continued to practice his unorthodox detective work.[406]

Another psychic detective, Janos Kele, worked for years in Hungary and Germany without ever accepting fees or rewards. His abilities were tested by Prof. Hans Dreisch at Leipzig University who pronounced him a "classic clairvoyant." He was also successfully tested by Dr. Karlis Osis, then at Duke University. According to Dr. Stephen Szimon, a deputy police chief in Hungary, Kele averaged 80% accuracy in the clues he provided for tracing missing persons.[407]

Today in the United States, a number of police officials have publicly credited clairvoyants who have helped them with difficult investigations. One of the most prominent of these seers is Marinus B. Dykshorn, a Dutchman, whose autobiography is titled, *My Passport Says Clairvoyant.*[408] Dykshorn's career spans three decades and three continents. He currently resides in the United States. For his psychic detective work he has twice been made an associate member of the Sheriffs Association of North Carolina. In May, 1971, he received a commission from Louis B. Nunn, the governor of Kentucky, as a Kentucky Colonel, "in consideration of outstanding achievement." Dykshorn's book contains ten notarized affidavits from individuals who have received benefit from his clairvoyant abilities. It is particularly interesting to note in his book the difficulties that he had getting researchers interested in testing his abilities, well after his practical successes had been acclaimed.

A psychic who has established lasting relationships with police authorities is Irene F. Hughes of Chicago. She is the head of an organization called the Golden Path where she has taught classes in psychic subjects and tests students interested in developing their own psychic abilities.[409] On the wall of her office, a plaque signed by three Chicago policemen expresses appreciation for the leads she has given in solving a number of cases. In one particular homicide case, Mrs. Hughes was able to provide police with the name and address of the murderer — adding that the case would take a long time to solve. It was, in fact, almost three years before the fugitive was found. According to crime reporter Paul Tabori, she is credited by police in Illinois with having helped to solve no less than fifteen murder cases.[410]

Other tested psychics who are known to have worked with police officials include Olaf Jonsson and Alex Tanous. Undoubtedly there are more who prefer to work quietly and without publicity. Police departments receive a regular stream of tips that allegedly come from psychic insights. Most of them simply do not prove useful. Nevertheless, this area deserves further exploration.

IRENE HUGHES

Paul Tabori writes of the Viennese Criminological Association meeting he attended in the early 1930s devoted to the question of "so-called occult phenomena" in police procedure and judicial investigation. Many learned academics voiced the opinion that clairvoyance, telepathy, and even hypnosis were too unreliable to be used with any advantage in police and judicial work. Equally insistent, however, were lawyers and police themselves who stated that practice had proved the value of psi in certain investigations and that it was foolish to reject it simply because of experimental and theoretical difficulties.[411]

Great caution must be exercised in evaluating psi claims related to crime investigation. Skeptical Dutch researcher Piet Hein Hoebens, for example, was able to find major loopholes in claims regarding the Dutch clairvoyant Gerard Croiset — "the Mozart of Psychic Sleuths."[412] Newspapers throughout Europe acclaimed Croiset as a great psychic, at the time of his death in July 1980. This is particularly disturbing, since Croiset's abilities were attested to by W. H. C. Tenhaeff, a psi researcher at the University of Utrecht in the Netherlands, who had studied Croiset's alleged abilities for several decades. Hoebens' investigation strongly suggests either incredibly shoddy research or fraud on the part of Tenhaeff.[413]

With such a history, it is understandably risky for me to report psi crime investigations with which I am personally acquainted. Yet, for some years I have been monitoring Kathlyn Rhea, a psi practitioner now living in San Rafael, California. Author of *Mind Sense* and *The Psychic Is You*, she is well known for her work with police departments.[414,415] She has been active on well over one hundred cases.

One case in particular provides evidence that Kathlyn Rhea was directly instrumental in locating a missing body. I personally obtained complete corroboration from the law enforcement officials involved. The case occurred several years ago in Calavaras County, California, in the foothills of the Sierra Nevada gold country. An elderly man, Mr. Russell Drummond, had been camping with his wife in the county. He was reported missing by his wife, after he left his campsite to use the latrine and never returned.

The local county sheriff organized a search party of some three hundred persons. However, after a two-week period of intensive combing through the adjacent areas, the searchers were unable to locate the body or any sign of what happened to Mr. Drummond. The sheriff therefore proclaimed that Drummond must have either left or been taken away from the county.

His wife was desperate at this point. Not only was she without her husband, but since his whereabouts was unknown she could not collect his pension or insurance.

Six months after the incident, Mrs. Drummond contacted Kathlyn Rhea. Mrs. Rhea sat down using her normal methods, which involved no profound altered state of consciousness. She simply dictated into a cassette recorder her impressions of what had happened to Mr. Drummond. She described in detail, in a tape lasting forty-five minutes, how he lost his sense of orientation and began wandering away from the campsite in an easterly direction. She described a gravel path near a small, chalet-like cottage, where there were trees and brush. There she described how he had a stroke and fell under one of the brush-like (madrone) trees in that area. She described him as still being under that brush, six months later, completely intact. This would be unusual for a body left in the woods for six months.

Mrs. Drummond took that tape to the new county sheriff, Claude Ballard, who had been elected during the intervening time. Based on his listening to the tape, Ballard acknowledged a general sense of the location described by Mrs. Rhea. He took his skeptical undersheriff with him to that potential site with the idea that if the location matched the description provided by Mrs. Rhea, he would then organize a new search

W. H. C. TENHAEFF

KATHLYN RHEA

party. In fact, her description was so accurate that Sheriff Ballard was able to walk right to the body and find it with no difficulty. According to undersheriff Fred Kern, the description provided by the tape was 99% accurate.

Another case involving Kathlyn Rhea, which I have personally verified, involved the murder of an Ohio woman. Rhea was approached by a local detective for information on this case, and she provided him with a detailed description of where the body could be found — in the country, on a gravel road near a bridge.

The case is full of several ironies. Based on this information, the detective visited a site where he thought the body might be found and was not successful. Being somewhat ill and unable to search further, he provided Kathlyn Rhea's description to the police. Simultaneously, some local Boy Scouts uncovered the body at another location which matched Rhea's description in major details. The sheriff's department, which had assumed jurisdiction over the case, took note that an accurate description of the body's condition and location had been turned in by this detective prior to the body's discovery. They detained him as a suspect.

Additional information developed by the detective, working with Kathlyn Rhea, was that the local police chief had actually committed this murder. Rhea suggested that fibers from the victim's clothing would be found in his police cruiser. Acting on this tip, investigators searched the car and did find fibers. The police chief was convicted of the murder and is now serving time in prison.

Journalism and Investigative Reporting

Just as ESP can ostensibly be used in crime investigation, there is a suggestion that it can be useful to the investigative reporter. At least one popular account describes such activity.[416]

History

Retrocognition is the apparent psi ability to see past events. Many popular claims in this area are made in connection with ostensible reincarnation. In some instances of documented xenoglossy, individuals have been able to speak dialects of languages that have not been spoken for centuries.[417] The most used application of this ability has been in connection with psychic archeology, in terms of interpreting the history of various artifacts through psychometry.[418]

Precognition in Business Management

Professors Douglas Dean and John Mihalsky at the Newark College of Engineering PSl Communications Project have spent ten years testing the precognitive abilities of more than five thousand businessmen. They had heard numerous stories of how fortunes were made by men whose intuitive decisions seemed to defy all logical considerations. In one series of studies they looked at company presidents who had doubled their company's profits during the last five years. They found that these individuals scored much higher in the precognitive tests than other executives. In fact, the ESP test seemed to be a much better indicator of executive success than other personality measurements. A number of companies have shown an interest in using this technique to screen applicants for management positions.[419]

One of the interesting outcomes of the PSI Communications work with executives was the high percentage of subjects (about 80%) who openly acknowledged a belief in ESP. When questioned further, the businessmen admitted that their belief was not based on either a familiarity with the scientific literature or an acquaintance with some psychics. These tough-minded individuals believed in ESP because they had seen it work in their own lives!

Another major endeavor in the business community is that of financial investments. Large investment companies typically spend millions of dollars in market research and investigation. In spite of all this effort, education and statistical science, major investment companies not infrequently fail to achieve results that equal the performance of the overall market.

In 1982, the *St. Louis Business Journal* conducted a special event with nineteen prominent stockbrokers and one psi practitioner, Mrs. Bevy Jaegers. Each participant was asked to select five stocks whose value would increase over a six-month period. Mrs. Jaegers, who has no professional training in corporate analysis, outperformed eighteen of the nineteen experts. During the test period, the Dow Jones Industrial Average fell eight percent. Jaeger's stocks were up an average of 17.2 percent. The only stockbroker who bettered her attained an average of 17.4 percent. Sixteen of the stockbrokers chose stocks that lost their value.[420]

Another psionic organization in the San Francisco Bay Area was Delphi Associates. This group was formed by Russell Targ and Keith Harary after leaving SRI International where they had been conducting remote viewing experiments, under government sponsorship, for several years. In a unique pilot study, Harary was able to predict the movement of silver futures accurately for nine consecutive weeks. Actually, he did not predict the price changes directly, rather he was asked to describe an object or scene that he was to view the following week. The objects were randomly coded by the investigators to correlate with movements in the price of silver. The success of this study, even though subsequent tests failed to yield the same extraordinary results, enabled Targ and Harary to financially sustain their continued business and research activities for a period of time. The organization has since disbanded — although not before Targ and Harary wrote a best-selling book describing their work, *The Mind Race.*

The applied psi activities of Uri Geller has been reported in business publications and in a biography which Geller co-authored with Guy Lyon Playfair.[421] Geller, of course, achieved notoriety for his unusual work as a psychic entertainer. In retrospect, it's fair to say that, for better or worse, he influenced the field of psychokinesis research. The term "mini-Gellers" now refers to ostensibly talented macro-PK subjects that have been reported in over a dozen countries — generally discovering their supposed talents after watching Uri Geller perform on television. However, for well over a decade, Geller has discontinued his work as a laboratory subject and has entered the world of psionics. News of Geller's more recent activities has surfaced in a national business publication, *Forbes Magazine*,[422] in a profile article following up on an earlier story which appeared in the *Wall Street Journal.* I will quote some portions from that article:

LEE PULOS, A CLINICAL PSYCHOLOGIST, RESTAURATEUR AND MANAGEMENT TRAINER. SKILLED IN THE USE OF SELF-HYPNOSIS AS A STATE CONDUCIVE TO APPARENT PSYCHIC INTUITION, PULOS BUILT HIS *SPAGHETTI FACTORY* BUSINESS UP TO A CHAIN OF TWENTY RESTAURANTS. (FROM *QUALITIES OF HIGH PERFORMANCE* VIDEOTAPE, COURTESY THINKING ALLOWED PRODUCTIONS)

"Big businesses," [Geller] says, "are beginning to listen to people who think they can deliver something with their sixth sense." What if some enterprising outsider were to hire a psychic to abscond with the secrets of IBM's new 1000K RAM chip? Or, if a Boone Pickens has the power to compel a Gulf or Cities Service to bow to the terms of his latest merger offer? What if he knows what the stock market or the gold market or the bond market will do in the next week or next month or next year?

It was Val Duncan, the late chairman of Britain's Rio Tinto Zinc Corporation that made him [Geller] see the possibilities that business could offer. Duncan tested Geller's ability to find minerals on Majorca and later put him in touch with the chairman of South Africa's Anglo Transvaal Mining Company. The Chairman spread out a map and said, "Tell me what you feel." Geller said, "I feel something here." And years later Geller said the company found coal on the spot. "That's when I learned I could do this for very big companies and profit myself also. For seven years I have been doing this and nobody knows anything about it."

Several oil companies . . . hired him. . . . Acting as a sort of airborne divining rod, Geller targeted 11 prospects, four of which he says proved out. Geller doesn't charge fees, he claims. He relies on people to pay him what his service is worth. In this case a percentage royalty. "It's a little percentage," he says, "but in oil a little is a lot."

Which companies he has worked for Geller won't say. "They do not want their name to be linked to the psychic, to the paranormal." His only really public venture to date is his success in bringing together Japan's Aoki Corporation and the U.S. Tishman Reality in a $500 million hotel, condominium shopping development near Disney World in Florida. Both John Aoki and John Tishman were personal friends. But Geller claims to be more than a mere go-between. "My role is that I predict the success of the venture." That's where the power lies in being able to predict future success.[423]

Above, Russell Targ (from *ESP, Clairvoyance and Remote Perception,* courtesy Thinking Allowed Productions). *Below,* Keith Harary (from *Practical Applications of ESP,* courtesy Thinking Allowed Productions)

Public Safety

After a mine disaster in Wales in which 144 people perished researchers collected reports from individuals who claimed to have had premonitions of the event. Of seventy-six reports, in twenty-four cases the percipient had actually talked to another person about the premonition before the catastrophe. Twenty-five of the experiences were in dreams.[424]

A study conducted by W. E. Cox indicates that precognition also operates on mass-awareness levels. Cox accumulated statistics on the numbers of passengers aboard twenty-eight railroad trains which were involved in accidents. These figures were found to be significantly less than the number of passengers on the same trains one week before or a few days after the accident. People somehow avoided the accident-bound trains. There were also fewer passengers in damaged and derailed coaches than would have been expected according to the figures for non-accident days. Cox hypothesized that many potential passengers were aware of the oncoming tragedy, but not on a fully conscious level.[425,426]

Several practitioners, such as Alan Vaughan, have suggested the use of ESP for safety inspection of such complicated systems as the space shuttle, nuclear reactors, or oil pipelines.[427]

About two months before the assassination of Robert Kennedy in June 1968, Alan Vaughan, then in Germany studying synchronicity at the Freiburg Institute for Border Areas of Psychology, began to develop a strong premonition that Kennedy would be assassinated. The event, he felt, was part of a complex archetypal pattern which he was tuned into, involving the killings of both JFK and Martin Luther King. Many coincidences and dreams began to support Vaughan's theory. On April 29 and again on May 28, Vaughan wrote letters to parapsychologists notifying them of his premonition and hoping that Kennedy could be warned. His letter was received by Stanley Krippner at the Maimonides Hospital Medical Center on the morning of June 4.[428] Subsequently, Vaughan's apparent precognitive abilities have been extensively tested.[429,430]

Ironically enough, at least three cases are on record regarding the accidental deaths of parapsychologists who did not heed the warnings of their psychic subjects.[431] Perhaps the subjects had not sufficiently demonstrated their reliability in the past.

Education

Modern American education is oriented toward the cultivation of the mental abilities charted by Benjamin Bloom et al. in their three taxonomies of educational goals: cognitive, affective and psychomotor. Psi functioning has not been included within this taxonomy. This "cultural vacuum" may have been responsible for the growth of many cults and organizations claiming to train psi and other related abilities. The mainstream educational community has yet to awaken to the challenge of assessing the

THERE ARE THINGS IN LIFE MUCH MORE VALUABLE THAN MONEY, ACCORDING TO MY DEAR FRIEND, JANA JANUS, WHO IS CURRENTLY RESIDING IN MOSCOW. FOR MANY YEARS, THE "CLOVER SPIRITS" HAVE SPONTANEOUSLY GUIDED HER TO RARE FOUR-LEAF CLOVERS. SHE HAS LITERALLY FOUND THOUSANDS OF THESE, AND OTHERS ARE ABLE TO DO SO IN HER PRESENCE. SHE IS "CLOVER RICH."

need and potential for parapsychological education within the school system. One area of obvious potential application and some apparent results is in special education for the blind.[432,433] Some steps toward the evaluation of popular programs for psi cultivation are found in my second book, *Psi Development Systems.*[434]

Creativity in Art, Literature and Music

Trance experience is sometimes accompanied by artistic production that seems beyond the experience and talent of the conscious mind. Well-known examples of this phenomena are the literary productions of the Brazilian spiritist Chico Xavier, the artistic productions of Luiz Gasparetto (another Brazilian spiritist) and the classical musical creations of the English medium, Rosemary Brown.[435] In each of these cases, mediums claim that the artistic creations are produced with the aid of discarnate entities.

Agriculture and Pest Control

The use of magical rituals as an adjunct to agriculture certainly dates back to prehistoric times. It eventually became formalized in the Eleusinean mystery religion of ancient Greece. In contemporary society, some practitioners use "radionics" instruments as devices for focusing concentration and eliminating pests and disease from crops.[436] Many anecdotal accounts, such as those of Arthur M. Young, attest to the success of this method. Controlled experimental investigations have not been published.

ALAN VAUGHAN

Athletics and Sports

Anecdotal accounts, collected by Michael Murphy and Rhea White, suggest that peak moments of psychic perception, psychokinesis, and synchronistic flow are associated with dramatic levels of athletic performance.[437] Murphy, founder of the Esalen Institute in Big Sur, California, has developed a database suggesting that athletes develop extraordinary control of the human body — powers comparable to that those reported of yogis, mystics, and saints.[438]

Finding Lost Objects

Tribal shamanism, perhaps the world's second oldest profession, cultivated practitioners who were noted for their ability to find lost objects. Adrian Boshier reported that tests for this skill are a normal part of shamanistic training in South Africa. An experimental model for testing this ability in natural situations has been designed by Patric Geisler, a psi researcher and anthropologist.[439] Popular advertisements in *Fate* magazine present many claims of this ability.

Scientific Discovery

At the beginning of the twentieth century the theosophist clairvoyant C. W. Leadbeater published a book called *Occult Chemistry* in which he attempted to use psi perception to determine the structure of molecules.[440] In so doing, he anticipated the discovery of deuterium, an isotope of

LUIZ GASPARETTO IN TRANCE OSTENSIBLY POSSESSED BY THE SPIRIT OF VINCENT VAN GOGH

hydrogen. Later, similar attempts were reported by French researcher Z. W. Wolkowski who conducted successful tests with a sensitive, Raymond Reant.[441] In the realm of astronomy, remote viewing experiments have reportedly provided unexpected data that was later confirmed by satellite probe.[442] These reports have been very controversial.

Weather Prediction and Control

The ability to control the weather is reputed to occur among shamans of many tribal groups and has reportedly been observed in the American Indian culture as well as the Tibetan. Several ostensible instances with an American practitioner have also been documented by me and are included in the earlier discussions of UFOs and macro-PK.[443] Followers of the psychiatrist Wilhelm Reich claim to have developed control over the weather through the use of mysterious "cloudbusters" that are said to manipulate "orgone energy."[444]

Animal Training and Interspecies Communication

Sir Arthur Grimble, who served as the British resident commissioner for the Gilbert and Ellis Islands in the South Pacific, reports witnessing a scene in which a native shaman entered a dream state in order to "call porpoises." After a period of time he awoke from his sleep and announced to the tribe that the porpoises were coming. The village of about one thousand individuals rushed down to the beach eagerly expecting a rare feast, and Grimble documents that he observed an entire flotilla of porpoises swim onto the beach, passively offering themselves to the natives:

> They were moving toward us in extended order with spaces of two or three yards between them, as far as the eye could reach. So slowly they came they seemed to be hung in a trance. It was as if their single wish was to get to the beach.[445]

The years since the first publication of *The Roots of Consciousness* have seen the birth and growth of an international movement to foster and study new forms of communication between humans and cetaceans. In 1976, I became a participant-observer of this movement. An organization, called the New Frontiers Institute, operating in a San Francisco suburb had developed some unusual methods for psychic work akin to traveling clairvoyance occurring during a mutual hypnotic induction.[446] Members of the group included a chiropractor, a housewife from Argentina, an astrophysicist, a psychiatrist and an accountant. I found them to be amiable and was willing to engage in group hypnotic experiences with them. Several such sessions ensued.

MICHAEL MURPHY (FROM *TRANSFORMING THE HUMAN BODY* VIDEOTAPE, COURTESY THINKING ALLOWED PRODUCTIONS). RHEA WHITE, FOUNDER OF THE PARAPSYCHOLOGY SOURCES OF INFORMATION CENTER (2 PLANE TREE LANE, DIX HILLS, NEW YORK 11746) MAINTAINS A COMPUTERIZED DATABASE OF ARTICLES AND BOOKS RELATING TO PSI PHENOMENA.

During these experiences, after a lengthy hypnotic induction, we engaged in a form of what might be called *group fantasy* — or if more than fantasy were involved, perhaps *group astral projection,* or, if any of this could be objectively verified, *group traveling clairvoyance.* Typically, we would "fly" to the top of the north tower of the Golden Gate Bridge. After, checking to make sure we were all "there," we would "fly" off on further adventures — each of us describing our experiences verbally to the others.

On one such occasion, we decided to visit Marine World, a Bay Area wildlife amusement park, to see whether we could communicate with a dolphin. As the conversation progressed, the following points emerged:

1) We were with a female dolphin.
2) She was sick and cantankerous.
3) Her name was "Dee" or began with "D."
3) She was in a separate tank without a mate.
4) She did not wish to cooperate with the trainers.
5) She wanted us to help her escape.
6) She wanted to travel with us *astrally.*

The issue of helping a dolphin escape was not entirely absurd — as such an instance had recently been reported in the news. We all felt a great deal of empathy for "Dee," but we were not prepared to engage in illegal activities in her behalf. We offered her another solution. If this ostensible communication were real, we would visit "Dee" at Marine World. If we could somehow objectively demonstrate some type of human-dolphin telepathic communication, the knowledge of this event might provoke other humans into rethinking our entire relationship with dolphins in a way that would greatly benefit "Dee" and others of her species. Meanwhile we would use whatever psychic means we had to help "Dee" and to heal her.

This particular session had a remarkable quality in terms of both its vividness and the group coherence. Immediately after awakening from the group hypnotic experience, we phoned Marine World to see if they had a dolphin that resembled "Dee." We were, fortunately, put through to one of the dolphin trainers who told us that there was a female dolphin who matched our description. Her name was "Dondi." Intrigued by our story, he invited us to Marine World for a private visit, to meet Dondi and for some informal testing of our possible communication with her.

During our first session at Marine World, the trainers had placed Dondi in a tank with five other dolphins. We were asked simply to see if we could identify which of these she was. Five of the six of our group succeeded in this — although there was some discussion among us and these could not be thought of as independent observations (which would be of relevance for statistical analysis).

The dolphin trainers also told us that, from their own experiences with these very intelligent animals, they had little doubt of the possibility of telepathic communication between cetaceans and humans. In fact, a recent issue of the dolphin trainer's professional newsletter had just featured an article about one of the most honored of their profession, Frank Robson. A New Zealander, Robson was knighted by Queen Elizabeth for the remarkable work described in his book *Thinking Dolphins — Talking Whales.* He had achieved heroic status by saving a pod of whales that had become stranded on a sandbar in the South Pacific. Robson described this success, and many other experiences as resulting from a type of telepathy. Thus the dolphin trainers at Marine World were most willing to work with the New Frontiers Institute group to explore further the possibility of human-dolphin telepathy.

Over a period of months, members of the group frequented the dolphin tanks at Marine World, paying particular attention to Dondi. Her condition improved markedly — to the point where instead of being sick, cantankerous and uncooperative, she had become the star performer in Marine World's dolphin show.

In May of 1977, a pilot attempt was made to assess the effectiveness of the New Frontiers Institute work with dolphins. The entire experiment was videotaped. In addition, detailed audiotaped records of Dondi's behavior were made by several independent observers (including me, on this occasion) and were later transcribed. The final results were suggestive of the possibility that Dondi was responsive, over 50% of the time, to telepathic instructions.

Another interspecies explorer, and international organizer of the movement, is Wade Doak of Whangerei, New Zealand, author of ten books on natural history and wildlife, including *Dolphin Dolphin* and *Encounters With Whales and Dolphins.*[447,448] Doak maintains that his exploration of interspecies communication began with a simple incident that took place on his catamaran in the waters off of New Zealand. An American visitor sitting on the deck suddenly and unexpectedly experienced an altered state of consciousness which he described as being like "a pinball machine in my head." Moments later, a dolphin jumped out of the water within a few feet of the American and looked at him right in the eye.

That evening, the American reported a strange dream. In the dream, Wade Doak was in the water surrounded by dolphins who were swimming about him in a figure-eight pattern. The dolphins were making a strange noise that sounded like "tepuhi."

Doak was a naturalist and filmmaker by training who had extensively studied communication patterns among the species inhabiting South Pacific coral reefs. In his explorations, he had encountered many exotic forms of communication among the various species — including the native, micronesian tribes. He suspected that his friend's dream might represent a form of "biological communication" between humans and dolphins.

In order to explore this hypothesis further, he visited a shaman of the New Zealand Maori people. The Maoris call themselves the "people of the dolphin" and legend has it that their tribe was originally guided to the New Zealand islands in past ages by the dolphins themselves. The shaman believed that this dream was not a mere fantasy. He pointed out that "tepuhi" is the Maori word for the sound that dolphins make when they blow water from their blowholes.

Pressing matters further, Doak launched an expedition to contact dolphins in the open waters off New Zealand. He equipped his catamaran with film equipment. However, no dolphins were sighted for several weeks. Finally, on one occasion, some dolphins were sighted in the distance.

RICHARD GIERAK, FOUNDER OF THE NEW FRONTIERS INSTITUTE

Doak jumped into the water and, while underwater, made the sound "tepuhi." At that moment, his film crew was fortunate enough to capture an extraordinary event. Circling the boat, dozens of dolphins simultaneously jumped into the air. Then they swam around Doak in a figure-eight pattern — a confirmation of the dream communication. Doak has since spent the intervening years continuing his efforts to communicate with and even live with dolphins in the open waters. He is also the founder of Project Interlock, an informal, international network of individuals who are engaging in similar experiences.

Intuitive Consensus

The Center for Applied Intuition was founded by William Kautz, who has also been a senior research scientist in the area of mathematics and computer science at SRI International. He developed a method called *intuitive consensus*:

> Essentially, the method of intuitive consensus consists of a careful preparation of the questions to be answered to eliminate ambiguity and vagueness; posing the questions independently to a team of four or more "intuitives," followed in each case by a dialogue until clear and detailed answers are obtained; then analysis and comparison with one another and existing knowledge to form a consensus of intuitively derived information. The team approach permits the occasional errors and discrepancies in the information provided by individual team members to be almost completely eliminated. The questioning cycle may be repeated one or more times in order to resolve ambiguities or contradictions, usually traceable to erroneous preconceptions or unintended vagueness in the questions and to otherwise reduce the noise level in the intuitive communication

WADE DOAK

channel. The final consensus takes the form of specific hypotheses amenable to test by ordinary scientific means. In some cases it provides new perspectives and new ideas about the problem under study.[449]

Actually, the method employed by Kautz is very similar to the "Delphi method" which is used in future forecasting. In this method, various experts are questioned about potential future phenomena which fall under their expertise. The primary difference is that Kautz does not employ intellectual or academic experts; he employs people whose knowledge is derived from intuitive sources, which may be taken, in this case, as a possible euphemism for psi sources.

Kautz has been applying this method, under contract, for various business and professional organizations. He recently completed a study for a Japanese research institute on the "Future of Japan." He has also employed the method to develop scientific hypotheses related to such unsolved problems as the cause of earthquakes, and the cause of crib death in infants, as well as technical problems in genetic engineering.

While the results he has obtained in these studies are not inconsistent with the possibility of successful ESP functioning, I am personally not aware of any individual instances of Kautz' work which could be put forward as examples of success in this area that could be attributed to psychic functioning. Kautz' approach has been detailed in two books, co-authored with Melanie Brannon, *Channeling* and *Intuiting the Future*.[450]

Some Concluding Thoughts about Folklore

There is an ironic boundary separating the worlds of science and folklore. While folklore, as such, carries little or no scientific weight, science, without folklore, has little or no real meaning. This is most evident in looking at the writings of certain critics of psi research. Such critics will, from time to time, grudgingly acknowledge that some studies contain statistically anomalous effects. However, they point out that these effects are simply meaningless statistical correlations pointing to no recognizable phenomena or principles.[451,452]

They may be correct, but only in a rigidly limited sense. I believe they are wrong to eliminate all history and folklore from scientific consideration. For within the real-life, human context which history and folklore provide, the seemingly meaningless statistical correlations of psi research take on a kind of life and color.

In Section II, I have reported some personal experiences which seemed extraordinary to me: an apparent clairvoyant dream experience which led to my career in the

WILLIAM KAUTZ (FROM *THE INTUITIVE CONNECTION* VIDEOTAPE, COURTESY THINKING ALLOWED PRODUCTIONS)

media, observations of Ted Owens who attempted to demonstrate that he was half-alien, work with the New Frontiers Institute in ostensible telepathic communication between humans and dolphins, observations of Kathlyn Rhea who located a missing body for the Calavaras County sheriff. I have been personally struck by these experiences. Whether I interpret them in a skeptical manner or as examples of psychic functioning is a personal matter. It is not possible to make a rigorous scientific case either way, because there were too many uncontrolled variables in every case.

In Section III, after looking at how little psychology has to say directly about consciousness and how much it has to say about human error and folly, we will examine the evidence from psi research. The strongest evidence (which some knowledgeable scientists still reject) comes from experiments which bear the least relation to the world of everyday activity. If you choose to accept the data from these studies, they may make more sense to you in the light of material presented in Sections I and II and vice versa.

THE SCIENTIFIC EXPLORATION OF CONSCIOUSNESS

INTRODUCTION

THE BOUNDARY BETWEEN THE REALM OF FOLKLORE AND THE REALM OF SCIENCE is sometimes fuzzy. Section II contains some material that was included in the "Scientific Exploration" portion of the original edition of *The Roots of Consciousness* — studies on auras, healing, out-of-body experience, UFOs, reincarnation, macro-psychokinesis. These areas have been investigated by individuals with scientific training who have attempted legitimate experimental undertakings. Seventeen years ago, I was optimistic that these areas would be rapidly embraced by the scientific community. Today, the major reason for optimism is the increasing willingness of both skeptics and proponents to admit their uncertainty.

The fact that Section II has been labelled "folklore" should not be taken to mean that all the studies cited there are devoid of scientific value. The folklore of consciousness is the very meat upon which the scientific endeavor feeds and sharpens its teeth. However, at this point in time it would be premature to say that the meat has been digested and assimilated.

I would not judge myself harshly had I chosen to label this entire book as a description of the "folklore of consciousness" — since, after more than a hundred years, psi research has yet to emerge as a mature scientific discipline. Some sociologists of science have come to view the debate over psi's existence as more of a social process than a scientific one.[1]

Yet, from my perspective, the distinction between folklore and science is not entirely arbitrary. The primary guideline I have used emerges from the debate between psi researchers and their most conscientious critics. Over many decades, these two groups have faced off on the wrestling mat of

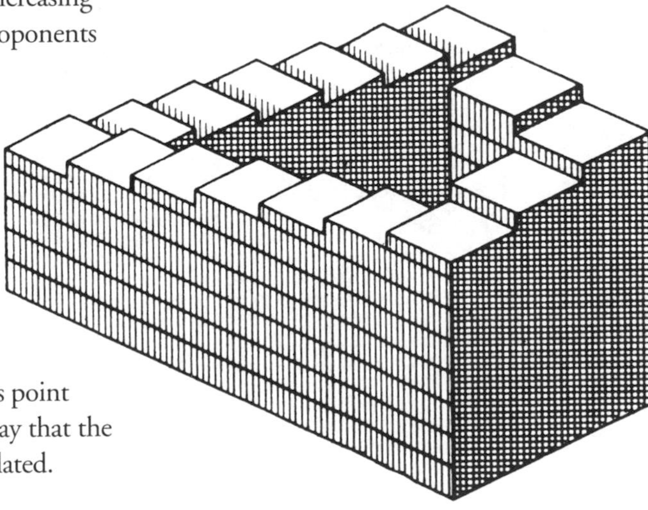

scientific debate. Both have had to cry "uncle" a few times. Out of this debate has emerged a kind of neutral territory. On one hand, the psi researchers admit that they have yet to establish their theoretical claims. On the other hand, the skeptics admit that they are unable to explain away some of the data. This process of scrutiny and debate, over many decades, has enabled the scientific endeavor to isolate certain lines of research as deserving of serious attention. Therefore, I feel comfortable including it in the realm of science rather than that of folklore. However, before we examine the best data of psi research, let us briefly examine relevant background material in the field of psychology.

WILLIAM JAMES

The Problem of Consciousness

It may seem ironic that a book titled *The Roots of Consciousness* has little to say about the field of psychology itself. The primary reasons for this situation is that in developing itself as a scientific discipline, psychology has moved away from the fundamental question of the human *psyche* in order to address more measurable, tangible issues that could properly be addressed by existing scientific methods.

Experimental psychology begins with the work of the mystical philosopher and scientist Gustav Theodore Fechner — whose theories on the soul life of plants, animals and humans have been outlined in Section II. In 1860, Fechner published *Elemente der Psychophysik.* The concepts he used and the problems he defined have become the foundation of experimental work in sensory experience since his time. He developed the psychological measure known as the *just noticeable difference* (*jnd*) which is the smallest observable difference between two stimuli, e.g., two different lights. He determined that the relationship between the intensity of a physical stimulus and the just noticeable difference at that intensity is a logarithmic one. In other words, the difference in intensity between two very bright lights will have to be much greater than the difference in intensity of two very dim lights in order for a *jnd* to be perceived.

Unlike the psychologists who followed in his footsteps, Fechner believed that he had discovered a relationship between the individual consciousness and the sublime universal soul. While Fechner's work was an effort to bridge experimental science and the world of mysticism, it also fell within the philosophy of mind-body dualism, espoused by Rene Descartes.

William James has been honored for generations as America's greatest psychologist. At the turn of the century, he outlined the foundations for the discipline of psychology that would include cognitive science, transpersonal psychology (the investigation of spiritual and religious experience), and psychical research. Consciousness was placed at the center of James' approach:

> The first and foremost concrete fact which everyone will affirm to belong to his inner experience is the fact that consciousness of some sort goes on. "States of mind" succeed each other in him. If we could say in English, "it thinks," as we say "it rains" or "it blows" we should be stating the fact most simply and with the minimum of assumption.[2]

In his classic text, *Principles of Psychology*, James adopted the accepted philosophy of dualism as the appropriate underpinnings of psychology. However, in one of his most brilliant works, the posthumously published *Essays in Radical Empiricism*, James shifted to a position of Berkeleyan idealism, which eliminated the material world and made consciousness the only reality. He argued that implicit in empiricism — the philosophy upon which science is based — is that we know the world only as it appears in our consciousness of it as mediated by our sensory systems.

Bishop Berkeley took this implication and used it to rule out the material world. Scottish philosopher David Hume, following Berkeley, carried this implication to its *reductio ad absurdum*, that all we know is a "flux of impressions," where they come from and where they are, we can never know.[3]

Building on these earlier empiricists, James formulated his "radical empiricism." Just as Berkeley denied the existence of matter, James denied the existence of both matter and consciousness:

> To deny plumply that "consciousness" exists seems so absurd on the face of it — for undeniably thoughts do exist....Let me then immediately explain that I mean only to deny that the word stands for an entity, but to insist most emphatically that it does stand for a function.[4]

James formulated a monism in which "there is only one primal stuff...of which everything is composed."[5] He called this stuff *pure experience*, which was prior to either material objects or consciousness. (See Appendix for a work-in-progress suggesting a rigorous mathematical approach to James' realm of pure experience.)

German experimentalist Wilhelm Wundt's approach to psychology was represented in America by E. B. Titchener (1867-1927), who gave it the name *structuralism*. According to Titchener,

> All human knowledge is derived from human experience...and there can be no essential difference between the raw materials of physics and the raw materials of psychology.[6]

Following Wundt's lead, Titchener used *introspection* — an approach to research which relied on verbal reports of internal mental states. Structuralists studied slices of consciousness that attempted to freeze a single moment or thin cross-section of the stream. They believed that, like chemists, they should search for the elements or basic building blocks of consciousness. This search proved unproductive. Consequently, the mainstream of American psychology, while originating with William James' concern with the fundamental role of consciousness, shifted direction dramatically.

Many psychologists in past decades behaved as if they were embarrassed by the very name *psychology*. They would have preferred that the discipline be called *behavioral science* and that they be referred to as *behavioral scientists* rather than as psychologists. Yet, for historical reasons, they dominated the discipline that came to be known as psychology. John B. Watson, who introduced behaviorism in 1912, defined psychology as a natural science that studied behavior, not consciousness, by observation and experimentation. In the beginning, Watson treated consciousness as an epiphenomenon but he later stated that

Behaviorism claims that "consciousness" is neither a definable nor a usable concept; that it is merely another word for the "soul" of more ancient times. The old psychology is thus dominated by a kind of subtle religious philosophy."[7]

Many behaviorists went so far as to argue, along with positivist philosophers, that mind or consciousness did not even exist — except as a concept used in popular language. From this sad perspective, the mind itself was considered a reification, a categorical error, a semantic confusion which did not, in any real scientific sense, exist.[8] Normal human cognition for much of this period was almost a "taboo" subject. Howard Gardner, a contemporary Harvard psychologist now engaged in resurrecting cognitive science, maintained that "it is difficult to think of this [behaviorist] phase as other than primarily negative and regressive."[9]

According to Karl Pribram, the neuropsychologist at Stanford University who has developed a holographic model of brain function, psychology has traveled full circle since the days of William James, so that consciousness is now a major topic of interest:

> The history of psychology in this century can be charted in terms of the issue that dominated each decade of exploration. Early studies on classical conditioning and Gestalt principles of perception were followed subsequently by two decades of behaviorism. In the 1950s information measurement took the stage to be supplanted in the 1960s by an almost frenetic endeavor to catalog memory processes, an endeavor which culminated in the new concepts of a *cognitive* psychology. Currently, the study of *consciousness* as central to the mind-brain problem has emerged from the explorations of altered and alternative states produced by drugs, meditation, and a variety of other techniques designed to promote psychological growth.[10]

The newest fad in psychology is *cognitive science* which defines "mind" as a system containing many components including sensory perception, memory, self-image, language, etc. By and large, cognitive scientists have used the information measurement and information processing approach to the brain-mind problem. The brain's *wetware* is viewed as akin to the hardware of computers and optical systems. Mental operations are analogous to programs and image constructions.

Karl Pribram (from *The Holographic Brain* videotape, courtesy Thinking Allowed Productions)

With the rise of *cognitive science,* consciousness is once again gaining legitimacy within psychology. But for most of the late nineteenth and twentieth century the scientific and philosophical investigation of consciousness has occurred within the tradition that started with psychical research.[11] The findings posed by psychical research, in my opinion, remain crucial for our understanding of the limits and nature of consciousness.

Psychical research has developed many new methods since the time of William James. During the past eight decades, many hundreds of research studies have been published attempting to measure the ostensible powers of the psyche under conditions of rigorous behavioral constraints.

Of course, we have learned more about the reach of the psyche than was known at the turn of the century. But not much more.

Eighty years of experimental progress has done little to convince the skeptical scientific community that direct psychic interactions with the environment, not mediated by the sensorimotor system, have been demonstrated. Lest we judge either science or psychical research too harshly, we will do well to understand the findings of psychology regarding the many types of error and folly to which the human mind is susceptible. Such an understanding can help us appreciate the innate conservatism of science in the face of psi research's extraordinary claims. It can also shed much light on the dynamics of the debate between proponents and skeptics of various psychic claims.

To Err Is Human[12]

To err is human. Understanding the mechanisms by which humans repeatedly make errors of judgment has been the subject of psychological study for many decades. Why do people disagree about beliefs despite access to the same evidence, and why does evidence so rarely lead to belief change? Psychological research has examined numerous risks of assessing evidence by subjective judgment. These risks include information-processing or cognitive biases, emotional self-protective mechanisms, and social biases.

All of these factors play a major role in both sides of the debate between proponents and opponents of psychic phenomena. No analysis of the controversies surrounding the nature of the human spirit, and its propensity for greatness, would be complete without a realistic look at the human proclivity for folly.

The Psychology of Cognitive Biases

The investigation of cognitive biases in judgment has followed from the study of perceptual illusions. Our understanding of the human visual system, for example, comes in large part from the study of situations in which our eye and brain are "fooled" into seeing something that is not there. In the Muller-Lyer visual illusion, for example, the presence of opposite-facing arrowheads on two lines of the same length makes one look longer than the other. We generally do not realize how subjective this construction is. Instead we feel as if we are seeing a copy of the world as it truly exists. Cognitive judgments have a similar feeling of "truth" — it is difficult to believe that our personal experience does not perfectly capture the objective world.

With a ruler, we can check that the two lines are the same length, and we believe the formal evidence rather than that of our fallible visual system. With cognitive biases, the analogue of the ruler is not clear. Against what should we validate our judgmental system?

One of the basic errors typical to intuitive judgments is called *the confirmation bias.* If you hold a theory strongly and confidently, then your search for evidence will be dominated by those attention-getting events that confirm your theory. People trying to solve logical puzzles, for example, set out to prove their hypothesis by searching out confirming examples, when they would be more efficient if they would search for *disconfirming* examples.[13] It seems more

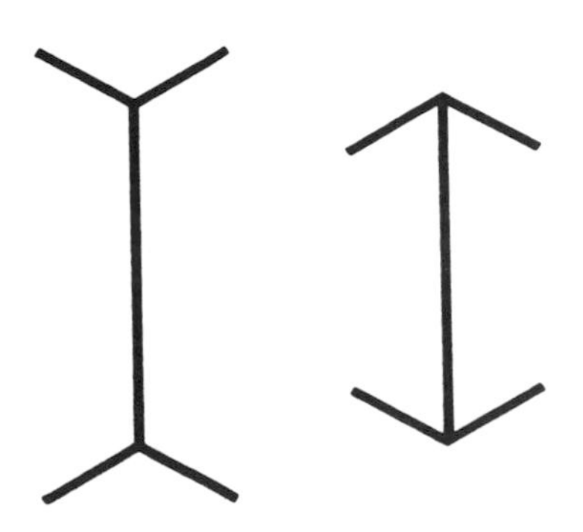

natural to search for examples that "fit" with the theory being tested, than to search for items that would disprove the theory. With regard to psychic phenomenon, this bias would explain why "skeptics" always seem to find reasons for doubting alleged instances of telepathy or clairvoyance, while "believers" continually find new instances to support their beliefs.

People in virtually all professions (except horse-racing handicappers and weather forecasters, who receive repeated objective feedback) are much more confident in their judgments and predictions than their performance would justify. One of the few ways to temper this overconfidence is to explicitly ask people to list the ways that might be wrong — for, unless prodded, we will only consider the confirmatory evidence.[14]

A dramatic example of *the confirmation bias* is the problem of the "self-fulfilling prophecy."[15] This now-popular phrase refers to the way our own expectations and behavior can influence events. Especially well known is the study by Harvard psychologists Rosenthal and Jacobson entitled *Pygmalion in the Classroom.*[16] Teachers were given false information on the expected achievement of some of their students. Based on the expectations created by this information, the teachers went on to treat the randomly selected "late bloomers" so differently that these students scored especially highly on subsequent achievement tests. Similar situations may occur between employers and their employees, between doctors and their patients, and between psychotherapists and their clients.

In scientific research, the *expectancy effect* suggests that experimenters' hypotheses may act as unintended determinants of experimental results. In other words, experimenters may have obtained the predicted results because they expected their subjects to behave as they did — even if the theory they were testing had little or no validity.

Although originally fraught with controversy, the existence of interpersonal expectancy effects is no longer in serious doubt. In 1978, Rosenthal and Rubin reported the results of a meta-analysis of 345 studies of expectancy effects.[17] This meta-analysis demonstrated the importance of expectancy effects within a wide variety of research domains including reaction time experiments, inkblot tests, animal learning, laboratory interviews, psychophysical judgments, learning and ability, person perception, and everyday situations or field studies.

Experimenter expectancy effects are a potential source of problems for any research area, but they may be especially influential in more recent research areas lacking well-established findings. This is because the first studies on a given treatment technique are typically carried out by creators or proponents of the technique who tend to hold very positive expectations for the efficacy of the technique. It is not until later that the technique may be investigated by more impartial or skeptical researchers, who may be less prone to expectancy effects operating to favor the technique.

People with strong pre-existing beliefs manage to find some confirmation in all presentations. The *biased assimilation* of evidence relevant to our beliefs is a phenomenon that seems true of others, but difficult to perceive in ourselves. Consider a classic social psychological study of students' perceptions of the annual Princeton-Dartmouth football game.[18] Students from the opposing schools watched a movie of the rough 1951 football game and were asked to carefully record all infractions. The two groups ended up with different scorecards based on the same game. Of course, we see this in sports enthusiasts and political partisans every day. Yet, the students used objective trial-by-trial recording techniques, and they still saw different infractions if they were on different sides.

Another typical error of intuition is our tendency to *overgeneralize* and *stereotype*. People are likely to make strong inferences about groups of people based on a few examples with little regard to the question of whether these sample were chosen randomly or whether they are representative of the larger group.[19] In one particularly unsettling set of experiments, people were willing to make strong inferences about prison guards and about welfare recipients, on the basis of one case study, even when they were explicitly forewarned that the case study was *atypical*.

At the heart of some possible intuitive misunderstandings about apparent psychic events is a widespread confusion about *randomness*. People believe the "laws" of chance are stricter in specific instances than they actually are. Processes occurring by chance alone can appear to be extremely usual. For example, the roulette wheel at Monte Carlo once landed on "black" twenty-eight times in a row. While the odds of this event, taken by itself, would seem extremely unlikely — and almost impossible to predict — in the larger picture of roulette gambling, such rare events are inevitable.

Another fallacy is to assume that, when a segment of a random sequence has strayed from the expected proportion, a corrective bias in the other direction is expected. In the roulette wheel, "red" is no more likely to have been selected after twenty-eight blacks in a row than after one black. To assume otherwise has been called *the gambler's fallacy*.[20]

When people see what appear to be patterns in a random sequence of events, they search for more meaningful causes than chance alone. Gamblers and professional athletes are known to become superstitious and attribute the good or bad luck to some part of their behavior or clothing.

Even lower animals such as pigeons have been shown in the laboratory to develop their own form of superstitious behavior.[21] When food is delivered according to a random schedule, the pigeons at first try to control the delivery by pecking on the food dispenser. The pigeons can be in the middle of any action at the moment the food comes, since there is no relation between its action and the food. Yet, they will continue to repeat the action that happened to occur at feeding time — and eventually their efforts will be "rewarded" by more food. This strengthens the "superstitious" behavior, and it is kept up because it appears to be successful.

An example of our human tendency to discover patterns in random data is the "hot hand" phenomenon in professional basketball. The hot hand phenomenon is the compelling, yet illusory, perception held by both fans and players that some players have "hot streaks" such that a successful shot is likely to be followed by another successful shot, while a failure is likely to be followed by another failure. University players were asked to place bets predicting the results of their next shot. Their bets showed strong evidence of a belief in the "hot hand" but their performance offered no evidence for its validity.

Researchers also examined the shooting records of outstanding NBA players and determined that the hot streaks of successful shots did not depart from standard statistical predictions based on the overall probability of success for each person individually.[22]

The most serious, common error in our understanding of probability is the *overinterpretation of coincidence*. In order to decide whether an event or collection of events is "unlikely," we must somehow compute a *sample space* — a list of all the other ways that the occasion could have turned out. Then we must decide which outcomes are comparable.

Before we attribute unusual and startling events to synchronicity or clairvoyance, we would do well to consider the many surprising ways that events can intersect within the *sample space* of our activities. It seems more like a miracle than a chance event when we encounter a familiar friend on a street corner in a distant city. We would never have predicted this in advance. The probability of this intersection of elementary events — being in another city, on a certain street, and meeting this familiar person — is indeed small. But the intersection of a *union* of elementary events — being in some other city, meeting some friend — is not so unlikely.[23]

In most spontaneous cases of apparent psychic abilities — such as precognitive dreams of disaster, or crisis apparitions followed by the death of a relative — there is no way of determining the likelihood of the event happening simply by chance. Therefore, spontaneous cases — such as those reported in Section II — can not be proof either of the existence or nonexistence of psychic events. Yet all people develop intuitive judgments about the nature of reality based on personal experience.

The Illusion of Self-Awareness

One source of overconfidence in our own judgments is the belief that we can always search our minds for the evidence and reasoning upon which these judgments are based. We sometimes mistakenly believe that we know whether we are biased and emotionally involved or evaluating objectively. Psychological studies indicate that this compelling feeling of self-awareness regarding our decision processes is exaggerated. Psychologists have found it surprisingly easy to manipulate preferences, and choices without the awareness of the actor.[24]

In a study of self-deception, subjects were told that certain uncontrollable indices diagnose whether they have a heart that is likely to cause trouble later in life. They were then given an opportunity to take part in a phony "diagnostic task" which supposedly indicated which type of heart they had. The task was painful and unpleasant. Subjects who believed that continuing with the task indicated a healthy heart reported little pain and continued with the task for a long period. Those who believed that sensitivity to pain indicated a healthy heart found themselves unable to bear the discomfort for more than a minute.

Some of the participants were aware that they "cheated" on the diagnosis. Those who were not aware of their own motivation expressed confidence that they had the healthy type of heart. These people were not deceiving the investigator, they were deceiving themselves. Continuing with the painful task could not have caused their heart to be of a certain type; but in order for them to believe the diagnosis, they had to remain unaware that they were controlling the outcome.[25]

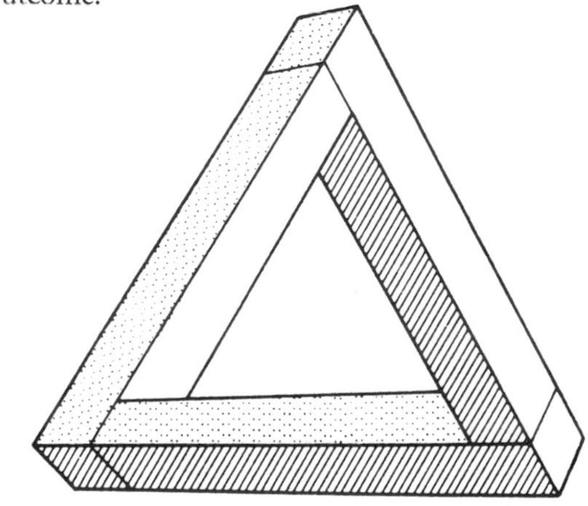

The Illusion of Control

We humans actively distort our perceptions in order to see what we *want* to see. Psychologists consider it normal behavior to distort our images of reality in order to enhance our own self-image.[26] In fact, the inability to create such *protective distortions* may lead to depression.

Depressives seem willing to accept that they do not have control over random events. On the other hand, normals maintain an *illusion of control* over chance events that have personal relevance.[27] Anecdotal examples from gambling are easy to generate: Dice players believe that their throwing styles are responsible for the high numbers or low numbers and Las Vegan casinos may blame their dealers for runs of bad luck.[28] People will wager more before they have tossed the dice than after the toss but before the result is disclosed — although the odds have not changed.[29] People believe their probability of winning a game of chance is greatest when irrelevant details are introduced that reminded them of games of skill. Allowing players to choose their own lottery numbers, or introducing a "schnook" as a competitor made people more confident in their likelihood of winning — without changing the chance nature of the outcome.

While depressives appear to be less vulnerable to this illusion, normal people will see themselves "in control" whenever they can find some reason. In a study of mental telepathy, researchers found that when subjects were able to choose their own occult symbol to send, and when the sender and receiver were able to discuss their communicative technique, they believed that they were operating at a success rate three times the chance rate. But when they were arbitrarily assigned a symbol and had no particular involvement in the task, they believed that they were operating at about the chance rate.

WHICH FACE OF THIS CUBE IS NEAREST YOU?

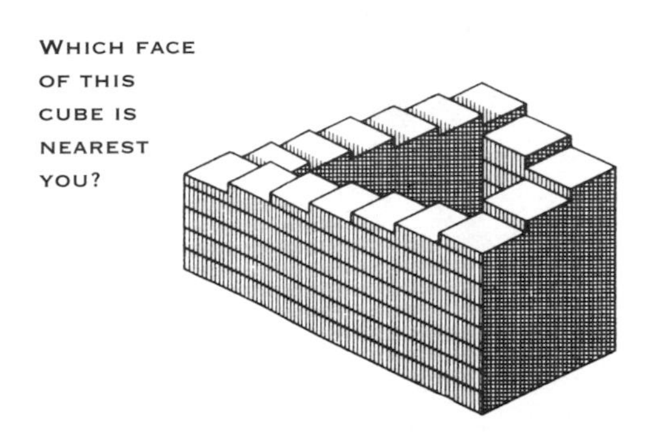

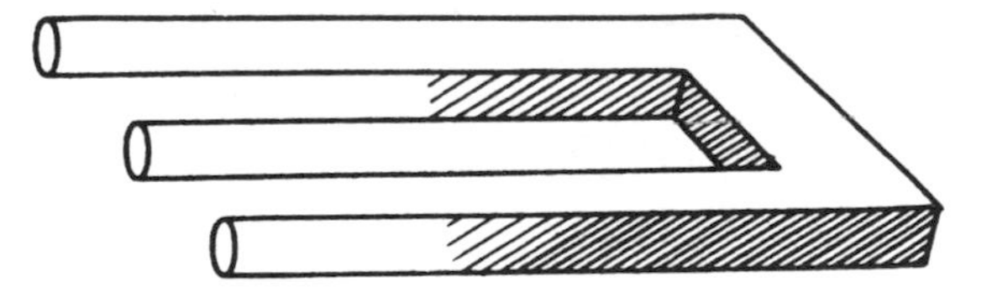

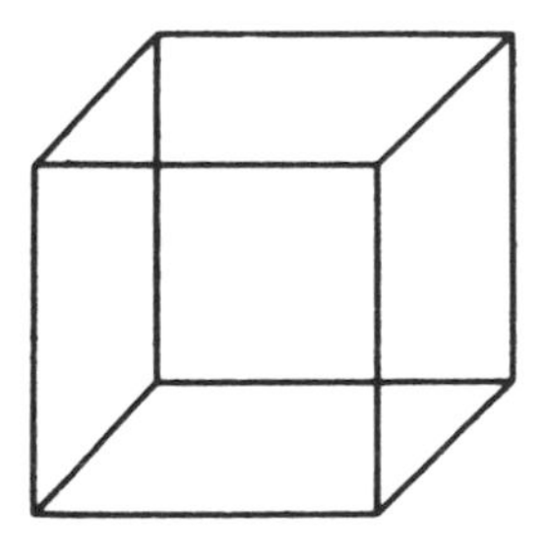

Actual scores did not deviate from chance throughout the entire experiment.[30]

A similar experiment used a psychokinesis task to test this hypothesis. Subjects' beliefs in their ability to influence the movement of a die did vary with active involvement in the task and with practice at the task.[32]

The Need to Be Consistent

One of the most powerful forces maintaining our beliefs in spite of others' attacks, our own questioning, and the challenge of new evidence — is the need to maintain *cognitive consistency* and avoid *cognitive dissonance.*

Modern social psychology came to public consciousness with the development of Leon Festinger's *theory of cognitive dissonance* which explains apparently irrational acts in terms of a general human "need" for consistency.[32]

In a dramatic field study of this phenomenon, Festinger and two colleagues joined a messianic movement to examine what would happen to the group when the "end of the world" did not occur as scheduled. A woman in the Midwestern United States who claimed to be in contact with aliens in flying saucers had gathered a group of supporters who were convinced that a great flood would wash over the earth on December 21, 1955. They made great sacrifices to be ready to be taken away by the flying saucers on that day. They also suffered public ridicule for their beliefs. Festinger hypothesized that if the flood did not occur and the flying saucers did not arrive, the members of the group would individually and collectively feel great dissonance between their beliefs and the actual events.

He felt that the members of the group had three alternatives: they could give up their beliefs and restore consonance; they could deny the reality of the evidence that the flood had not come; or they could alter the meaning of the evidence to make it congruent with the rest of their belief system.[33]

Public commitment made it unlikely that the members of the group would deny their beliefs. Yet, the existence of the unflooded world was too obvious to be repressed or denied. Therefore, the psychologically "easiest" solution was to make the evidence congruent with the prior beliefs. No flying saucers arrived, no deluge covered the earth, but a few hours after the appointed time, the communication medium received a message: the earth had been spared due to the efforts of the faithful group. The "disconfirmation" had turned into a "confirmation."[34]

Overcoming discomfort and actually considering the truth of a threatening idea does not necessarily lead to a weakening of our commitment. In a study of the reaction of committed Christians to scholarly attacks on the divinity of Christ, researchers found that only those who gave some credence to the evidence became more religious as a result of their exposure to the attacks. Only when they thought about the evidence did they become sufficiently distressed to resolve the dissonance by strengthening their beliefs.[35]

A similar situation may exist with regard to skeptics of psi research. For example, a recent (and highly criticized) report prepared for National Research Council of the National Academy of Sciences concludes that "The Committee finds no scientific justification from research conducted over a period of 130 years for the existence of parapsychological phenomena."[36] That this conclusion reflects the strengthening of previously held beliefs, in the face of threatening evidence, is suggested by many lines of argument. The committee's review, for example, was restricted to four selected areas of research conducted over the previous twenty years. And, one of the background papers prepared for the committee by an eminent Harvard psychologist ranked the methodological rigor of psi experiments very favorably in comparison to other areas of psychological research.[37]

The Effect of Formal Research

> What I do wish to maintain — and it is here that the scientific attitude becomes imperative — is that insight, untested and unsupported, is an insufficient guarantee of truth, in spite of the fact that much of the most important truth is first suggested by its means.

The research previously cited demonstrates the problems that arise when intuition replaces logic as the arbiter of truth. Intuitive processes based on personal experience sometimes seem designed as much for protecting both our sense of self-esteem and our prior opinions as for generating accurate predictions and assessments.

Organized science can be thought of as an extension of the ways that humans naturally learn about the world — with added procedures and methods designed to protect against the biasing effects of prior theories, salient evidence, compelling subsets of evidence and other natural pitfalls that beset all humans. The process of relying strictly upon personal experience often leads to certain violations of the protections against error that science affords.

In order to predict and control their environment, people generate hypotheses about what events go together and then gather evidence to test these hypotheses. If the evidence seems to support the current belief, the working hypothesis is retained; otherwise it is rejected.

Science adds quantitative measurement to this process. The procedures and measurements can be explicitly recorded and the strength of the evidence for different hypotheses can be objectively tallied. A key difference between intuitive and scientific methods is that the measurements and analyses of scientific investigations are publicly available for criticism and review, while intuitive hypothesis-testing takes place inside one person's mind.

When can we rely upon folklore, philosophy, spiritual tradition or personal experience without bias? These realms are our only guidance in many areas of life, but they can never be *decisive* when pitted against objective quantitative evidence. Informal examination of theories developed through personal experience or exposure to tradition is subject to flaws both in the gathering and in the analysis of the data. We cannot be "blind" to our theories when collecting the data, and we always know whether each data point collected supports or weakens the evidence for a theory. Without careful consideration for research design, people inevitably tend to collect samples of data biased in favor of the theories they wish to confirm.

Intuitive self-knowledge of the type required for a wide variety of higher mental functions requires a healthy respect for our natural human biases of attention and memory. Only if we are aware of these biased processes as they occur, can we begin to know when to trust our intuitive judgments.

While scientists and scientific methods are susceptible to errors of judgment, good research is designed to minimize the impact of these problems. Formal research methods are not the only or necessarily best way to learn about the true state of nature. But good research is the only way to ensure that real phenomena will drive out illusions.

The story of the "discovery" of N-rays in France in 1903 reveals how physics, the "hardest" of the sciences, could be led astray by subjective evaluation. This "new" form of X rays supposedly could be detected by the human eye in a nearly darkened room. The best physical scientists in France accepted this breakthrough. Within a year of its original "discovery" by Prof. R. Blondlot, the French Academy of Science had published nearly a hundred papers on the subject.[38]

However, in 1904 the American physicist Robert Wood visited Blondlot's laboratory and discovered, by secretly changing a series of experimental conditions, that Blondlot continued to see the N-rays under circumstances that Blondlot claimed would prevent their occurrence. When

Wood published his findings,[39] it became clear that the French scientists had believed so strongly in N-rays that they had virtually hallucinated their existence. Good research can disconfirm theories, subjective judgment rarely does.

An essential aspect of scientific research that is usually neglected when we rely on personal experience is the need for experimental control. Control is used in an experiment to ensure that the only thing that differs between the "present" and "absent" conditions is the particular variable of relevance to our hypothesis.

The need for such control is well illustrated in the problems of medical experimentation. When new types of surgery come along, physicians sometimes have good, humanistic reasons for violating scientific conditions in providing treatment to experimental patients. Instead they may offer the new surgeries to the patients who would seem to benefit the most. The results of such tests often seem impressive when compared to the survival rates of those who do not receive the surgery. However, those receiving the surgery start out differently on health variables than those who do not. They know they are receiving special treatment, and are cared for by staff who also understand this situation. When such uncontrolled field trials are compared with randomized experimental trials, it turns out that about 50% of surgical innovations were either of no help or actually caused harm.[40]

Medical research also demonstrates the necessity for *placebo* controls. When patients are given a pill of no medical value and told that they are participating in research on a new drug (but not told that they are in the "control group"), a substantial proportion will improve simply because of their belief in the possible efficacy of the drug. Therefore, all new drugs or treatments must be compared to a *placebo* to test whether they are of greater value than simple suggestion.

The analogue to the *placebo effect* in industrial research is the *Hawthorne effect*— named after a classic investigation into methods of improving worker efficiency at the Hawthorne plant of the Western Electric Company.[41] The researchers found that *every* alteration in working conditions led to improved efficiency — not because the changes in conditions affected productivity but because the attention from the investigators improved the workers' morale. (This study itself, however, has been cited so often as to have become *psychological folklore.* It is not clear that it holds up under careful experimental scrutiny.)

Similar problems make it extremely difficult to conduct valid research into psychotherapy. Control groups must offer supportive attention without the actual psychotherapy in order to test the effects of a particular therapeutic procedure.

Modern psi research is almost entirely an experimental science, as any cursory look through its journals will demonstrate. Articles published in the *Journal of Parapsychology* or the *Journal of the Society for Psychical Research* explicitly discuss the statistical assumptions and controlled research design used in their studies. Most active psi researchers believe that the path to scientific acceptance lies through the adoption of rigorous experimental method.

Psi researchers have amassed a large literature of experiments, and this compendium of studies and results can now be assessed using the language of science. Discussions of the status of psi research hypotheses can be argued on the evidence: quantified, explicit evidence. Our ability, as a culture, to objectively evaluate this evidence is itself a test of the scientific method as a valid tool for understanding the nature of consciousness.

The critical-thinking ability of believers and nonbelievers in psychic phenomena was examined in two studies conducted by James Alcock, a noted skeptic, and his associate Laura Otis. In the first study, believers and skeptics were given Watson and Glaser's *Critical Thinking Appraisal Scale* as well as Trodahl and Powell's *Dogmatism Scale*. Skeptics showed a significantly higher level of critical thinking ability and significantly less dogmatism than believers.

This result might lead some readers who consider themselves to be "believers" to question their own commitment to critical thinking. One might also question whether such a finding is the result of an *expectancy effect*. As many skeptics are *believers-in-the-negative*, they can be expected to show a bias like other believers.

The second study was carried out to evaluate the critical thinking ability of believers and skeptics on a task dealing with the psychokinesis. "Believers" and "skeptics" were asked to critically evaluate either a research article on psychokinesis or a similar article on pain tolerance. It was anticipated that believers would show a bias in favor of the psychokinesis article; however, results indicated that believers and nonbelievers were equally critical of the psychokinesis article.[42] This finding, particularly since it was conducted by skeptics, lends support to the notion that individuals who have accepted the psi hypothesis are capable of critical, scientific evaluation in this area.

Extrasensory Perception (ESP)

Introduction

Although many who read this book, like me, have few personal doubts that something like extrasensory perception is real, the effort to establish ESP as a scientific fact has been a continuous struggle, the outcome of which still remains uncertain.[43] Many subjects whose demonstrations had originally convinced researchers from the British Society for Psychical Research were later detected using bogus means to dupe these eminent scientists.[44] Fascinated by their few successes, researchers continued undaunted in the midst of failures, criticism, and detected frauds.

J. B. Rhine's Early Research at Duke University

Perhaps the most publicized early experiments were those published by Dr. Joseph Banks Rhine in 1934 in a monograph entitled *Extra-Sensory Perception*, which summarized results from his experiments at Duke University beginning in 1927. Although this work was published by the relatively obscure Boston Society for Psychic Research, it was picked up in the popular press and had a large impact throughout the world. While earlier researches had been fruitful, they were generally neither as systematic nor as persistent as Dr. Rhine's studies.

These experiments used shuffled decks of *ESP cards* with five sets of five different symbols on them — a cross, a circle, a wavy line, a square and a star. This method reduced the problem of chance-expectation to a matter of exact calculations. Furthermore, the cards were designed to be as emotionally neutral as possible to eliminate possible response biases caused by idiosyncratic preferences. However, other studies have shown that emotionally laden targets can also work without impairing statistical analysis.

Rhine describes his early work with one of his more successful subjects, Hubert E. Pearce, a graduate divinity student:

The working conditions were these:

observer and subject sat opposite each other at a table, on which lay about a dozen packs of the Zener cards and a record book. One of the packs would be handed to Pearce and he allowed to shuffle it. (He felt it gave more real "contact.") Then it was laid down and it was cut by the observer. Following this Pearce would, as a rule, pick up the pack, lift off the top card, keeping both the pack and the removed card face down, and after calling it, he would lay the card on the table, still face down. The observer would record the call. Either after five calls or after twenty-five calls — and we used both conditions generally about equally — the called cards would be turned over and checked off against the calls recorded in the book. The observer saw each card and checked each one personally, though the subject was asked to help in checking by laying off the cards as checked. There is no legerdemain by which an alert observer can be repeatedly deceived at this simple task in his own laboratory. (And, of course, we are not even dealing with amateur magicians.) For the next run another pack of cards would be taken up.[45]

The critical reader will find several faults with this experiment. First, as long as the subject is able to see or touch the backs or sides of the cards, there exists a channel of sensory leakage through which the subject might receive information about the face of the cards. Several critics reported that this was why they were able to obtain good scoring results. Second, there was no adequate safeguards against legerdemain. For example, what would prevent the subject from making small markings on the cards with his fingernails in order to identify the cards later on? It almost seems as if the optimism of the experimenter could mitigate against sufficiently careful observation.

Furthermore, was it really possible for one experimenter to maintain sufficient concentration to insure that the subject did not cheat? Experience of other researchers has sadly shown this is quite doubtful. Perhaps Rhine did utilize other safeguards. If so he could be (and was) fairly criticized for not adequately reporting his experimental conditions, although other experiments in his monograph were admittedly better controlled. Finally, there was no mention of any efforts to guard against recording errors on the part of the experimenter. One can hardly expect the cooperation of the subject, who may have a personal interest in the outcome, to be an adequate control against experimenter mistakes.

As Rhine's positive results gained more attention, arguments of this sort began to proliferate in the popular and scientific literature. It is much to Rhine's credit that he encouraged such criticism and modified his experiments accordingly. In 1940, J. Gaither Pratt, J. B. Rhine, and their associates published a work, titled *Extra-Sensory Perception After Sixty Years*, which described the ways in which the ESP experiments had met the thirty-five different counter-hypotheses that had been published in the scientific and popular press.

The areas of criticism Rhine and Pratt focused on in 1940 included: hypotheses related to improper statistical analysis of the results; hypotheses related to biased selection of experiments reported; hypotheses dealing with errors in the experimental records; hypotheses involving sensory leakage; hypotheses charging experimenter incompetence; and, finally, hypotheses of a general speculative character. In each case, Rhine and Pratt pointed to experimental evidence to counter the hypotheses.

Many prominent mathematicians in the field of probability who have made a detailed investigation have approved his techniques. In fact, in 1937 the American Institute of

JOSEPH BANKS AND LOUISA E. RHINE (COURTESY FOUNDATION FOR RESEARCH ON THE NATURE OF MAN, DURHAM, NORTH CAROLINA)

Statistical Mathematics issued a statement that Rhine's statistical procedures were not in the least faulty.[46] In most experiments, both significant and chance results were reported and averaged into the data.

Between 1880 and 1940, 145 empirical ESP studies were published using 77,796 subjects who made 4,918,186 single trial guesses. These experiments were mostly conducted by psychologists and other scientists. In 106 such studies, the authors arrived at results exceeding chance expectations.[47]

High scores due to inaccurate recording of results had been reduced to an insignificant level by double-blind techniques in which both subject and experimenter notations were made without knowledge of the scores against which they were to be matched. Errors were further reduced by having two or more experimenters oversee the matching of scores. Furthermore, original experimental data had been saved and double checked for mistakes many times by investigators. Tampering with these original records was prevented by having several copies independently preserved.

Sensory cues were impossible in many tests of clairvoyance because the experimenter himself and all witnesses did not know the correct targets. In other tests, the cards were sealed in opaque envelopes, or an opaque screen prevented the subject from seeing the cards. Often the experimenter and the subject were in completely different rooms.

Those who charged the experimenters with incompetence failed to find any flaws in several experiments (although rarely, if ever, are these early studies cited as evidential today, in an era of stricter experimental controls). In cases of inadequate reporting, Rhine indicated that further data would always be supplied upon request. In several cases, experimenter fraud would have had to involve the active collusion on the part of several teams of two or more experimenters. Critics who claimed that the results came only from the laboratories of those with a predisposition to believe in ESP were also ignoring at least six successful studies gathered from skeptical observers.

Other criticisms generally claimed that ESP could not exist because of certain philosophical assumptions about the nature of the universe or scientifically uninformed assumptions of what ESP would be like if it did exist. Rhine argued forcefully that such assumptions were scarcely sufficient cause to dismiss the carefully observed experimental data.

Of the 145 experiments reported in the sixty year period from 1880 to 1940, Rhine and Pratt selected six different experimental studies of ESP that they believed were not amenable to explanation by any of the counter-hypotheses offered by critics of psi research at that time.[48]

One of the more carefully controlled studies was the Pearce-Pratt series, carried out in 1933 with Dr. J. Gaither Pratt as agent and Hubert Pearce as subject. In these experiments, the agent and his subject were separated in different buildings more than a hundred yards apart. Pratt displaced the cards one by one from an ESP pack at an agreed time without turning them over. After going through the pack, Pratt then turned the cards over and recorded them. The guesses were recorded independently by Pearce. In order to eliminate the possibility of cheating, both placed their records in a sealed package handed to Rhine before the two lists were compared. Copies of these original records are still available for inspection. The total number of guesses was 1,850 of which one would expect one-fifth, or 370, to be correct by chance. The actual number of hits was 558. The probability that these results could have occurred by chance is much less than one in a hundred million.[49]

Criticisms of ESP Research

After the publication of *ESP After Sixty Years,* both the quality and quantity of criticism of ESP research declined until the mid-1970s (when a new wave of still-ongoing criticism was launched by the Committee for the Scientific Investigation of Claims of the Paranormal). That is not to say, however, that psi research met with general acceptance in the United States or in other countries. The work of the psi researchers was simply ignored by many universities and the major scientific publications. The public guardians were not then ready for ESP.

In August 1955, *Science* carried an editorial on ESP research by Dr. George R. Price, a chemist from the University of Minnesota, stating that scientists had to choose between accepting the reality of ESP or rejecting the evidence.[50] Price had carefully studied the data and he frankly admitted that the best experiments could only be faulted by assuming deliberate fraud, or an abnormal mental condition, on the part of the scientists. Price felt that ESP, judged in the light of the accepted principles of modern science, would have to be classed as a miracle (this judgment, as we will point out later, is ill-founded). Rather than accept a miracle, he suggested accepting the position of the eighteenth-century philosopher, David Hume, who said those who report miracles should be dismissed as liars.

Similar criticisms were published by Prof. C. E. M. Hansel. Regarding the Pearce-Pratt experiment, he suggests that after Pratt had left him, Pearce departed from the university library, followed Pratt to his office, and looked through the fanlight of Pratt's door thus observing the target cards being recorded by Pratt.[51] While it is true that Hansel exposed the defect in the experimental design of having left Pearce alone in the library, the structure of Pratt's office would have made it impossible for Pearce to see the cards even if he had taken the great risk of staring through the fanlight of Dr. Pratt's door.[52] In subsequent experiments psi researchers have generally (but, inevitably, not always) eliminated such defects.

Official recognition of the experimental competency of psi researchers did not come until December of 1969 when the American Academy for the Advancement of Science granted affiliate status to the researchers in the Parapsychological Association. Recent decades have shown authoritative scientific voices displaying a new willingness to deal with the evidence for ESP. In the "letters" column of *Science* for January 28, 1972, there appeared a brief note from Dr. Price titled "Apology to Rhine and Soal," in which Price expressed his conviction that his original article was highly unfair to both S. G. Soal and J. B. Rhine.

Ironically, Price's apology seventeen years after his original editorial was partially premature. In 1978, psi researchers found that S. G. Soal, a British mathematician who also reported significant ESP results, had fraudulently manufactured his data.[53]

Other criticisms relating to repeatability, fraud, statistical inferences, experimental design and interpretation of data have continued. In fact, psi researchers closely scrutinize each other's work and have often been their own most thorough critics (making it rather easy for would-be debunkers to seize upon their criticisms as grounds for discrediting the entire field). As a response to criticism, psi researchers have slowly, sometimes erratically and sometimes steadily, improved the quality of their experiments while continuing to obtain data which they believe is anomalous.[54] Skeptics, meanwhile, while continuing to reject the psi hypothesis, acknowledge that some of the research deserves careful scrutiny from the mainstream scientific community.

Unconscious ESP

One of the first theories about the nature of ESP was put forward by Frederick Myers, author of the 1903 classic *Human Personality and Its Survival of Bodily Death*, when he associated psychic phenomena with the workings of the *subliminal* mind, below the limits of consciousness. Studies in which ESP signals are registered by the body's physiological processes even when the subject is unaware of the message support the concept of unconscious ESP. For example, in a series of studies conducted by E. Douglas

John Palmer argues that, while psychic researchers have not proven the existence of psi, they have established a scientific anomaly that cannot be explained away by skeptics. (Photograph taken by Richard Broughton)

Dean, subjects were hooked up to a plethysmograph. Increases or decreases in blood and lymph volume, resulting from emotional responses, are measured by this instrument. A telepathic agent in another room then concentrated on different names, some of which were known to be emotionally significant to the subjects. The results indicated changes in the blood volume which significantly correlated with the emotionally laden target messages.[55] This finding was confirmed in a second series of studies conducted by Dean and Carroll B. Nash at St. Joseph's College in Philadelphia. Most of the subjects were totally unaware of the changes in their blood supply which were responding to the target material.[56]

A similar study was conducted by Charles Tart in which subjects were hooked up to a plethysmograph, an electroencephalograph, and a device for measuring galvanic skin response. The agent in this experiment was periodically given a mild electric shock. The subjects did not know that they were being tested for ESP, but rather were told to guess when a "subliminal stimulus" (sensory stimulation below the threshold of conscious awareness) was being directed to them. The subjects' hunches failed to correlate to this disguised target. However, their physiological measurements showed abrupt changes when the shocks were administered to the agent in another room.[57]

Dream Telepathy

Frederick Myers noted in the early years of psychical research that the workings of the subliminal mind were most visible in such phenomena as dreams, trance states, hypnosis, and states of creative inspiration. In fact, a large proportion of the reported cases of ESP occurred while the percipient was in such altered states of consciousness.

An important series of studies on the nature of ESP in dreams was carried out by a team of researchers at Maimonides Hospital in Brooklyn, New York. Using equipment which monitored brain waves and eye movements, the investigators could determine when subjects were dreaming. By waking the subjects at these times they were able to obtain immediate reports of the dream contents. Earlier in the day, in another room, the telepathic senders had concentrated on target pictures designed to create a particular impression.

Independent judges compared the similarity subject's responses displayed to all of the actual targets in each series and found evidence for nocturnal telepathy and precognition (when targets were not chosen until the following day) of the actual targets used.[58,59]

In addition to these careful experiments, there were some interesting one-time studies.

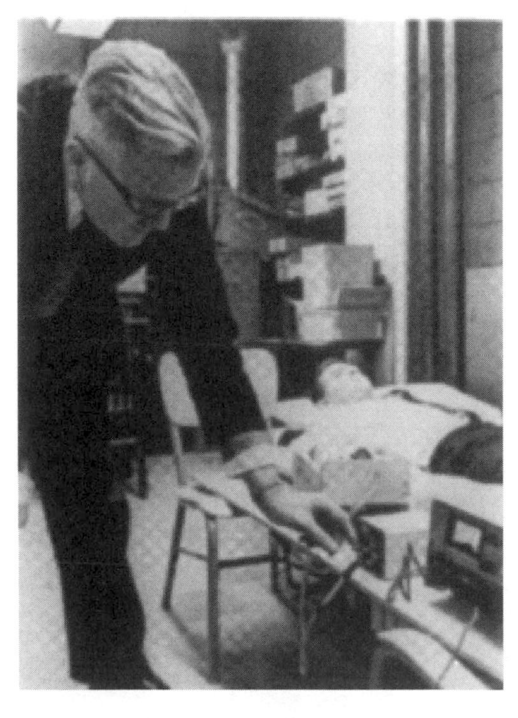

E. DOUGLAS DEAN CONDUCTING A PLETHYSMOGRAPH STUDY

In one such test, telepathic transmission was obtained by having about two thousand persons attending a Grateful Dead rock concert focus on a color slide projection image and attempt to send it to the dream laboratory forty-five miles away in Brooklyn. Many of these individuals were in altered states of consciousness from the music and the ingestion of psychedelic drugs. This test proved successful.[60]

Psychologist David Foulkes at the University of Wyoming, in consultation with the Maimonides team, attempted unsuccessfully to replicate the dream studies. Critic C. E. M. Hansel, a psychologist at the University of Wales in England, attributed the failure to tighter controls against fraud in the Wyoming experiments, whereas dream researcher Robert Van de Castle from the University of Virginia, one of the subjects in both the Wyoming and the Maimonides experiments, stressed the debilitating effect of the skeptical attitude of the Wyoming team.[61]

In 1985, Yale University psychologist Irvin Child published a review of the Maimonides dream studies in the *American Psychologist*.[62] Child's basic point was that this research was basically sound. The experiments have received little or no mention in the pertinent psychological literature. When these studies were reviewed, Child claims that they were so severely distorted as to give an entirely erroneous impression of how they were conducted. Child used the example of the dream research to illustrate the general point that books by psychologists purporting to offer critical reviews of psi research do not use the scientific standards of discourse prevalent in psychology.

IRVIN L. CHILD, PROFESSOR EMERITUS OF PSYCHOLOGY, YALE UNIVERSITY

Hypnosis and ESP

Milan Ryzl, a chemist who defected to the United States from Czechoslovakia in 1967, developed a hypnotic technique for facilitating ESP which, although it has not been successfully replicated, attracted attention and may yet prove fruitful. Ryzl's technique involved the intensive use of deep hypnotic sessions almost daily for a period of several months. The first stage of these sessions was to instill confidence in his subjects that they could visualize clear mental images containing accurate extrasensory information. Once this stage was reached, Ryzl concentrated on conducting simple ESP tests with immediate feedback so that subjects might learn to associate certain mental states with accurate psychic information. Subjects were taught to reject mental images which were fuzzy or unclear. This process, according to Ryzl, continued until the subject was able to perceive clairvoyantly with accuracy and detail. Finally, Ryzl attempted to wean the subject away from his own tutelage so that he or she could function independently. While still in Czechoslovakia, Ryzl claimed to have used this technique with some five hundred individuals, fifty of whom supposedly achieved success.[63]

Other studies have shown heightened ESP in states of physical relaxation or in trance and hypnotic states.[64,65,66,67] In fact, the use of hypnosis to produce high ESP scores is one of the more replicablе procedures in psi research. A particularly notable series of experiments were described in 1910 by Emile Boirac, rector of the Dijon Academy in France, which produced what he described as an "externalization of sensitivity." When the hypnotist placed something in his mouth, the subject could describe it. If he pricked himself with a pin, the subject would feel the pain. The most striking experiments were those in which the subject was told to project his sensibility into a glass of water. If the water was pricked, the subject would react by a visible jerk or exclamation.[68] This phenomenon has been repeated by the Finnish psychologist, Jarl Fahler.[69]

In a similar experiment performed by one of Boirac's associates, blisters were raised on the skin of a hypnotized subject, simply by pricking a photograph of the subject's hand.[70]Boirac, as well as Soviet investigators, have reported the ability to induce a hypnotic trance simply through telepathic concentration directed toward their subjects.[71,72]

In 1969, Charles Honorton and Stanley Krippner reviewed the experimental literature of studies designed to use hypnosis to induce ESP. Of nineteen experiments reported, only seven failed to produce significant results. Many of the studies produced astounding success. In a particularly interesting precognition study, conducted by Fahler and Osis with two hypnotized subjects, the task also included making confidence calls — predicting which guesses would be most accurate. The correlation of confidence call hits produced impressive results with a probability of 0.0000002.[73]

In 1984, Ephriam Schechter reported an analysis of studies comparing the effect of hypnotic induction and nonhypnosis control procedures on performance in ESP card-guessing tasks. There were twenty-five experiments by investigators in ten different laboratories. Consistently superior ESP performance was found to occur in the hypnotic induction conditions compared to the control conditions of these experiments.[74]

Hypnosis typically involves relaxation and suggestion in an atmosphere of friendliness and trust. We do not know which of these factors, or combination of factors, accounts for heightened psi scores. Schechter himself, as well as other psi researchers, has been reluctant to conclude that hypnosis does facilitate psi performance. He noted

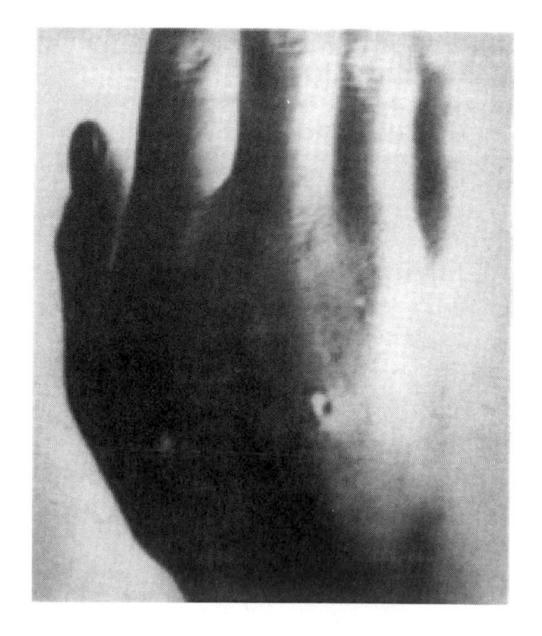

"EXTERIORIZATION OF SENSITIVENESS." A PHOTOGRAPH OF THE SENSITIVE WAS TAKEN AND THE NEGATIVE WAS THEN HELD BY HER A FEW MOMENTS. THE OPERATOR, WITH A PIN, SCRATCHED THE HAND ON THE NEGATIVE. INSTANTLY THE SENSITIVE EJACULATED WITH PAIN, AND A SMALL RED SPOT APPEARED ON THE BACK OF HER HAND. THIS RAPIDLY GREW INTO THE BLISTER SHOWN IN THE ABOVE PHOTOGRAPH.

"THE CONDUCTIBILITY OF PSYCHIC FORCE." THE TWO GLASSES WERE CONNECTED BY A COPPER WIRE. WHEN THE EXPERIMENTER PINCHED THE AIR-ZONE ABOVE THE WATER GLASS NEAREST HIM, OR PLUNGED HIS FINGER OR PENCIL INTO IT, THE SUBJECT IMMEDIATELY REACTED. THIS REACTION DISAPPEARED IF THE CONNECTION BETWEEN THE GLASSES WAS REMOVED. THESE RESULTS WERE UNDOUBTEDLY SHAPED BY THE BELIEF SYSTEMS OF EXPERIMENTER AND SUBJECT.

that the studies were not designed to control for an expectancy effect.[75] In a retrospective critique of research on altered states of consciousness and psi, St. Johns University psychologist Rex Stanford argued that many alternative explanations of heightened psi effects were not controlled for in the research studies. Stanford called for more rigorous process oriented research to determine why hypnosis and other altered states enhance psi scores — if, in fact, they actually do.[76]

Exceptional ESP Laboratory Performers

PAVEL STEPANEK

Western researchers who traveled to Prague to personally investigate Milan Ryzl's hypnotic training program were able to test one of his better subjects, Pavel Stepanek. During a long period of experimental investigations, Stepanek proved to be one of the most successful subjects ever tested. More than twenty studies with him have now been published.[77]

What we still do not know is whether Stepanek always had this ESP ability or whether it developed as a result of Ryzl's training. There was a period of time during which Stepanck's scores did drop down to chance levels and then jumped up again after a hypnotic session with Ryzl.

In a recently published book titled *How Not to Test a Psychic,* skeptic Martin Gardner reanalyzed the tests with Stepanek, offering detailed hypotheses as to how the results obtained for over a decade by several independent experimental teams may have resulted from cheating by Stepanek.[78]

BILL DELMORE

In a study with an exceptional subject, Bill Delmore, confidence calls were made using a deck of ordinary playing cards as the target. The technique used was a "psychic

MILAN RYZL TESTING HIS STAR SUBJECT, PAVEL STEPANEK. THE TARGET IS INSIDE A TRIPLE-SEALED ENVELOPE. (COURTESY MILAN RYZL)

MARTIN GARDNER, AUTHOR OF *HOW NOT TO TEST A PSYCHIC, FADS AND FALLACIES IN THE NAME OF SCIENCE,* AND OTHER SKEPTICAL BOOKS (COURTESY MARTIN GARDNER)

shuffle" in which the experimenters randomly select a predetermined order which the subject must match by shuffling the target deck. In each of two shuffle series, with fifty-two cards in a series, Delmore made twenty-five confidence calls — all of which were completely correct.[79] The probability of such success is only one in fifty-two. Other studies with Delmore have also produced extraordinary results.

Delmore seems not to use an altered state of consciousness as a means of gaining psychic information. In fact, the research in altered states points to other variables which are really more significant. For Delmore, a congenial atmosphere is important.

Skeptical statistician Persi Diaconis, who observed some informal tests with Delmore which amazed a group of Harvard faculty and students, hypothesized that the results were due to a set of complicated maneuvers familiar to magical practitioners.[80]

Uri Geller

In 1974, Harold Puthoff and Russell Targ at SRI International (then Stanford Research Institute) in Menlo Park, California, reported on experiments conducted with the Israeli psychic performer Uri Geller. Their results covered studies taking place over an eighteen month period:

> In the experiments with [Uri] Geller, he was asked to reproduce 13 drawings over a week-long period while physically separated from his experimenters in a shielded room. Geller was not told who made any drawing, who selected it for him to reproduce or about its method of selection.
>
> The researchers said that only after Geller's isolation — in a double-walled steel room that was acoustically, visually and electrically shielded from them — was a target picture randomly chosen and drawn. It was never discussed by the experimenters after being drawn or brought near Geller.
>
> All but two of the experiments conducted

with Geller were in the shielded room, with the drawings in adjacent rooms ranging from four meters to 475 meters from him. In other experiments, the drawings were made inside the shielded room with Geller in adjacent locations. Examples of drawings Geller was asked to reproduce included a firecracker, a cluster of grapes, a devil, a horse, the solar system, a tree and an envelope.

Two researchers — not otherwise associated with this research — were given Geller's reproductions for judging on a "blind" basis. They matched the target data to the response data with no errors, a chance probability of better than one in a million per judgment....

In another experiment with Geller, he was asked to "guess" the face of a die shaken in a closed steel box. The box was vigorously shaken by one of the experimenters and placed on a table. The position of the die was not known to the researchers.

Geller provided the correct answer eight times, the researchers said. The experiment was performed ten times but Geller declined to respond two times, saying his perception was not clear....[81]

Because of the publicity which these studies received, they have been highly criticized by skeptics.[82] These criticisms are largely speculations as to how Geller might have cheated. The SRI researchers have responded by maintaining that the experimental conditions were such that the alleged cheating could not have taken place. Nevertheless, psi researchers in general are reluctant to defend Geller.

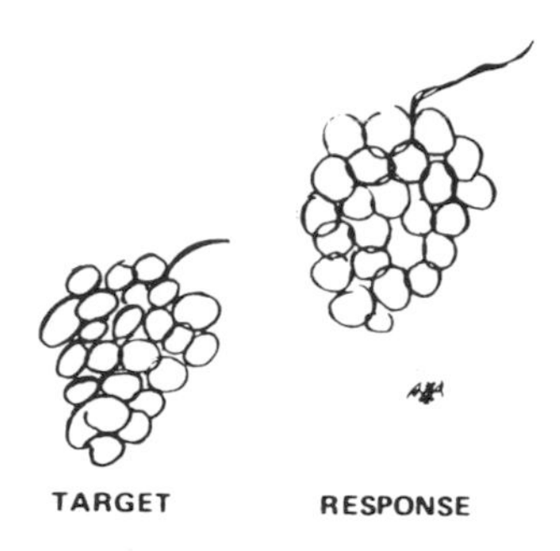

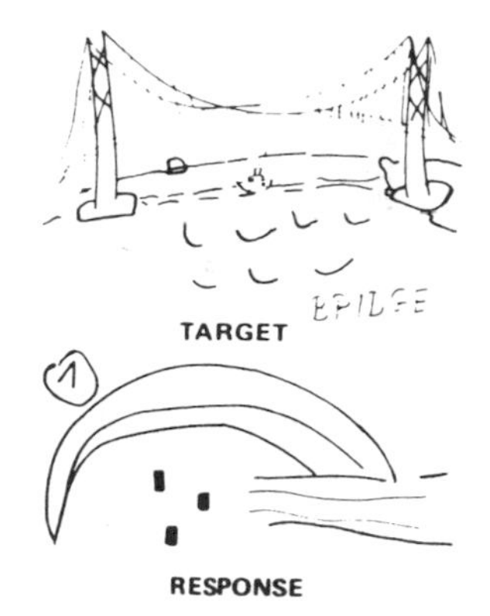

SOME TARGETS AND RESPONSES FROM ESP TESTS WITH URI GELLER AT STANFORD RESEARCH INSTITUTE

Ganzfeld Research

A very thoughtful approach toward investigating ESP was reported in 1974 simultaneously by investigators in New York and Texas.[83,84] These researchers hypothesized that the reason high scoring occurred in altered states (e.g., the Maimonides Hospital ESP dream research) was that the normal, waking mind was less active at these times; thus, there was less mental "noise" covering up the signals coming through the subliminal mind. To test this theory, they utilized a *ganzfeld* technique of covering the eyes of their subjects with halved ping pong balls so that the visual field was seen as solid

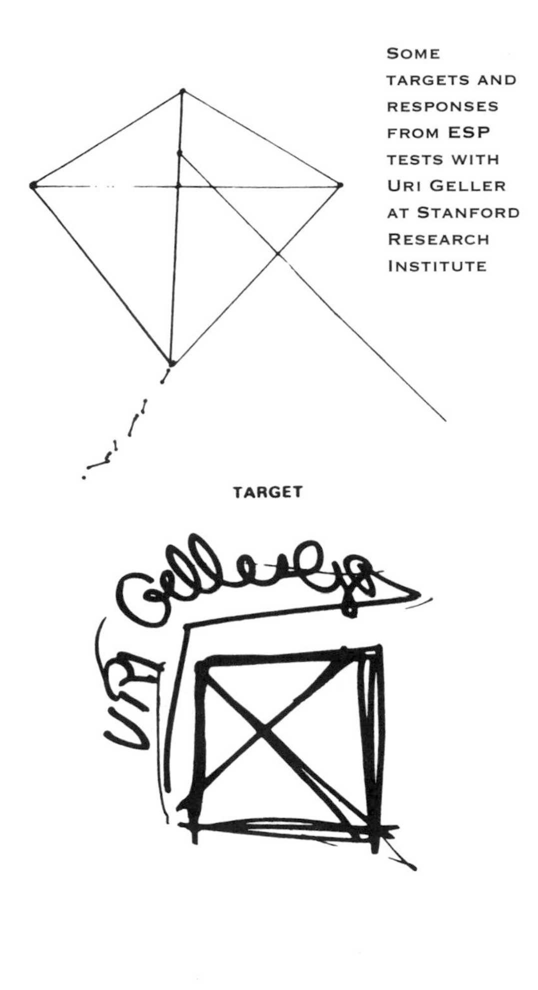

Some targets and responses from ESP tests with Uri Geller at Stanford Research Institute

white. A constant auditory environment was provided by either a white noise generator or a tape of the seashore. Under these conditions, with a constant sensory input, psi signals were expected to be easier to perceive. Subjects were put into this condition and asked to free-associate out loud while their responses were put on to magnetic tape. In another room, the telepathic sender chose, at random, a set of slides to look at and try to send to the subject. After the experiment, the subject was asked to guess which of the View-Master reels, of a group of four, had been the target. The subject's taped responses were also independently judged. The qualitative results of this procedure were often striking, and statistical results also proved impressive.

In 1985, a meta-analysis of twenty-eight psi ganzfeld studies by investigators in ten different laboratories found a combined *z* score of 6.6, a result associated with a probability of less than one part in a billion.[85] Independently significant outcomes were reported by six of the ten investigators, and the overall significance was not dependent on the work of any one or two investigators. Moreover, in order to account for the observed experimental results on the basis of selective reporting, it would have been necessary to assume that there were more than four hundred unreported studies averaging chance results. Part of this problem was addressed by British psychologist Susan Blackmore's survey of unreported ganzfeld studies. Seven of these nineteen studies (37%) yielded statistically significant results.[86] This proportion was not appreciably lower than the proportion of published studies found significant.

In evaluating the ganzfeld database, Harvard psychologists Monica Harris and Robert Rosenthal compared it in quality to research in biofeedback.[87] This is not to say, however, that the studies were flawless. Several critics found methodological problems with these studies.[88,89,90] In fact, for every ganzfeld study reporting significant evidence of psi communication, there has been at least one critical review or commentary.

By their nature, ESP studies must eliminate opportunities for sensory cueing. An exception occurred in some ganzfeld studies when the subject was asked to choose which target picture had been "sent" by another person or agent. When slides held originally by the sender were shown to the receiver, finger smudges or other marks could theoretically have served as cues.

Honorton has shown, however, that ganzfeld studies which eliminated this type of cue yielded at least as many significant psi effects as the studies with poorer controls.

In response to the various criticisms which had been addressed to the ganzfeld studies, Honorton and seven associates reported on a series of eleven new experiments at the 1989 convention of the Parapsychological Association.[91] These studies were conducted using an automated testing system which controlled random target selection, target presentation, the blind-judging procedure, and data recording and storage. Targets were recorded on videotape and included both video segments (dynamic targets) and single images (static targets). In all, 243 volunteer receivers completed 358 psi ganzfeld sessions. The success rate for correct identification of remotely viewed targets was statistically highly significant. The likelihood that these results could have been obtained by chance was less than one in ten thousand. The results were consistent across the eleven series with eight different experimenters.

A number of other interesting correlations were noted. The success rate for sessions using dynamic targets was significantly greater than those with static targets and accounted for most of the successful scoring. Significantly stronger performance occurred with sender/receiver pairs who were acquainted than with unacquainted sender/receiver pairs. Furthermore, comparison of the outcomes of these eleven automated ganzfeld studies with a meta-analysis of the original twenty-eight direct hits ganzfeld studies indicated that the two sets were consistent on four dimensions: (1) overall success rate, (2) impact of dynamic and static targets, (3) effect of sender/receiver acquaintance, and (4) impact of prior ganzfeld experience.

This new series of eleven ganzfeld studies provides a model for psi research in that it combined exacting methodological rigor with a consideration of the humanistic considerations essential for psi performance. Participants were all informed that they were free to bring a friend or family member to serve as their sender.[92] Additionally, the researchers encouraged participants to reschedule their session rather than feel that they had to come in to "fulfill an obligation" if they were not feeling well or were experiencing personal problems.

Researchers greeted participants at the door when they arrived for their session and attempted to create a friendly, informal social atmosphere. They offered coffee, tea, or soft drinks. The experimenter and other staff members engaged the participant in conversation during this period. If the session involved a laboratory sender, time was taken for the sender and participant to become acquainted. The experimental protocols included many other details to help subjects feel comfortable using psi faculties.[93]

The possibility of sensory cueing was eliminated by the use of the automated target selection system. Both sender and receiver were kept in separate sound attenuated chambers during each experiment. Only the sender was aware of the target's identity. The use of a videotape display system prevented potential cues that might result from manual handling of target pictures. Dozens of details in the experimental protocols addressed every criticism which has been made of the experimental procedures. All electronic circuits, for example, were carefully monitored to exclude the possibility of even subliminal sensory leakage. All of the data from every test with every subject was reported using statistical tests which were specified in advance.

The use of a large number of subjects and the significance of the outcome using subjects as the unit of analysis, rules out

JOHN BELOFF

subject deception as a plausible explanation. The automated protocol had been examined by several dozen psi and behavioral researchers, including two well-known critics of psi research. Some participated as subjects, senders, or observers, and all expressed satisfaction with the handling of security issues and controls.

In addition, two experts on the simulation of psi ability have examined the system and protocol. Ford Kross has been a professional mentalist for over twenty years. He is the author of more than a hundred articles in mentalist periodicals, and has served as secretary/treasurer of the Psychic Entertainers Association. Mr. Kross has stated:

> In my professional capacity as a mentalist, I have reviewed Psychophysical Research Laboratories' automated ganzfeld system and found it to provide excellent security against deception by subjects.[94]

Similar comments were made by Daryl Bem, professor of psychology at Cornell University. Professor Bem is well known for the development of *self-perception theory* in social and personality psychology. He is also a member of the Psychic Entertainers Association and has performed for many years as a mentalist. He visited PRL for several days and was a subject in one series. In a book review of *Advances in Parapsychology, Vol. 5* published in *Contemporary Psychology*, Bem made the following statement about the Honorton-Hyman debate over the psi ganzfeld studies:

> (For what it's worth, I find Honorton's conclusion that there is a significant and nonartifactual effect in the Ganzfeld data more persuasive than Hyman's more pessimistic conclusion.) Apparently parapsychological data will remain a projective test for all of us.[95]

The researchers claim that analysis has shown, contrary to the assertions of certain critics, that the ganzfeld psi effect exhibits "consistent and lawful patterns of covariation found in other areas of inquiry."[96] The automated ganzfeld studies display the same patterns of relationships between psi performance and target type, sender/receiver acquaintance, and prior testing experience found in the earlier ganzfeld studies, and the *magnitude* of these relationships is consistent across the two data sets. The impact of target type and sender/receiver acquaintance is also consistent with trends in spontaneous case studies, linking ostensible psi experiences to emotionally significant events and persons.

Skeptic Ray Hyman and psi researcher Charles Honorton stated in a joint communique regarding the status of the *ganzfeld* studies:

> ...the best way to resolve the controversy...is to await the outcome of future ganzfeld experiments. These experiments, ideally, will be carried out in such a way as to circumvent the file-drawer problem, problems of multiple analysis, and

> the various defects in randomization, statistical application, and documentation pointed out by Hyman. If a variety of...investigators continue to obtain significant results under these conditions, then the existence of a genuine communications anomaly will have been demonstrated.[97]

The researchers claim that they have presented a series of experiments that satisfy these guidelines. No single investigator or laboratory can satisfy the requirement of independent replication, but the automated ganzfeld outcomes are quite consistent with the earlier psi ganzfeld studies and the psi researchers believe that the burden of proof is on the critics to show why these findings should not be accepted.

The automated ganzfeld studies show an overall success rate slightly in excess of 34%. A *power analysis* by University of California statistician Jessica Utts shows that for an effect this size, the investigator has only about one chance in three of obtaining a statistically significant result in an experiment with fifty trials. Even with a hundred trials, which is unusually large in ganzfeld research, the probability of a significant outcome is quite small.

The Experimenter Effect

For some time psi researchers have been suggesting that the failure of some investigators to replicate their findings was due to attitudes and expectations, conscious or unconscious, which were communicated through subtle sensory or psychic channels to their subjects.[98] A number of psi research projects have been designed to study the factors that Harris and Rosenthal identified as contributing to the experimenter expectancy effect.

For example, one project compared the effects of a warm and cold social climate on ESP scores. All of the subjects had the same instructions and the same long ESP task. For half, there was a friendly, informal conversation with the experimenter for a quarter of an hour before the orientation began, and the experimenter made encouraging remarks during the breaks. The other half were treated formally and rather abruptly. The experimenter began the orientation immediately and also made discouraging remarks during the breaks. Results clearly confirmed the hypothesis. ESP scores of subjects treated warmly were significantly higher than mean chance expectation; scores of subjects treated coldly were significantly below mean chance expectation.[99]

Judith Taddonio followed Rosenthal's classic expectancy effects design in an experiment with two series. Her experimenters were six undergraduates with previous practice in conducting psychological experiments. All felt neutral toward ESP but agreed to help her when she told them that a particular ESP method needed checking out. Three were told that prior findings with this method could not fail. The other three were told that Taddonio's colleagues were worried because the method seemed to elicit only psi-missing. All experimenters used the same materials and method.

Both in the first series and in the second, subjects of experimenters with high expectations made ESP scores above chance and subjects of experimenters with low expectations made ESP scores below chance. In each series, the difference was significant.[100]

A study conducted at the University of Edinburgh in Scotland suggests that the attitude of the experimenter regarding the existence of ESP correlates the results of that person's research.[101] Ironically, one of psi researcher's least successful experimenters is from the University of Edinburgh, psychologist John Beloff. Beloff is a respected researcher who has consistently failed to confirm several psi experiments that have done well elsewhere.[102,103,104] Yet, in spite of his inability to confirm psi studies in his own laboratory, Beloff has remained convinced of the legitimacy of the research conducted by others. If he is a psi-inhibitory experimenter, it is not likely that his belief system is the inhibiting factor.

In 1988, psychologist Gertrude R. Schmeidler conducted an analysis of psi research studies testing for expectancy effects. Her conclusions were:

> Psychological research on the experimenter effect has shown higher scores with a warm than with a cold experimenter climate and with an experimenter who expects high rather than low scores. Eight experiments, comprising 12 series, tested for the experimenter effect in psi. Nine of the 12 series had significant results, all in the predicted direction. Six other experiments tested a related hypothesis: that psi experimenters would be self-consistent in obtaining results like their prior ones. Four of these had significant results, all in the predicted direction.[105]

Of course, effects of this sort have long been recognized in psychology, where they have been attributed to the desire of subjects to fulfill the expectations of the experimenters. Psychologists usually guard against this type of "artifact" by designing *double-blind* studies in which the experimenters are kept unaware of which subjects are in the test and control groups.

However, the psi experimenter effect actually has revolutionary implications for normal psychological research. The following study was reported by Prof. Hans Kreitler and Dr. Shulamith Kreitler of the Department of Psychology at Tel Aviv University in Israel:

> The first experiment dealt with the effect of ESP on the identification of letters projected at subliminal speed and illumination. The second experiment dealt with the effect of ESP on the direction of perceived autokinetic motion (i.e., of a stationary point of light in a dark room). The third experiment dealt with the effect of ESP on the occurrence of specific words and themes in the stories subjects tell to TAT (Thematic Apperception Test) cards.
>
> In all these three experiments the subjects did not know that ESP communications were "sent" to them, the "senders" never met the subjects and "senders" were naive in the sense that they were not particularly interested in parapsychology, were unselected, and did not get any training for the experiments. The precautions undertaken against any sensory contact between "senders" and subjects were highly complex and included the spatial separation of "sender" and subject (they were in two different soundproof rooms with another room between them), the decentralization of information about the experiment among different people, strict randomization of all stimuli and sequences, the use of experimenters who were disbelievers in ESP, etc.
>
> The results show that in every experiment there was a significant effect due to the ESP communication....The effect...was particularly pronounced with regard to responses with an initially low probability of occurrence.[106]

Ironically, while scientists in many other fields question the reliability of experiments in psi research, it may well be that psi effects underlie many controversial studies in other areas of science which have proved difficult to repeat — for example, physicists' attempts to measure gravity waves.

The Sheep-Goat Effect

The hypothesis that the attitude the experimenter takes can affect ESP scores also applies to ESP subject attitudes. Dr. Gertrude Schmeidler, working at Harvard University and at the City College of New York, divided her subjects into "sheep," who believed that ESP might occur in their experiment, and "goats," who did not. Her studies, which were conducted over a nine-year period and have since been replicated, showed an unquestionable difference between the "sheep" whose scores fell above chance expectation and "goats" who scored below chance levels. The phenomenon of *psi-missing* is thought to be a psychological effect in which psychic material is repressed from consciousness.[107,108]

In a review of seventeen experiments testing the hypothesis that subjects who believed in ESP would show superior ESP performance compared to subjects who did not believe in ESP, psychologist John Palmer found that the predicted pattern occurred in 76% of the experiments, and all six of the experiments with individually significant outcomes were in the predicted direction.[109] These findings suggest an overall statistical significance for this effect.

It is important to realize, however, that the sheep-goat studies do not necessarily distinguish those who believe in ESP from those who do not. In most studies, the "sheep" were not "true believers"; they merely accepted the possibility that ESP could occur in the test situation. On the other hand, many of the "goats" were willing to accept that ESP could occur between people who loved each other, or in certain times of crisis; but they rejected all possibility that ESP would manifest for them in their particular test situation.[110]

Psi-Missing

One would expect that if a person had ESP the level of performance on an experimental ESP test should be higher than chance expectation. Some individuals (and some groups in specific experiments) on the other hand have been found to score significantly *below* chance. This is termed *psi-missing*. It is important to note that this phenomenon does not necessarily indicate a lack of ESP since that situation would be associated with *nonsignificant* scores. Rather, *psi-missing* can be viewed as an expression of psi in a way that produces a result different from one's conscious intent. The percipient shows sensitivity to the targets' identity but tends to make calls at odds with this sensitivity.

Psi-missing is thought to be in part a consequence of negative elements in mood and attitude or aspects of personality; these elements may cause *psi-missers* to focus their ESP inaccurately.[111] While observations of *psi-missing* are much less common than those of *psi-hitting*, the former are sufficiently numerous to suggest that there is an effect here to be explained.[112]

Critics argue that the occurrence of so-called *psi-missing* confirms their view of reported data as statistical freaks. These below-chance scores are seen as nothing else than the negative tail of the normal distribution of random guessing scores. Two points count against this interpretation. First,

GERTRUDE SCHMEIDLER

under this "normal curve" model the incidence of *psi-missing* data should be the same as that of *psi-hitting* results, yet in fact the former is much lower than the latter. Second, the occurrence of *psi-missing* seems to be correlated with certain psychological variables; this should not be the case if properly controlled ESP tests entail purely random guessing.[113]

ESP and Personality Traits

Beginning in the early 1940s numerous attempts have been made to correlate experimental ESP performance with individual differences in subjects' personality and attitudinal characteristics. A series of studies with high school students in India by B. K. Kanthamani and K. Ramakrishna Rao has given further insight into the personality traits associated with psi-hitters and psi-missers. The following adjectives summarize the results of their work.[114]

Positive ESP Scores	**Negative ESP Scores**
warm, sociable	tense
goodnatured, easy going	excitable
assertive, self-assured	frustrated
tough	demanding
enthusiastic	impatient
talkative	dependent
cheerful	sensitive
quick, alert	timid
adventuresome, impulsive	threat-sensitive
emotional	shy
carefree	withdrawn
realistic, practical	submissive
relaxed	suspicious
composed	depression-prone

These particular traits are not surprising, in that people who frustrate themselves in the course of their other affairs are quite likely to behave the same way with regard to psi. It is much harder to define the personality of someone who expresses no ESP ability and whose scores will always approximate chance. For example, many people who indicate a fair amount of spontaneous ESP experience, and even professional psychics (whom I would assume have at least some ability) often do not score well in a laboratory.

Extraversion/Introversion

Extraversion is a personality type in which one's interests are directed outward to the world and to other people. This is contrasted with *introversion* in which one's interests are more withdrawn and directed toward the inner world of thoughts and feelings.

Gertrude Schmeidler, building on earlier studies by John Palmer and Carl Sargent, reviewed thirty-eight experiments involving the relationship between ESP performance and standard psychometric measures of introversion/extroversion. She found that extraverts scored higher than introverts in 77% of these experiments. In twelve of these

K. RAMAKRISHNA RAO, DIRECTOR OF THE FOUNDATION FOR RESEARCH ON THE NATURE OF MAN, DURHAM, NORTH CAROLINA

studies, the difference between introverts and extraverts was statistically significant.[115,116]

Effects of Different ESP Targets

Robert L. Morris who holds the Arthur Koestler Chair of Parapsychology at the University of Edinburgh in Scotland has proposed that each target be viewed as having both physical and psychological characteristics. The psychological characteristics seem to be more salient for psi research subjects than the physical. Morris has also suggested that researchers consider not only the targets themselves, but also the systems to determine and display the targets.[117]

The nature of the test situation and the target material itself is likely to affect ESP scores. Some people prefer material which involves other human beings on a feeling level. Other subjects who do well with ESP cards show little psychic skill outside of the laboratory. The technical name for scoring well on some kinds of targets and not on others is *the differential effect* and seems to follow a trend relating to emotional preferences, attitudes, and needs.

For example, in a test conducted by Jim Carpenter at the University of North Carolina, Chapel Hill, with male college students unknown to the subjects, some of the ESP cards had sexually arousing pictures drawn on them. The subjects showed a greater ability at guessing the ESP symbols on these cards than on the regular cards. In another study with a female patient in psychotherapy, an ESP test was given using words which were emotionally potent for her. Half of them were of a traumatic nature and half of them were of a pleasant nature. In this test, she showed *psi-missing* for the traumatic words and *psi-hitting* on the emotionally positive targets. This test was conducted by Martin Johnson at Lund University in Sweden.[118]

ROBERT L. MORRIS

Psi-mediated Instrumental Response

One of the most ingenious theories regarding the role of psi in everyday life was developed by Rex Stanford, who is currently teaching in the psychology department of St. John's University in Jamaica, New York. Stanford developed the concept of the psi-mediated instrumental response (PMIR) to explain non-intentional psychic experiences. For example, there is the story about a retired army colonel who found himself unconsciously getting off of the subway in New York at the wrong exit and then running into the very people he was intending to visit. Is it possible that his response of getting off the subway was triggered by ESP?

To test this hypothesis, experiments were designed to see if subjects would use ESP in a situation in which they would be rewarded for it, although they did not know they were being tested. In one such experiment, students in a psychology class were given an essay-type exam with the answers to half the questions sealed in opaque envelopes which were handed to them with the exam. They were told that the envelope contained carbon paper which would make copies of their answers. The experimenters thought that the students would use ESP as well as other means in order to do well on the examination. In fact, the students did better on the questions which were answered in the sealed envelopes. Furthermore, in a study where sealed answers were incorrect, the students did poorer on the corresponding questions. This study was conducted by Martin Johnson at Lund University in Sweden.

Another study has indicated that subjects who use the PMIR to avoid unpleasant situations and to encounter favorable situations, also score better than average on tests of conscious ESP.[119]

The PMIR model and research program have not been addressed by outside critics. However, Stanford himself has abandoned the model because he found its "psychobiological" or cybernetic assumptions to be untenable. Other psi researchers, such as John Palmer, feel that the PMIR model need not be abandoned, although modifications are necessary.[120]

Stanford's Conformance Behavior Model

Stanford calls his new approach the "conformance behavior" model of psi.[121] He challenges the "psychobiological" or "cybernetic" assumptions of PMIR, which assume that psi abilities are similar to other sensory-motor functions. Stanford questions whether ESP is analogous to normal perceptual or cognitive processes, since it occasionally manifests itself through unconscious motor behavior (such as when one misses a subway train and then encounters the person one is going to see in the first place). For this reason, Stanford sees psi as "dispositional" in character. He no longer assumes any communication of information across a channel.

As further support for his view, Stanford cites evidence that psi success is independent of task complexity. If psi were akin to normal sensory-motor skills, Stanford argues, one would expect deterioration in psi performance when doing a multicomponent task.

As evidence of the complexity independence of ESP, Stanford cites a 1940 study showing no deterioration in performance when a subject had to integrate information from two extrasensorily perceived targets from that obtained when the information was contained on a single target.[122] Had the subject been cheating, and thus relying on normal senses, one could anticipate deteriorating performance in the more complex task.

Stanford interprets psi events as the conformance behavior of "random event generators" (such as quantum mechanical REGs or human brains) to the needs of a "disposed system" (typically, the subject in a psi experiment, or the agent or percipient in a spontaneous case). In order for such conformance behavior to occur, the REG must produce events that are "unequally attractive" to the disposed system. Further, *labile* systems characterized by a great deal of random fluctuation should produce more conformance behavior than more deterministic systems.

In principle, the concept of *conformance behavior* does away with the distinction between ESP and PK. As some support for

REX STANFORD

this view, we note that studies have shown no decrease in PK success when the PK target is a complex, multiprocess REG as opposed to a simple, single-process REG. However, we are getting ahead of our story.

Precognition

ESP is generally divided into telepathy, i.e., extrasensory communication between two minds; clairvoyance, i.e., extrasensory perception at a distance, without the mediation of another mind; and precognition, which is ESP across time into the future. There is still some controversy as to whether telepathy actually exists, or whether it is simply another form of clairvoyance. However, precognition, a most unusual ability in terms of our conventional notions of time and free will, is a rather well-established ESP phenomenon. In fact, precognition tests afford some of the best evidence for ESP, since sensory leakage from a target which has not yet been determined is impossible. For example, in early studies with Hubert Pearce, the subject was able to guess what the order of cards in a pack would be after it was shuffled at the same high rate of scoring (up to 50% above chance levels) as in clairvoyance tests.

While many people tend to reject ESP because it seems to contradict the classical laws of science, precognition is even harder to swallow for exactly the opposite reason — it seems to imply a completely mechanical, predetermined universe. Ironically, it is this determinism which violates the sensibilities of twentieth-century science. In fact, precognition is very difficult to prove, although its alternatives are not exactly palatable.

For example in the precognitive card-guessing studies, one might say that the subject psychokinetically caused the order of the cards to conform to his guesses. Or perhaps, more reasonably, the experimenter, using his clairvoyance subconsciously, determined the subject's guesses and (with possible PK influence) shuffled the cards accordingly. It is impossible for precognitive experiments to rule out the possibility of contamination by other forms of psychic interaction. The methodological difficulty in distinguishing different types of extra-sensory transmission and reception had led researchers to use the more general term psi.[123]

HELMUT SCHMIDT

There is evidence to suggest that precognition actually does occur — with all of its ramifications regarding time and free will. Among the most sophisticated tests for precognition were those designed by Dr. Helmut Schmidt, a physicist now associated with the Mind Science Foundation in San Antonio, Texas. Subjects in his experiments were asked to predict the lighting of one of four lamps which was determined by theoretically unpredictable, radioactive decay. Schmidt gives us the following description of his apparatus:

The target generator consists of a radioactive source (strontium 90), a Geiger counter, and a four-step electronic switch controlling the four lamps [see illustration]. The strontium 90 delivers electrons randomly at the average rate of ten per second to the Geiger counter. A high frequency pulse generator advances the switch rapidly through the four positions. When a gate between the Geiger counter and the four-step switch is opened, the next electron that reaches the Geiger counter stops the switch in one of its four positions (whichever one it happens to be in when the electron registers) and illuminates the lamp corresponding to that position.[124]

In precognition experiments, the subject makes his guess before the apparatus makes its random selection of a target. The results of these experiments were automatically recorded and the device was frequently subjected to tests of its true randomness. The instrument can also be modified for experiments in clairvoyance and psychokinesis. In all three modes of psi testing with the Schmidt device, significant results have consistently been obtained.[125,126,127] Many other studies also show precognition.

Schmidt's studies have come under close scrutiny by skeptics. In 1981, psychologist C. E. M. Hansel suggested that Schmidt's experimental designs were not adequate to prevent cheating by Schmidt himself.[128] Fellow skeptic, psychologist Ray Hyman responded to Hansel's critique by pointing out that a charge of possible fraud is "a dogmatism that is immune to falsification."[129] Both points have some merit, and Schmidt himself has in some subsequent studies (to be covered in the upcoming discussion of psychokinesis research) collaborated with other researchers to minimize the possibility of experimenter fraud. Perhaps the most cogent critique of Schmidt's research is that randomization checks on the instrumentation could have been conducted in a manner which more closely simulated actual experimental conditions.[130] There is no data to suggest that this methodological weakness actually contributed to artifactually inflated psi scores. There is also, unfortunately, no way in which such control tests can be designed to be immune from possible psychic influences!

In 1989, Charles Honorton and Diane C. Ferrari reported a *meta-analysis* of *forced-choice precognition* experiments published in the English language between 1935 and 1987. "Forced choice" experiments are those, such as Schmidt's, in which the ESP percipient is asked to select among a limited number of choices — as opposed to "free-response" experiments in which the percipients' responses are not limited. These studies involve attempts by subjects to predict the identity of target stimuli selected randomly over intervals ranging from several hundred milliseconds following the subject's responses to one year in the future. Some 309 studies reported by sixty-two investigators were analyzed. Nearly two million individual trials were contributed by more than fifty-thousand subjects. Study outcomes were assessed in terms of overall level of statistical significance and effect size. There was a reliable overall effect with chance probability less than 10^{-25}. Thirty percent of the studies (by forty investigators) were statistically significant at the 5% level. A ratio of forty-six unreported studies averaging null results would be required for each reported study in order to reduce the overall effect to non-significance. No systematic relationship was found between study outcomes and eight indices of research quality. Effect size has remained essentially constant over the survey period, while research quality has improved substantially.

Four moderating variables were significantly associated with study outcome: (1) Studies using subjects selected on the basis of prior testing performance show significantly larger effects than studies involving unselected subjects; (2) subjects tested individually by an experimenter show significantly larger effects than those tested in groups; (3) studies in which subjects are given trial-by-trial or run-score feedback have significantly larger effects than those with delayed or no subject feedback; (4) studies with brief intervals between subjects' responses and target generation show significantly stronger effects than studies involving longer intervals. The combined impact of these moderating variables appears to be very strong. A nearly perfect replication rate is observed in the subset of studies using selected subjects, who are tested individually and receive trial-by-trial feedback.[131]

An interesting study comparing real-time ESP tests with tests of precognition was conducted by Charles T. Tart, a psychologist at the University of California, Davis. Using

CHARLES TART

a measure of information rates, Tart analyzed fifty-three studies of present-time ESP and thirty-two studies of precognitive ESP — all of which used a forced-choice test model. A striking and robust performance difference was found: present-time ESP worked up to ten times as well as precognitive ESP in forced-choice tests.

Tart suggested three possible explanations for this finding: (1) Real-time ESP may be emotionally more acceptable than precognition; (2) perhaps present-time ESP and precognition are two basically different processes, with inherently different characteristics; and, (3) something about the nature of time itself attenuates ESP performance that extends into the future.[132]

A SUBJECT PRESSES A BUTTON RECORDING A GUESS ON ONE OF THE AUTOMATED TESTING DEVICES DEVELOPED BY HELMUT SCHMIDT. THERE IS A PROBABILITY OF ONE IN FOUR THAT THE SUBJECT WILL SCORE CORRECTLY BY CHANCE ALONE. (COURTESY FOUNDATION FOR RESEARCH ON THE NATURE OF MAN)

A study on a single individual, Malcolm Bessent, who has a history of success in laboratory precognitive tasks, suggests that the barriers to precognition may, indeed, be psychological.[133] Bessent completed a thousand trials in a computer-based experiment comparing precognition and real-time target modes. A diode-based electronic number generator (RNG) served as the target source. Target mode was randomly selected at the outset of each ten-trial run and was unknown to Bessent until the completion of each run. Bessent's task was to identify the actual target from a judging pool of four graphic "card" images presented on a computer graphics display.

Based on Bessent's prior research history, two formal hypotheses were tested by researcher Charles Honorton: (1) Bessent would demonstrate statistically significant hitting in the precognitive target mode, and (2) his precognitive performance would be significantly superior to his performance on real-time targets. Significance criteria were specified in advance. Both hypotheses were confirmed. Bessent's success rate in the precognitive target mode was 30.4%. This is reliably above the 25% chance level. Real-time performance did not exceed chance expectations.

As is customary in psi research, various rival hypotheses including sensory cues, faulty randomization, data-handling errors, data-selection bias, multiple analysis, and deception were assessed. Honorton found them to be inadequate explanations of the beyond-chance result.

This was the fourth precognition experiment with Bessent, each involving a different methodology and each yielding a statistically significant outcome. The combined result is highly significant with a chance probability of less than one in a billion.

An ingenious experiment, designed and conducted by Dean Radin, using himself as subject, attempted to explore the hypothesis that precognition entails the ability to see probability wave, to see probable rather than actual futures.

Radin designed a computerized Random Event Generator that would, in effect, change the probabilities of the various targets with each trial. A conventional precognition hypothesis would suggest a greater than chance number of hits for the correct target (regardless of *a priori* probability). Radin suggested that if precognition involved probable futures, the incorrect responses might still match the targets which were given a high *a priori* probability of being selected by the computer on a given trial — even if they were not ultimately selected. This hypothesis was confirmed.[134]

How unfortunate that, while the hypothesis would seem to have enormous ramifications for our understanding of precognition, there is no known way to distinguish the outcome of Radin's experiment from the possibility that he simply used his own psi abilities, if they exist, to confirm his favorite hypothesis.

One of the most rigorous and successful series of precognitive studies has been conducted by Brenda Dunne and colleagues at the Princeton Engineering Anomalies Research program. The Princeton group used a free-response, *remote-viewing* procedure which was developed by physicists Harold Puthoff and Russell Targ at SRI International (working in conjunction with research subjects Ingo Swan and Pat Price).[135] Thus far, 336 experimental trials have been conducted in which randomly selected targets are not chosen or visited until the percipient's responses have been recorded.[136]

The targets are real-life locations that are actually visited by an experimental agent acting something like a telepathic sender. The experimental subjects or percipients are asked to report any and all imagery which comes to them during the testing period. Then percipients are asked a series of thirty questions about the target which are to be answered yes or no. These questions include such items as: Is the target outside? Is it man-made? Does a single object dominate the scene? Are animals present? Is it colorful? Are there any loud sounds? A statistical analysis then compares the subjects' responses both to the actual target and to the descriptor ratings for all the other targets in the target pool. The results are summarized:

> Effects are found to compound incrementally over a large number of experiments, rather than being dominated by a few outstanding efforts or a few exceptional participants. The yield is statistically insensitive to the mode of target selection, to the number of percipients addressing a given target, and, over the ranges tested, to the spatial separation of the percipient from the target and even to the temporal separation of the perception effort from the time of target visitation. Overall results are unlikely by chance to the order of 10^{-10}.[137]

Psychokinesis

Rhine's Early Studies

In 1934, several months prior to publishing his famous paper on Extra-Sensory Perception, Dr. J. B. Rhine received a visit from a young gambler. After comparing notes on conditions for success in psychic testing, he remarked that similar conditions seemed to favor his luck in gambling. Furthermore, he claimed that he himself was sometimes able to exercise a mind-over-matter effect on dice-throwing games. While belief in such an influence on dice was both common and ancient, until then it had not been deemed a serious problem for scientific study. Rhine discovered that preliminary experimentation would be quick, easy, and inexpensive. The results proved encouraging enough to warrant further research.

Experiments continued during the next decade using protocols that systematically eliminated bias from unbalanced dice. The dice were placed in special cups, so subjects could not use special tricks to throw them. Still later, the dice were placed in electrically driven rotating cages and were also photographed automatically in order to eliminate experimenter error. In general, the tests entailed asking the subjects to will the fall of the dice with selected target faces showing. Numerous throws were made in succession for each target before another target was chosen.

By the end of 1941, a total of 651,216 experimental die throws had been conducted. The combined results of these experiments pointed to a phenomenon with 10115 to 1 odds against chance occurrence. Nevertheless, Rhine hesitated to publish his results. The scientific world was still reacting emotionally to his announced proof of ESP, and he felt no need then to raise eyebrows by announcing another unorthodox discovery.[138]

In 1942, with most of the staff at the Parapsychology Laboratory called away to war, continued experimentation in PK proved difficult. At this time, Rhine went over the records of earlier experiments so conducted that an analysis of position effects could be made, similar to the decline of high ESP scoring toward the end of experimental sessions, detected a few months earlier. If the above chance results had been caused by probability, artifacts, or illegitimate means, one would expect that the distribution of hits would be consistent throughout the experiment and would not decline.[139]

The results of this survey indicated that there were more hits near the beginning of each run of twenty-four die throws. There were also more hits during the earlier runs of each experimental session which would typically last for ten runs. These results were not expected or even considered by the experimenters and subjects at the time of the experiments. The odds against such distribution occurring by chance were about a hundred million to one. This evidence of a presumably psychological effect, similar to that noted with ESP, made a case for psychokinesis strong enough to warrant publication. The first of the papers appeared in the *Journal of Parapsychology* in 1943.[140] Many others followed.

In 1946 a study was published that pitted the psychokinetic skills of veteran gamblers against those of divinity students.[141] In this contest atmosphere, both groups scored well above chance expectations.

PK with Random Number Generators (RNGs)

The random number generator experiments pioneered by Helmut Schmidt, previously described as tests of precognition, have also been used extensively as tests for psychokinesis. As mentioned earlier, it seems theoretically impossible to clearly distinguish between psychokinesis and precognition in quantitative research. Generally, the tests for psychokinesis are those in which the experimenter instructs the subject to will or intend that a particular target be selected by the RNG. This reduces the possibility that the subject could be using precognition, but does not eliminate the possibility of experimenter precognition. The hypothesis of *intuitive data sorting* suggests that the subject might use precognition to start the RNG at the exact time required to match the preselected PK target sequence.

One interesting version of the Schmidt RNG studies involved the cooperation of Robert Morris and Luther Rudolph at Syracuse University, with the experimental protocol published prior to the beginning of the experiment.[142,143] These studies, which used prerecorded targets, were reviewed by skeptic James Alcock in a report for the National Research Council's study on methods for enhancing human performance. Alcock, who had access to the pertinent raw data, admitted that this study was much better executed than other studies by Schmidt and merited further replication attempts.[144,145]

An elegant and sophisticated research program involving Random Event Generators has been underway for over ten years at Princeton University under the aegis of Dean Emeritus of Engineering, Robert Jahn, and the management of psychologist Brenda Dunne. Other staff members include psychologists Roger Nelson and Angela Thompson, electrical engineer John Bradish, and physicist York Dobyns.

In the formal test series, generation rates of either one hundred or one thousand per second are used, and each trial comprises two hundred binary samples. The count data are permanently recorded on a strip printer as well as being entered on line into computer memory. The subject receives immediate feedback via electronic displays which show the number of trials, the number of hits in the last trial, and the average number of hits since some predetermined starting point. The REG and the on-line IBM PC/AT computer independently calculate the mean of each trial and the standard deviation for every block of fifty trials.

J. B. RHINE CONDUCTING A PK EXPERIMENT USING DICE IN A MECHANICAL DICE TUMBLER. (COURTESY FOUNDATION FOR RESEARCH ON THE NATURE OF MAN)

The equipment can be run in one of two modes, either manual or automatic. In the former case, the machine will generate a trial only when a switch is pressed; while in the automatic mode, once started, the machine will automatically initiate a block of fifty trials.[146]

There are two types of procedure, either "volitional mode," in which case the subject chooses whether to aim for a high score (PK+) or a low score (PK-) in a given run, or "instructed mode" where some kind of random process determines which way the subject is to aim. There are also baseline runs interspersed ("in some reasonable fashion," the nature of which is unspecified) with the PK runs; in this case the subject is to exert no influence, so that these will serve as a randomization check. The choice of volitional/instructed mode and automatic/manual mode are "normally left to the preference of the operators (subjects), but they are encouraged to undertake additional series employing the other modes for comparison."[147]

The formal data base consists of well over 750,000 trials (or 150,000,000 binary digits) carried out on two different machines by thirty-three different subjects over a period of nearly ten years.[148] Typically, a session had three types of trials: high-aim, low-aim, and a control series. Pooled, overall results of the high- and low-aim trials are clear: significantly higher scores for high aim; significantly low scores for low aim. The outcome of the control trials was illuminating. Here the subjects were told not to try to influence the REG. Presumably, they hoped their data would be "normal." In

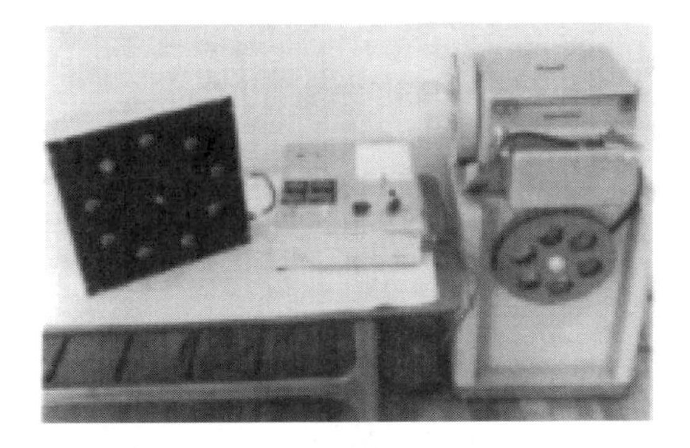

TOP RIGHT, PK TEST EQUIPMENT (COURTESY HELMUT SCHMIDT)

BOTTOM RIGHT, REG TESTING AT THE PRINCETON ENGINEERING ANOMALIES RESEARCH PROGRAM fact, although the mean score was almost identical with the theoretical mean, the distribution was unique: a statistically significant surplus of scores at the precise theoretical mean. When control, high-aim, and low-aim series are pooled, the distribution is what would be expected by chance.[149] This result, which may be the consequence of the subjects' intentions to "achieve a baseline" in the control condition, shows the difficulty of conducting psi experiments with a true *control group.*

Interestingly, both earlier and later calibration trials nicely conformed to a normal statistical curve, showing that the equipment was probably operating properly. If psi is operating in the ostensible control tests, then the critics' demands for additional control tests seems naive.

Ironically, comparable results were obtained using a pseudo-random noise source (i.e., based on both computer algorithms and prerecorded targets). This finding makes it difficult to interpret the Princeton data in terms of a conventional psychokinesis hypothesis. How can one psychokinetically influence targets which have already been determined? For this reason, the Princeton team does not attempt to describe their research program as a test of PK, but simply refers to the program as a study of anomalous man-machine interactions.

The Princeton team has gone to great lengths to try to ensure that their equipment is unbiased. Internal circuits are continually monitored with regard to internal temperature, input voltage, etc. Successive switching of the relationship between the sign of the noise and the sign of the output pulse on a trial-to-trial basis was done to provide a further safeguard against machine bias. Results were automatically recorded and analyzed. Extensive tests of the machine's output and its individual components were also carried out at times separate from the test sessions. The provision of baseline trials interspersed with test trials provided a randomization check which overcame some of the weaknesses of Schmidt's procedure.

Psi researcher John Palmer has drawn attention to the fact that there is no documentation regarding measures to prevent data tampering by subjects. This is of concern since the subject was left alone in the room during the formal sessions along with the REG.[150]

In evaluating these studies, skeptic James Alcock claimed that only one subject (Operator 10) accounted for virtually all the significant departures from chance in the Princeton studies.

Noting that details regarding precautions against subject cheating were not specified, Alcock stated:

> I am not trying to suggest that this subject cheated; I am only pointing out that it would appear that such a possibility is not ruled out. Had the subject been monitored at all times, such a worry could have been avoided or at least reduced.[151]

The Princeton team has chosen a policy of keeping the identity of all experimental subjects anonymous — among other reasons, in order to eliminate motivation for subjects to cheat. However, the fact that Subject 10 contributed considerably more to the database than any other subject suggests that this individual was either a member of the experimental team or someone who had become a close friend of the experimenters. As such, Subject 10 might well have had access to information which would make it possible to tamper with the data recording system.

In response to the criticisms of Palmer and Alcock, the Princeton researchers have prepared a detailed analysis of the equipment, calibration procedures and various precautions against data-tampering.[152]

According to the researchers, the automated and redundant on-line recording of data preclude data-tampering — as does the protocol requirement that the printer record be on one continuous, unbroken paper strip. It would appear that all necessary precautions have been taken, short of submitting subjects to constant visual observation. The subjects are submitted to intermittent visual observation which the researchers believe is sufficient to control against tampering with the equipment, given their particular setup.[153]

In further response to Alcock's critique, the Princeton team conducted further analyses of the data which show that the anomalous RNG effects were contributed by most of the subjects, and were not dependent upon the scores of Subject 10. Several other subjects, who participated in fewer experimental trials, actually had scores with greater chance deviations. By analyzing the data from only the first series of 7,500 trials (1,500,000 binary digits) from each subject, it was possible to level the influence that Subject 10 exerted on the database. In this analysis, with each subject carrying an equal weight, the results were significantly beyond chance.[154] Another analysis was conducted which eliminated all of the data from Subject 10. This, too, was statistically significant.[155]

DEAN RADIN

A comprehensive meta-analytic review of the RNG research literature encompassing all known RNG studies between 1959 and 1987 has been reported by Radin and Nelson, comprising over eight hundred experimental and control studies conducted by a total of sixty-eight different investigators.[156] The probability 597 experimental series was $p < 10^{-35}$, whereas 235 control series yielded an overall score well within the range of chance fluctuation. In order to account for the observed experimental results on the basis of selective reporting (assuming no other methodological flaws), it would require "file drawers" full of more than fifty thousand unreported studies averaging chance results.

Some people seem to produce data in random number generator (RNG) experiments that display idiosyncratic patterning that appears to be consistent from one run to the next. To explore the idea of person-unique signatures, Dean Radin, working at Princeton University, used a powerful, new "neural network" computational technique that is proving to be adept at discovering weak patterns in noisy data.

Neural networks are a form of parallel processing based upon research about how the brain encodes and processes information. The power of these networks rests upon the discovery that when numerous elementary processing units are richly interconnected under the right conditions, they can automatically learn to associate arbitrarily complex inputs with arbitrarily complex outputs. Information processing in these networks takes place in the interactions among large numbers of artificial neurons. Learning takes place by changing the interconnection strengths between neurons.

The study involved training a network to associate data with given individuals, then observing whether the trained network could successfully identify these people based upon new data. Two sub-datasets were required for each person: One was used to train the network and the other was used to see whether the trained network could transfer its knowledge to new data. Thus, each series of fifty runs was split in half, using the first half as the *training set* and the second half as the *transfer set.* Results showed that these networks were able to learn to associate data with thirty-two different individuals, then, in statistical terms, successfully transfer that knowledge to new data.[157]

PK Placement Studies

A number of researchers have conducted studies designed to determine whether naked human intention could affect the movement of moving objects. The most recent version of this approach is the database of studies using a *random mechanical cascade* at Princeton University's Engineering Anomalies Research program. The experimental apparatus allows nine thousand polystyrene balls to drop through a matrix of 330 pegs, scattering into nineteen collecting bins. As the balls enter the bins, exact counts are accumulated photoelectrically, displayed as feedback for the operator and recorded by a computer. Subjects are asked to concentrate on shifting the mean of the developing distribution of balls to the right or left, relative to a concurrently developing baseline distribution. Over three thousand experimental runs have been conducted with twenty-five individuals. The results are significantly beyond chance expectation.[158]

The Princeton University researchers note that virtually all of the statistically significant results have come from a deviation of the balls to the left of the baseline. This, they claim, cannot be attributed to any known physical asymmetry in the system.

Chinese Reports of Psychokinesis Associated with ESP

Reports of psi research in China claim that certain subjects showed consistent success in ESP tests.[159,160,161] A report from the Chinese Academy of Sciences states that if the ESP response was incorrect there was no change at the target, but that, in more than seven hundred trials when the ESP response was correct, there was always an accompanying PK effect at the target location. These effects included clouding of X-ray or photographic film, or pronounced changes in the records of photoelectric tubes, thermoluminescence docimeters, or biological detectors.[162] It should be noted, however, that the quality and reliability of reports of Chinese research is very inconsistent. At least one set of observers believes that some reported results were produced by sleight-of-hand.[163]

Scott Hubbard, Edwin May and Harold Puthoff at SRI International in Menlo Park, California, searched for such changes using a detector that the Chinese had found most sensitive — a photomultiplier tube. ESP targets were slides of scenes from the *National Geographic.* There were four subjects with six sessions each. The pooled sessions showed significant ESP success. The researchers then correlated ESP scores with four measures of photomultiplier output: low-amplitude increase and decrease and high-amplitude increase and decrease in a number of pulses. One correlation was significant: the correlation with increase in the number of high-amplitude pulses, as the Chinese had reported.[164]

PK Metal-Bending

A renewed interest has recently developed in large-scale (macro) PK effects, particularly metal-bending. The most extensive research on metal-bending has been conducted by physicist John Hasted at the University of London's Birkbeck College. His subjects were mostly adolescents who had developed an interest in metal-bending upon exposure to the public performances of Uri Geller. They were asked to bend or deform latchkeys or bars of aluminum alloy without touching them. The specimens were attached to resistive strain gauges or (in later work) piezoelectric sensors. Signals from these devices were then amplified and registered on chart recorders.

Actual bending was observed in only a minority of sessions; however, anomalous signals frequently appeared on the chart records — from sensors separated up to several feet from each other. This led Hasted to hypothesize an unknown form of conduction of electrical charge from the subjects' bodies through the atmosphere to the sensors.

Hasted claimed that the subjects had no opportunity to interact directly with the chart recorder. Furthermore, he employed dummy loads along with electrical shielding of the test channels to minimize global electrical artifacts.[165]

Psi researchers have been rather reluctant to accept Hasted's findings. In part, this is because macro-PK effects remain very controversial. In part, as enumerated by psi researcher John Palmer, it is because Hasted's research procedures would benefit from additional refinements:

> Even if one grants the paranormal origins of the signals, Hasted's methodology makes it difficult to draw valid conclusions about their nature, including whether or not they truly represent strain. Use of an inadequately fast chart recorder, failure to adopt proper principles of experimental design, and failure to use statistical analyses are the most serious problems. In particular, it is impossible to distinguish basic physical characteristics of the phenomena from those correlated with preferences, attitudes, etc., of the subject or experimenter.[166]

JOHN HASTED

In general, "non-touching" is considered an essential prerequisite control for a variety of possible conventional influences in PK metal-bending research. However, some studies were conducted which allowed touching of the target specimens and, yet, still merit some scientific consideration.

One test used by Hasted employed a brittle alloy bar that supposedly could not be bent to a particular angle of deformation in less than a certain known time. When excessive force was applied, it simply broke. The only way to bend it is to apply a small force slowly over time, which produces bend by a process known as *creep*. Hasted has reported bending of such alloys in well under the minimum time thought to be possible using a *creep* process.[167]

Charles Crussard and J. Bouvaist, two French metallurgists whose research was funded by a metals company, took the following experimental measures in PK studies with a magical performer, Jean-Paul Girard:

(1) All dimensions of metal strips or rods were measured before and after bending; (2) the microhardness of the metal was measured at several points before and after bending; (3) residual strain profiles (measures of crystalline structure) were examined; (4) electron micrograph analyses of the fine structure of ultrathin foil specimens were often made; (5) analyses of the chemical composition at various places along the strip or rod were made. Additional precautions included consultations with magicians, video recording of trials, and the marking of test specimens.

Crussard and Bouvaist described eight of twenty trials conducted with Girard. The specimens were bars of aluminum alloys, stainless steel cylinders, and Duralumin plates. During the trials, Girard was allowed to touch and hold the specimens, while at all times being observed by the experimenters. Bending was observed in four of the specimens. Structural changes inconsistent with physical bending were found for a stainless steel cylinder and Duralumin plate.[168]

Since Girard is a conjurer, researchers are cautious in interpreting the above results. In his 1985 evaluation for the U.S. Army, psi researcher John Palmer reached the following conclusion with regard to this report:

> Only in the case of the bending of one of the aluminum bars do the controls as

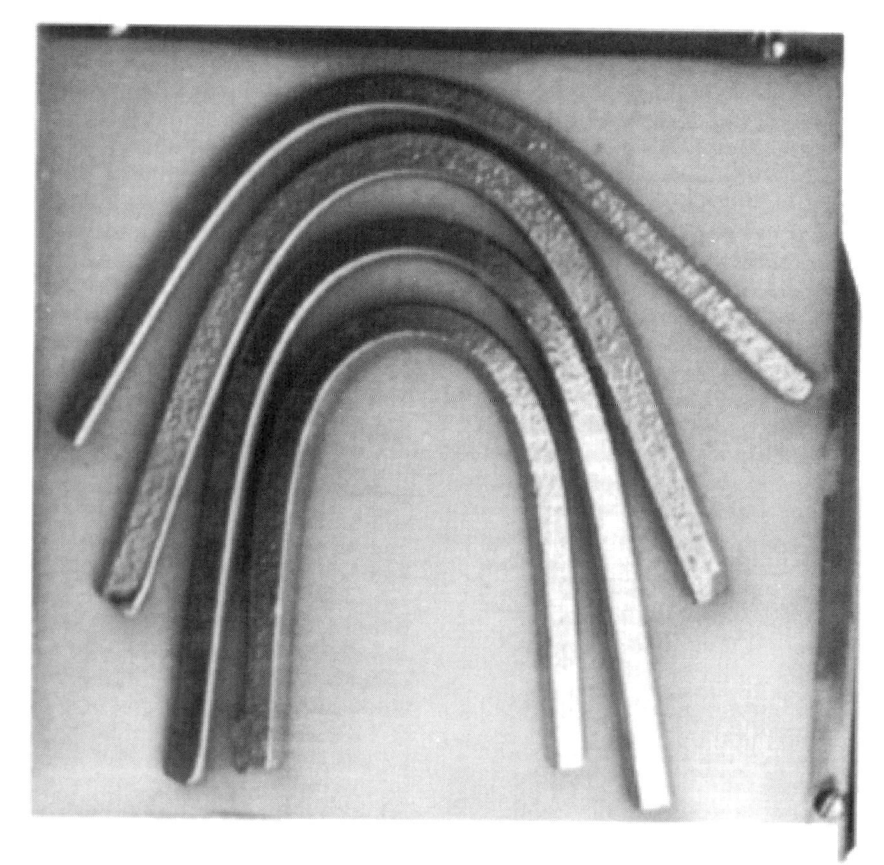

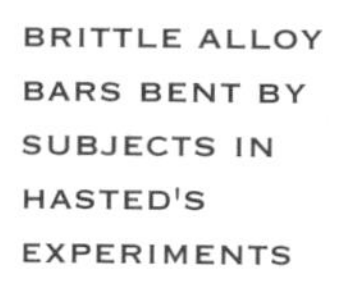

BRITTLE ALLOY BARS BENT BY SUBJECTS IN HASTED'S EXPERIMENTS

reported seem to completely rule out the possibility of Girard substituting previously deformed specimens for the test specimens. Nonetheless, the assumptions that must be made to explain away these results seem rather farfetched.[169]

Bio-PK

A number of studies are suggestive of the possibility that conscious intention can influence the growth and movement of biological targets. One of the first of these studies involved the one-celled protozoan, *paramecium.* The organism was centered under the cross hairs of a microscope, and it moved with significant frequency into the randomly selected quadrant of the field.[170] Water fleas observed under a microscope turned in the randomly selected direction, i.e., either right or left, with greater frequency than these crustaceans turned in the opposite direction.[171] Significant results were also obtained in mentally directing ants to carry away matches on the selected side of a wooden slide.[172] Carroll B. Nash conducted an experiment in which bacterial growth was psychokinetically accelerated and retarded according to the intentions of randomly selected college students.[173] In two separate studies, the growth of fungus was less when an attempt was made to mentally retard it than was the growth in the controls.[174,175] Physicist Elizabeth Rauscher conducted a study with biochemist Beverly Rubik in which bacteria exposed to an antibiotic grew more rapidly in a sealed tube surrounded, but not touched, by the hands of psychic healer Olga Worrall than in tubes not treated by her.[176] (Further experimental work in the area of healing is reported in the section on potential applications of psi.)

In a study following up on his earlier finding correlating students' intentions with bacterial growth, Carroll B. Nash conducted another study looking at mutation rates. He put suspensions of *E. coli* into nine tubes, in a three-by-three arrangement, for each of fifty-two subjects. He randomly designated one set for rapid mutation into another strain, one set for inhibition of mutation rate, and one set for control. He arranged that the subjects would know the instructions but the student experimenters would be blind. The rapid mutation tubes showed significantly more growth than the inhibition tubes. The promotion tubes had nonsignificantly more growth than the controls; the inhibition tubes had significantly less.[177]

In a computer-automated study, Charles Pleass and N. Dean Dey, at the University of Delaware, tried to have subjects use PK to speed or to slow the swimming of algae. They measured swimming speed by the Doppler effect of laser light, which does not affect algae adversely. Each experimental run was preceded by a control run. A run consisted of the collection of a thousand data points and ordinarily took thirty seconds. Each subject participated in ten runs. Data analysis showed differences in speed, in the anticipated direction, between scores for PK and control runs, and also indicated changes with the subject's mood and with differences in the instructions.[178]

CARROLL B. NASH

Psionics — Potential Practical Application of Psychic Awareness

Even for individuals who accept the reality of psi ability, the question of its practical applications remains controversial. Can these powers be used for detrimental purposes? What are the limits of psychic ability? Can psychic abilities be brought under institutional or government control? Are we opening another Pandora's Box?

It is difficult to answer these questions based on the scant amount of solid, scientific evidence we have so far accumulated. Certainly more inferences can be obtained by drawing upon the history, the literature, and the folk wisdom of psi.

Many fields of human endeavor today include both a scientific/scholarly branch and a branch devoted to practical application. Examples include education, physics, psychology, music, drama, agriculture, sociology, political science, genetics, chemistry, rhetoric and athletics. In fact, in each case, the practical applications of a discipline preceded the science and scholarship. Certainly the same is also true of psi research. Folklore and history suggest that the practical application of psi abilities is a major interest of shamanism — one of the world's oldest professions. I sometimes use the term "psionics," taken from the science fiction literature,[179] to describe the applied branch of psi exploration.

Some psi researchers and other scientists have been known to argue that there really is no proper applied branch of psi exploration.[180,181,182] Some say that psi cannot be applied until its existence has been proven to the scientific community. Others, who accept the scientific proofs of psi's existence, argue that it cannot be applied because experimental tests have shown it to be too weak and unreliable. Still others argue that, even if psi is not always reliable, it cannot compete in effectiveness with the technology of today's modern world. Furthermore, in a real-life setting, there is no way to properly distinguish genuine psi functioning from other possible interpretations. If psi exists, its practical application in a consistent, reliable manner still awaits many future developments. Nevertheless, there are some potentially promising beginnings.

Casino Gambling Simulation

Researchers attempting to apply ESP to casino gambling have hoped that eventually they might be able to use psi-missing to their advantage. In a series of studies reported by Robert Brier and Walter Tyminski several statistical techniques were used to apply ESP choices to gaming.[183] One of these was the majority-vote technique.

Suppose, for example, we know a given subject can consistently score 52% on an ESP test where chance expectation is 50%. Although such scoring can be extremely significant over several thousand trials, there are many situations where one could not afford to be wrong 48% of the time. In this situation, scores can be strengthened, although the procedure is somewhat slower. Suppose we make ten guesses of each target instead of one; then we determine our final guess using the majority vote of those ten guesses. This is essentially the same repetition principle a radio-communications engineer uses when a signal is obscured by noise. If we made a thousand guesses at the target, with 52% ESP, chances are very high that a majority-vote would yield the correct answer — unless one's ESP were working in a negative direction.

Working with actual casino games as their targets, Brier and Tyminski utilized a majority-vote technique so bets were determined by the vote of five different guesses. Furthermore, they divided the gambling situation into test runs and play runs. The guesses were made well in advance of the betting situation and were recorded carefully on scoring sheets. During the test runs of twenty-five bets, the experimenters calculated whether the subject was psi-hitting or psi-missing. If the subject was showing positive psi, the play-run bets were predicted according to the regular majority vote. However, if negative-psi was indicated, the play-run bets were then predicted in the opposite direction of the majority vote. If scores in the test run were close to chance, no attempt was made to predict the play-run bets.

This technique proved remarkably successful, from an experimental point of view, in the six test situations reported. Its promise of psi applications seems very encouraging. However, readers are not encouraged to rush out and try to apply this technique right away. The statistical procedures can be quite complex, particularly when one takes into account position effects and scoring declines. Attention must also be paid to creating a psychological atmosphere for the subject conducive to ESP testing.

James Carpenter's experiment resulted in the successful identification of the binary (Morse) code equivalent to the word "PEACE."[184] Later, Puthoff et al. reported encouraging results based on a more efficient statistical averaging procedure.[185]

In one series of studies, Robert Mihalsky and E. Douglas Dean looked at company presidents who had doubled their company's profits during the last five years. They found that these individuals scored much higher in the precognitive tests than other executives. In fact, the ESP test seemed to be a much better indicator of executive success than other personality measurements.

Given the challenge of applying remote viewing protocols to practical ends, Harold Puthoff worked with a group of parents hoping to raise money for an alternative school for their children. They undertook a thirty-trial series in which remote viewing was used to predict the daily outcomes of a commodities market variable (which was then successfully traded in the market).

The technique employed was an ARV (associational remote viewing) procedure. Several remote viewers were asked to describe (free-response) a target object to be shown them at the close of the following day, the selection of that object to be determined by that day's market activity (e.g., if market up, an apple; if down, a pencil). The task of the remote viewing judge is to determine from the viewers' transcripts the likely feedback object, and hence (in advance) the associated market movement. The sequence in detail was: (a) remote viewers generate transcripts; (b) without reference to the transcripts, two objects are selected and labeled (by use of a random number generator) market-up, market-down objects; (c) a judge determines a consensus vote as to which of the two objects is being described (and the associated

HAROLD PUTHOFF

market-movement prediction is passed on to a trader); (d) at the close of the following market day the actual "ground-truth" market-movement object is shown the viewers for feedback, closing the loop.

Seven parents interested in raising funds for the school volunteered as remote viewers. After an evening's instruction on the SRI remote viewing protocols, a series was begun. The number of remote viewing trials per person over the entire series ranged from a maximum of thirty-six (six-pilot, thirty-market trials) to a minimum of twelve. Consensus judging yielded a result of 21/30 (70.0%), significant at $p < 2.2 \times 10^{-2}$, and a series of profits/losses at about $1,000 - $2,000/trial, netting more than $25,000 profit for the entire series.[186]

In a second series, using the same formal method, when they were trying to make money for themselves, Puthoff and his colleagues were unsuccessful.

Possible Psi Healing

Bernard Grad's Research at McGill University

In 1959, a Canadian researcher, Dr. Bernard Grad, was introduced to a Hungarian refugee named Oskar Estebany who claimed that some form of healing energy emanated from his hands. Estebany had been a cavalry officer in the Hungarian army before the 1956 uprising and originally discovered his healing abilities in treating the army horses. In a series of ingenious experiments with Estebany, Grad provided the scientific foundation for the existence of psychic healing.

His first experiments were with laboratory mice whose backs had been deliberately wounded by carefully removing an area of skin. The areas of these wounds were measured over an eighteen-day healing period. The treatment consisted of Mr. Estebany's holding the caged mice between his hands for twenty minutes twice daily. One control group received similar handling from medical students who did not claim to have unusual healing ability, while the mice in another control group simply remained in their cages without handling at all. The experiment was carefully controlled so that the individuals who cared for the mice and measured their wounds did not know which of the test groups they were in. A total of three hundred mice were used in one experiment, which was eventually published after several pilot studies. This experiment showed significantly faster wound healing in the mice treated by Mr. Estebany.[187] It was difficult to maintain that the mice were susceptible to the power of suggestion. Rather, the experimenters felt that there was some sort of healing emanations coming from Estebany's hands.

In his next experiment, Grad used barley seeds which were treated with a saline solution. Sterile and sealed in bottles under vacuum, the solution was normally used for intravenous infusion of humans. The healer merely held one sealed bottle in his hands for thirty minutes. A control bottle of solution was not "healed." The seeds were soaked in the solution and allowed to dry for forty-eight hours. Then they were baked in an oven just long enough to injure, but not kill them. Twenty seeds were planted in each of twenty-four pots — with identical soil, temperature, and humidity conditions. During the test period no person knew which seeds had been given the treated water. Estebany himself had no contact with the barley seeds. However, after the conclusion of the experiment it was found that those pots with seeds which had been watered from the bottles treated by the healer had more plants growing in them and the plants were also taller.[188]

In a third experiment, Grad attempted to determine if he could get effects from other subjects. In fact, he hypothesized that if a psychic healer could cause greater plant growth, perhaps treatment by psychiatric patients would inhibit growth. A similar experimental technique was used with three subjects. One of these was psychiatrically normal, the second was a hospitalized depressed neurotic, and the third a hospitalized depressed psychotic patient. A control group of plants received no treatment at all. The results of this experiment were very intriguing. The plants treated with the solution held by the normal subject showed greater growth than either the control or the depressed subjects. This effect was statistically significant and the "normal" subject also claimed to feel some sort of flow through his hands during the experiment. One of the depressed subjects was so amused with the experiment that her mood picked up as soon as she was asked to hold the bottle of saline solution. Her plants grew consistently, but not significantly larger than the control plants. The third subject stayed in an unhappy mood throughout the experiment. The seeds treated from the bottle of solution which she held showed less growth than the untreated control group.[189] This effect was small in Grad's experiment, but we will see that it has been confirmed by other studies.

Another experiment was performed testing the effect of healing on the rate at which thyroid goiters developed in mice whose diet was deficient in iodine and contained thiouracil. It was found that mice in cages held for two fifteen-minute periods each day by the psychic Estebany developed goiters at a slower rate and recovered from them at a faster rate than mice in the control group, which were given the same amount of warmth as that produced by the healer's hands.[190]

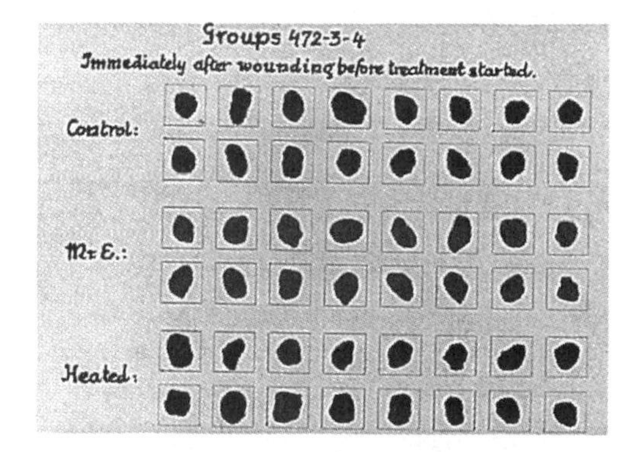

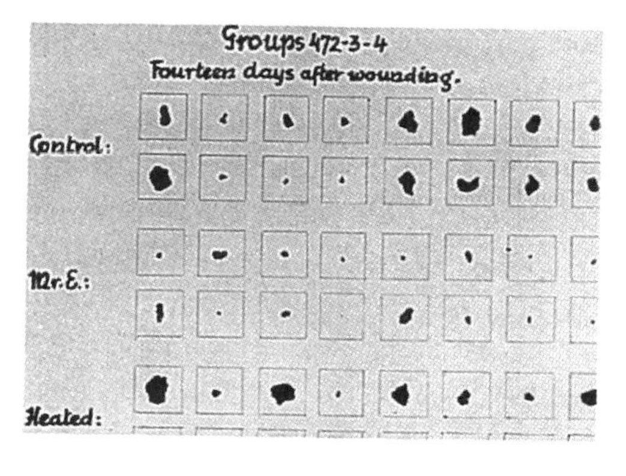

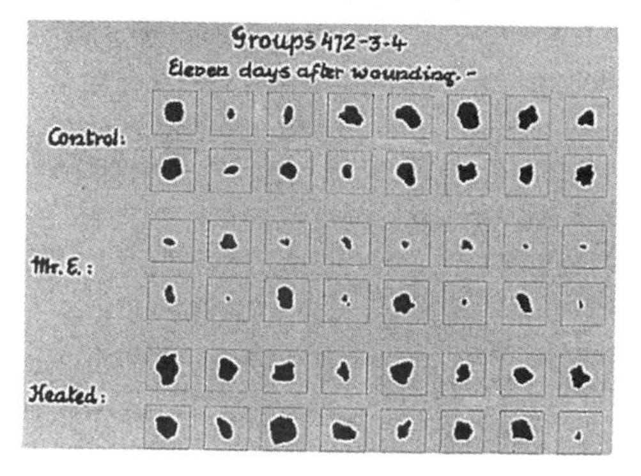

ABOVE LEFT: WOUND SIZE IN MICE IMMEDIATELY AFTER REMOVING A PORTION OF THE SKIN. THE CENTER GROUP RECEIVED PSYCHIC HEALING TREATMENT (COURTESY BERNARD GRAD).

BOTTOM LEFT: WOUND SIZE AFTER ELEVEN DAYS

TOP RIGHT: WOUND SIZE AFTER FOURTEEN DAYS

Critics have faulted Grad's studies as well as other work in the area of ostensible psychic healing for two primary reasons. Studies did not adequately shield for the possibility of conventional physical influences (i.e., heat and electromagnetic radiation). Also, researchers measuring the results were often not blind as to which animals were in the test condition and which were controls.

Conceptual Replications of Grad's Research

One experiment examined the healing effect on wounds that were surgically administered on human subjects by a medical doctor using a skin biopsy instrument and local anaesthetic. This procedure created a relatively consistent wound size upon incision for all subjects.

This study eliminated the influence of suggestion and the expectation of healing, as well as the placebo effect, by utilizing a double-blind design. Neither the doctor nor the laboratory technician knew whether any particular subject was in the experimental or control group. In fact, none of these individuals even were informed, until after the data had been collected, that this experiment was a test of psychic healing. A special laboratory had been constructed which kept the healing practitioner separated from all other experimental personnel, including the subjects themselves, who simply placed their arms through a hole in the wall of the laboratory room. They were told that the study was designed to measure electrical conductivity of the body. The noncontact healing treatments lasted for five minutes each day, for sixteen days. The healing practitioner was located behind the wall through which the subjects placed their arms. Circumstances were exactly the same for both treatment and control groups with the exception that the healing practitioner was not present during the sessions with control subjects.

After sixteen days, thirteen of the twenty-three "treatment" group subjects were completely healed — whereas none of the "nontreatment" group subjects was completely healed on this day. Statistical differences in wound size between the healing and control group were significant when measurements were made on both the eighth and sixteenth days after the original wounds.[191]

After the statistical analysis had been completed, the experimenter interviewed the subjects, the healing practitioner, the medical doctor and the laboratory technician independently. These interviews confirmed that almost all essential experimental protocols were maintained throughout the study.

Although this is one of the best healing studies on record, it may be faulted on several grounds. The experimental protocols were violated when the healing practitioner was unable to schedule afternoon sessions. Thus, all of the experimental subjects were scheduled in the morning, while control subjects were scheduled in the afternoon. While it is unlikely that the time of treatment, in itself, would have produced a differential healing effect, a significant difference still exists insofar as there was no human being near the arm of the control subjects during the supposed treatment period. Thus, there was no effective experimental control for effects such as electromagnetic radiation that may have emanated from the healing practitioner. While one may claim that a non-contact therapeutic touch (NCTT) effect was observed, there is no reason to assume that this was a psychokinetic effect as opposed to one that might be explainable in terms of conventional physical interactions.

Carroll B. Nash, who directed the parapsychology laboratory at St. Joseph's College in Philadelphia, performed an experiment with nineteen psychotics in which each held a bottle of glucose solution. When poured on suspensions of yeast cells the solutions were found to have a slight inhibiting effect on the growth of this organism as compared with controls in which the glucose had not been subjected to treatment.[192]

The demonstrated existence of effects caused by the laying on of hands still left unanswered many questions regarding the mechanisms of this phenomenon. What is it about the hands of a healer that can affect wound healing or plant growth?

This question was on the mind of the biochemist, Sr. M. Justa Smith, who, in 1970, invited Mr. Estebany to her laboratory at Rosary Hill College in Buffalo, New York. Sister Smith's research had focused on enzymes — large protein molecules which act as catalysts, speeding up biochemical reactions such as those associated with wound healing and growth of tissue. Her research had shown that the reactivity of enzymes, which were treated by a strong magnetic field, was increased,[193] and she wondered if Estebany's hands might imitate this effect.

In her laboratory, Estebany held sealed test tubes of enzymes in his hands while Sister Smith's assistants tested their reactivity every fifteen minutes, using an infrared spectrophotometer. They found that the enzymes behaved as if they had been exposed to a magnetic field of thirteen thousand gauss. This is a very strong field when one considers that the magnetic field of the earth is only about one-half of a gauss. Further testing with a magnetometer, however, revealed that there was no unusual magnetic field around Estebany's hands.[194]

In another experiment with Mr. Estebany, Dolores Krieger, a nursing instructor at New York University, measured the hemoglobin levels of sixteen patients who were treated with laying on of hands for fifteen minutes three times daily. Over a six-day period, the patients showed an average increase of 1.2 gm. of hemoglobin per 100 cc. of blood. There was no increase in the level of patients who did not receive the healing treatment.[195]

Further testing of the water treated by Mr. Estebany showed distinct spectrophotometric differences from untreated water. Bernard Grad passed infrared light through both experimental and control saline solutions and revealed a difference in percent transmission between 2,800 and 3,000 millimicrons. But he called it an error since the region was outside the

Plant growth affected by psychic healing treatment (courtesy Bernard Grad).

instrument's specifications.

Douglas Dean, as reported in his 1983 doctoral dissertation research at the Saybrook Institute, replicated this observation on an instrument designed for that particular region of infrared.[196]

Dean used a double-blind study and tested the healer, Olga Worrall, holding distilled water bottles for five, fifteen, and thirty minutes. Larger 2,700 millimicron infrared bands appeared with longer time of holding. Dean also replicated these results in London at Kings College with healer Rose Gladden.

Hand-held and imagining "higher consciousness" gave a large band, whereas imagining "magnetizing" gave control values. Boiling the healer-treated water to steam and condensing it back to water seemed not to boil the healer effect out.

In a study conducted by Stephen Schwartz, et al. of the Mobius Society in Los Angeles, fourteen practitioners were tested — each of whom employed a personal variation of the Laying-on-of-Hands/Therapeutic-Touch processes. Through standard techniques of infrared spectrophotometry, sterile water samples in randomly selected vials evidence alteration of infrared (IR) spectra after being proximate to the palms of the hands of practitioners. A variation in the spectra of all treated samples compared with all control samples was observed in the 2.5 - 3.0 micrometer range, as predicted.[197]

Environmental factors including temperature, barometric pressure, and variations dependent on sampling order, do not appear to explain the observed infrared spectrum alteration.

A series of experiments in psychic healing was conducted by Graham and Anita Watkins at the Foundation for Research on the Nature of Man in Durham, North Carolina, who attempted to find out whether psychics would be able to cause mice to awaken more quickly from ether anesthesia than would normally be expected. Altogether thirteen different subjects were used for this experiment. Three of these subjects were members of the laboratory staff who claimed no special healing ability or significant psychokinetic ability in general. The remaining ten subjects either had claimed to have healing abilities or had performed well on a psychokinetic test under controlled laboratory conditions. In some of the experiments, the subjects were in the same room with the mice they were attempting to revive. In another experiment, the mice were in one room and the subjects were in an adjoining room viewing them through a one-way glass.

SUBJECT'S VIEW OF THE MICE THROUGH A ONE-WAY GLASS. EITHER THE RIGHT OR LEFT SIDE WAS THE TARGET, THE OTHER MOUSE SERVING AS A CONTROL. FUNNELS IN THE BACKGROUND WERE USED TO GUIDE THE MICE ONTO THE PHOTOCELL PLATFORM WITHOUT TOUCHING THEM. (COURTESY GRAHAM WATKINS)

The results of this experiment were highly significant overall. Thirty-two runs were performed with twenty-four trials in each run. In each trial the subject was presented a mouse to revive, and a control mouse which was simultaneously anesthetized.

The control mice averaged 30.43 seconds and the experimentals 25.36 seconds to revive from the ether. The probability that this result was due to chance is less than one in a million. Only one of the talented subjects scored at a chance level as did all three of the laboratory staff. The nine remaining talented subjects scored extremely well.[198]

One unusual finding of this experiment was that the subjects failed to produce a significant effect when they were assigned a random target series instead of using one target location (right or left) throughout a half-run of twelve trials. This apparent failure could be explained, however, if the psychic effect which was causing the accelerated waking of the mice did not immediately dissipate when the subject ceased to concentrate, but rather lingered on for a certain period of time. This was suggested to the experimenters by the fact that when the subjects were asked to change target sides at the end of a half-run, approximately thirty minutes was required between halves to insure a successful second half. This was seen in a number of preliminary runs in which this interval was varied between five minutes and an hour.

To test this hypothesis, the Watkins conducted another experiment in which the subject was asked to leave the building upon completing the first half-run, and the second half began immediately with a pair of mice being placed on the table as though the healer were still present in the adjoining room. In this experiment, the mice on the side of the table that had previously been the target side continued to revive faster — even though no healer was concentrating on them. This "linger effect" was found to be at least as reliable as the main healing effect.[199]

A replication of this study was conducted by Roger Wells and Judith Klein using four "healers." Eight experiments, each consisting of twenty-four trials, were conducted. Seven of the eight were in the expected direction, one being independently significant with a chance probability level less than one in a hundred. There was an average difference of 5.52 seconds between the time the experimental mice awoke and the time the control mice awoke. Of the total 192 trials, 110 were hits.[200]

The Transpersonal Imagery Effect

Researchers in psychoneuroimmunology have determined that individuals can use mental imagery to influence their own physiological states. This process is sometimes called *preverbal imagery.* William Braud and Marilyn Schlitz, of the Mind Science Foundation in San Antonio, Texas, recently reported on a series of studies attempting to assess whether one can use mental imagery to influence the physiological state of another person at a distant location.[201] This process is referred to as *transpersonal imagery*, a concept which, as defined by psychologist Jean Achterberg, "embodies the assumption that information can be transmitted from the consciousness of one person to the physical substrate of others."[202]

Braud and Schlitz developed an objective, quantitative methodology for the study of transpersonal imagery within the framework of experimental psychology. In a typical experiment, Person A (the *influencer*) was instructed to use mental imagery in order to induce physiological changes in Person B (the *subject*), who was isolated in a distant room. The expected physiological effect was assessed by measuring the spontaneous electrical skin resistance responses of Person B. During some experimental sessions, Person A was instructed to produce imagery, during thirty-second recording

WILLIAM BRAUD

"epochs," for increased relaxation in Person B. In other sessions, the mental imagery from Person A was directed toward arousal. At other specified moments in time, according to a random pattern, Person A was asked to refrain from generating any relevant imagery. Some experimental sessions included sequences with both calming and arousing influence, as well as control periods.

The distance (twenty meters or more) between the two rooms used in an experiment, and the presence of several intervening closed doors and corridors, isolated the participants from possible sensory interaction. Additionally, verbalization of any information regarding the imagery/nonimagery schedule by the influencer or the experimenter was not allowed during the experimental sessions. There were no active microphones in either room through which participants could communicate. The headphones through which the participants in the two rooms received required auditory information were attached to independent electrical circuits so that possible "crosstalk" between two sets of headphones was eliminated (i.e., it was impossible for one person's headphone to function as a microphone for the other person's headset).

Throughout an experimental session, the subject sat in a comfortable armchair in a dimly illuminated, closed room. The subject was instructed to make no deliberate effort to relax or become more active, but rather to remain in as ordinary a condition as possible and to be open to and accepting of a possible influence from the distant influencer whom he or she had already met. The subject remained unaware of the number, timing or scheduling of the various influence attempts, and was instructed not to try to guess consciously when influence attempts might be made. The subject was asked to allow his or her thought processes to be as variable or random as possible and to simply observe the various thoughts, images, sensations, and feelings that came to mind without attempting to control, force, or cling to any of them.

The influencer sat in a comfortable chair in front of a polygraph in another closed room. The polygraph provided a graphic analog readout of the concurrent electrodermal activity of the distant subject. Each change in imagery was signaled to the influencer by an auditory signal that could not be heard by the distant subject. Immediately before each signal, the experimenter exposed a card to the influencer containing an instruction for the upcoming recording "epoch." During control periods, the influencer attempted not to think about the subject or about the experiment, and to think of other matters. During influence periods, the influencer used the following strategies (alone or in combination) in an attempt to influence the somatic activity of the distant subject.

1. The influencer used imagery and self-regulation techniques in order to induce the intended condition in *himself or herself*, and imagined a corresponding change in the distant subject.

2. The influencer imagined the *other person* in appropriate relaxing or activating settings.

3. The influencer imaged the desired outcomes of the polygraph pen tracings — i.e., imagined few and small pen deflections for calming periods and many and large deflection for activation periods.

There were rest periods, ranging in duration from fifteen seconds to two minutes in the various experiments, between the thirty-second recording epochs. During those periods, the influencer was able to rest and to prepare for the upcoming epoch.

In order to eliminate the possible influence of common internal rhythms and to remove the possibility that the influencer and the subject just happened to respond at whim in the same manner and at the same times, it was necessary to *formally assign* to the influencer specific times for engaging in imagery; such assignments had to be truly random and, of course, could not be known to the subject (lest the subject self-regulate his or her own physiology on the basis of such knowledge, in order to confirm the expectations of the experimenter).

Evaluation of whether the influencer's imagery influenced the subject's somatic activity was carried out on a session-by-session basis, and involved a determination of the proportion of somatic activity in the prescribed direction which occurred during the influence periods, relative to its occurrence during control periods.

The experimental design guaranteed that the effect could not be attributed to conventional sensorimotor cues, common external stimuli, common internal rhythms, or chance coincidence. Polygraph readings were scored on a blind basis and were eventually computer-automated in order to prevent recording errors or "motivated misreadings" of the records.

The subjects were not told when or how many influence attempts would be made, nor was the experimenter aware of the influence/control epoch until all preliminary interactions with the subject had been completed.

Additional experimental precautions were taken to prevent progressive (time-based) errors. Equipment was allowed to warm up for fifteen to twenty minutes prior to the beginning of a session and therefore had become thermally stable before the experiment began. The use of randomly counterbalanced design prevented any possible progressive effects from contributing differentially to influence versus control epochs.

In all, 337 persons participated in these experiments: 271 served as subjects, sixty-two as influencers, and four as experimenters.

The subjects and influencers were unselected volunteers with no apparent motive for trickery. However, even if a subject were motivated to cheat, such an opportunity was not present. Cheating would have required knowledge of the session's influence/control epoch sequence and of the precise starting time for the session. These requirements were eliminated.

Braud and Schlitz concluded, based upon overall statistical results of thirteen experiments, that the transpersonal imagery effect is a relatively reliable and robust phenomenon. In fact, under certain conditions, the magnitude of the transpersonal imagery effect compared favorably with the magnitudes reported for self-regulation effect. The ability to manifest the effect is apparently widely distributed in the population. Sensitivity to the effects appeared to be normally distributed among the 271 volunteer subjects tested in these experiments.

In addition to responding physiologically in a manner consistent with the imagery of the distant influencer, subjects often reported subjective responses which correspond to the influencer's images. Sometimes these reports were of relatively vague feelings of relaxation or activation. However, there were also reports of extremely *specific* thoughts, feelings, and sensations which strikingly matched the imagery employed by the influencer. For example, a subject reported spontaneously that during the session he had a very vivid impression of the influencer coming into his room, walking behind his seat, and vigorously shaking the chair; the impression was so strong that he found it difficult to believe that the event had not happened in reality. This session was one in which the influencer had employed just such an image in order to activate the subject from afar.

MARILYN SCHLITZ (FROM *BIOLOGICAL PSYCHOKINESIS* VIDEOTAPE, COURTESY THINKING ALLOWED PRODUCTIONS)

Subjects sometimes spontaneously reported mentation which corresponded closely to that of the influencer or the experimenter, even when that mentation was incidental and not employed consciously as part of an influence strategy. For example, at the beginning of one session, the experimenter remarked to an influencer that the electrodermal tracings of the subject were very precise and regimented and that they reminded him of the German techno-pop instrumental musical group, Kraftwerk. When the experimenter went to the subject's room at the end of the session, the subject's first comment was that early in the session, for some unknown reason, thoughts of the group Kraftwerk had come into her mind. The subject could not have overheard the experimenter's earlier comment to the influencer. Such correspondences were not rare.

Proper Scientific Controls for ESP Experimentation[203]

To some extent, psi research — which legitimately falls within the domain of psychology, has suffered by borrowing the methods of psychology. Psychological methods, in general, have not been designed to test for scientific anomalies and extraordinary claims. While there can be little doubt that psi research studies are generally far more carefully scrutinized than other studies in psychology, this, in itself, is not sufficient reason for supposing that psi's existence has been scientifically established. If anything, a close examination of the methodological rigor used in psi research might tend to create doubts about the validity of a great deal of research in psychology — as well as other behavioral and social sciences.

The usual logic is to suppose an ESP effect only when all other alternatives have been ruled out — not by post hoc analysis, but by design of the experiment. The reason for this reasoning is that ESP is not an established process; it is only a label that we assign to certain anomalous findings. To verify the anomaly it is necessary to absolutely rule out sources of error.

Randomization

In theory, if ESP targets are chosen "at random," there is no logical way that the next target in a sequence can be predicted. In a table of random numbers, all of the numbers occur with a frequency approximating chance expectation. Furthermore, all sequences of numbers (i.e., doublets, triplets, etc.) also occur with a frequency approximating chance. We can be certain that a table of random numbers meets the various statistical tests that have been devised for randomness. However, since such tables are published and generally available in libraries, there is always a risk that "random" target sequences could be predicted by someone who obtained access to the random number table being used in a particular study. (A standard precaution against this possibility in psychological and psi research is to randomly select the entry point into a random number table.)

Psi researchers can generate their own random number sequences, without reference to a published table of random numbers, by using an electronic or mechanical Random Event Generator (REG). The best of these devices rely on random sources which are quantum mechanical in nature, such as electronic white noise or radioactive decay. Due to the uncertainty principle in quantum mechanics, the output of such devices in theoretically unpredictable; they are thus the most random sources known to nature.

Ironically, it is entirely possible, however, that a truly random source will provide a short-term output that fails the statistical tests for randomness mentioned above. Thus, a genuinely random target sequence can be problematic if it mimics the properties of a non-random sequence. A test subject who receives trial-by-trial feedback in such a situation might make inferences that happen to match the random output. Beyond-chance scoring in such an experiment would not necessarily be due to ESP. Therefore, the ideal experiment must not only be derived from a true random source, it must also meet post hoc tests for randomnicity. Such tests will also detect more serious sources of bias that may develop within electronic or mechanical apparatus designed to generate random events.

Non-random target sequences may not necessarily, in themselves, account for an alleged ESP score. The main concern is that subjects will learn the characteristics of previous target sequences, and use this information to infer the characteristics of future target sequences. This, of course, requires periodic feedback on targets. Or, subjects may simply have a personal bias toward some targets that, coincidentally, matches the patterns that emerge in a non-random sequence.[204] (In fact, this can occur in target sequences which are truly random, as well, creating a — generally short-term — false impression of beyond chance, psi scoring.) This can be detected if the features of the target sequence are matched or correlated with non-random patterns in the subject's calls — other than those intended for each target. (It is entirely possible that a subject's successful ESP calls will be unrelated to the non-random features in a target sequence.) Whether or not a non-random feature can account for an alleged ESP effect, it is a disquieting situation when departures from randomness are severe, and of unknown origin. If a critical procedure such as randomization breaks down, this raises the possibility that there were other breakdowns, or failures to carry out the experimental plan. At least one case of experimenter fraud (involving British mathematician S. G. Soal) was traced through a detection of non-random target sequences.

Originally, in early ESP studies, randomization was achieved through manual shuffling of card decks. It is now generally agreed that this is not adequate. It could be argued that informal randomization invalidates any ESP experiment. If unbalanced decks are used, then psi-like artifacts can easily emerge.[205]

Randomization procedures, or REG test machines, must be thoroughly described in experimental reports. If random number tables are used, the application procedures must be fully described. This is particularly important when untrained assistants are asked to generate target sequences.

Electronic random event generators should employ a switching system to correct for possible systematic bias. However, there is always the possibility that the output bias could, in some manner, correlate with the switching sequence itself. This would occur if the REG bias oscillates. Is there any way to prevent this occurrence? Schmidt, at a 1974 research meeting at the Foundation for Research on the Nature of Man, suggested doing so by incorporating the Rand Corporation random number tables into such a switching system.

When targets are obtained from a random event generator (REG) rather than from a predetermined sequence, then it is critically important to include control runs in a systematic fashion (i.e., a counterbalanced sequence of control and experimental runs). In spite of this importance, there is no way to theoretically ensure that such control runs are immune from psychokinetic influence. This uncertainty places an absolute limit on the degree of precision which is possible in psi research using REGs. Undoubtedly, it will also provide sufficient grounds for some skeptics to refuse to accept any REG data as evidential of psi functioning.

As REGs are continually being modified or replaced, there is not much standardization. This makes it all the more important to introduce systematic controls, especially as a guard against short-term generator bias.

One means of systematically controlling for generator bias is to randomly pair control and experimental trials.[206] This is quite easily accomplished when one has an REG interfaced to a computer.

To what extent can one make allowances for non-random target sequences and salvage an experiment which is flawed in this respect? This is an extremely important question because: (a) pure randomness is an ideal which can never truly be reached in the real world;[207] and (b) valid random procedures may, in fact, produce target sequences which in retrospect do not appear random — i.e., at Monte Carlo when black came up thirty-two times in a row. In other words, a genuinely random sequence of sufficient length will have many subsequences which do not appear random.

Failure to record actual target sequences is a severe shortcoming in any ESP experiment.

A crucial issue in examining target sequences generated by a REG for ESP experimentation is whether non-random sequences may have been produced by PK. There is no way possible to control for this factor other than by using pseudorandom REGs or by referring to a random number table. (Even then some sort of macro-PK might theoretically be involved.)

During PK experiments with REGs, control tests must check for temporal stability of the random sources during the course of the experiment. Such randomness tests should be conducted in the actual experimental environment with all peripheral equipment attached. As a precaution against PK influence on the control tests, experiments could be designed with various blinds to prevent both subjects and experimenters from knowing when and how these control tests will be run. Ultimately, of course, at least one experimenter who designs the study will have to know the arrangements. A further control would be to have this individual be someone who has no particular history of manifesting psi in REG testing situations. (However, all of these controls will ultimately lead to uncertainties.)

Control studies must also specify the physics and constructed parameters of the experimental apparatus to assess the possibility of environmental influences.

It should be mentioned that some Schmidt generators have been tested by generating sequences of over a million trials, and have shown no evidence of either short- or long-term bias.[208] Hence the problem is not a severe one with a well-designed generator which has been thoroughly tested.

Sensory Leakage

A standard rival hypothesis to the hypothesis of ESP is that sensory leakage occurred and that the receiver was knowingly or unknowingly cued by the sender or by an intermediary between the sender and receiver. As early as 1895, psychologists described "unconscious whispering" in the laboratory and were even able to show that senders in telepathy experiments could give auditory cues to their receivers quite unwittingly.[209,210,211] Ingenious use of parabolic sound reflectors made this demonstration possible. Many researchers in the early years of experimental psychology and psi research gave early warnings on the dangers of unintentional cueing.[212,213,214,215] The subtle kinds of cues described by these early workers are just the kind psychologists have come to realize mediate experimenter expectancy effects found in laboratory settings.[216]

In designing experiments to prevent sensory leakage, experimenters cannot assume that there are no tricksters present among the subjects. Precautions must be taken that would prevent the most skilled of tricksters or magicians from succeeding in obtaining normal sensory information about the targets.

Experimental reports must clearly describe the relative location of subjects and targets.

If visual targets are within proximity of the subject, they must be in an opaque container, unopened until the subject's responses have been recorded. The opaqueness of the container should be objectively assessed and the container should be kept well out of the subject's reach. Subjects must not be present at any time while the target materials are being prepared.

If the subject is allowed access to the container, it must be made "fraudproof." This is no easy task. Chemist George Price suggested, in his 1955 *Science* article advocating the presumption of fraud, that a metal container be used with a cover welded on and photomicrographs taken of the welds.[217] Even with such precautions a clever subject, using advanced technology, under unobserved circumstances could devise ways of penetrating such a device. For day to day research, such precautions are not practical. Thus, it is simply better not to allow the subjects to have unobserved access to containers with target materials.

Particular care must be taken with using computerized REG devices that subjects not remain unobserved with the computer device. James Davis, of the Institute for Parapsychology in Durham, North Carolina, has observed that a subject who has access to the computer, knows the data format, and has sufficient programming knowledge, might subvert experimental precautions.[218]

In addition, it is also unclear as to whether computers might provide sensory cues regarding target information. This might be in terms of subtle audio or electromagnetic signals that could aid a sensitive individual in distinguishing different targets.

James Davis, psi researcher who was instrumental in detecting experimenter fraud in 1974 at the Foundation for Research on the Nature of Man in Durham, North Carolina.

In telepathy experiments one must exclude the possibility that the subject learns about the targets indirectly, through cues from the agent. Thus, any individual with information regarding the targets must not be within auditory or visual range of the subject. This often means that putting subject and agent in adjacent rooms is insufficient. Sound isolation must also be insured. (Electromagnetic shielding between rooms can also prevent leakage through the use of radio transmission between agent and percipient — an issue which may more appropriately come under the heading of cheating.) In some buildings, sounds travel quite readily between distant rooms. Furthermore, any communication between subject and agent with regard to timing of trials (i.e., intercom or knocking on the walls) may also inadvertently contain sensory information regarding ESP targets. If "ready" signals are used, they should operate only from the percipient's room to the agent's room, so that cueing is eliminated.

Precognitive testing, where targets are not selected until some time after the subject's calls are recorded, allows the tightest control over conditions that might otherwise contribute to sensory leakage.

Experimenter cueing can be eliminated by keeping experimenters blind as to target order. This means that experimenters cannot administer their own ESP tests; or, if they do, they must be shielded for all sensory contact with subjects. In order to insure experimenter blindness with regard to targets, experimenters should have no sensory contact with individuals who are aware of the target order. Otherwise, it is possible to hypothesize a chain of nonverbal communication.

In free-response experiments with independent judges, it is also essential that the judges be shielded from all sensory cues just as if they were subjects. In addition, judges must be provided with no sensory cues whatsoever regarding the order of the various calls made by subjects. Such cues must be edited out of any transcripts provided to judges. If such cues are provided, judges may succeed in time-ordering both the target and the response sets, thus contaminating the judging process with additional logically derived information.

Providing feedback in a "closed deck" target pool situation reduces, in effect, the degrees of freedom of the final target (even if the total pool is unknown to the subject) and thus may be a source of experimental contamination. This may occur if subjects avoid producing imagery related to any targets for which imagery has already been provided.

Handling cues are also best avoided. Both judges and percipients may detect creases, marks, smudges, temperature differences or other artifacts that result if actual targets have been handled and then mixed in with targets from a pool for judging. Handling cues may also result when targets placed in envelopes are opened and then resealed or placed in new envelopes, as has sometimes been done.

In studies where ESP scoring is correlated with other factors, such as personality variables, it is important that subjects be given no feedback on their ESP scores prior to other testing. Otherwise, it is possible that the personality score is influenced by feedback from the ESP test.

Subject Cheating

Fraud is encouraged in studies where incentives are offered (i.e., employment, publicity) for high ESP or PK scores. Some researchers deal with this issue by never working with well-known "psychics" and by insisting that all subjects participate anonymously in their studies. These precautions, however, are not sufficient to preclude fraud.

During an eight-month period in 1983 and 1984 researchers at the Department of Psychology, University of Edinburgh, in Scotland, conducted twenty experimental sessions with a teenager who claimed to possess macro-PK abilities. Although many phenomena were apparently observed, none of these manifested under well-controlled conditions. Eventually, researchers suspected fraud and set up hidden cameras which succeeded in revealing blatantly fraudulent activities. When confronted with this evidence of his deception, the subject denied that his activities has been fraudulent. However, several months later he confessed that he was a practicing magician "who had wished to see if it were possible for a magician to pose successfully as a psychic in a laboratory."[219]

While, ultimately, he did not succeed in fooling the researchers, he did manage to take up approximately sixty hours of their time. Psi researchers are somewhat vulnerable to this type of invasion, because — in attempting to establish conditions conducive to the alleged phenomena they wish to investigate — they attempt to establish good rapport and thus avoid treating experimental subjects with suspicion.

Human nature is often unpredictable; this is sometimes the case with regard to fraudulent and criminal activity, especially when it occurs in the absence of apparent motive. Some subjects may simply get a kick out of fooling experimenters (especially when researchers have claimed fraudproof conditions). In one clever incident, a research study conducted at Harvard University, the agent and percipient were a hundred feet apart, with four closed doors separating their rooms. The student subjects used a confederate hidden in one of the rooms to aid in passing a signal. The system was so successful that the percipient was able to guess the correct color of a deck of playing cards, on all fifty-two attempts.[220] The researcher, George Estabrooks, was initially fooled, even though he knew that cheating would be attempted.

A basic precaution, when subjects are provided with feedback regarding targets, is that subjects' calls be clearly recorded and submitted to the experimenter before feedback is provided.

When REGs or other testing equipment is used, precautions must be carefully instituted to prevent subjects from resetting counters or, in any other uncontrolled manner, manipulating the parameters of the device.

In developing precautions against radio transmission of target information to an ESP percipient, care must be made to objectively assess electromagnetic shielding of laboratory rooms at the time of the experiment. Often shielding characteristics change over time, as modifications are made to experimental rooms (e.g., for insertion of cables or ventilation ducts).

The possibility of such cheating is obviously lessened when controls are such that no friends or associates of the percipient are allowed access to ESP target information (i.e., by participating as an agent in an ESP test).

Another form of control against the possibility of cheating is to have the subject and the agent observed (or videotaped) continuously during the experimental period. Some experimenters have attempted to control for fraud by locking subjects into experimental rooms during testing. Naturally, this control would be of little success if the subject is skilled at picking locks.

Readers interested in developing a more detailed knowledge of the many ways in which fraud may be committed to create the illusion of psychic functioning will want to study the literature of professional magic.[221]

Recording Errors

The possibility of unconscious errors in the recording of experimental data has been observed in psi research since the 1930s.[222,223] In a meta-analysis of 139,000 recorded observations in twenty-one psychology studies, Harvard researcher Robert Rosenthal found that about 1% of all observations were in error. Of the errors committed, twice as many favored the hypothesis as opposed it.[224] Thus, such errors are a very real, yet small (i.e., 0.33%), factor in studies involving manual recording of data.

When recording ESP and PK target and response sequences, automated equipment, which cannot be tampered with, provides the best insurance against recording errors. If targets and responses are recorded manually, this must be done by individuals who are blind (or unaware) as to the correct targets in order to preclude unconscious errors.

When automated equipment is used, it is critical to good experimental design that automated equipment be subject to periodic tests, during the course of an experiment, to ensure that the equipment functions as it was intended to function.

When computers are used to record psi targets and responses, it is important that the paper printout be kept in its original continuous condition, in order to prevent misplacement of some records. In the event that computer printout paper becomes severed, the paper sheets should be prenumbered. Another precaution would be to keep duplicate records in magnetic storage format.

Classification and Scoring Errors

When different experimental conditions (i.e., high-aim and low-aim PK tests) are built into the design, automated recording equipment should be programmed to carefully distinguish between conditions.

When experimental subjects are divided into various classifications, this must not be done on a post hoc knowledge of the subjects' psi scores. The classification must be blind and would ordinarily be completed before psi testing is conducted. Ideally, the basis for classification into high and low scorers should be made public in full detail to the research community before any testing begins. Fixed response alternatives or a preplanned scoring system are essential for such classifications.

In all aspects of data handling, such as computing various statistical parameters from raw data, automated analysis is preferable. If analysis is done manually, it is advantageous that the scorer or statistician be blind as to the various experimental conditions and hypotheses — in order to avoid inadvertent bias.

Statistical/Methodological Violations

Some critics claim that experiments should compare psi scores with scores obtained in some sort of control conditions, rather than simply with the expected statistical means. The reason for this insistence is not that the predictions of probability theory are being called into question — but rather because a control group may tend to benefit from any otherwise undetected, non-psi sources of information related to methodological flaws, recording errors or sensory leakage. The problem with this approach, however, is that it is not possible to devise a control condition that would be identical to the experimental condition in every way except that it would eliminate possible psi communication.

In situations where subjects were instructed not to use psi, as in the Princeton RNG studies, it was possible (and seems probable) that subjects used PK to dampen the variability of the RNG so that scores were unusually close to the theoretical mean. This dilemma would seem to place a limit on the degree of precision which is possible to attain in psi research.

Trial-by-trial feedback, given in studies using a "closed" ESP target sequence (e.g., a deck of cards) violates the condition of independence used for most standard statistical tests. A clever card-counter, such as those who sometimes use the system to win at Blackjack in Las Vegas, can increase the certainty of their guesses on the final cards in a deck. Independence, of course, is not violated when using an "open" sequence in which targets are chosen from a much larger universe. In such situations, card-counting types of strategies would be useless.

Multiple responses for a single target cannot be evaluated using statistical tests that assume independence of responses. In psi research this error is known as the *stacking effect.* It often occurs with informal classroom or media tests of ESP. If there is a sufficient number of responses, the data can be analyzed by the *Greville method,* which accounts for the *stacking effect.* While the *stacking effect,* however, is a theoretical possibility, empirical tests have shown that

JESSICA UTTS

for multiple data in typical forced-choice ESP tests, it makes little practical difference whether the results are analyzed by the usual binomial formula, or by the more appropriate Greville method.[225,226] The situation is quite different in remote-viewing and other free-response experiments where the number of targets is generally small. It then becomes critical to control for response bias.

In remote-viewing experiments, violations of independence have arisen even when a single judge is asked to rank a small number of targets against an equal number of responses. The judge may, under these conditions, be influenced in assigning a rank or rating to a given target by the memory of how he or she assigned ranks or ratings to other targets.

The assumptions used in most statistical tests are violated if optional stopping is used. This could occur by limiting the number of experimental trials at the experimenter's option (particularly after receiving feedback as to success rates) or by optionally limiting the number of experimental subjects. The Princeton method of converting free-response information to binary data avoids this criticism.[227]

Statistical tests are not accurate if researchers are free to censor data which does not support their hypotheses. This sometimes occurs by the use of post hoc decisions as to whether a study will be reported as an informal, preliminary demonstration or as part of an experiment. Ideally, all formal psi experiments should be registered, in advance, specifying the total number of trials, runs, subjects, etc. Then all data from those experiments should be reported, regardless of the outcome.

All planned statistical tests must be announced in advance to enable the research community to clearly distinguish between the main analysis and post hoc analysis. When multiple analyses are used, statistical tests must take into account the increased possibility of a Type I error — i.e., the greater likelihood that one of the many tests will attain a probability level of less than .05.

University of California statistician Jessica Utts has pointed out that psi researchers often place too great an emphasis on the probability level of experiments.[228] This is a mistake, as the validity, magnitude, and reliability of a possible psi effect have very little to do with probability levels. When a great many trials are involved (as in the Princeton RNG studies), a very weak effect can yield probability significance levels that are astronomical. This misunderstanding also results in much confusion with regard to replication in psi research. Utts recommends the use of confidence intervals and power analyses in order to determine the sample sizes necessary to attain significant results in particular experimental designs.

Reporting Failures

Experiments must be reported in sufficient detail so that other researchers may attempt independent replication. If this is not possible within the context of a journal article, sufficient detail must be made available in unpublished documentation. All of the details regarding protocol should be included in such reports. When space does not allow publication of full details, provisions must be made for independent release of experimental details to interested researchers.

Experimenter Fraud

Fudging or "tidying" of data, or outright fraud by experimenters is a factor that, given the history and controversies of psi research, must be accounted for in experimental planning and reporting. Within the past two decades, two major psi researchers were caught by their colleagues in acts of fraud.

While it is impossible to absolutely preclude experimenter fraud in any field of science, it is possible to design studies in such a fashion that if fraud were to occur it could not logically be attributed to a single deviant scientist, but would rather have to be explained as the result of a conspiracy among members of a research team. The basic methodology for accomplishing this level of safeguard is to ensure that researchers work together in teams, always observing and double-checking each other's work.

Other methodologies for preventing experimenter fraud involve strategies such as sharing of data with outside research teams, establishing various double- and triple-blinds so that researchers not possess data necessary to commit fraud.

Evaluating Psi Research

How are we to evaluate the evidence for psychic interactions? Psi research offers a mass of experimental evidence that *prima facie* requires us to take psi seriously. If we apply the same standards used to judge non-controversial claims in the behavioral and social sciences, we would have little choice but to accept that extra-sensorimotor interactions have been experimentally established.

The critics argue that psi claims, if accepted, would dramatically change our self-image and world view. Therefore, they claim, we must use extraordinary care in evaluating the data. As extraordinary claims require extraordinary evidence, it is reasonable to use a higher standard in looking at psi. They are correct; and psi research, after more than a century, has yet to meet such a higher standard.

Often the skeptics' arguments go even further and assert that psi research is a pseudoscience that mimics the methods of science, but has no real subject matter (since psi does not exist). More than a hundred years of research, they assert, have failed to produce solid scientific evidence. Therefore, they maintain that scientists are entitled to ignore psi research, or even formally disaffiliate psi researchers from scientific organizations.[229]

The claim that we should abandon an empirical discipline is, itself, an extraordinary claim, and, as such, requires extraordinary evidence. For the skeptics to make this claim stick, it would be incumbent upon them to demonstrate that the beyond-chance findings of psi research can reasonably be attributed to various artifacts, fraud, or other conventional hypotheses. At the current time, the skeptics are much farther from this goal than the psi researchers are from establishing psi. Thus, according to Trevor Pinch, a sociologist of science:

> Psi will not lie down and die; neither will it stand up and be counted.[230]

Psi researchers, by adopting the methods, procedures, and institutions of orthodox science, have been quietly attempting to gain the acceptance and approval of the scientific community, by increasing their rigorous adherence to scientific method and conducting new experiments that meet all published criticisms. Most scientists generally ignore these new studies until, eventually, critics develop new lines of attack against them.

Critics also reserve for themselves the right to take a great deal of time in examining studies for flaws. Psychologist Ray Hyman, a prominent skeptic, once suggested with regard to Helmut Schmidt's PK studies with random number generators that, although he could find no serious flaws in these studies, he would like to have another twenty-five or thirty years to examine them more carefully.[231] This strategy leaves the critics grinning in confidence, for it is highly unlikely (but not impossible) that any scientific study (or series of studies) would be found "completely flawless" after twenty-five or thirty years of further technological progress.

Psi researchers respond to this approach by claiming that potential flaws in their studies do not suffice to disqualify their findings. They claim that the burden is on the skeptics not only to find flaws but, additionally, to demonstrate that these flaws could have contributed somehow to the beyond-chance results they have obtained. Otherwise, they argue, the alleged flaws may simply be trivial. To support this argument, they point to meta-analyses in which studies that were free of particular flaws showed as strong a psi effect as other studies in which the flaws had not been eliminated.

In response, the critics have suggested that the ostensible psi effects may have been contributed by artifacts other than the particular methodological flaws that were being examined in the meta-analysis comparisons. The only way to be certain that this is not occurring, they claim, is for experiments to be completely flawless.

Psi researchers counter by claiming that the demand for completely flawless experiments is, for good reason, not required in any other field of science — it is scientifically sterile:

> . . . resort to the "dirty test tube" metaphor provides an unrestricted license for the wholesale dismissal of research findings on the basis of vague and ad hoc "weaknesses." Whole domains of research are dismissed through allusions to "inadequate documentation," "inadequate controls," "overcomplicated experimental setups," and "lack of experimental rigor." Yet if the measure of good scientific methodology is its capacity to rule out plausible alternatives, such attributions are clearly inappropriate. No scientific experiment is so pristine that it can withstand the efforts of a sufficiently determined critic, but the fact remains that by any reasonable standard the methodology of many successful psi experiments is fundamentally sound.[232]

In effect, the psi researchers are accusing their critics of unscientifically taking a position that is *unfalsifiable*, e.g., no matter what evidence is produced in favor of a psi effect, skeptics will come up with some argument in favor of dismissing it. Skeptics counter by claiming that the psi hypothesis, itself, is *unfalsifiable* and therefore unscientific. They claim that no matter how many experiments fail to obtain positive results, or how many studies are shown to be faulty, researchers will claim that they have not disproven psi's existence.

Thus, the convoluted arguments reach deep into the philosophy of science, causing us to question almost every premise of ontology and epistemology. However, the tone of the debate between proponents and opponents of psi's existence is not always polite. Name-calling is common enough that outside observers have come to assume that there exists "an extraordinarily annoying 'tendency' to sloppiness among parapsychological experimenters."[233] Of course, name-calling occurs on both sides of the debate, as exemplified by William James' earlier use of the term "ignoramus." Such characterizations, however, are often unfair.[234,235] Several independent critics have concluded that psi experiments are comparable to research studies in the field of psychology.[236,237] The problem of image is one that has plagued psi research from its very beginnings. Yet, even at the turn of the century, the level of critical intelligence among researchers affiliated with the Society for Psychical Research was extremely high, according to William James (whom many still regard as America's greatest psychologist):

> According to the newspaper and drawing-room myth, soft-headedness and idiotic credulity are the bond of sympathy in this Society, and general wonder-sickness its dynamic principle. A glance at the membership fails, however, to corroborate this view. The president is Professor Henry Sidgewick, known by his other deeds as the most incorrigibly and exasperatingly critical and skeptical mind in England....Such men as Professor Lodge, the eminent English physicist and Professor Richet, the eminent French physiologist, are among the most active contributors to the Society's *Proceedings*; and through the catalogue of membership are sprinkled names honored throughout the world for their scientific capacity. In fact, were I asked to point to a scientific journal where hard-headedness and never-sleeping suspicion of sources of error might be seen in their full bloom, I think I should have to fall back on the *Proceedings of the Society for Psychical Research*. The common run of papers, say on physiological subjects, which one finds in other professional organs, are apt to show a far lower level of critical consciousness.[238]

There is, of course, another perspective. For example, Eric Dingwall, a British anthropologist who has spent sixty years affiliated with psychical research, eventually left the field in exasperation with the low standards he found and convinced that irrational, popular occult groups were thriving on misinformation propagated by psychical researchers.[239]

The fact — that, after more than a century of inquiry, psi researchers have yet to firmly establish the phenomena which they purport to study — cannot help reflecting on the quality of the research itself (or the researchers themselves) in comparison to work in other related disciplines. Yet, considering the low level of funding for psi research, one might well argue that both the quality and quantity of research studies have been surprisingly high.

The strategy of psi researchers is to obtain from their critics the most complete description possible of the methodological requirements that would suffice to prove psi's existence. However, in order to meet these requirements, the field of psi research will undoubtedly itself require levels of funding comparable to that available for other fields of science in which a high level of methodological rigor is standard practice. Skeptics wish to see experiments independently replicated in many laboratories, using tamper-proof, automated equipment whose parameters are well understood. Psi researchers seem to be closing in on this goal and, perhaps, with additional levels of funding, they will over the next several decades begin to attract the serious attention from mainstream science which they feel they deserve.

Since before the turn of the century, breakthroughs have periodically appeared to be just around the corner, but somehow they never seem to have arrived. This situation led William James to comment, when writing in 1909 about the first twenty-five years of psychical research:

> It is hard to believe, however, that the Creator has really put any big array of phenomena into the world merely to defy and mock our scientific tendencies; so my deeper belief is that we psychical researchers have been too precipitate with our hopes, and that we must expect to mark progress not by quarter-centuries, but by half-centuries or whole centuries.[240]

The history and sociology of science shows that the arguments surrounding psi research are not atypical.[241,242] Even in the hard sciences, it is possible to find experimental evidence that is compelling to some scientists but contested by others. The outcomes of experiments can always be contested by challenging one or more of the many assumptions upon which those outcomes rely. Such arguments in most cases are resolved fairly quickly by means of further experimentation and argument.[243] In most cases the process of criticism will come to an end. One feature of psi research is that consensus never seems to emerge.

While the cycle seems like pointless repetition to some, my own opinion is that the debate, over the generations, has increased in both subtlety and sophistication. This is the best progress we can expect until such time as a decisive case can be made. The debate is an ancient one. Perhaps it will be finally solved in our lifetime; however, until then, the wisest course is to tolerate the ambiguities, uncertainties and multiplicities of perspective.

This is not always easy. Often, we are tempted to seek the comfort that comes from certainty. Were I to do so, I would echo, as a starting point, the words of William James:

> In psychology, physiology, and medicine, wherever a debate between the mystics and the scientifics has been once for all decided, it is the mystics who have usually proved to be right about the facts, while the scientifics had the better of it in respect to the theories.[244]

Science today is incomplete in many respects. We lack a unified theory of the physical forces. We lack a theory of consciousness itself. Our ability to integrate psi (if it exists) into our scientific world view is extremely limited until we can develop adequate theories of these fundamental constituents of the universe.

In the seventeen years since the original publication of *The Roots of Consciousness*, there has been considerable development in the area of grand unified field theory. While a final solution to this basic problem in physics has not arrived, the outlines of what such a solution will look like seem to be emerging. Section IV discusses progress in the area of consciousness and psi research with particular focus on a potential theoretical environment for integrating consciousness and psi with grand unified field theory.

WILLIAM JAMES, LEFT, AND A FRIEND AT JAMES'S SUMMER HOME, ABOUT 1887.

THEORIES OF CONSCIOUSNESS

INTRODUCTION

IN A CLASSIC 1962 ESSAY, MICHAEL SCRIVEN, A PHILOSOPHER OF SCIENCE (and one of my mentors), argued that psi research was a field that had amassed a substantial body of data for which there was no adequate theory. He contrasted it with the field of psychoanalysis which had a substantial theory that lacked adequate empirical support.[1] In recent years, some psi researchers have taken issue with this point of view — claiming that they are, indeed, actively engaged in producing theories of psi.[2,3] In particular, as Gertrude Schmeidler has demonstrated in her book, *Parapsychology and Psychology,* parapsychologists generally operate on the theoretical premise that psi is a psychological function.[4]

In fact, since 1962, there have been many more efforts to develop theories of psi than can be presented in Section IV. While data has accumulated to support some theories, they are all inherently unsatisfactory. Psychological theories fail to account for the uncanny transfer of information which psi implies. Physical theories which deal with the issue of information transfer generally fail to address crucial psychological issues. The bottom line is simply this: A theory of psi cannot be complete without a theory of consciousness — and this is what we lack. Yet we are slowly making progress.

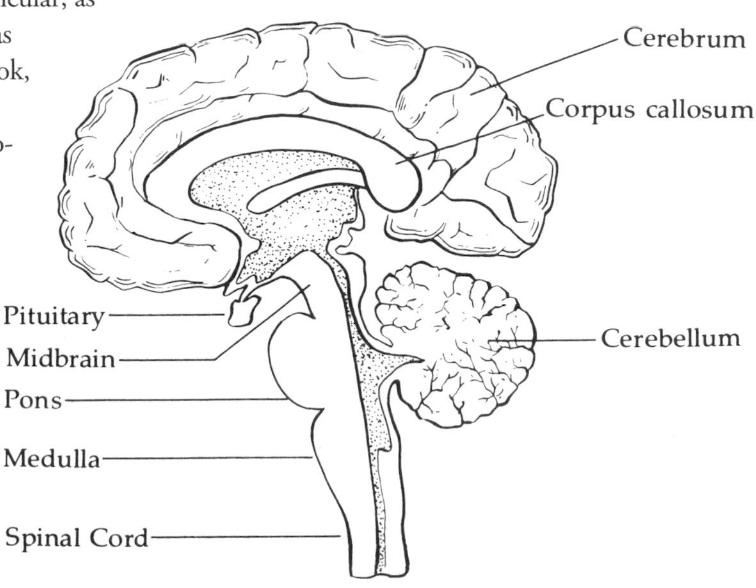

MICHAEL SCRIVEN (FROM *EXPLANATIONS OF THE SUPERNATURAL*, COURTESY THINKING ALLOWED PRODUCTIONS)

THE BIOLOGICAL PERSPECTIVE

In the Introduction, I discussed the evolution of organized matter from the photon through particles, atoms and molecules to living cells which begin to differentiate in structure and function forming a wide variety of tissues and organs that play a specialized function in the human body. It is reasonable to assume that all these levels of organization including the whole human being, play a role in *shaping* consciousness. Particularly important are the nervous system, comprising brain and spinal cord, and the endocrine system, comprising a number of ductless glands that secrete hormones into the bloodstream. Many biological scientists today implicitly believe that these structures not only *shape* consciousness, but are actually the source of conscious awareness. This view is known as the *biological identity theory.*

The Nervous System

Neuron cells are the principle units of the nervous system. Their function is to conduct nerve impulses transmitting information. The twelve billion neurons in our bodies vary greatly in size and shape; however, they all have two general parts: a *cell body* and *fibers.* The cell body contains structures that keep the neuron alive and properly functioning. The neural fibers are of two classes: *dendrites* stimulated by neighboring neurons or physical stimuli; and *axons,* which transmit impulses to other neurons or to an effector, such as a muscle or gland.

The process by which pulses transmit across the neural membrane is electrochemical. The pulses are caused by rapid and reversible changes in the permeability of the membrane to certain ions. The resulting

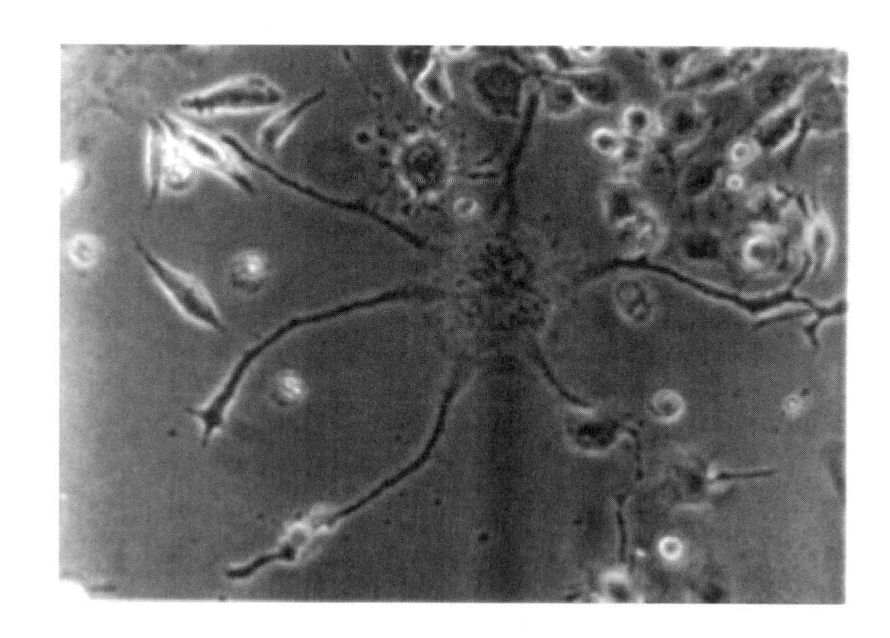

MULTI-POLAR NEURON. CIRCLE INDICATES A SYNAPSE. (COURTESY ANGELA M. LONGO)

flows of ions across the membrane give rise to electrical impulses, which can be detected and recorded with various instruments. The size of the nerve impulses and the speed with which they travel are unique to each particular neuron and do not relate to the strength of the stimuli that initiated them. Firing thresholds will vary with time from neuron to neuron depending on many factors; however, once the threshold is reached, the electrochemical changes that cause the impulse proceed to completion. Therefore, information about any stimulus is carried by (1) the frequency of nerve firing and (2) by the number of particular fibers carrying impulses, and not by the strength of any single impulse. This, incidentally, is the same on-off principle by which information is coded in a digital computer. Some nerves transmit as many as one thousand impulses each second.

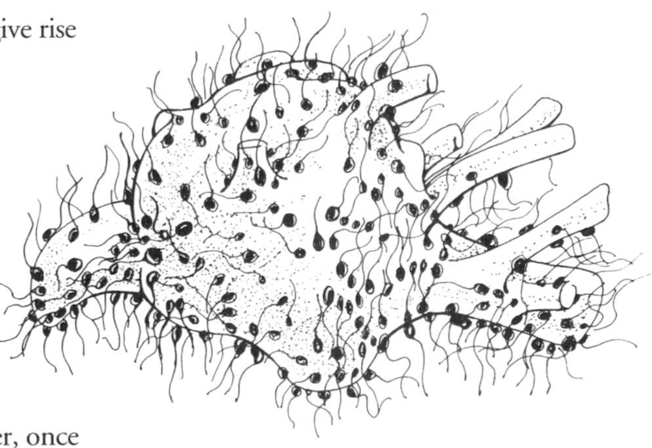

NEURON CELL BODY WITH SYNAPSES FROM OTHER NEURONS

Neurons are stimulated to fire by either sensory receptors or other neurons. Nerve impulses are transmitted from one neuron to another or from a neuron to a muscle or gland across an important gap known as a *synapse.* The whole region including the *bouton* on the end of the axon on one neuron, the gap, and the post-synaptic membrane of the adjoining cell, can be called the *synaptic region* (the circled area in the multi-polar neuron photograph). Information is transmitted across the synaptic gap by enzymes delicately released from little spheres in the bouton called *vesicles.* The information is received at the *postsynaptic membrane,* which is generally either excited or inhibited by these chemicals depending again on

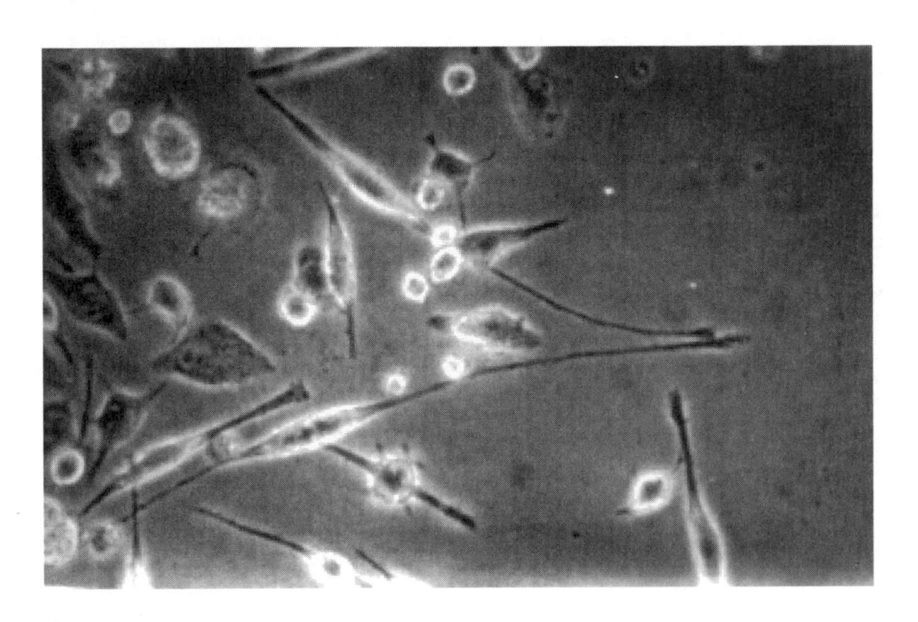

BIPOLAR NEURONS (COURTESY ANGELA M. LONGO)

many factors, such as the particular combination of enzymes transmitted across the synapse or the interaction with the electromagnetic environment around the body.

If the post-synaptic membrane is stimulated by an inhibiting neurotransmitter, its firing threshold will become higher. An excitatory neurotransmitter will lower the firing threshold of a given neuron, causing it to fire more often. The actual firing threshold of a neuron is variable and is often determined by the combined influence of hundreds of synapses. Thus the synaptic aspect of neural transmission is not an all or none affair, and may be thought of as the analog or continuous aspect of the human bio-computer. Some nerves actually loop back upon themselves to form reverberating circuits which may be the neural basis for memory storage.

The nervous system itself is quite complex and may be divided into several different structures.

The *peripheral* nervous system comprises those neurons or parts of neurons that lie outside the bony case formed by the skull and the spine. The somatic nerves of this system mediate the sensory inputs and muscle movements that we are consciously aware of during waking hours.

The *autonomic* part of the peripheral system regulates many functions — such as the heart rate, blood pressure, endocrine and digestive processes of which we are not normally conscious, but which can be brought under conscious control through bio-feedback and yoga techniques. The *sympathetic* aspect of the autonomic system generally comes into play when we experience strong emotions, while the *parasympathetic* system tends to be active when we are calm and relaxed. The cell bodies of the autonomic nervous system, as well as of the sensory nerves of the somatic system, gather together in *ganglia* alongside the spinal column, and at other points in the body. The cell bodies of somatic motor-nerve fibers, however, are located inside the central nervous system.

The central nervous system is organized into two principle parts, the spinal cord and the brain. The spinal cord serves as a conduction path to and from the brain and also as an organ for effecting reflex action. The brain seems to play an important role in all the complex activities constituting consciousness — thinking, perception, learning, memory, etc. The three main structures of the brain are known as the *hindbrain*, the *midbrain*, and the *forebrain*.

Within the hindbrain lie the *cerebellum*, the *pons*, and the *medulla*. These neural centers regulate breathing, heartbeat, motor coordination, posture, and balance. They are also involved in mediating nerve impulses from the body to the higher brain centers.

The midbrain contains numerous nerve fiber tracts and neural centers regulating body changes in response to visual and auditory stimulation.

The forebrain has reached its greatest development in humans and other highly evolved animals, such as porpoises. It comprises the *cerebrum*, which is covered by the *cerebral cortex*, the *thalamus*, and a group of closely related structures forming the *limbic system*. These parts of the brain mediate our inner mental and emotional processes.

The sensations in your mind are mapped out on the cerebral cortex of your brain, which mediates your conscious sensory and motor functions, as well as complex perceptual processes.

One method of researching cerebral functioning has been to electrically stimulate the exposed cortex of human subjects, under local anesthesia, who could then report on their experiences.[5] By stimulating certain areas, various types of sensations, movements and thought patterns can be evoked. Another method of research is to observe the functioning of individuals who have had portions of their brain removed or damaged. Especially in the case of young children, removing a portion of the brain does not seem to impair the functioning of the mind.[6]

One important line of research has indicated that the two hemispheres of the cerebral cortex function differently. The speech areas of the human cortex are almost always located on the left hemisphere, regardless of whether the person is right- or left-handed. Several researchers have suggested that the mind's logical and linear functions are associated with the left hemisphere, while the more kinesthetic, preverbal, intuitive properties of consciousness derive from the right hemisphere.[7] The particular functions each hemisphere assumes may vary with different individuals. However, the capacity for two uniquely different modes of consciousness within each individual seems well established. Important differences also seem to exist between the intellectual cortex and other deeper, emotional layers of the forebrain.[8]

SENSORY-MOTOR MAPPING ON THE HUMAN CORTEX.

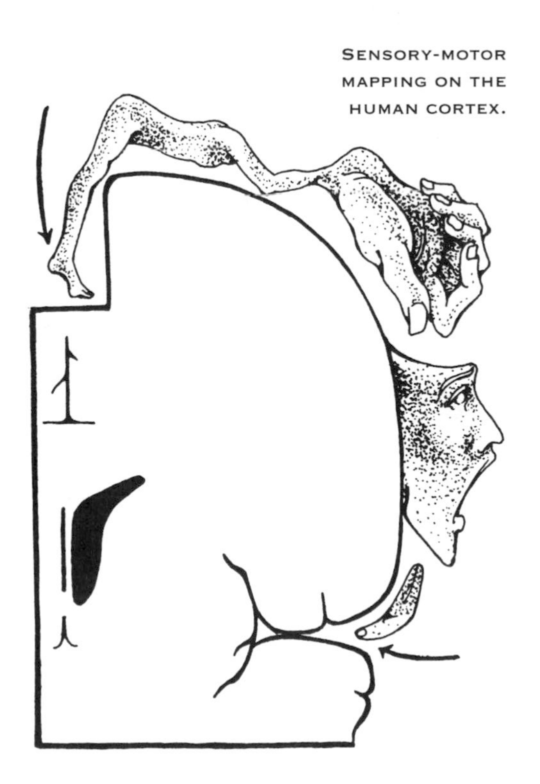

Those parts of the brain most attuned to the body's needs and emotional states are the *limbic system* and the *hypothalamus.* The hypothalamus is a bundle of nerve bodies, about the size of a peanut, located just above the roof of the mouth. It contains several centers that mediate the excitement and inhibition of the hunger, thirst, and sexual drives, as well as emotional arousal. The activity of these centers is in turn regulated by such factors as hormones in the blood and signals from other parts of the brain, including the cortex. Certain areas in the hypothalamus and limbic system, when stimulated, can be a source of enormous pleasure for the body.

In conjunction with the *reticular activating system,* the hypothalamus is also involved in the mediation of sleep and arousal states.

By attaching electrodes to the skin of the head, psychologists are able to measure the electrical activity of the brain as a whole. Brain waves thus measured can generally be correlated with different states of consciousness ranging from the alert waking state, to drowsiness, hypnagogic imagery, meditation, sleep, and dreaming. Individuals can learn to control their brain waves, and also their internal states of consciousness, through techniques providing them with immediate feedback on their physiological state. Researchers suggest that there may be no biological functions that cannot be brought under conscious control in this fashion.[9] Many individuals are able to develop this control through simple techniques of yoga, hypnosis, and meditation.

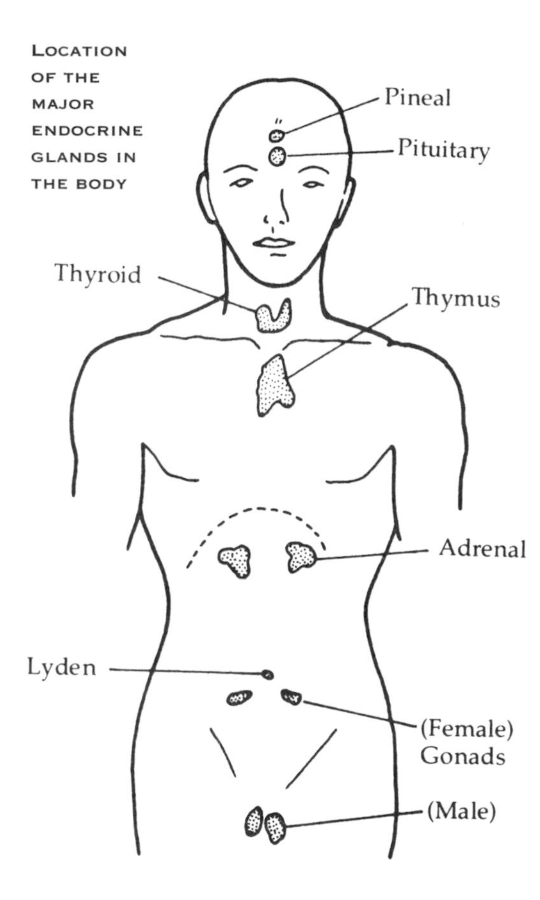

LOCATION OF THE MAJOR ENDOCRINE GLANDS IN THE BODY

The Endocrine System

The *endocrine* system, which comprises glands secreting powerful hormones into the bloodstream, is one of the most interesting areas of autonomic functioning. Our personality and character is profoundly affected by our hormone balance. The major endocrine glands are the *pituitary* and *pineal* glands in the brain, the *thyroid* and *parathyroids* in the throat, the *thymus* gland located near the heart, the *adrenal* glands, and the *sexual* glands. To a lesser degree, other parts of the body, including neurons, also secrete hormones into the bloodstream. The endocrine system is self-regulating in that hormone secretion from any gland is activated in part by other hormones in the bloodstream. The hypothalamus also plays an important role in stimulating certain hormone secretions from the pituitary gland.

The pituitary is often called the "master gland" because it secretes a number of hormones that stimulate or inhibit secretion in the other glands of the body. It also produces hormones that regulate the growth rate of children and awaken the sexual glands at puberty.

The pineal gland produces several substances including a hormone known as *5-hydroxytryptamine* or *serotonin.* Serotonin is of the same chemical series of indole alkaloids that includes psychedelic drugs such as LSD-25, psilocybin, DMT, and *bufotenine.*[10,11] The exact mechanism by which serotonin might affect consciousness or behavior is not well understood by scientists today. Research findings are paradoxical as serotonin is known to affect different parts of the body and brain in

LEFT, SEROTONIN; *RIGHT,* PSILOCYBIN.

different ways, depending on the proportions and combinations of other hormones and enzymes present during the interaction. Generally speaking, serotonin is recognized as a neural inhibitor in the brain. The stores of serotonin in the brain are depleted by *reserpine*, a tranquilizer, and augmented by *iproniazid*, a mood elevator.[12] Large amounts are present in the limbic system and the hypothalamus. Smaller concentrations occur in the cortex and the cerebellum. Ablation of the nerve network in the brain called the *raphe* system, which contains considerable amounts of serotonin, is known to produce permanent insomnia. The ingestion of serotonin is unlikely to affect the central nervous system as it does not cross the blood-brain barrier. If it did, its main result would be to put one to sleep. Most of the serotonin in the brain is in the reticular activating system where it plays an important role in the sleep-wake cycle. When serotonin levels in the r.a.s. rise, the brain goes into deep sleep.[13] Other studies have shown greatly increased amounts of serotonin in the brains of psychotic patients. According to biologist John Bleibtrau, "Bananas and plums abound in serotonin; so do figs, and among species of figs none is richer in serotonin than the *ficus religiosa*, known in India as the Bo tree, under which the Buddha reportedly sat when he became enlightened."[14] Thus, the hormone produced by the pineal gland makes possible emotions, perception, sleep and wakefulness, and orientation to conventional reality.

The thyroid gland produces a hormone known as *thyroxin*, which controls the metabolic rate at which the body produces energy. Whether a person is slow and sluggish or extremely active is influenced by this hormone. (Occult systems often associate this gland with the throat chakra).

The hormones produced in the thymus gland regulate the process by which the body learns to differentiate its own proteins from foreign substances which may be harmful to it. By this process *antibodies* are manufactured that react only against invading *antigens* and not to the myriad similar substances necessary to the body. One could think of the thymus gland as being closely related to the body's sense of organic identity.

The adrenal glands, located in the back of the body above the kidneys, secrete the hormones *epinephrine* and *norepinephrine*, which are related to states of strong emotion. The sympathetic nervous system can stimulate the adrenal glands and the action of the adrenal hormones produced generally intensifies the actions of the sympathetic system throughout the body. It helps mobilize sugar into the blood and makes more energy available to the brain and muscles. It stimulates the heart to beat faster and also constricts the peripheral blood vessels, thus raising blood pressure.

The sex glands or gonads are the testes in men and the ovaries in women. The hormones they produce are responsible for the physical changes that take place during puberty — the onset of menstruation, growth of the breasts, voice changes and beard and body hair growth.

It is important to recognize that the complex activity of manufacturing the hormones and enzymes, which regulate both neural transmission and the endocrine system, is guided by the subtle programming coded into the genetic structure of each cell in the body. One can view these three modes of physiological functioning as communication systems. Neural transmission provides rapid communication for the whole body — requiring fractions of a second for feedback. The endocrine system provides inter-organ, slow communication — requiring minutes to hours for feedback. While the genetic structure can be seen as an organism-environment communication system requiring many generations for feedback.

It is recognized that manufacturing protein substances within the cells is guided by the DNA codes; however, scientists have yet to find a satisfactory explanation for the development of tissues, organs, and whole organisms.[15]

Melanin: The Organizing Molecule

Building on the "reflexive universe" model of Arthur M. Young (to be presented at the end of Section IV), physician Frank Barr hypothesizes that *neuromelanin*, a complex category of light-and-sound-absorbing molecules, is responsible for our experience of a continuum of mental states.[16] It is the molecule, he claims, that coordinates interactions between the endocrine and nervous systems. Barr summarizes his theory:

> *Neuromelanin* — through 1) its photon-phonon-(exciton)-(soliton) interactions; 2) its semi- (and possibly super-) conductive capacities; 3) its cation exchange flow; 4) its continuous free radical signal; 5) its neuroglial direct current; 6) its potentially diverse covalent modifications; 7) its potential to trigger reversible enzyme cascade amplifications; etc. — could precisely regulate the neuroendocrine system. By meticulous phase-timing, neuromelanin could coordinate the synthesis, release, uptake, destruction, modification, and/or recycling of the various neuroamines and peptides throughout the brain.[17]

The Temporal Lobe Factor in Psychic Experience

Psychologist Michael A. Persinger of Laurentian University in Canada states that, whether psi experiences are real or imagined, the *temporal lobes* of the brain play a significant role in mediating such experiences. Deep within the temporal lobes are the *mesiobasal* structures, specifically the *hippocampus* (often referred to as the gateway to memory) and the more anterior, *amygdala* (the mediator of affect and meaning).

The temporal lobes have diverse structures and multiple functions including memory, the sense of self in space and time, the attribution of meaning and emotional significance, audition, organization of complex visual patterns, smell, and language.

Persinger suggests that psi information signals are carried on extremely low electromagnetic frequencies to which temporal lobe structures are sensitive. He describes his approach to understanding psychic functioning in the temporal lobes:

> The deep structures of the temporal lobes are the most electrically unstable portions of the human brain. This instability is really a sensitivity, due to the microcircuitry of the neurons; it allows the phenomena of declarative memory and its consolidation to occur.[18] However, there are consequences to this sensitivity. The temporal lobe structures are prone to electrically active foci....Local and paroxysmal discharges can even be produced by specific memories and biofrequency (extremely low frequency) magnetic fields that penetrate brain tissue.
>
> The contribution of temporal lobe processes to psi phenomena have two important implications. Firstly, the phenomenological characteristics of psi experiences, especially spontaneous ones, should be dominated by the functions of the temporal lobes. Such evidence is clearly seen in the propensity for spontaneous psi experiences to involve visuoauditory modalities, dreams (modulated via the hippocampus), and intense affect (the amygdala) that attributes the experience with intense, personal meaningfulness.[19] Secondly, the electrical lability means that many other stimuli could both compete for neural substrates that facilitate psi experiences and stimulate psi-like experiences, that is generate pseudo-psi or quasi-psi.[20]

Persinger also notes that no other brain

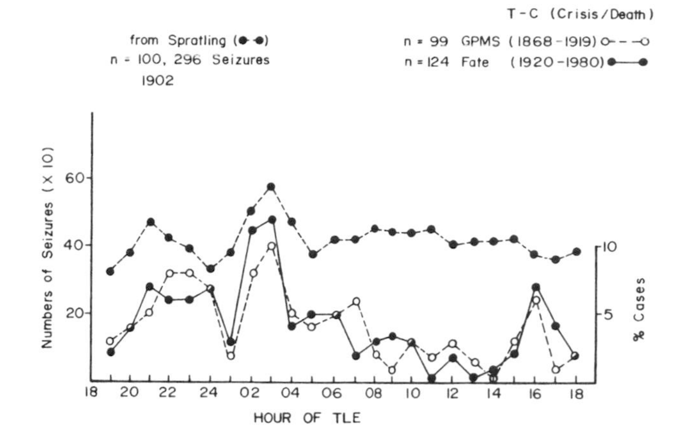

HOURLY DISTRIBUTIONS OF THE NUMBERS (LEFT VERTICAL AXIS) OF EPILEPTIC SEIZURES (MINIMALLY INFLUENCED BY MEDICATION) COMPARED WITH THE PERCENT HOURLY DISTRIBUTION (RIGHT VERTICAL AXIS) OF TELEPATHIC-CLAIRVOYANT EXPERIENCES CONCERNING DEATH OR CRISIS TO SIGNIFICANT OTHERS. TWO SEPARATE TYPES OF CASES ARE COMPARED: THE GURNEY-PODMORE-MYERS-SIDGWICK COLLECTIONS (GPMS; VERIFIED) AND THE FATE REPORTS (UNVERIFIED.)

condition simulates spontaneous psi experiences as closely as temporal lobe epilepsy. This disorder is associated with brief paroxysmal electrical discharges within the mesiobasal regions of the temporal lobe. If the discharge remains within one lobe and does not propagate to motor regions, there are no convulsions. An observer might not realize that the person is experiencing a seizure.

However, there are often experiential phenomena that are associated with such discharges which resemble the major manifestations of spontaneous psi experiences. These include deja vu, depersonalization, out-of-body types of experiences, a sense of a presence, time distortions, an internal "hearing and knowing," anxiety or panic, experiences of floating or falling, shapes in the peripheral visual field (especially the upper quadrant), and complex visual "hallucinations."[21] Electrical stimulation studies have demonstrated that these experiences are specific to temporal lobe structures.[22,23]

People who have chronic electrical discharges within temporal lobe structures also develop a behavioral pattern which overlaps with the profile of persons interested in psychic and "new age" matters.[24] These patterns include: a widening of affect, such that unusual events acquire special personal meaning; an interest in philosophy and mysticism; a sense of personal destiny; episodes of delusions; and a desire either to record one's experiences or to communicate one's beliefs.[25]

Following up on his interest in geomagnetic effects upon consciousness (which will be discussed further), Persinger has assembled a body of data suggesting a marked similarity between the diurnal distribution of limbic epilepsy and psi experiences. The number of temporal lobe seizures (with observable motor activity) were plotted for each one hour interval from a population of about 100,000 events collected before anticonvulsants were introduced into medicine.[26] Seizures were most prominent between 0200 and 0400 hours local time, with a secondary peak around 2200 hours.

For comparison, note the percentage of total cases per hour for all of the histories of spontaneous telepathy concerning death and crises to significant others from the Society for Psychical Research collections that contained the hour of the occurrence (open circles).[27,28] In addition, Persinger collected similar cases that contained this information as reported in *Fate* magazine. A statistical analysis demonstrated no significant difference between the well-documented SPR collections and the less-documented *Fate* cases — suggesting the possibility of a similar mechanism affecting their occurrence.[29]

Peak displays of spontaneous experiences concerning death and crises to significant others occurred between 0200 and 0400 hrs, with a secondary peak around 2100 to 2300 hours. However, unlike the epileptic events, there was increased incidence of ostensible psi experiences around 1600 hours.

The partial similarity of the hourly distribution of the incidence of both epileptic episodes and ostensible psi experiences is an example of the commonality of the two phenomena. They appear to exist along a continuum of temporal lobe lability or sensitivity. They may both involve local microseizuring that generate experiential phenomena without overt motoric displays. However, Persinger claims that "it would be incorrect to assume that psi experiences are a form of limbic epilepsy."[30] One must also take into account that normal microseizuring occurs every night, during the dream or REM (rapid eye movement) state.[31] The most important difference from the perspective of psi research, of course, is the trigger that evokes the experience.

Persinger has verified the existence of a temporal lobe continuum of activity in normal individuals who show no signs of epilepsy or abnormal personality.[32] The more frequent the number of temporal lobe signs a person reports, Persinger suggests, the more likely he is to report spontaneous psi experiences and to score well in laboratory tests of psi.[33]

ALBERT P. KRUEGER, EMERITUS PROFESSOR OF BIOLOGY, UNIVERSITY OF CALIFORNIA, BERKELEY.

The Ecology of Consciousness

An interesting new area of science concerns electrostatic interactions between biological organisms and the environment. I have already indicated that the electrochemical nature of neural transmission plays an important role in mediating information-transfer throughout the body. Now we will take a look at some of the more subtle extensions of our biological functioning:

Our bodies are influenced — in ways often overlooked — by the existence of small ions in the atmosphere. The research of scientists such as Albert P. Krueger are sometimes dismissed as insignificant in the face of gross environmental pollution; however, they seem to show important implications for consciousness:

> Air ion formation begins when enough energy acts on a gaseous molecule to eject an electron. Most of this energy comes from radioactive substances in the Earth's crust, and some from cosmic rays. The displaced electron attaches itself to an adjacent molecule, which becomes a negative ion, the original molecule then becoming a positive ion...natural gas or water molecules cluster about the ions to form small air ions of four types: $H+(H_2O)n$, $(HaO)+(H_2O)n$, $O_2(H_2O)n$ and $OH-(H_2O)n$, where *N* is a small number.
>
> In normal clean air over land, there are 1500 to 4000 ions/cubic centimeter. But negative ions are more mobile and the

earth's surface has a negative charge, so negative ions are repelled from the earth's surface. Thus the normal ratio of negative to positive ions is 1.2 to l.

Man often encounters very low concentrations of ions, and modern city life increases the ratio of positive to negative small air ions. A 14-day study in 1971 by B. Maczynski (*lnt. J. Biometeor*, vol. 15, p. 11) in an office containing four people showed that the small air ion concentration dropped as the day went on, falling on average to only 34 positive ions and 20 negative ions/cm^3. And a test at a light industry area of San Francisco by J. C. Beckett (*J. Amer. Soc. Heating, Refrig, and Air Cond.*, vol. 1, p. 47) showed a small ion count of less than 80 ions/cm^3. In both cases the number of physiologically inert large ions rose considerably — apparently small ions react with dust and pollutants to form large ions.

People travelling to work in polluted air, spending eight hours a day in offices or factories, and living their leisure hours in urban dwellings, inescapably breathe ion-depleted air for substantial portions of their lives. There is increasing evidence that this ion depletion leads to discomfort, enervation and lassitude, and loss of mental and physical efficiency. This syndrome appears to develop quite apart from the direct toxic effects of the usual atmospheric pollutants. It occurs in the absence of such pollutants, in the "clean" air of rural schools or libraries which happen to be ion-depleted due to special factors which remove ions, such as stray electrical fields. On the other hand, evidence is accumulating that substantial increases in ions can have highly beneficial effects, from relieving the pain of burns to promoting plant growth.[34]

Experiments have shown that negative ions promote the healing rate of animals with severed peripheral nerves, skin lacerations, burns, and post-operative discomfort. They are known to greatly enhance cell proliferation, and under certain circumstances they are known to raise the critical fusion frequency threshold (the point at which a flickering light appears constant) in humans and decrease visual reaction time.[35]

In several instances both positive and negative ions are shown to have similar effects. High doses of either type of ion have been shown to be lethal to bacteria. High densities of negative or positive ions increase, on the other hand, the maze-learning ability of rats. Low concentrations of positive and negative ions are known to produce fewer alpha frequency brain waves in human beings. High concentrations of ions tend to disrupt alpha frequencies in a more variable fashion. In rats, varying outputs of ions in either polarity will produce measurable changes in urine, defecation, sleeping period, respiration rate, and attacks on the aluminum foil ground plate used to generate the ions. In general, oddly enough, the lowest ion concentrations were the most effective in evoking (or provoking) such changes.[36]

Particularly interesting is Kreuger's demonstration of the effects small air ions have on the levels of serotonin in the blood and in the brain. He has shown that in mice positive ions raise blood levels of serotonin, and negative ions depress them.[37] In these rodents' brains, low dosages as well as high dosages of both negative and positive ions produced significant decreases in serotonin — as compared to normal atmospheric levels.[38] This disparity can be accounted for by the fact that serotonin does not cross the blood-brain barrier. (You will recall the important role brain-serotonin plays in mediating many facets of consciousness.) Negative ions are also known to play a role in speeding up plant growth and in increasing resistance to influenza.[39]

Research from Israel dramatically illustrates the link between atmospheric ionization, physiological levels of serotonin, and consciousness. In many parts of the world, observers have noted that certain "winds of ill repute" have a discomforting effect upon individuals — the Santa Ana winds in Southern California, the Chinook winds in Canada, the Mistral winds of France, the Zonda winds of Argentina, the Sirocco winds of Italy, and the Sharav or Chamsin winds of the Near East.[40] Symptoms such as sleeplessness, irritability, tension, migraines, nausea and vomiting, scotoma (diminished vision), amblyopia (dimness of vision), and edemata (swelling of tissue) have been noted. These symptoms resemble the effects of hyper-production of serotonin. In weather-sensitive people, urinary serotonin output showed a steep rise two days before the onset of the Sharav winds in Israel. They remained high the following day and dropped only after the winds began. In addition to increase in positive ionization, the salient meteorological features of these winds are a rapid rise in temperature and a decrease in humidity. These factors by themselves, however, fail to account for the physiological changes noted. The negative psychological and physiological effects are attributed to the rise in the ratio of positively charged ions in the atmosphere preceding the onset of the winds.[41] It is interesting to note in this connection that the word doldrums has two dictionary meanings: (1) dullness; a state of listlessness and boredom, (2) a part of the ocean near the equator abounding in calms, light winds, and squalls.

On the other hand, in locations where (-) air ion densities are relatively high, such as near waterfalls, the general effect of the environment is tranquilizing and conducive to good health.[42] It is no wonder then that scientists in the know, such as Dr. Albert Krueger in Berkeley, use air filters and negative ion generators to restore the environment around them to its natural unpolluted and electrostatically balanced state.[43]

Stepping into Krueger's laboratory in the Life Science Building at the University of California, Berkeley, and breathing deeply was like all of a sudden being out in the crisp, clean air of a mountain wilderness.

Closely related to the electrostatic and ionic phenomena of the biosphere are electromagnetic phenomena that also play an important role in the ecology of consciousness.

The magnetic field of the earth extends around the planet like a large donut and is probably created by the flow of molten metals in the earth's core.[44] The average intensity of this field is about 0.5 gauss and it pulses at frequencies ranging from 0.1 to 100 cycles per second. The predominant frequency range of magnetic pulsations, known as the Schumann resonance, is around 7.5 cycles per second. Several researchers have suggested that this resonance in the geomagnetic and electrostatic field has an effect upon the human nervous system — and upon consciousness itself.[45]

The Schumann resonance is an effect due to the fact that an electromagnetic wave (traveling at the speed of light, 186,000 miles a second) goes around the earth's 25,000-mile circumference around 7.5 times a second. Perhaps it is useful to think of the 7.5 c.p.s. brain wave frequency as the boundary between alpha waves and theta waves. If that frequency predominates in your brain waves, you are generally in the hypnogogic or hypnopompic state just on the border of waking up or falling asleep. The theta wave is frequently observed in the EEG patterns of experienced meditators, who must pass through the Schumann resonance portal without falling asleep.

The field of the earth is about a thousand times weaker than the field from a small horseshoe magnet. The reported effects of such weak magnetic fields include altered cellular reproduction, plant growth and germination, orientation to direction, amplitude of motor activity, and enzyme activity. Of particular interest is the work of Dull and Dull, which showed a striking correlation between incidents of human illness and death during periods of sharp geomagnetic disturbances (such disturbances are often related to solar-storm activity).[46] Another study conducted by Robert Becker and his associates at the Veterans Administration Hospital in Syracuse, New York, showed a positive correlation between days of geomagnetic intensity and the number of persons admitted to a psychiatric hospital.[47]

Professor Michael Persinger, of the Psychophysiology Laboratory at Laurentian University, hypothesizes that the extremely low frequency (ELF) Schumann waves may serve as a carrier for psi information. He points out the near impossibility of shielding against such waves, requiring no less than "an underground bunker surrounded by several inches of steel."

Noting that ELF waves propagate more easily from midnight to 4:00 A.M., and that they are easier to transmit from west to east rather than east to west, Persinger surveyed the ESP literature for any correlations. His findings were as he predicted. Telepathy and clairvoyance do show a tendency to peak roughly between midnight and 4:00 A.M. There is also a slight tendency for the telepathic agent to be west of the percipient rather than to the east. To clinch his argument, Persinger observes that fewer psi experiences are reported during periods of geomagnetic disturbance. Such disturbances also impair the propagation of ELF waves.[48,49]

Several investigators have shown that humans are sensitive to slight variations of magnetic intensity. Once accustomed to distinguish between the presence and absence of a weak magnetic field, subjects in several experiments were asked to walk back and forth over a given area without knowing whether an artificial magnetic field had been activated. Under these conditions, the subjects were extremely accurate in guessing whether the current was in operation.[50,51] This sensitivity is offered as a partial explanation for the effectiveness of dowsers in finding water:

> Water filtering through porous media produces electric currents through electrofiltration potential and concentration batteries. If the medium is sufficiently conducting, and the current of the soil is sufficiently high, then there exists at the surface of the soil a small magnetic anomaly.[52]

The precise channels by which the human body detects magnetism are still a matter of speculation. However, we know that most biological processes are based on chemical interactions, which can be accounted for, in the last resort, by the interactions of atomic nuclei and electrons. In one study with dowsers, using strict experimental controls and a double blind,[53] weak magnetic fields were shown to cause measurable changes in the electrical skin potential.[54]

Another study was conducted in which future astronauts spent up to ten days in a special chamber free of magnetic fields. During this time, no serious psychological or physiological deviations were reported — although some of the findings have remained classified. It was found, however, that the subjective perception of general brightness was lower under the non-magnetic condition — thus implying a magnetic effect upon the visual cortex. Soviet studies, in addition, have determined that weak magnetic fields can effect the direction-finding orientation of birds, fish, and insects.[55,56,57] Research with honey bees shows that they are sensitive to fields of one gamma, i.e., several thousand times weaker than the earth's 1/2 gauss

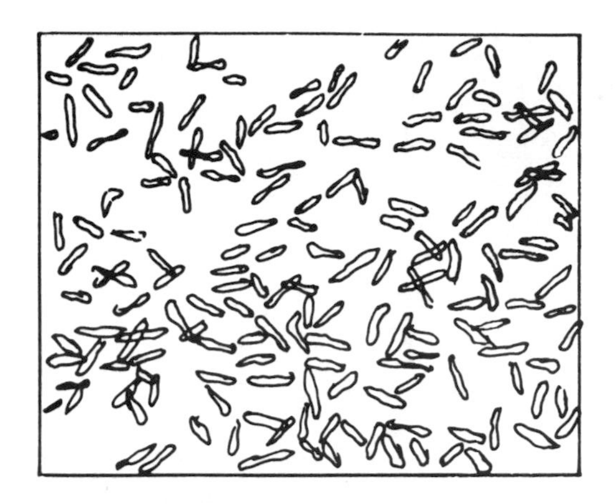

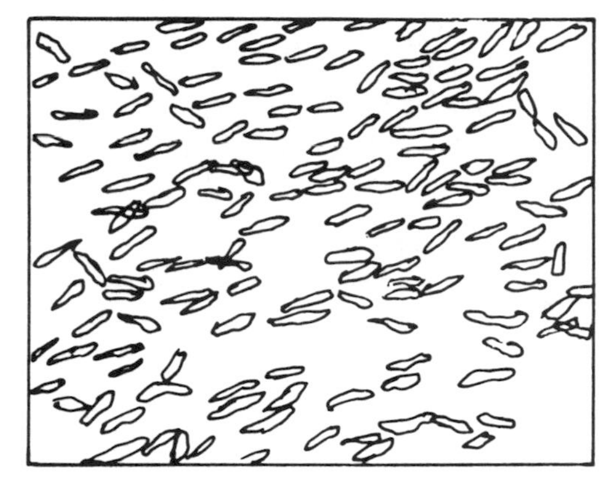

ABOVE: RANDOM MOTION OF UNICELLULAR ORGANISMS IN ABSENCE OF AN ELECTROMAGNETIC FIELD.
BELOW: ORIENTED MOVEMENT OF UNICELLULAR ORGANISMS DUE TO ELECTROMAGNETIC FIELD OF 5-7 MHZ. MOTION IS PARALLEL TO ELECTRIC LINES OF FORCE. (AFTER PRESSMAN, *ELECTROMAGNETIC FIELDS AND LIFE,* P. 163)

field. Homing pigeons may rival honey bees in sensitivity. Other studies have shown that germs and viruses are sensitive to the slightest departure of the earth's magnetic field from the average — this is reflected in reproduction rates and in genetic changes. For example, exposure to magnetic fields causes resistance to penicillin in certain strains.[58]

Sister M. Justa Smith, Ph.D., a biochemist associated with the Roswell Park Memorial Institute in Buffalo, New York, has shown that strong magnetic fields affect the reactivity of certain enzymes in the human body. These enzymes can act as a catalyst to speed up the body's natural healing processes; and, in fact, Sister Smith observed that psychic healers do exert a non-magnetic effect on the enzyme similar to the magnetic field.[59] Such studies have left scientists with a firm conviction that magnetic fields play an important role in the body's healing and immunological processes.[60]

The world map shows the variations in the intensity of the earth's geomagnetic field. Movement of high and low centers varies very slowly with time — the rate of this movement is measured in feet per year. The center of lowest magnetic intensity on the planet (25 gauss) is in Brazil right over Rio de Janeiro.[61]

(In terms of psychic consciousness, it is interesting to note that Spiritism has flourished in Brazil, in spite of opposition from the Catholic Church, perhaps more than in any other nation. Brazilian spiritists, synthesizing modern European, native Indian, and African culture, number over a third of Brazil's population and comprise powerful interest groups with their own elected representatives in the national legislature. There are entire towns in Brazil composed solely of spiritists.)[62]

The areas of greatest geomagnetic intensity center near the poles where readings are found in the .60-.70 gauss range. Spacecraft at the altitudes and latitudes of the usual near-earth orbits are generally not exposed to magnetic fields lower than those in Brazil. However, spaceflights more than about one-sixth the distance to the moon enter a magnetic environment near-zero in intensity.[63] It is still uncertain precisely how these variations of magnetic field will affect the consciousness of astronauts, as scientists are just beginning to explore the interactions of electromagnetism on the mind and body.

For nearly thirty years doctors in Austria,

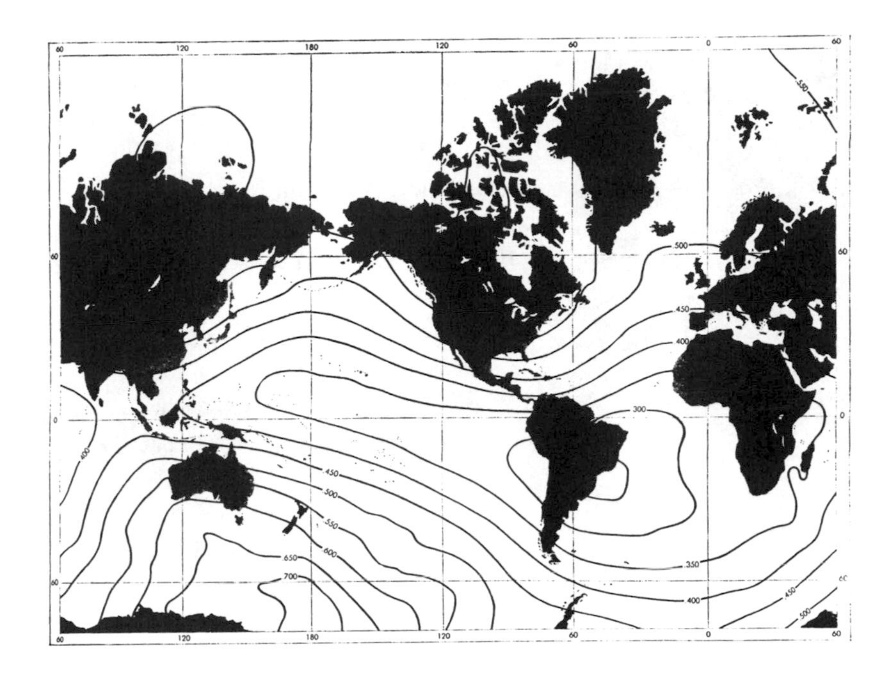

From "The Airborne Magnetometer," by Homer Jenses, Copyright © June 1961 by Scientific American, Inc. All rights reserved.

West Germany, and the former Soviet Union have used a therapeutic technique known as *electrosleep* to cure a wide variety of psychological problems related to insomnia. A weak electric current (just enough to cause a tingling sensation) is passed through the head by attaching electrodes over the closed eyes and over the mastoid process (behind the ears). This induces an altered state of consciousness, and eventually sleep.

More than five hundred articles about electrosleep have been published in the Russian literature, and a number of sophisticated studies in Western Europe have produced evidence that the therapeutic process is effective. However, American clinicians have remained skeptical about all electronic therapeutic processes, which have long been associated with medical quackery. (The unfortunate exception to this is *electroshock therapy* where powerful current — 70 to 130 volts — jolts through a patient's brain causing convulsions, memory loss, temporary relief of depression and other symptoms. No one is sure how or why it works.)

In the last few years, American researchers have shown a new interest in electrosleep. A number of favorable research papers have been presented using electrosleep with humans and animals. Improvements have been shown in cases of insomnia as well as in removing neurotic and psychotic symptoms.[64] The exact mechanisms are still unknown, but it is quite clear, as we have already pointed out, that electromagnetic brain fluctuations are involved in the basic rest and activity cycle.

The problem of bio-electromagnetic interactions is much more intrinsic than the comparatively simple question of brain activity. The enormous role that light plays in our daily lives is so obvious that we ordinarily overlook it. The most dramatic responses to light can be observed in plants, upon which we are dependent for oxygen and nutrition. The Swedish naturalist Carolus Linnaeus (1707-1780) first noticed that various flowers opened at different hours and could actually be used as a clock.

SIR JOHN ECCLES, NOBEL LAUREATE NEUROPHYSIOLOGIST AND MIND-BODY DUALIST

LINNAEUS FLOWER CLOCK:

6 A.M.	Spotted Cat's Ear opens
7 A.M.	African Marigold opens
8 A.M.	Mouse Ear Hawkweed opens
9 A.M.	Prickly Sowthistle closes
10 A.M.	Common Nipple Wort closes
11 A.M.	Star of Bethlehem opens
noon	Passion Flower opens
1 P.M.	Childing Pink closes
2 P.M.	Scarlet Pimpernel closes
3 P.M.	Hawkbit closes
4 P.M.	Small Bindweed closes
5 P.M.	White Water Lily closes
6 P.M.	Evening Primrose opens

In nineteenth-century Europe, formal gardens were sometimes planted to form a clockface, with flowers in each bed blossoming at a different hour. One could tell the time to within a half hour by glancing at the garden.[65]

We wake and sleep according to cycles of light and darkness. Furthermore, our adrenal hormones, pineal hormones (such as serotonin), and our sexual hormones all follow a twenty-four hour circadian production cycle which changes with the seasons according to the amount of available sunlight. Reflect for a moment on how much your consciousness is affected by sunlight and artificial light in your environment in a church or temple...in the forest on a bright afternoon...in the moonlight...by the flickering firelight a lamplit room...just after sunset...or in the dark.[66] One of the things I love to do is get up early in the morning, several hours before sunrise. From a hilltop, I can silently watch the gentle conquest of darkness as the earth turns and the birds, insects and the hormones flowing in my own blood are all part of the music — the planetary rotation raga. (The Hindu musicians understood this perfectly well when they composed music to be played at different times of day.)

In Robert O. Becker's opinion, electromagnetic fields have enormous implications for understanding consciousness. He suggests that the analog-synaptic aspect of the central nervous system is regulated in part by electromagnetic interaction with the environment.[67] His research relating geomagnetic disturbances to psychiatric admission rates has already been cited. In other studies he has indicated that geomagnetic disturbances affect the behavior of patients on a psychiatric ward, and that magnetic fields also have an effect on human reaction time.[68,69]

CHALLENGES TO THE BIOLOGICAL IDENTITY MODE

Ever since its eloquent expression in the philosophy of Rene Descartes, dualism has been a feature of Western philosophy and cultural thought. While most physiologists implicitly subscribe to the materialistic, biological indentity model of consciousness, many of the most prominent members of the field have opted for a clean-cut dualism. Wilder Penfield, the Canadian neurosurgeon whose experiments of electrical stimulation of the brain were instrumental in developing our knowledge of cortical functioning, ended a renowned scientific career by renouncing the biological identity principle:

> In the end I conclude that there is no good evidence, in spite of new methods, such as the employment of stimulating electrodes, the study of conscious patients and the analysis of epileptic attacks, that the brain alone can carry out the work that the mind does. I conclude that it is easier to rationalize man's being on the basis of two elements than on the basis of one.[70]

Some neurophysiologists such as Sherrington, Eccles, and Sperry have proceeded further in stating that mind can act on brain directly.[71,72,73] They have not specified, however, what they mean by mind, nor by what mechanism mental organization can influence brain function. This is the basic problem of dualism. Nevertheless, support for the dualistic position has come from the logician and philosopher of science, Karl Popper, who summarizes the crux of the argument against a materialistic biological identity model:

> [Materialists suggest] that consciousness is nothing but inner perception, perception of a second order, or perception (scanning) of an activity of the brain by other parts of the brain. But [they] skip and skim over the problem why this scanning should produce consciousness or awareness, in the sense in which all of us are acquainted with consciousness or awareness; for example, with the conscious, critical assessment of a solution to a problem. And he never goes into the problem of the difference between conscious awareness and physical reality.[74]

The monist materialist can respond — as philosopher Thomas Hobbes did in refuting Descartes' dualism — that there is no reason why matter should *not* be capable of thinking. This formulation is correct as far as it goes. If we conceive of matter vaguely at the start, we cannot deny it the faculty of thought. But this essentially destroys the mechanistic world view; in addition to the classical properties of extension and motion, an entirely different sort of property is now being ascribed to matter. The mechanistic claims of materialism are thereby fundamentally changed, raising severe problems for conventional physical notions.

Some leading physicists have gone even further in their dissolution of the idea of matter. Under the influence of Ernst Mach, a physicist who believed neither in matter nor in atoms, and who proposed a theory

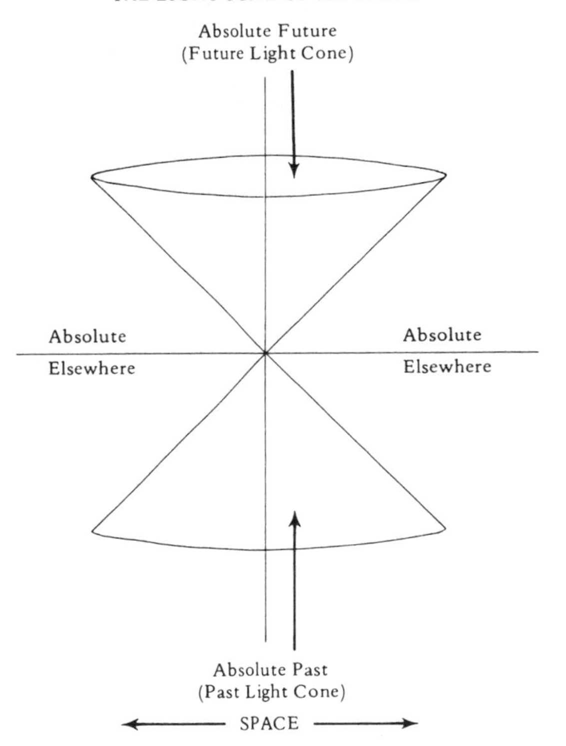

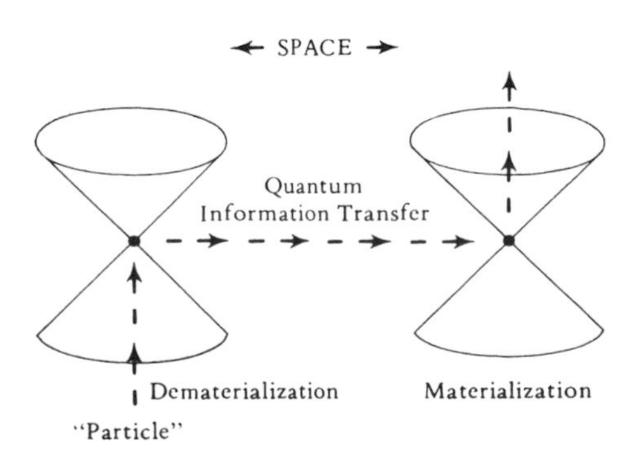

ALBERT EINSTEIN WITH INDIA'S GREAT MYSTICAL POET RABINDRANATH TAGORE. PHOTOGRAPH TAKEN AT A MEETING OF NOBEL LAUREATES, NEW YORK, 1940

of knowledge reminiscent of William James' radical empiricism, idealistic interpretations of quantum mechanics have been put forward. As Bertrand Russell has eloquently stated:

> It has begun to seem that matter, like the Cheshire Cat, is becoming gradually diaphanous until nothing of it is left but the grin, caused presumably, by amusement at those who still think it is there.[75]

CONSCIOUSNESS AND THE NEW PHYSICS

Space-Time According to Einstein

The special theory of relativity, formulated by Albert Einstein in 1905, is based on the experimentally confirmed idea that the velocity of light is the same universal constant, $c = 3\text{x}10^{10}$ cm./sec., for all observers who move uniformly in straight lines relative to each other. Consequently, Einstein's genius deduced that events which are simultaneous to one observer are not simultaneous to a second observer.

Furthermore, moving clocks run slow. Moving measuring sticks contract in length along the direction of motion. Energy is equivalent to mass — i.e., $E = mc^2$. And the mass of a particle increases to infinity as the velocity approaches that of light. Einstein's results have been confirmed many times in physics laboratories.

Like all scientific facts, these results presuppose that the observers are in a common state of consciousness whose legitimacy is determined by their agreement or social contract. The legitimacy accorded any scientific theory is a sociological matter. In fact, one interpretation of quantum physics is that physical reality does not objectively exist independent of the participating observers.

Physicists use a simple geometric picture of the flat spacetime of special relativity called a "Minkowski diagram." Relativity unites space and time into a unified "four dimensional space-time continuum" in which time appears in the distance formula with a sign different from the sign of space. Events are conceived of as points on the Minkowski diagram. The history of a sequence of events is described by a curve or path on the Minkowski diagram called a *world line.* Each event is the origin of a future light cone and a past light cone. World lines that are everywhere inside the light cones are called *time-like* and describe the history of particles moving at velocities less than the velocity of light. World lines that are everywhere on the light cones are called *light-like* and describe the histories of real photons, neutrinos and gravitons that move at exactly the velocity of light. World lines that are everywhere outside the light cones are called *space-like* and would correspond to *tachyonic* processes happening faster than the velocity of light.

Space-like processes, if they exist, could be in two or more widely separated places at the same time. Furthermore, these space-like processes allow the effect to precede the cause for some observers and not for others. They are not allowed in classical physics but are acceptable in quantum physics according to some interpretations. Quantum transitions or "quantum jumps" may be thought of as space-like processes.

Folded Space

Some psi researchers have attempted to use the concept of curved spacetime to eliminate some of the apparent paradoxes involved in psi phenomena. Psychologist Gertrude Schmeidler has suggested that the universe may contain an extra dimension that permits "topological folding" to occur so that two regions which are widely separated in an Einsteinian universe might be in immediate contact, much as two points on a towel which are normally quite a distance apart may be adjacent when the towel is folded.[76] Thus, apparent instances of ESP across great distances might be explained by postulating that the persons involved are somehow in close proximity in the "folded" space.

FRED ALAN WOLF (FROM *PHYSICS AND CONSCIOUSNESS* VIDEOTAPE, COURTESY THINKING ALLOWED PRODUCTIONS)

Physicist John Archibald Wheeler (a man with pronounced antipathy toward psi research) has theorized that, at a microscopic level, quantum effects might tear the fabric of spacetime, producing a structure involving *wormholes.* He speculated that such wormholes could connect pairs of oppositely charged particles such as electrons and positrons.[77] Wheeler's hypothetical structure is sometimes called the "quantum foam." Such wormholes may exist on a macroscopic scale and, in some cases, rotating black holes may give rise to a "tunnel" or shortcut to another region of spacetime. Physicist Fred Alan Wolf has implicitly suggested (in a cartooned text called *Space, Time and Beyond*) that such wormholes may provide the connections needed to explain psi phenomena over long distances or temporal intervals.[78]

Wolf, himself, has become one of the most prolific and articulate writers interpreting the complexities of theoretical physics to a general audience — particularly those interested in psi and consciousness.[79,80,81,82] His book, *Parallel Universes,* is probably the best popular explanation of Everett and Wheeler's "many worlds" interpretation of quantum mechanics.[83]

Multidimensional Spacetime

Multi-dimensional models of spacetime have been proposed by physicist/psi researchers Russell Targ, Harold Puthoff and Edwin May. They propose that ordinary four-dimensional Minkowski spacetime may be the "real" part of an eight-dimensional complex spacetime.[84,85]

And eight-dimensional models of spacetime to account for psi have also been proposed by physicist Elizabeth Rauscher.[86,87,88,89] She suggests that *soliton* waves in a complex multidimensional space might serve as possible psi signals, as they would be able to propagate over large "distances" with little attenuation. She asserts that signals that appear to be superluminal in four-dimensional spacetime may be subluminal in eight-dimensional spacetime. She also contends that the problem of causal loops arising from backward causal chains need not arise in eight-dimensional spacetime. Rauscher suggests that any space-time dependence that exists for psi effects may be accounted for in terms of signal propagation velocities in complex spacetime. However, it is not clear that Rauscher's theory can be tested by this method unless some means of measuring the complex coordinates are provided; otherwise, they simply constitute free parameters that may be adjusted at will, rendering the theory incapable of falsification.

A more comprehensive and sophisticated hyperspace model, developed by Saul-Paul Sirag, is summarized in this section under the heading of "unified field theories" and developed further in the Appendix.

The EPR Effect and Bell's Theorem

Recent theoretical developments in quantum theory known as the EPR effect (named for Einstein, Podolsky, and Rosen's 1935 paper on the quantum connection between spatially separated systems), now formulated in a theorem by John S. Bell (called Bell's Theorem), allow for an instantaneous effect between any two places in the physical universe.[90,91,92] There is no violation of Einstein's theory of relativity because the effect does not require the propagation of energetic signals. The confirmation of this principle of *nonlocality* suggests that psi phenomena, if they exist, need not be in conflict with the established laws of science.

The prejudice of classical causality says that an event can only be influenced by other events that are in its past light cone. Events in the future light cone and outside the light cone in the "absolute elsewhere" are said not to influence the event of interest. Classical causality does work on the statistical level in which we average our observations over sets of events. Almost all of the measurements of atomic physics are adequately described by the statistical limit of the quantum principle.

However, both general relativity and quantum theory in the form of Bell's Theorem show that classical causality is not correct in principle on the level of individual events.[93,94] Recent experiments by John Clauser at the University of California, Berkeley, and Alain Aspect at the University of Paris, show that classical causality is violated for individual atomic events. (Local causes operate within the velocity of light.) These experiments measure the simultaneous arrival of two photons at spatially separated detectors.[95,96] The two photons originate from the same atom. Bell's theorem enables one to calculate what the rate of simultaneous arrival should be if the statistical predictions of quantum theory are correct. It also enables one to calculate the rate of simultaneous arrival if physical reality is objective and locally causal for the individual photons.

The experiments of Clauser and Aspect contradict the rate of photon coincidences predicted on the basis of an objective and locally causal reality. The measured rate agrees with the prediction of ordinary quantum theory. This means that physical reality either is not subject to the principle of local causation or does not objectively exist independent of the observers who participate in its creation.

Bell's Theorem and the related experiments may have importance for the understanding of personal human experience. The human brain stores and processes its information at the level of single organic molecules and is a single macroscopic quantum system. Acts of consciousness may be viewed as incorporating quantum events.

The illusion of the classical scientific paradigm that is shattered by the quantum principle is the assumption that there is an immutable objective reality "out there" that is totally independent of what happens in consciousness "in here." Quantum theory forces a new kind of logic in science that is still mathematical and disciplined. The Nobel prize physicist Eugene Wigner of Princeton has repeatedly written that consciousness is at the root of the quantum measurement problem.[97]

All classical measurements, including classical measurements of quantum processes of the type considered by Heisenberg in his "microscope" that leads to the uncertainty principle, involve the actual flow of energy and momentum in order to convey information. For example, Heisenberg reasons that the position of an electron must be measured by means of a second particle, e.g., a photon, that must collide with the electron in order to get the information on the electron's position. The fact that action is quantized in units of Planck's constant, $h\ 10^{-27}$ erg-sec., implies uncontrollable minimal energy and momentum transfers between photon and electron in the collision. The result of Heisenberg's thought experiment is that it is impossible to predict the simultaneous values

NICK HERBERT (FROM *CONSCIOUSNESS AND QUANTUM REALITY* VIDEOTAPE, COURTESY THINKING ALLOWED PRODUCTIONS)

of both the position and the momentum of the electron with complete certainty. The only way to gain knowledge of the uncertainties is to repeat the experiment many times under "identically prepared" conditions. These kinds of classical measurements of quantum processes are fundamentally statistical.

Josephson proposes that there may be another level of measurement that transcends the limitations of Heisenberg's uncertainty principle.[98] He says that this limitation is perhaps only a "reflection of the kinds of observation we can make," and that "the physical description of the world would change radically if we could observe more things." Einstein was also firmly convinced that there was another way to knowledge, but his refusal to accept the "telepathic" implications that he saw so clearly in his EPR effect prevented him, like Moses, from seeing the promised land. Thus, Einstein's autobiographical notes contain this remark about the EPR effect:

> There is to be a system which at the time t of our observation consists of two partial systems S_1, and S_2, which at this time are spatially separated....If I make a complete measurement of S_1, I get from the results...an entirely definite Ψ-function Ψ_2 of the system S_2. The character of Ψ_2

> then depends upon *what kind* of measurement I undertake on S_1....One can escape from this conclusion only by either assuming that the measurement of S_1 (telepathically) changes the real situation of S_2 or by denying independent real situations as such to things which are spatially separated from each other. Both alternatives appear to me entirely unacceptable.[99]

It is very interesting to note here that the Ψ function referred to by Einstein is the standard quantum probability function, referring to the mathematical probabilities which underlie the subatomic interactions of the physical world (i.e., Schrodinger's Wave Function). At least one physicist has commented on the possible synchronicity that this physical term may be very relevant in the psi effect of consciousness researchers.[100]

Physicists have actually developed a number of possible conceptual strategies for integrating the EPR effect and Bell's Theorem. Physicist Nick Herbert, in his book *Quantum Reality,* describes eight possible interpretations: (1) There is no underlying reality; (2) reality is created by observation; (3) reality is an undivided wholeness; (4) there are actually many worlds; (5) the world obeys a non-human kind of reasoning; (6) the world is made of ordinary objects; (7) consciousness creates reality; (8) unmeasured quantum reality exists only in potential.[101] Each of these interpretations poses its own paradoxes. Given Bell's Theorem and the EPR effect, all of them must allow for nonlocal (or superluminal) interactions.

EVAN HARRIS WALKER

The Implicate Order

The *nonlocal* nature of the state vector collapse, as described above, suggests that particles of matter are not accurately describable as separate, localized entities. Rather, seemingly isolated or separate particles may be intimately connected with one another and must be seen as parts of a higher unity.

Physicist David Bohm has referred to the universe as a "holomovement," invoking an analogy to a hologram (a three-dimensional photograph in which the entire picture is contained in each part). Bohm has termed the world of manifest appearances the "explicate order" and the hidden (nonlocal) reality underlying it the "implicate order." He also proposes a new mode of speaking, which he calls the rheomode, in which "thing" expressions would be replaced by "event" expressions.[102]

In contrast with theories such as Evan Harris Walker's and Saul-Paul Sirag's, the implicate order theory lacks a specific mathematical formulation from which testable predictions may be derived. On the other hand, the implicate order theory is consistent with and provides a good philosophical underpinning for the testable observational theories, such as those of Mattuck and Walker.

Observational Theories

Physicist Evan Harris Walker's *observational theory* equates the conscious mind with the "hidden variables" of quantum theory.[103,104] Walker notes that, due to the necessarily nonlocal nature of such hidden variables, quantum state collapse by the observer should be independent of space and time; hence, psi phenomena such as telepathy should be independent of space-time separation.

Noting that the conventional view in physics is to deny that the paradoxes of quantum mechanics have implications beyond the mathematical formalisms, Walker defines his theory:

> The measurement problem in Quantum Mechanics has existed virtually from the inception of quantum theory. It has engendered a thousand scientific papers in fruitless efforts to resolve the problem. One of the central features of the controversy has been the argument that characteristics of QM imply that an observer's thoughts can affect an objective apparatus directly, which in turn implies the reality not only of consciousness but of psi phenomena. I have written several papers saying that such a feature of QM is not a fault, but rather represents a solution to problems that go beyond the usual purview of physics. Thus, I have developed a theory of consciousness and psi phenomena that arises directly from these bizarre findings in QM, findings now supported by specific tests of the principles of objective reality and/or Einstein locality.[105]

Walker specifies channel capacities for various "regions" of mental activity. He calculates the rate for "dataprocessing of the brain as a whole at a subconscious level" (S) to be equal to 2.4×10^{12} bits/sec. The data rate for conscious activity (C) is equal to 7.5×10^{8} bits/sec., and the channel capacity of the "will" (W) is equal to 6×10^{4} bits/sec.

Walker's derivation of the above rates is based on the assumption that electron tunneling across synapses is the basis for the transmission of impulses across synapses and that the large-scale integration of brain activity is also mediated by electron tunneling.

Copenhagen physicist Richard Mattuck has proposed an observational theory which builds on the work of both Helmut Schmidt and Evan Harris Walker. He asserts that PK results from the restructuring of thermal noise through the action of mind, involving a decrease in entropy. His hypothesis is "not of the 'Maxwell demon' type" as "it does not operate by selection of states of individual molecules, but rather by the selection of *macroscopic* pure states."[106] Using the example of a moving ball, Mattuck notes that, as its velocity is distributed about its current mean due to thermal noise, an observer can select increasingly higher velocity states. This selection may be made in steps, resulting in possible incremental increase in velocity by the ball.[107]

Unified Field Theory and Consciousness

A hyperspace model of consciousness has been developed by interdisciplinary scholar Saul-Paul Sirag, at the Institute for the Study of Consciousness in Berkeley and San Francisco's Parapsychology Research Group. Further details of Sirag's work-in-progress are presented in the Appendix. In my estimation, this work (while incomplete) represents the most advanced model available linking consciousness at a deep level with physical reality. I have been closely associated with Sirag since before he began this work in 1974, when he was a research associate at the Institute for the Study of Consciousness (ISC) in Berkeley. Frankly, after years of detailed discussions with him, I still find it very difficult to comprehend his model. I have included it as an Appendix to the revised edition because I believe that Sirag may well be speaking the language of the future in consciousness research. Here is the story of the development of Sirag's approach:

Arthur Young, the founder of the Institute (whose own "reflexive universe" model is presented next in this section), asked Sirag to work out the algebraic group

SAUL-PAUL SIRAG (FROM CONSCIOUSNESS AND HYPERSPACE VIDEOTAPE, COURTESY THINKING ALLOWED PRODUCTIONS)

structure of the rotations of the tetrahedron. Young also encouraged Sirag to study the works of Sir Arthur Eddington, the physicist who was famous for producing a nearly incomprehensible unified field theory, which purportedly unified gravity and electromagnetism as well as general relativity and quantum mechanics. The key to this unification was also group theory. Sirag was impressed by the fact that, although Eddington's work had been neglected for decades, the central importance of group theory for unified field theory had become established by recent physics.

Eddington's unification was based upon the four-element group called the Klein group K_4. Eddington thought of this group as describing the structure of the most elemental measurement: seeing whether or not two rigid rods are the same length. He regarded group theory as the solution to the mind-matter duality problem. His solution can be stated in this way: Insofar as the mind can know matter, it has a group structure isomorphic to that of matter.

Eddington's "structuralist" approach found support from an unexpected quarter for Sirag when he came upon Piaget's work on the structure of the acquisition of knowledge by children. Eddington had declared K_4 to be the primary group structure of the acquisition of physical knowledge by professional physicists because of his use of K_4 to describe the fundamental structure of measurement. Piaget found, by testing children in precisely contrived situations, that K_4 was also the basic structure of children's acquisition of physical knowledge. Piaget's names for the four elements of K_4 are well known: identify, negation, collaterality and reciprocity.

The problem, for Sirag, was that K_4 as a mathematical group structure had not offered sufficient complexity to capture the richness of theoretical physics since the time of Eddington. He assumed that there had to be a much larger group structure. He was intrigued with the possibility that a larger, finite group structure called S_4 (with subgroup K_4) was the right path to unification of mind and matter. This idea took many years to mature.

In 1977, Sirag published a short piece in the prestigious British science magazine, *Nature*, that was both a criticism of and a tribute to Eddington's mass ratio derivation.[108] Sirag was very impressed by Eddington's use of epistemological principles as a clue to unify gravity and electromagnetism, and his attempt to account for the fundamental pure numbers in physics by purely epistemological reasoning. Eddington's program was too ambitious to be carried out directly, Sirag thought, so as a kind of half-way measure, he tried to reduce the number of pure numbers to be accounted for by judicious combinatorial reasoning. This kind of reasoning led to a rather extensive paper, "Physical Constants as Cosmological Constraints" published in 1983.[109]

In this paper Sirag showed that the physical constants determine the large-scale structure of the universe in such a way that the present-day scale factor — the "radius" can be calculated, as well as the age and the density, and various other cosmological properties. Sirag hypothesized the age of the universe to be thirty-two billion years. This differs markedly from the usual statements of ten to twenty billion years. These numbers are really based on the measurement of Hubble's constant which Sirag has calculated as fifteen kilometers per second per megaparsec (which implies a closed universe), while the usual "measurement" is fifty to one hundred in the same units, implying an open universe. Presumably the Hubble telescope will settle the issue, although it is likely to take many years for scientists to settle the various technical problems necessary before they can agree on a result. (Should Sirag's predictions prove correct, he could be considered a possible Nobel Prize candidate.)

Additionally, Sirag presented a finite-group-algebra unification model in January 1982 at the American Physical Society meeting in San Francisco under the title, "Why There Are Three Fermion Families."[110] This work is particularly significant as physicists have recently confirmed that there are indeed exactly three families of subatomic matter particles, as Sirag had predicted.[111] An Associated Press article on the discovery quotes Nobel Laureate physicist Burton Richter, director of the Stanford Linear Accelerator Center, as saying that the major mystery remaining is "why God chose three families instead of one or nine or 47."[112] Burton had apparently not read Sirag's paper, as this is precisely the issue Sirag has addressed.

In his various published works, Sirag claims to have developed new solutions for some of the most fundamental problems in all of science: the age and size of the universe and the number of basic subatomic building blocks. The predictions which he has made in these areas stand to be either confirmed or refuted in the coming decades. It is from this theoretical work that his mathematical theory of consciousness has emerged. While models of consciousness are far more difficult to verify or falsify than models of the physical universe, the logic of developing a model of consciousness from advanced views of physical reality is quite compelling. Whether or not Sirag's particular models are confirmed, it seems possible that a successful physical-mathematical solution to the mind-matter problem may eventually develop from the type of ambitious program which Sirag has developed.

Sirag's model of consciousness, as presented in the Appendix, could be called a Pythagorean approach to consciousness, since Sirag's strategy is to look to mathematics for an appropriate structure to describe the relationship between consciousness and the physical world. He finds that unified field theories of the physical forces depend fundamentally on mathematical structures called *reflection spaces*, which are hierarchically organized in such a way that an infinite spectrum of realities is naturally suggested.

This situation is natural because mathematicians have discovered that the hierarchical organization of reflection spaces also corresponds to the organization of many other mathematical objects — e.g., catastrophes, singularities, wave fronts, and contact structures, error correcting codes, sphere packing lattices, and, perhaps most surprisingly, certain regular geometric figures including the Platonic solids.

It is generally believed by physicists working on unified field theory that space-time is hyperdimensional, with all but four of the dimensions being invisible. The reason for this invisibility is a major subject of research. Beside space-time dimensions, there are also other *internal* (or invisible) dimensions called *gauge* dimensions. The reality of these gauge dimensions is also a topic of controversy and research. In Sirag's view, both the extra space-time dimensions and the gauge dimensions are real. This provides scope for considering ordinary reality a substructure within a hyperdimensional reality. This idea has, of course, been suggested before — e.g., it is implicit in the Cave Parable of Plato. The difference in Sirag's approach is that the structure of the hyperspace is defined directly by the properties of physical forces.

A further innovation in Sirag's approach is that his version of unified field theory embeds both spacetime and gauge space in an algebra whose basis is a finite group. This group, which directly models certain symmetries of particle physics, is a symmetry group of one of the Platonic solids — the octehedron. Thus, it is a mathematical entity contained in the reflection space hierarchy. In fact, the reflection space corresponding to the octehedron is seven-dimensional and is also a superstring-type reflection space, so that a link with the most popular version of unified field theory is provided.

The central postulate of Sirag's paper is that this seven-dimensional reflection space is a universal consciousness, and that individual consciousnesses tap into this universal consciousness. This implies that the high level of consciousness enjoyed by humans is due to the complex network of connections to the underlying reflection space afforded by a highly evolved brain.

Moreover, the hierarchy of reflection spaces suggests a hierarchy of realms (or states) of consciousness. Each realm would correspond to a different unified field theory with different sets of forces. In fact, the seven-dimensional reflection space is contained in an eight-dimensional reflection space, and contains a six-dimensional reflection space, so that there would be a realm of consciousness directly "above" ordinary reality, and a realm of consciousness directly "below" ordinary reality. In principle, the relationship between the different forces in these different realms could be worked out in detail, so that precise predictions could be made.

Sirag believes that this hierarchy of realms of consciousness is analogous to the spectrum of light discovered in 1864 by James Clerk Maxwell in his electromagnetic theory of light, which unified the forces of electricity and magnetism. Maxwell had no way of directly testing his theory, which proposed the reality of frequencies of light both higher and lower than that of ordinary light. He boldly proposed the existence of invisible light, simply because his equations contained the higher and lower frequencies.

Similarly, in the unification of all the forces, we can expect something new to be described, which could be the analog of light. Sirag proposes that this new thing be consciousness, and that since the mathematics of the unification gives reflection space a central role, the hierarchy of reflection spaces suggests a hierarchy of realms of consciousness.

Evaluating Implications of the New Physics

One of the most fundamental developments in the past two decades has been the experimental confirmations of the principle of nonlocality in quantum mechanics and the realization of the importance of that principle for a theory of psi phenomena. If nothing else, this breakthrough strongly suggests that psi phenomena, if they exist, need not be in conflict with established laws of science.

At present, theories regarding psi are somewhat premature for two reasons. We still lack a reliable database and repeatable psi effects upon which a theory might be constructed and refined. We also lack a comprehensive theory of consciousness itself, upon which a theory of psi must, inevitably, be built. Thus, many of the theories discussed represent mere presentations of "theoretical environments" in which more testable theories might be constructed.[113] Sirag's "work in progress" as presented in the Appendix represents the beginnings of a venture which, if successful, will run a course of many generations.

A note of caution may be appropriate at this point. While I have been focusing on the relationship between physics and consciousness, this is only a short step from the issue of physics and mysticism. It is in this realm that many physicists themselves, as well as scholars of mysticism, feel that physics can have little to say. Ken Wilbur, for example, firmly maintains that the attempt to prove the reality of mystical experience by resorting to scientific arguments does a great injustice to genuine mysticism which is self-supporting and timeless. Whereas scientific theories are in constant flux.[114] This is an important point; however, it is also premature to assume that physics will never develop permanent and complete answers. After all, physics is based upon mathematics, and that field does seem to have developed some permanent solutions.

ARTHUR M. YOUNG (FROM *SELF AND UNIVERSE*, COURTESY THINKING ALLOWED PRODUCTIONS)

The Reflexive Universe

I feel fortunate that one of my mentors has been Arthur M. Young, an iconoclastic genius who invented the first, commercially licensed helicopter and later became a philosopher of cosmology and process theory. Many thousands of lives have been saved as a result of his revolutionary invention. Yet, the helicopter was only a tangible by-product of Young's deeper, lifelong search for a philosophy that could integrate human consciousness with the physical, biological, and social sciences. This is a far bolder endeavor than the search for a grand unified field theory in physics. It is a project whose completion may well take many generations, perhaps even millennia. I believe that Arthur Young's models stand as landmarks along this great journey.

Young's work cannot be considered a theory in the strict scientific sense. It is larger than a theory; it is a model of reality that goes beyond science.[115] The potential value of such a world view, paradigm or model for the scientific endeavor is heuristic: it suggests new avenues of inquiry. In this sense, Young's approach has been an inspiration to a generation of scholars working on the leading edge of consciousness exploration — including Kenneth Pelletier, Stanislav Grof, Saul-Paul Sirag, and Frank Barr.

As an anchor point for understanding Young's cosmology, we can begin with the formula for the volume of the Einstein-Eddington Universe, the boundary region of what physicists call the hypersphere. It is $2\pi^2r^3$. This is also the formula for the volume of a torus (donut) with an infinitely small hole. It is in the torus topology that Young sees a possible answer to the philosophical problem of the individual (or part, or microcosm) versus the collective (or whole, or macrocosm); in a toroidal universe, a

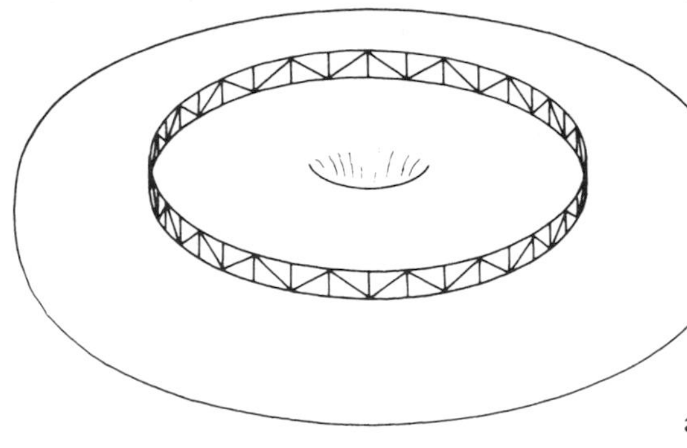

We must, however, bear in mind that the volume of the torus is three-dimensional and is kin to the surface of the four-dimensional hypersphere of Einstein and Eddington.

Suppose you had to draw a map on the surface of a torus so that all of the bordering countries would be distinguished by differences in color. On an ordinary surface, say a plane or sphere, such a map would require no more than four colors. The sphere ($4/3\pi r^3$) to Young is analogous to structure in the universe. Later we shall show how a "cycle of action" divides the sphere from the torus which is, in Young's scheme, analogous to universal process. It requires seven different colors to create a map on the surface of a torus. Therefore, Young reasoned, there might be seven stages to process just as there is a fourfold division to *structure*.

This inspiration is affirmed somewhat by ancient myths and cosmogonies. The Hindu, Zoroastrian, Japanese and Genesis creation myths all describe a seven-stage process. There are also seven rows to the periodic table of elements. Taking these cues, Young

part can be seemingly separate and yet connected with the rest.

If we think of the fence in the diagram below as separating the inner from the outer, the torus provides a paradigm that permits us to see a monad as both separate from the rest of the universe by the fence — and still connected to everything else through the core. The core of the torus, with its infinitely small hole, is for Young a representation of inner consciousness.

Young points out that magnetic fields, vortices and tornados all have the toroidal form. The vortex is, in fact, the only manner in which a fluid can move on itself. Thus it is a very suitable shape for the universe to have.

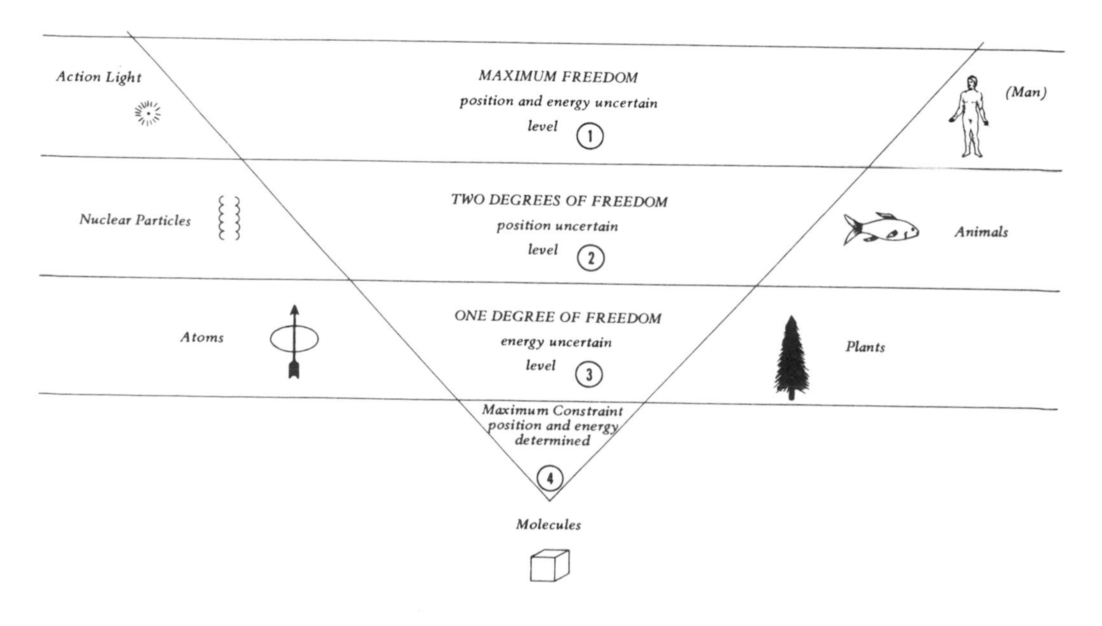

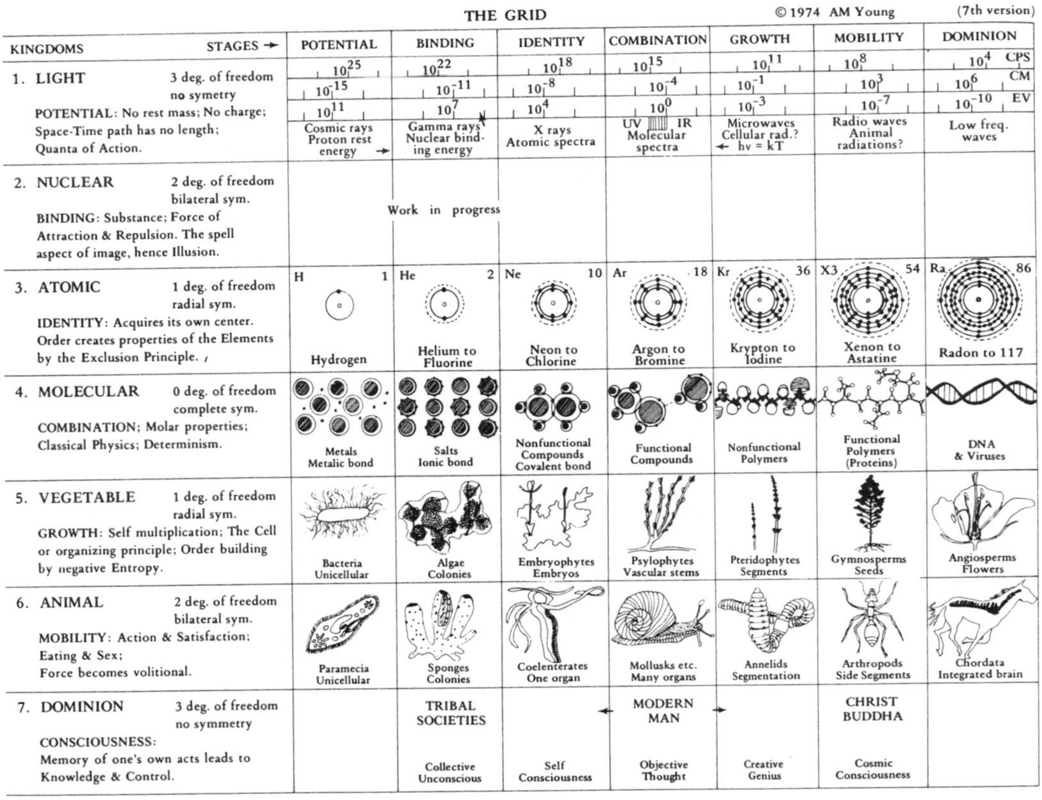

THE GRID © 1974 AM Young (7th version)

KINGDOMS STAGES →	POTENTIAL	BINDING	IDENTITY	COMBINATION	GROWTH	MOBILITY	DOMINION
1. LIGHT 3 deg. of freedom no symetry POTENTIAL: No rest mass; No charge; Space-Time path has no length; Quanta of Action.	10^{25} 10^{-15} 10^{11} Cosmic rays Proton rest energy →	10^{22} 10^{-11} 10^{7} Gamma rays Nuclear binding energy	10^{18} 10^{-8} 10^{4} X rays Atomic spectra	10^{15} 10^{-4} 10^{0} UV IR Molecular spectra	10^{11} 10^{-1} 10^{-3} Microwaves Cellular rad.? ← $h\nu = kT$	10^{8} 10^{3} 10^{-7} Radio waves Animal radiations?	10^{4} CPS 10^{6} CM 10^{-10} EV Low freq. waves
2. NUCLEAR 2 deg. of freedom bilateral sym. BINDING: Substance; Force of Attraction & Repulsion. The spell aspect of image, hence Illusion.		Work in progress					
3. ATOMIC 1 deg. of freedom radial sym. IDENTITY: Acquires its own center. Order creates properties of the Elements by the Exclusion Principle.	H 1 Hydrogen	He 2 Helium to Fluorine	Ne 10 Neon to Chlorine	Ar 18 Argon to Bromine	Kr 36 Krypton to Iodine	X3 54 Xenon to Astatine	Ra 86 Radon to 117
4. MOLECULAR 0 deg. of freedom complete sym. COMBINATION; Molar properties; Classical Physics; Determinism.	Metals Metalic bond	Salts Ionic bond	Nonfunctional Compounds Covalent bond	Functional Compounds	Nonfunctional Polymers	Functional Polymers (Proteins)	DNA & Viruses
5. VEGETABLE 1 deg. of freedom radial sym. GROWTH: Self multiplication; The Cell or organizing principle; Order building by negative Entropy.	Bacteria Unicellular	Algae Colonies	Embryophytes Embryos	Psylophytes Vascular stems	Pteridophytes Segments	Gymnosperms Seeds	Angiosperms Flowers
6. ANIMAL 2 deg. of freedom bilateral sym. MOBILITY: Action & Satisfaction; Eating & Sex; Force becomes volitional.	Paramecia Unicellular	Sponges Colonies	Coelenterates One organ	Mollusks etc. Many organs	Annelids Segmentation	Arthropods Side Segments	Chordata Integrated brain
7. DOMINION 3 deg. of freedom no symmetry CONSCIOUSNESS: Memory of one's own acts leads to Knowledge & Control.		TRIBAL SOCIETIES Collective Unconscious	Self Consciousness	← MODERN MAN → Objective Thought	Creative Genius	CHRIST BUDDHA Cosmic Consciousness	

divided all of nature into seven stages of process or evolution. The diagram on page 370 illustrates the seven kingdoms of Young's "reflexive universe" arranged in an arc, on four levels according to their relative degrees of uncertainty.

This chart symbolizes the mythological descent of spirit into matter and the corresponding ascent of matter into spirit. The greatest amount of constraint and symmetry occurs in the molecules' crystalline structure. This kingdom is most subject to science's deterministic laws, and thus is most predictable. Both atoms and plants possess radial or two-dimensional symmetry. They have two degrees of constraint and one degree of freedom, which constitutes their ability to store and release energy, within certain boundaries, without any specified prompting from without. Animals have bilateral symmetry along one dimension. Young believes that electrons and protons also possess symmetry along one dimension.

The experiments of Lee and Young, who discovered that chirality, or "handedness," characterizes nuclear particle reactions, suggests this possibility. Young points out that "handedness" requires bilateral symmetry. Heisenberg's principle states that we are uncertain of the position and the momentum of the nuclear particle. Young states that this principle applies to the animal as well. Thus, both nuclear particles and animals possess one degree of constraint and two degrees of freedom.

The first kingdom, which Young refers to as light (or action), and the seventh kingdom, of which humanity is a part, theoretically possess complete asymmetry and complete freedom. A photon released at a certain point could be anywhere within a radius of 186,000 miles a second later.

Furthermore, since observation annihilates a photon, it cannot be predicted. Although light has no rest mass, when it is annihilated it can create electrons and protons which do have mass. It has no charge, yet the particles it creates do. In fact, for a pulse of light, time does not exist. Clocks stop at the speed of light. Thus, mass, energy, and time are born when the photon condenses into a particle. This is the first step in the process that engenders the universe.[116]

Young regards action as the primary constituent of the universe, and other measures such as force (including gravity), energy and even time as *derived* parts of a whole which manifests as action. He also introduces the notion of purpose or intention into his scheme. The *principle of least action* is that light always follows the precise path that gets it to its destination in the shortest possible time. Planck himself observed that this principle expresses, "an explicitly teleological character."

> Thus the photons which constitute a ray of light behave like intelligent human beings: Out of all the possible curves they always select the one which will take them most quickly to their goal.[117]

Leibnitz, who discovered this principle, believed himself to have found evidence for an ubiquitous higher reason ruling all of nature. This characterization of light is the one exception to the exclusion of purpose from science. Purpose, associated with the quantum of action, becomes the keynote of Arthur Young's theory. He draws on a rich, although often discarded, tradition in science and philosophy.

> ...as Whitehead pointed out in his *Function of Reason*: "Scientists, animated by the purpose of proving they are purposeless, constitute an interesting subject for study."[118]

Young points out that *2ph* is the quantum of uncertainty. Thus we have a fundamental relationship between purpose and uncertainty, confirmed by the fact that *h* contains an angle, *2p,* which according to Eddington (the physicist from whom Young derives the greatest inspiration), is a phase dimension. For Young, the *2p* represents choice. Uncertainty then is not so much a limitation upon science as the positive introduction of purpose and choice and therefore free will.

Essentially then, a light pulse is a piece of uncertainty, and it is possible to account for the chain of effects that it can produce. If it is of a high frequency, it can become a nuclear particle, a proton, or an electron. Some uncertainty will become mass (or certainty). Another step combines nuclear particles into atoms with a further loss of uncertainty, followed by still more at the molecular stage. Nevertheless, there still remains enough uncertainty and choice of timing (phase dimension *2p*) in certain large molecules, within narrow temperature ranges, to extract energy from the environment and build organizations that emerge as life.

Referring to Young's "grid," one notices that each of the seven kingdoms is divided into seven substages.

The turning point of the arc is the middle of the fourth substage of the molecular, or fourth, kingdom. The fifth substage of the molecular kingdom represents the non-functional (covalently bonded) polymers such as cellulose, celluloid, rayon, nylon, dacron, etc. Young maintains that the distinguishing properties of these polymers is that they grow, like cells, in chains or a series of links. The growth of polymers reflects an ability to store order — to drain energy from the environment. This is an example of negative entropy and a prelude to the living kingdoms which follow the turn of the arc.

This turn marks the beginning of consciousness in Young's theory — although clearly not anthropomorphic consciousness. The amount of indeterminacy here is very small indeed, but it is such that it enables the molecule to *use the laws of determinism* to build more complex structures and processes with even greater freedom. The 90° turn in the arc is a change in direction that symbolizes this freedom. Thus, the uncertainty which is unconscious on the left side of the arc achieves ever greater degrees of voluntary control on the right side of the arc. Self-control, as such, is generally not recognized in classical physics. But, as was shown in the astrology section, Young assigned to it the measure formula T^3, the third derivative of position which is equivalent to the rate of change in acceleration.

A logically elegant feature in Young's scheme is the way basic characteristics of each of the seven kingdoms or stages (see the notated keywords on the grid) apply in an analogous fashion to the corresponding seven substages within each stage. Thus, the chain polymers in the fifth substage of the molecular kingdom have the property of growth referred to above which is characteristic of the fifth or plant kingdom. Furthermore, plants often consist of the polymers cellulose and lignin; so the fifth-stage growth involves the fifth substage chemical. The ionic bonding in the second substage of the molecular kingdom is characterized by the binding potential of the subatomic particles of the second kingdom. And, in fact, these particles are actively involved in ionic bonding. A third example is the principle of mobility that manifests in the sixth substage of the molecular kingdom, via the stretching proteins — actin and myosm — as well as in the sixth kingdom of animals. Actin and myosin are involved in the muscular movements of animals. Numerous examples are evident throughout the grid.

One of the major characteristics of the fifth substage of the animal kingdom is a hierarchical series of organs from the head to the tail, through a segmented structure. The earthworm is a typical example. This segmented organization occurs in the fifth substage of the molecular and plant kingdoms as well. In the sixth substages of these kingdoms, the structural property involves side chains attached to the main segmented structure. This is evidenced in protein amino and side chains, the branches of gymnosperms, and the jointed feelers and antennae of arthropods.

While recognizing the importance of DNA genetic material in the organization of intercellular structure, Young shares the doubt previously expressed that the DNA code can account for the hierarchy and diversity of organs. Furthermore, he thinks that animal instinct cannot be explained by DNA. To account for this type of extra-cellular organization, he postulates an organizing field. Young suggests that the corresponding organizing principle in the fifth substage of human beings (genus) is related to the awakened *kundalini* concept of the yogis.[119]

APPENDIX

Consciousness: A Hyperspace View

by Saul-Paul Sirag

Introduction

Ordinary reality, objectified by the methods of measuring space, time and matter, is a subrealm of a larger reality. This is an ancient idea — at least as ancient as Plato's cave story in which prisoners are chained in such a way that they identify themselves with their own shadows on the cave wall. It seems clear that Plato meant to imply that the larger reality is hyperdimensional — i.e., although we tend to identify ourselves with our 3-d bodies, there is a higher dimensional realm in which we are higher dimensional beings of which our 3-d bodies are mere shadows. This interpretation of the cave parable is augmented by Plato's motto for his Academy: "Let no one enter here without geometry" (cf. Hinton, 1904, 1980).

The idea that reality is hyperdimensional is entertained today by physicists attempting to unify all the physical forces in a unified field theory. It is reasonable to suppose that the detailed description of this hyperdimensional reality will yield a theory of *consciousness.*

Consider the following aspects:

1. The unification of electricity and magnetism worked out by James Clerk Maxwell (1831–1879) entailed the first satisfactory theory of light: i.e., light is an electromagnetic wave. This theory introduced the unexpected idea that visible light is only a tiny part of the electromagnetic spectrum. Subsequent discovery of radio waves and X rays confirmed this theory. Analogously, we can expect something new to come out of the unification of all the forces. We should not be too surprised to see that this something is consciousness and that the unified theory provides a basis for a "spectrum" of states of consciousness.

2. In philosophy, consciousness is usually discussed in the context of the mind-body problem: Are the basic entities of the world mind-like or body-like, or some mixture of mind-like and body-like entities? The defining characteristic of body-like is assumed to be extension in space. The defining characteristic of mind-like is assumed to be sensation. So the question

becomes: Can the world be constructed from sensations alone, extensions alone, or some mixture of both? There are other possible views; for example: both sensations and extensions are derived from different combinations of some neutral entity (cf. Russell, 1954).

The extension-alone school (materialism) maintains that consciousness is an epiphenomenon of the complex structures of the brain. The sensation-alone school (idealism maintains that consciousness (or a universal mind) is the ultimate reality and that the material world, as well as the existence of individual minds is a construct within the universal mind. The sensation-and-extension school (dualism) maintains that consciousness is an aspect of reality separate from, but somehow interacting with, the material world. The chief criticism against dualism is that an entity which is not extended in space cannot interact with matter (which is extended in space). The view that consciousness and matter arise from different combinations of some neutral set of entities is called neutral monism.

It is easy to imagine that a hyperspace view of reality will entail a reevaluation of the traditional categories of the mind-body problem.

3. The unification of all the forces is possible only if a theory can be constructed unifying general relativity (Einstein's theory of gravity) with quantum mechanics. The main problem is that these two theories appear to be incompatible because general relativity is a deterministic theory, whereas quantum mechanics employs a fundamentally nondeterministic description of measurement. A possible means of reconciliation is suggested by the fact that quantum mechanics itself has a deterministic side as well as the well-known nondeterministic side. The reconciliation of these two opposed aspects of quantum theory is called the quantum measurement problem. To state this problem clearly, a brief description of quantum theory is necessary.

A quantum system is represented by a vector (called the state vector) which rotates in an abstract space (which may very well be infinite-dimensional). Note: a *vector* is an arrow-like entity with both length and direction. The rotation of the state vector is deterministic in the sense that if its position is known at one time, its position at another time can calculated. (It is just like a clock hand except that a clock hand is rotating in a 2-d space, whereas the quantum state vector

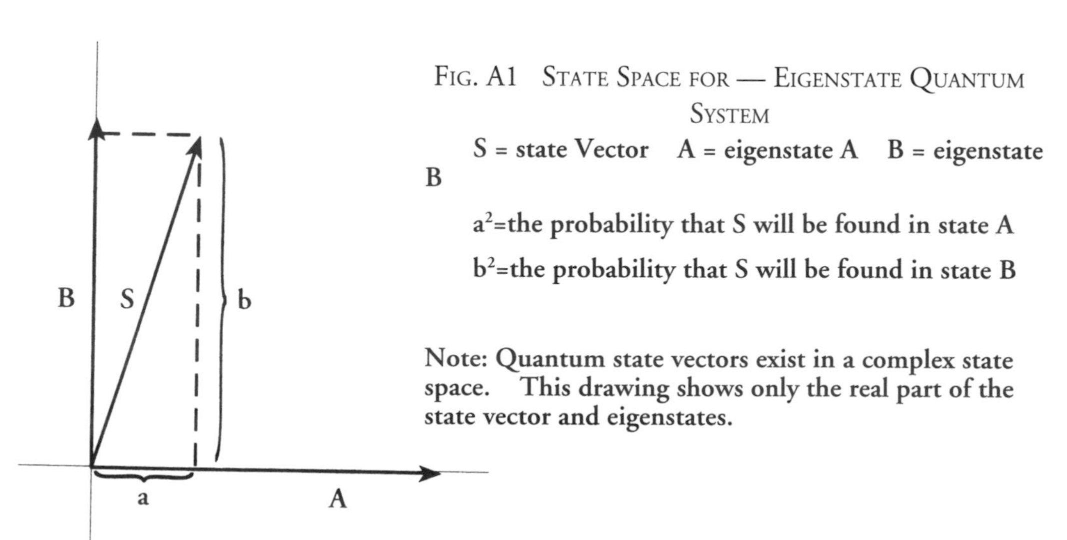

FIG. A1 STATE SPACE FOR — EIGENSTATE QUANTUM SYSTEM

S = state Vector A = eigenstate A B = eigenstate B

a^2=the probability that S will be found in state A

b^2=the probability that S will be found in state B

Note: Quantum state vectors exist in a complex state space. This drawing shows only the real part of the state vector and eigenstates.

might be rotating in a hyperspace.) As long as no *measurement* is made on the system, the state vector keeps rotating smoothly in the state space according to a deterministic equation called Schroedinger's equation; but, as soon as a measurement is made, the state vector immediately jumps to a position (an *eigenvector*) corresponding to an allowed value (an *eigenvalue*) of the particular measurement that is being made. This jump is called the collapse of the *wave function* (another name for the state vector), but it is more appropriately described as the projection of the state vector onto an eigenvector belonging to a measurement. Most important: It is completely undetermined which eigenvector is projected out of the state vector by the measurement; however, we can calculate the probability of this projection, and we can verify this probability by repeating the measurement over and over or by making simultaneous measurements on a multitude of similarly prepared systems. Each type of measurement (e.g., a position measurement or a momentum measurement) has its own set of allowed states (eigenvectors) which belong to allowed values (eigenvalues). The eigenvectors corresponding to a type of measurement provide a *coordinate system* for the space in which the state vector rotates (state space). Each type of measurement provides such a coordinate system.

Note: Since a coordinate system is imposed on a vector space arbitrarily (by choosing a measurement), we will consider a coordinate system as equivalent to a point of view (cf. Weinberg, 1987).

A crucial question is: Under what conditions do different types of measurement provide identical coordinate systems for the state space? And what is the consequence of this coincidence?

The answer is somewhat abstract: To each type of *measurement* corresponds an *operator* which acts on the state vector and on the state space in which the state vector lives. Of all the vectors of the state space, let us pick out only those which do not change direction, but change only their length under the action of the operator. These vectors are the eigenvectors belonging to the operator. The factor by which an eigenvector changes length is the eigenvalue belonging to the eigenvector. (e.g., if the operator doubles the length of the eigenvector, the eigenvalue is 2). Now it is a theorem of pure mathematics that *commuting* operators have the same eigenvectors. Note: If $AB = BA$, where A and B are operators, then we say that A and B *commute*. Thus, in this case, A and B provide the same coordinate system (the same set of eigenvectors for the state space on which they act).

The consequence for quantum theory is that any two types of measurement represented by commuting operators can be made in any order. Thus, in unified field theory, we are looking for a complete set of commuting operators whose eigenvalues are the complete set of "simultaneous" eigenvalues of the world. As we shall see, these operators are expected to be the basis elements of *a maximal commutative subalgebra of a Lie algebra* (and these terms will be defined in due time).

If two types of measurement are represented by *noncommuting* operators ($AB \neq BA$), the order in which the measurements are made affects the outcome of the measurements.

This is the basis of *Heisenberg's Uncertainty Principle.* For example, *position* and *momentum* are measurements corresponding to *noncommuting* operators, so that $PQ - QP = hl\,(i\,2\pi)$, where P is the momentum operator (in the x direction), Q is the position operator (in the x direction), i is the square root of minus one, and h is *Planck's constant,* which is the very small quantity $\approx 10^{-27}$ erg seconds.

This implies that the uncertainty in position times the uncertainty in momentum is always greater than or equal to Planck's constant — which is the usual way the uncertainty principle is stated.

(Note: The smallness of Planck's constant accounts for the fact that we didn't notice it until we investigated atomic phenomena. If we measure ordinary-sized objects, the ordinary measurement errors are much larger than the fundamental uncertainty due to Planck's constant.)

The most peculiar thing in quantum theory is not, however, the Uncertainty Principle. Rather it is the Measurement Problem (cf. Wheeler and Zurek, 1983), which we state as follows: How can we reconcile the deterministic evolution of the state vector with the random projection of the state vector onto a measurement eigenvector?

The problem becomes more acute when we realize that measuring devices themselves are ultimately composed of quantum entities so that, in principle, there is a single state vector **V** which corresponds to the combination of the measured system and the measurement device. This state vector **V** must rotate deterministically like any other quantum state vector. The projection of **V** onto an eigenvector is induced by a measurement of the measuring device by another measuring device (which, for the time being, is considered outside the quantum system). Of course, this second measuring device can be combined with the above system into an even larger quantum system and treated similarly. In fact, an indefinitely long chain of such devices can be hooked together. The most interesting question is: What is the ultimate observer? Is it the apparatus on the physicist's laboratory table? Is it the physicist's eyes? His optic nerves? His brain? His consciousness? (see Fig. A2)

According to the mathematician John Von Neuman and the physicist Eugene Wigner (cf. Wigner, 1979), the best solution to the measurement problem is that ultimately consciousness projects the state vector onto the eigenvector. However, if one adopts the Wigner interpretation of quantum theory, the *nonlocality* of the state vector as implied by Bell's Theorem (cf. d'Espagnat, 1976,1979 1983; Herbert, 1985) suggests that this consciousness which projects the state vector must be a *universal consciousness.*

Of course, many other solutions to the measurement problem have been proposed. None of them, however, has been universally accepted by physicists. And all of them are as strange, in their own ways, as the Wigner proposal. For example, Heisenberg (1958) proposed that the state vector is essentially a mental entity in the sense that it describes not the state of the physical world but rather the state of our knowledge of the physical world.

We should expect that a view of unified field physics, in which hyperspace is considered real, would throw new light on the quantum measurement problem. And we should not be surprised if consciousness is involved in this picture.

4. *Cosmology,* the study of the large-scale structure of spacetime, presents a special challenge to unified field theory. On the large scale, gravity is the dominant force, so that the unification of gravity with the other forces should have implications for cosmology. Moreover, it is believed that the forces become unified only in extremely high-energy interactions, such as occurred shortly after the big bang explosion of the universe. Thus it is hoped that unified field theory can provide clues to the origin of the universe. It

Fig. A2

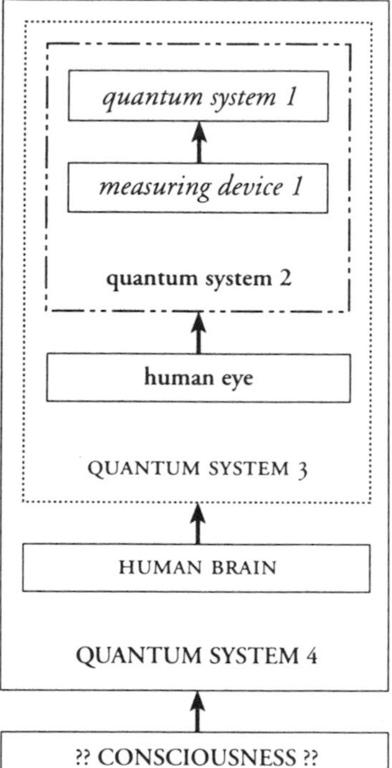

should be clear from the discussion above that quantum cosmology will entail a state vector for the universe as a whole. In this case, the problem of having an observer outside the universe to project the state vector onto eigenvectors becomes acute. The keyword here is *outside,* and it should not be surprising to find that a hyperspace view of reality provides a solution to this problem.

5. Among physicists working on *unified field theory,* it is widely believed that the deepest aspect of the world is the *symmetry principle* which we state (cf. Weinberg, 1987) as follows: The rule describing how the state vector evolves does not change even though the direction of the state vector can be altered by the changes in point of view (i.e., coordinate system).

These changes in point of view which leave the evolution rule unchanged are called *symmetries,* and together they make up a mathematical structure called the *symmetry group* of the world. The symmetry principle severely restricts the form of the rule of state-vector evolution. In fact, it is believed that is possible to specify the rule uniquely by specifying the symmetry group. Thus, in order to unify all the forces, we must answer the question: What is the symmetry group of the world? Since each element of this symmetry group corresponds to a change in point of view, it is clear that the symmetry principle requires the basic rule of evolution to be independent of point of view (i.e., coordinate-system independent).

Some of these symmetries are spacetime symmetries, such as the fact that the rules of physics are independent of the time and place and orientation of the measuring equipment. It is rather amazing that from these quite plausible requirements we can derive the following laws: conservation of energy, conservation of linear momentum and conservation of angular momentum. Moreover, the rules of Einstein's theory of special relativity can be derived from the postulate that the laws of physics must remain invariant under any coordinate system change which does not stretch the vectors in spacetime. The constancy of the speed of light c is built in here by the assumption that c is the intrinsic conversion factor necessary to put space and time into the same coordinate system — ct is a distance, if t is time. In the last several decades it has become increasingly clear that similar symmetry statements can be made about vectors in the state spaces of quantum mechanics. These are called "internal" symmetries in order to distinguish them from the symmetries of spacetime.

Symmetry and Groups

Because of the central role that symmetry plays in the unified field theory, it is necessary to present a more detailed description of symmetry and symmetry groups.

We define a *symmetry* as a change in an object which leaves some property of the object unchanged (or *invariant*). We use (rather loosely) the name of the change to identify the symmetry. For example, we say that a sphere is rotationally symmetric because its shape does not change under any rotation. We say that a cylinder is axially symmetric, because it does not change its appearance under a rotation about one particular axis (cf. Weyl, 1949, 1952; Elliott and Dawber, 1979).

The set of all symmetry transformations of an object forms a mathematical structure called a *group,* the most important feature of which is that one transformation followed by another transformation is equivalent to a third transformation. This property of the group is called *closure.* There are three other necessary properties:

1. *Associativity:* $a(bc) = (ab)c$ where a, b, c are group elements

2. *Identity elements:* $ea = ae = a$ where a is any group element and e is the identity element of the group

3. *Inverse elements:* $aa^{-1} = a^{-1}a = e$ where a is any group element and a^{-1} is the inverse of a; i.e., every element has an inverse.

There are many extra properties which a group might possess, the most important of which is

4. *Commutativity:* $ab = ba$ where a, b are group elements.

In other words, the order in which we perform all the transformations is irrelevant. We call such a group commutative. A noncommutative group may have a commutative subgroup. If this commutative subgroup commutes with all the group elements, it is called the *center* of the group.

Groups come in two types: *continuous* groups and *discrete* groups. A continuous group is also a space. Thus, a continuous group possesses not only the above algebraic properties which make it a group but also the geometric properties which make it a space. This means that the elements of the group are also the points of a space.

A discrete group is not a space (or is a 0-dimensional space if one insists in calling it a space). However, it may be a subgroup of a continuous group, i.e., a set of points set at intervals in a space. The simplest example is the group of integers, which can be viewed as a set of discrete points on the real number line. The identity element is 0; the inverse elements are the negative integers. The integers are closed and associative under the operation of addition. Moreover, the integers form a subgroup of the real numbers (viewed as a group under addition). The integers form an infinite but countable set. In fact, we ordinarily use the positive integers to do our counting.

There are also discrete groups which have a finite number of elements. These groups are called *finite groups.* The most important finite groups are called *symmetric* groups. A symmetric group S_n consists of all the *permutations* (reorderings) of n objects. The group property of closure arises out of the fact that a permutation of a permutation yields another permutation. For each n, there exists a permutation group of $n!$ (pronounced *n factorial*) elements, because there are $n!$ ways to permute (rearrange) n objects. For example, there are $4! = 4 \times 3 \times 2 \times 1 = 24$

ways to permute four objects. It should be noted that S_n is a *noncommutative* group, except in the two trivial cases $n = 1$ or 2.

If the space of a continuous group is *smooth,* the group is called a *Lie group* (named for the Norwegian mathematician Sophus Lie, 1842–1899). The real number line mentioned above is the simplest example of a Lie group. The best examples of Lie groups are groups consisting of rotations of vectors in vector spaces (which could, for example, be the state spaces of quantum theory).

One must be very careful to distinguish between two spaces here: (1) the space which the group is and (2) the space on which the group acts. One must also keep in mind that the group can act on itself as a transformation group. In this special case of self-action, (1) and (2) are the same space. This is hardly an unimportant technical detail. For as we will see, the action of the Lie symmetry group on itself corresponds to *force fields,* whereas the action of the group on an "outside" space corresponds to *matter fields.* And this is the fundamental basis of the distinction between these two types of field.

The set of all *rotations* of a sphere is a very useful example of a Lie group. There is a continuous infinity of rotations of an ordinary sphere (called a *2-sphere,* with the label S^2, because it is a 2-d space, coordinatized by latitude and longitude, let us say). These rotations themselves form a 3-d space, called *the 3-sphere mod plus or minus 1,* with the label $S^3/\{\pm 1\}$. The group name for this space is *SO(3),* which means the set of all *special-orthogonal* 3-by-3 *matrices* (cf. Schutz, 1980; Poor, 1981).

Note: A *matrix* is a rectangular array of numbers. The word *orthogonal* refers to the fact that the matrices rotate without stretching the vectors on which they act; the word *special* implies that volume remains invariant under the action of the matrices. Thus, the 3 in the label *SO(3)* refers not to the 3-d space of the group itself but to the dimensionality of the space on which the group acts, i.e., the 3-d vector space in which the sphere S^2 is embedded as the set of all unit-length 3-d vectors.

In general, the set of n-by-n special orthogonal matrices is a group called *SO(n)* which acts on n – *dimensional* vector space and in so doing rotates on $n - 1$ dimensional sphere S^{n-1}.

The importance of *SO(3)* may be realized by considering its finite subgroups which are the symmetry groups of various *polyhedra* inscribed in the sphere S^2 on which *SO(3)* acts. (cf. Weyl, 1952; DuVal, 1964; Slodowy, 1983). Each element of the group leaves the polyhedron invariant — i.e., looking the same (see Fig. A3).

The finite subgroups of *SO(3)* are:

1. All of the *cyclic* groups c_n (n = any integer) are commutative groups which rotate n-sided pyramids. We can imagine these pyramids as inscribed in S^2 with the base inscribed in the "equator" and the apex at the "north" pole.

2. All the *dihedral groups* d_n (n = any integer) are (except for $n = 1$ or $n = 2$) noncommutative groups which rotate n-sided *oranges.* We can imagine such an orange as S^2 partitioned into n sectors with the two vertices at the two poles. The noncommutativity of the symmetry group arises from the fact that a flip-over of the orange, exchanging the "north" and "south" poles, followed by a clockwise rotation through angle x, is different from a clockwise rotation through angle x followed by a flip. (To make this concrete, try it on a beach ball with different colored segments.)

3. The three *regular polyhedral* groups, which are symmetry groups of the *five*

platonic solids: the *tetrahedron,* the *octahedron,* the *cube,* the *icosahedron* and the *dodecahedron.* (Note: The octahedron and cube have the same symmetry group; likewise for the icosahedron and the dodecahedron. All three groups are noncommutative.)

T is the 12-element *tetrahedral* group, which we depict as the set of symmetries of the tetrahedron inscribed in S^2.

O is the 24-element *octahedral* group, which we depict as the set of symmetries of the octahedron inscribed in S^2. This is also the symmetry group of the cube, since the six faces of the cube correspond to the six vertices of the octahedron and the eight faces of the octahedron correspond to the eight vertices of the cube.

(The noncommutativity of *O* is easily seen by rotating a cube with six different colored faces. Or simply visualize the following: The three axes about which a cube rotates could be labeled — according to aeronautical terminology — yaw, pitch and roll. Then imagine you are the pilot of an airplane which is flying upright and level due west, and consider a pitch downward through 90° followed by a yaw to the left through 90°; you are now flying due south and with your right wing pointed to the ground. Now go back to the original starting position and reverse the order of the two maneuvers. You will "end up" with your nose headed straight to the ground!)

I is the 60-element *icosahedral* group, which we depict as the set of symmetries of the icosahedron inscribed in S^2, as well as the symmetry group of the dodecahedron, since the 12 faces of the dodecahedron correspond to the 12 vertices of the icosahedron, and the 20 faces of the icosahedron correspond to the 20 vertices of the *dodecahedron.*

The finite groups c_n, d_n and the regular polyhedral groups *T, O* and *I*

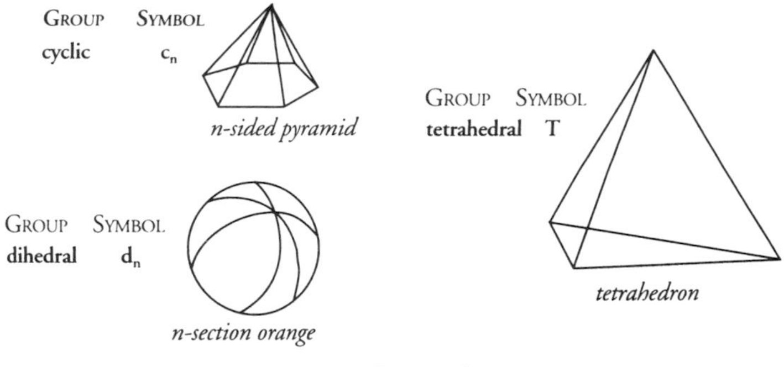

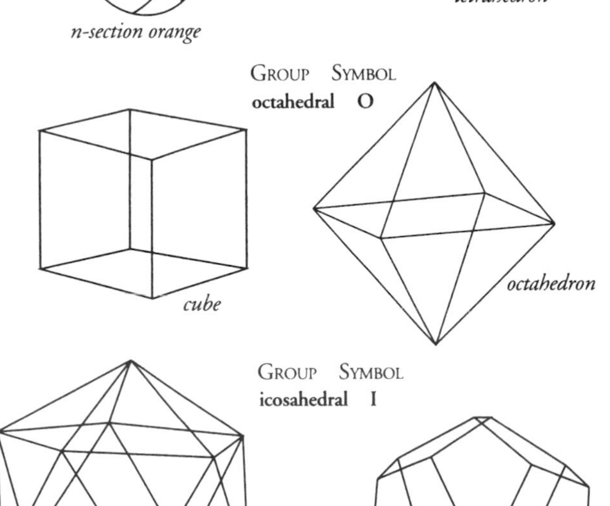

FIG. A3

Symmetry figures embedded in S^2 correspond to finite subgroups of SO(3). The regular figures (all sides equal) are the five Platonic

form the complete set of finite subgroups of *SO(3)* (cf. Weyl, 1952). There is a corresponding set of finite subgroups of the Lie Group *SU(2)*. This group is the set of all *special-unitary* 2-by-2 matrices: the elements of *SU(2)* rotate the vectors in the *complex* 2-d vector space on which the matrices act. In other words, *SU(2)* acts on the 2-d complex space (4-d) real space) just like *SO(3)* acts on 3-d real space. In fact, *SU(2)* as a space itself is called the *double cover* of *SO(3)*. This is because *SU(2)* as a space is the sphere S^3, whereas *SO(3)* is $S^3/\{\pm 1\}$. This means that *SO(3)* can be derived from *SU(2)* by pairing up positive and negative elements of *SU(2)* to form elements of *SO(3)*. Geometrically, this would be like considering the "north" and "south" poles of a sphere as the same element (cf. Penrose, 1978).

Thus each *x*-element finite subgroup of *SO(3)* has a 2*x*-element double cover in *SU(2)*. These finite subgroups of *SU(2)* can be called *polyhedral double groups*. For example, the 24-element octahedral group *O* has a 48-element double cover in *SU(2)* called the *octahedral double group OD*, and *O* can be derived from *OD* by pairing up positive and negative elements in *OD*. We can write: $O = OD/\{\pm 1\}$.

Another important class of finite groups is *reflection groups* (Coxeter, 1973), the most useful examples of which are the *Weyl groups* of Lie groups. A Weyl group acts on the *reflection space* h^* of the Lie algebra of the Lie group. The concept of a reflection space plays an important role in our theory, so it is necessary to describe this idea, charming in itself, in some detail. It will perhaps interest the reader to know ahead of time that the observable quantities (eigenvalues) of a quantum system, mentioned previously, exist in a reflection space, so that we could also call this space *eigenvalue space*.

An ordinary *mirror* seems to transfer objects from one side of the mirror to a similar space on the other side, seemingly changing the direction of objects as if the space in front of the mirror had simply flipped through the mirror. Imagine a two-sided mirror that actually could transfer objects, without changing their size, in both directions — the space in front of the mirror would be flipped, without stretching, behind the mirror, and vice versa, while the 2-d space of the mirror itself is left unmoved. This action of a 2-d mirror plane on a 3-d vector space is called a *reflection* in mathematics. And the idea generalizes to any number of dimensions, provided the mirror of an *n*-dimensional vector space is a plane of one dimension less, called a *hyperplane*. (For example, a line is the mirror hyperplane of 2-d space, and a 3-d plane is the mirror hyperplane in a 4-d space.) Only in this way can the mirror be two-sided. Just as a line has two sides in an ordinary plane but an infinity of "sides" in a 3-d space, so a 2-d plane is 2-sided in 3-d space but infinite-sided in 4-d space. Notice how these considerations provide some intuition about hyperspace.

Since we are by now accustomed to the idea of group elements acting on a vector space, we should not be surprised to learn that a reflection, as defined above, is a group element, and that a reflection is its own inverse element. That is, if *r* is a reflection, the *r* following *r* has the same end effect as doing nothing to the vector space. So we can write: $r^2 = 1$, where 1 is our symbol for the identity element (which means "do nothing").

Suppose we have two or more mirror planes intersecting each other. At the point where all the mirrors intersect, one vector (a *mirror vector*) is defined for each mirror, so that the vector corresponding to a mirror points away from the mirror *orthogonally* (at a right angle). There would, of course, be a negative mirror vector attached to the back of each mirror at the same point. We have pictured our mirrors as living in a vector space, and we usually imagine a vector space as having some set of *basis vectors* to provide a *coordinate system* for the vector space. Moreover, we customarily make this coordinate system rectangular by having the basis vectors orthogonal to each other, such as in the familiar *x–y–z* axis system for a 3-d vector space. For any vector space, however, there is an infinity of coordinate systems, not all of them rectangular. To set up a coordinate system, we need one basis vector for each dimension of the space, but these basis vectors need not be orthogonal (or even all of the same length). (see Fig. A4)

FIG. A4 THE A_2 REFLECTION SPACE

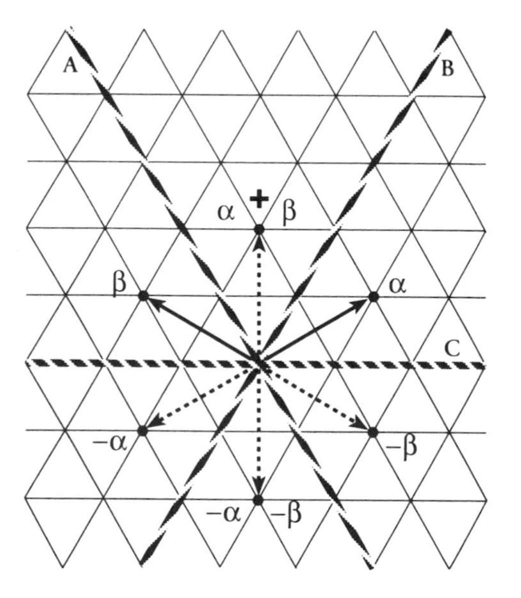

A and B are basic mirrors (set at 60° to one another).

C is a mirror generated by reflection in either A or B.

α and ß are basis roots (set at 120° to one another).

The other four roots are generated by reflections in the mirrors. The nodes (crossing points) are the weight lattice. Each node is a different weight.

Note: The root lattice is a sub-lattice of the weight lattice. The large dots are part of the root lattice.

We can now define a *reflection space* as: a vector space whose basis vectors are mirror vectors.

This implies that all the reflection activity possible in such a space is generated by the basic mirrors attached to the basis vectors. All the other mirrors are derived by reflection in these basic mirrors. Each basic mirror defines a reflection in these basic mirrors. Each basic mirror defines a reflection group element and all the elements of the reflection group are generated by (i.e., are combinations of) the reflections defined by the basic mirrors.

It is known (cf. Coxeter, 1973) that in setting up reflection spaces, the angles between the basic intersecting mirrors can only be: 60°, 45°, or 36° (except in the 2-d case where the angles can be 180°/*p*, where *p* is any integer larger than 2). Thus all possible reflection spaces can be defined by a graph (called a *Coxeter graph* or *Dynkin diagram*) for each reflection space. The nodes in each graph stand for the basic mirrors.

The most important reflection spaces have basic mirrors which are all set at 60° to each other. The *A-D-E* Coxeter graphs are displayed in Figure A5.

Thus, we have two infinite series: A_n and D_n (corresponding to two types of *n*-dimensional reflection spaces, and the "exceptional series" consisting of the three graphs E_6, E_7 and E_8 (corresponding to three exceptional reflection spaces of dimensions 6, 7 and 8).

Not only do these graphs classify reflection spaces, but they define a *nesting* of lower dimensional reflection spaces within the higher dimensional reflection spaces. This is because a lower-rank graph can

Fig. A5 A–D–E Coxeter Graphs

Lie algebra label	Coxeter graph	Total number of mirrors
A_n	0–0–0–. . . –0 (n nodes)	$(n^2 + n)/2$
D_n	0–0–0–. . . –0 (n nodes), with a node 0 attached below the second node	$n^2 - n$ (n greater than 3)
E_6	0–0–0–0–0, with a node 0 attached below the third node	36
E_7	0–0–0–0–0–0, with a node 0 attached below the third node	63
E_8	0–0–0–0–0–0–0, with a node 0 attached below the third node	120

always be derived by removing a node from a higher-rank graph. (The *rank* of a graph is the number of nodes, which is equal to the dimensionality of the reflection space defined by the graph.) For example, starting from E_8, we can remove appropriate nodes to create the following two hierarchies of nestings:

$E_8, E_7, E_6, D_5, D_4, A_3, A_2, A_1$;

$E_8, E_7, D_6, D_5, D_4, A_3, A_2, A_1$.

and many other hierarchies as well.

The nesting hierarchies imply that the lower dimensional reflection spaces are contained in higher reflection spaces. In fact, since a mirror in an n-dimensional reflection space is an $(n-1)$-dimensional plane, we can regard this mirror plane itself as a reflection space defined by a graph of $n-1$ nodes; and, in turn, each $(n-2)$dimensional mirror can be considered a reflection space defined by a graph of rank $n-2$; and so on, all the way down through the hierarchy of reflection spaces. In effect, each node in an n-rank graph (of *A-D-E* type) is a graph of rank $n-1$.

As one might guess, this hierarchy of reflection spaces is also a hierarchy of algebraic and topological structures, which are of great beauty and utility (cf. Gilmore, 1981).

As mentioned above, the reflection spaces are Lie algebra reflection spaces h^*. In fact, the *A-D-E* labels name various Lie algebras. Thus we can expect an intimate relationship between Lie groups, Lie algebras and Weyl reflection groups, since all three structures can be derived from the Coxeter graphs.

For example, given the A_2 Coxeter graph, we specify a 2-d reflection space (Fig. A4) with two basic mirrors (i.e., lines) intersecting at 60°. Call the mirrors A and B. The two mirror vectors, α and β, attached to these mirrors are angled at 120° to each other. These are the basis vectors of this reflection space. By reflection in the basic mirrors (and a third mirror generated from either of the basic mirrors), we generate four new vectors: $-\alpha$, $-\beta$, $\alpha + \beta$ and $-\alpha - \beta$. Thus we have six vectors attached in pairs to each side of three mirrors. These six mirror vectors, usually called *roots,* are *eigenvalue vectors* in the sense that the coordinates of each root are eigenvalues of an eigenvector belonging to two commuting operators h_1 and h_2, which for the basis of h, the *maximal commutative subalgebra* (the *Cartan subalgebra* of A_2. The *noncommuting* part of the Lie algebra requires a basis made of one eigenvector corresponding to each root; thus, the A_2 algebra is 8-d, with eight basis vectors: h_1, h_2 and six noncommuting eigenvectors.

Note: An element of an *algebra* is both a *vector* and *operator,* because an algebra has both additive and multiplicative properties. Imagine each element represented by a matrix. The addition of two matrices corresponds to the addition of two vectors. We represent the algebra acting on its own vector space by the multiplication of one matrix by another. In a *Lie algebra,* multiplication of elements X and Y is defined as the combination $XY - YX$, which is written via Lie brackets: $[X,Y]$. Thus, if we consider X an operator and Y a vector, we can write the *eigenvector equation* as:

$$[X,Y] = aY$$

where y is an eigenvector with eigenvalue a belonging to the operator X. In the case of A_2, for example, we have:

$$[h_1, h_2] = 0;$$

$$[h_2, h_1] = 0;$$

$$[h_1, e] = 2e;$$

$$[h_2, e] = -e.$$

Thus h_1 and h_2 are commutative and are eigenvectors of each other with eigenvalue 0. And e is an eigenvector of both h_1 and h_2, with eigenvalues 2 and -1, respectively. The root associated to e is a vector in the reflection space h^* with coordinates 2 and –1. There are five other eigenvectors followng the same pattern as e but with different eigenvalues, and thus corresponding to the other five roots.

In general, a Lie algebra of rank n is defined by an n-node Coxeter graph, such that there is an n-dimensional Cartan subalgebra with n commuting operators as basis, and a set of noncommuting operators corresponding 1:1 to the roots in the reflection space defined by the Coxeter graph. In fact, the reflection space is called the *dual space* h^* of the Cartan subalgebra h. Moreover, h^* and h can considered two different views of the same space, which we now call $\mathbf{C}^n$, because we regard the Lie algebras as complex Lie algebras (cf. Humphreys, 1972).

Since quantum theory restricts observable quantities to eigenvalues of measurement operators, these quantities are to be found in some Lie algebra reflection space. For example, the first *grand unified field theory* proposed the A_4 reflection space as the unified eigenvalue space. This 4-d reflection space has: one dimension for electric charge, one for weak charge, and two for strong (color) charge (cf. Georgi, 1981,1982).

Note: A_4 is the label for a complex Lie algebra whose compact Lie group is *SU(5),* with Cartan subgroup T^4, so that the A_4 reflection space is also the reflection space of the group *SU(5).* In general, the Lie algebra A_n with reflection space $\mathbf{C}^n$ has a compact Lie group $SU(n + 1)$. Thus the first grand unified field theory is usually referred to as the *SU(5)* theory.

The term *grand unified theory* is really a misnomer because gravity is missing, and it is mainly the attempt to include gravity in unified field theory that has forced us to look to much larger Lie algebra reflection spaces. For example, the most celebrated version of *superstring theory* proposes to unify all the forces via the 16-dimensional reflection space $E_8 \times E_8$. This is possible only because the hierarchy of reflection space embeddings provides for an embedding of A_4 in E_8 (cf. Duff, 1986).

Since reflection spaces seem to be the key unification structures, we would do well to study them in more detail. One of the most striking facts of recent mathematics is a 1:1 correspondence between the finite subgroups of *SU(2)* mentioned above and the *A-D-E* series of Lie algebras (cf. McKay, 1980; Slodowy, 1983; Arnold, 1986). We list the most important aspects of this correspondence in Figure A6 (the terms will be explained in due course).

The *extended Coxeter graph* of a particular *n*-rank Lie algebra defines an infinite number of mirror planes in the *n*-dimensional reflection space of the Lie algebra. This is accomplished by starting from the *n* basis vectors defined by the ordinary Coxeter graph (cf. p. 335), and then constructing a new vector corresponding to the node marked with an asterisk. We lengthen all the other vectors by the factors indicated in the extended graph. These $n + 1$ vectors are now in *balance:* if they were force vectors, the total force would be zero.

These vectors-in-balance define a mirror plane which forms a closed *alcove* (called a *fundamental alcove*) with the basic mirror planes. An observer in the closed alcove would see the space of the alcove reflected in all the mirror walls of the alcove so that a tessellation of mirror-walled alcoves (called *Coxeter alcoves*) would be generated, thus filling the entire reflection space (cf. Coxeter, 1973; Bröker and Dieck, 1985) (see Fig. A5).

For example, the A_3 fundamental alcove is a tetrahedron. Put a candle in its center; then an infinity of candles in the centers of tetrahedral alcoves will be reflected in the four mirrored walls of the fundamental tetrahedron.

There are many uses for the *balance numbers* in the extended graphs. The sum of the balance numbers for a particular graph is called the *Coxeter number c.* We can calculate the *mirror number m* of intersecting mirrors in an *n*-dimensional reflection space as $m=nc/2$. We can check that for E_7 the number of mirrors is 63 = 7 x 18/2, because the Coxeter number is 18 = 1 + 2 + 3 + 4 + 3 + 2 + 1 + 2. Since there are two roots for each intersecting mirror, the number of roots is simply 126 = 7 x 18. And thus the *dimensionality* of the E_7 Lie algebra is the rank plus the number of roots: 133 = 7 + 126.

This is all on the Lie algebra side.

FIG. A6 THE MCKAY CORRESPONDENCE

LIE ALGEBRA	EXTENDED COXETER GRAPH	SU(2) SUBGROUP	CATASTROPHE GERM
A_n	1–1– . . . –1 (both ends joined to 1*)	$^c n + 1$	$A^{n+1} + B^2 + C^2$
D_n	1–2– . . . –2–1 (1 below first 2, 1* below last 2)	$^d n - 2$	$A(B^2 - A^{n-2}) + C^2$
E_6	1–2–3–2–1 (2 below 3, then 1*)	TD	$A^4 + B^3 + C^2$
E_7	1*–2–3–4–3–2–1 (2 below 4)	OD	$A^3 + AB^3 + C^2$
E_8	2–4–6–5–4–3–2–1* (3 below 6)	ID	$A^5 + B^3 + C^2$

What about the supposed correspondence with finite subgroups of *SU(2)?* The following relations are due to the mathematician John McKay (1980), so we call these finite groups *McKay groups.* For a given extended graph:

1. The sum of the squares of the balance numbers is the number of elements in the associated McKay group.

For E_7:

$$1^2 + 2^2 + 3^2 + 4^2 + 3^2 + 2^2 + 1^2 + 2^2 = 48$$

where 48 is the number of elements in the McKay group *OD.*

2. The balance numbers are the dimensions of fundamental vector spaces on which the McKay group acts. Such vector spaces are called *inequivalent irreducible representation* spaces (or *iirep* spaces). Thus the number of balance numbers *b* (i.e., the number of nodes in the extended graph) is equal to the number of iireps of the McKay group.

For E_7: the iirep dimensions of *OD* are: 1, 2, 3, 4, 3, 2, 1, 2.

3. The number of *classes* within the McKay group is also equal to *b.* (Note: Two group elements are in the same *class* if their action on a vector space differs only by a change of basis, or coordinate system, for the vector space.)

For E_7: the number of classes in the McKay group *OD* is eight.

4. If we make the McKay-group elements into basis vectors of a vector space, this vector space becomes an algebra called a *group algebra.* We can also make the classes into basis vectors of an algebra which is called the *center* of the group algebra. (Note: The center is a commutative subalgebra which commutes with the entire group algebra.) The dimensionality of the center is *b*, since *b* is the number of classes.

For E_7: the McKay-group algebra is **C**[*OD*], each of whose elements is a complex sum over the group elements *OD* — i.e., any element of **C**[*OD*] can be written as $c_1x_1 + c_2x_2 + \ldots + c_{48}x_{48}$, where c_1 through c_{48} are complex numbers, and x_1 through x_{48} are the elements of *OD.* There are eight classes in *OD.* Thus, if we partition the 48 elements into these eight classes and assign to each element of a class the same complex number, we will have an element of the 8-d center of **C**[*OD*].

5. Just as rectangular basis vectors define a grid structure whose vertices are called a "lattice" on a vector space, o the basis roots define a nonrectangular lattice (called a *root lattice* L_r) in the reflection space. The intersection points of the *alcoves* generated by the extended graph also define the vertices of a lattice, called a *weight lattice* L_w (see Fig. A4). In unified field theory, the roots correspond to eigenvalues of *force particles,* whereas the weights correspond the eigenvalues of *matter particles.* If we think of the reflection space (where these roots and weights reside) as a space acted on by an operator, the most important such operator is one changing from a weight basis to a root basis. This is called the *Cartan matrix.* The Cartan matrix has as columns coordinates of the basic roots and thus be derived directly from the Coxeter graph. In like manner, the *extended Cartan matrix* can be derived from the extended Coxeter graph (see Fig. A7).

The Cartan matrix can be defined as $C = -A + 2E$, where A is the *adjacency matrix* of

FIG. A7 E_7 & (EXTENDED) $^\wedge E_7$ COEXETER GRAPHS

(Where the underlined numbers are indexing numbers)

```
1   2   3   5   6   7        0   1   2   3   5   6   7
0 – 0 – 0 – 0 – 0 – 0        1*– 2 – 3 – 4 – 3 – 2 – 1
        |                                |
        0                                2
        4                                4
```

the graph, and E is the identity matrix consisting of ones written along the primary diagonal of the matrix.

Note: In the adjacency matrix A (of either the ordinary or extended graph), the component corresponding to the i'th row

FIG. A8 $^{\wedge}E_7$ (EXTENDED) CARTAN MATRIX

C = -A + 2E

	0	1	2	3	4	5	6	7
0	2	-1	0	0	0	0	0	0
1	-1	2	-1	0	0	0	0	0
2	0	-1	2	-1	0	0	0	0
3	0	0	-1	2	-1	-1	0	0
4	0	0	0	-1	2	0	0	0
5	0	0	0	-1	0	2	-1	0
6	0	0	0	0	0	-1	2	-1
7	0	0	0	0	0	0	-1	2

Note: If we remove the 0'th row and the 0'th column, we have the ordinary Cartan matrix for E_7.

and the j'th column has a 1 if the i'th node is adjacent to the j'th node, and 0 otherwise (see Fig. A8).

6. McKay (1980) has proved the amazing fact that eigenvectors (i.e., the vectors changing only in length) under the action of an extended Cartan matrix are the columns of *character table* of the McKay group associated to the extended graph.

Note: The character table of a finite group is a square array of numbers, which are the *characters* (i.e., the sums of the diagonal numbers) of the iirep matrices for each class of elements of the group. The rows of the character table correspond to iireps; the columns correspond to the classes.

McKay's theorem implies that the columns of the character table of the McKay group provide a rectangular basis for the n-dimensional reflection space embedded in the C[OD] center of dimension $n+1$ (see Figs. A9 and A10).

Let us look at these facts from the point of view of quantum mechanics in the context of unified field theory. We pick a Lie algebra which we hope encodes the basic symmetries of the world. The basis operators of its Cartan subalgebra h are a complete set of commuting operators. The eigenvectors correspond to eigenstates of the quantum system — the world. There are two types of eigenvalue in the reflection space h^*: force eigenvalues corresponding the root lattice and matter

FIG. A9 THE INTERSECTION OF E_7 WITH C[OD]

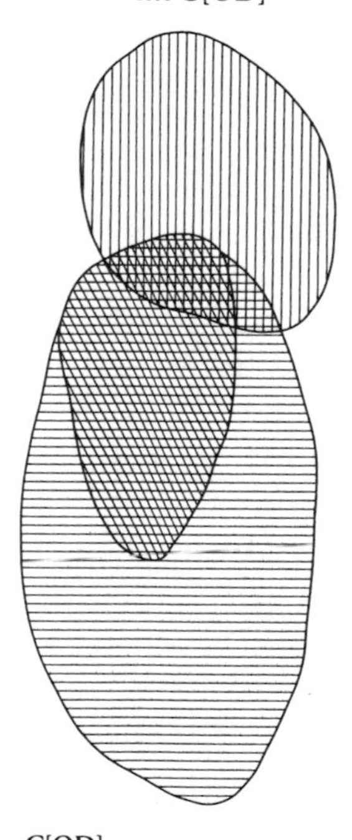

Fig. A10 The OD Character Table

(8 classes C_i & 8 iireps R_j)

	C_0	C_1	C_2	C_3	C_4	C_5	C_6	C_7
R_0	1	1	1	1	1	1	1	1
R_1	2	0	1	0	$\sqrt{2}$	-2	-1	$-\sqrt{2}$
R_2	3	-1	0	-1	1	3	0	1
R_3	4	0	-1	0	0	-4	1	0
R_4	2	2	-1	0	0	2	-1	0
R_5	3	-1	0	1	-1	3	0	-1
R_6	2	0	1	0	$-\sqrt{2}$	-2	-1	$\sqrt{2}$
R_7	1	1	1	-1	-1	1	1	-1

Note: The ordering of iireps accords with the indexing of the $\hat{E}_7$ (extended) graph (see Fig. A7). The first column, which contains the E_7 balance numbers, corresponds to the identity element.

eigenvalues corresponding to the weight lattice.

The Cartan matrix acts on reflection space h^* and transforms its weights into roots: i.e., matter eigenvalues into force eigenvalues. Thus it is a kind of super-operator. Think of the extended Cartan matrix as a super measurement operator whose eigenvectors are the classes of the McKay group associated to the Lie algebra. Eigenvectors should correspond to particle states. In this case (since the operator is a super-operator), we expect the eigenvectors to correspond to classes of observable particle states.

Is there an assignment of particle classes to the eight classes of the *OD* group? Actually, the most straightforward assignment is to the five classes of the *O* group. But since $O = OD/\pm 1$, the elements of *O* are pairs of *OD* elements; so the classes of *O* are derived from the classes of *OD*. The classes of *O* are equivalent to the classes of permutation *cycle pattern*. For example, there are six ways to permute four objects by *swapping* two and leaving the other two untouched. If we name the objects 1, 2, 3 and 4, the six such permutations can be written as: (1 2)(3)(4), (1 3)(2)(4), (1 4)(2)(3), (3 4)(1)(2), (2 4)(1)(3) and (2 3)(1)(4). These are called cycle patterns because the numbers in brackets are considered as a cycle — e.g., (1 2)(3)(4) is read as 1 goes to 2 and 2 goes to 1; 3 goes to 3; 4 goes to 4.

The five permutation classes partition the 24 elements of *O* into three classes of *even* permutations and two classes of *odd* permutations.

Note: An *even* permutation can be considered as a combination of an even number of *swaps;* an *odd* permutation is equivalent to an odd number of swaps.

The number of elements in the three even classes are: one, three and eight.

The number of elements in the two odd classes are: six and six.

If we call an even permutation *e* and an odd permutation *o,* we can write the general rules for combining permutation as:

$e \times e = e;$

$o \times o = e;$

$e \times o = o;$

$o \times e = o.$

We notice that this matches the quantum field rules for force bosons *b* and matter fermions *f*: *f* interacts with *f* by exchanging *b* ($f \times b = f$), while *b* interacts with *b* by exchanging *b* ($b \times b = b$).

This suggests that the class numbers one, three and eight correspond, somehow, to the numbers of force particles in each of three classes. Moreover, it would seem that the class numbers six and six correspond to the two classes of matter particle: quarks and leptons.

If we also partition the 24 elements of *O* into six *cosets* of the subgroup K_4, then three of the cosets correspond to even permutations, and the three remaining cosets correspond to three families of fermions, with two quarks and two leptons in each family.

We display the *O multiplication table* and define the 24 permutations with assignments to particle labels as Figure A11.

Since the fermion family structure is generally regarded as the deepest mystery of particle physics (cf. Georgi, 1982), we are justified in thinking that the finite group solution to this problem is a fundamental clue to the structure of unified field theory (cf. Sirag, 1982).

To test this idea further, we must construct an interpretation of the multiplication table which makes sense of the particle labels. First we note that every particle interaction fundamentally entails three particles. Such an interaction is represented by a vertex in a *Feynman diagram,* i.e., three lines meeting at a point. There are only two patterns to this vertex: two fermion lines and a boson line, three boson lines. This matches the odd-even structure of the *O* multiplication table, as mentioned above (see Fig. A12).

FIG. A11

Multiplication table for O (i.e. S_4):						**Permutation:**	**Particle**	**Coset:**
ABCD	**EFGH**	**IJKL**	**MNPQ**	**RSTV**	**WXYZ**	**Even:**	**Bosons:**	
ABCD	EFGH	IJKL	MNPQ	RSTV	WXYZ	**A** (1) (2) (3) (4)	identon	1st
BADC	FEHG	JIKL	PQMN	TVRS	XWZY	**B** (12) (34)	kleinon	(=K_4)
CDAB	GHEF	KLIJ	MNQP	VTSR	YZWX	**C** (13) (24)	kleinon	
DCBA	HGFE	LKJI	QPNM	SRVT	ZYXW	**D** (14) (23)	kleinon	
EGHF	IKLJ	ACDB	WZYX	MPNQ	RVST	**E** (124) (3)	familon	2nd
FHGE	JLKI	BDCA	XYZW	PMQN	TSVR	**F** (234) (1)	familon	
GEFH	KIJL	CABD	YXWZ	NQMP	VRTS	**G** (143) (2)	familon	
HFEG	LJIK	DBAC	ZWXY	QNPM	STRV	**H** (132) (1)	familon	
ILJK	ADBC	EHFG	RTSV	WYZX	MQPN	**I** (142) (3)	familon	3rd
JKIL	BCAD	FGEH	TRVS	XZVW	PNMQ	**J** (134) (2)	familon	
KJLI	CBDA	GFHE	VSTR	YWXZ	NPQM	**K** (123) (4)	familon	
LIKJ	DACB	HEGF	SVRT	ZXWY	ZMNP	**L** (243) (1)	familon	
						Odd:	**Fermions:**	
MQNP	RSVT	WZYX	ACDB	EFHG	ILKJ	**M** (24) (1) (3)	quark	4th
NPMQ	VTRS	YXWZ	CABD	GHFE	KJIL	**N** (13) (2) (4)	quark	
PNQM	TVSR	XYZW	BDCA	FEGH	JKLI	**P** (1234)	lepton	
QMPN	SRVT	ZWXY	DBAC	HGEF	LIJK	**Q** (1432)	lepton	
RVTS	WYXZ	MNPQ	IJKL	ADCB	EGFH	**R** (14) (2) (3)	quark	5th
STVR	ZXYW	QPNM	KLJI	DABC	HFGE	**S** (23) (1) (4)	quark	
TSRV	XZWY	PQMN	JILK	BCDA	FHEG	**T** (1342)	lepton	
VRST	YWZX	MNQP	KLIJ	CBAD	GEHF	**V** (1243)	lepton	
WXZY	MPQN	RTSV	EHFG	IKJL	ABDC	**W** (12) (3) (4)	quark	6th
XWYZ	PMNQ	TRVS	FGEH	JLIK	BACD	**X** (34) (1) (2)	quark	
YZXW	NQPM	VSTR	GFHE	KILJ	CDBA	**Y** (1423)	lepton	
ZYWX	QNMP	SVRT	HEGF	LJKI	DCAB	**Z** (1324)	lepton	

FIG. A12 FEYMAN DIAGRAMS
2 Types of Vertices

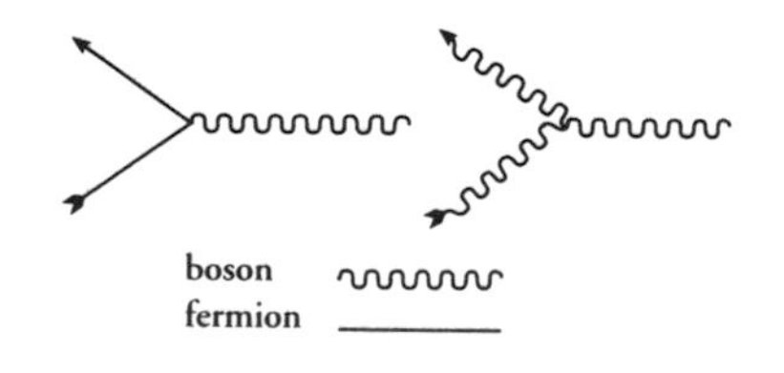

Electromagnetism

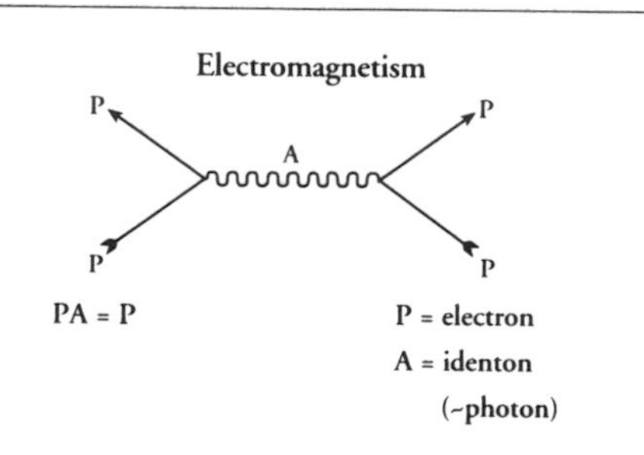

Weak Interaction

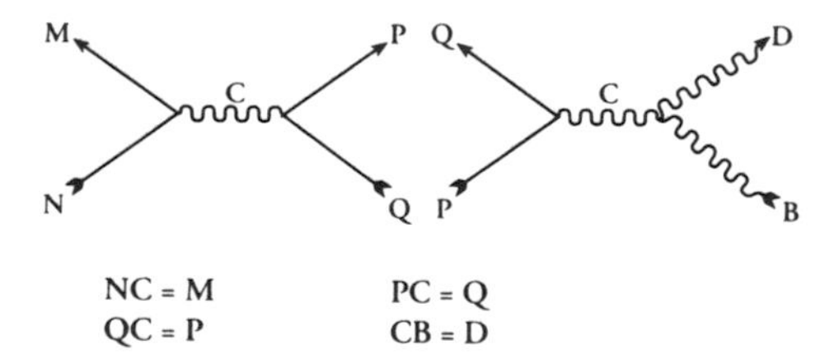

Q = electron neutrino
P = electron
M= up quark
N = down quark
C = kleinon (~weakon)
B = kleinon (~weakon)
D = kleinon (~weakon)

Feyman Diagrams display two types of vertices. The second type of vertex shows up in the weak interaction because weakons can interact with each other.

Actually A is somewhat different from a photon, so we call it an identon, because it corresponds to the identity element of the group. Also, B, C, and D are somewhat different from the weakon so we call them kleinons, after Felix Klein for whom the Klein group K_4, consisting of A, B, C, & D, is named.

We recover the standard gauge theory of these force particles by the embedding of U (1) x SU (2) x SU (3) in the algebra C[OD].

The three lines of any interaction vertex can be labeled: forward, lateral and forward; and the multiplication table can be interpreted so the *ab* = *c* means forward *a* and lateral *b* and forward *c* meet at a vertex. In short, forward x lateral = forward. When this rule is postulated, the deep structure of weak and strong interactions is displayed (see Fig. A13). To distinguish these *O* group bosons from the standard force bosons, we use the particle labels *identon, kleinon* and *familon,* rather than *photon, weakon* and *gluon,* respectively. The standard force bosons are recovered by embedding the gauge groups *U(1)* x *SU(2)* x *SU(3)* in the group algebra **C**[*O*], which is a subalgebra of **C**[*OD*].

It may seem that we have strayed far from the idea of a reflection space in defining particle classes corresponding to finite group classes. However, although the McKay group *OD* is not a reflection group, its factor group *O* is a reflection group. In fact, *O* is the reflection group of the A_3 reflection space. And A_3 can be used to unify the eigenvalue structure of the electromagnetic and strong color forces as emphasized by Georgi (1981) (see Fig. A14).

The exact interpretation of the relationship between *O* as a factor group of *OD* and *O* as A_3 reflection space has yet to be worked out. Since the eigenvalues associated with the measurement of particles reside in reflection space, it is reasonable to expect that it will be intimately related to the fundamental fact of quantum mechanics: only eigenvalues are observable.

Furthermore, because there seems to be no fundamental distinction between observation and what is observed, I propose that the reflection space (i.e., eigenvalue space) is universal consciousness.

Consciousness as Reflection Space

As we have seen, all of the magical structure of unified field theory is in the reflection space. Because of the McKay theorem, we can view this reflection space via the Lie algebra, or the

FIG. A13 THE STRONG INTERACTION MODELED BY THE FAMILON PRODUCTS OF THE O GROUP

MG = V
VJ = M
MF = S
SL = M
NL = Z
ZF = N
NJ = X
XG = N
NH = S
SK = N
NL = Z
ZF = N
FM = S
SL = M
NK = W
WH = N
NG = R
RJ = N
MK = Y
YH = M
MJ = Z
ZG = M
NG = R
RJ = N

LF = A
FL = A
JG = A
GJ = A
KH = A
HK = A

These products are derived from the multiplication table of the O group. They model the labeling of this diagram via the rule:
forward x lateral = forward.

The strong force binding the proton and neutron is mediated by the exchange of a pi meson, which changes the identity of these particles. According to the quark model, the pi meson (consisting of an up quark (u) and an anti-down quark ($\bar{d}$)) transfers an up quark (u) from the proton to the neutron, and a down quark (d) from the neutron to the proton. The quarks are bound to each other by gluons, each of which consists of a color and an anticolor.

Here the "gluons" are replaced by a complex of familons, which changes the family label of the quarks.

The standard theory of quarks and gluons is recovered by the embedding of the color gauge group SU (3) in C[OD], but here we're examining the underlying structure of OD itself via = OD/±1.

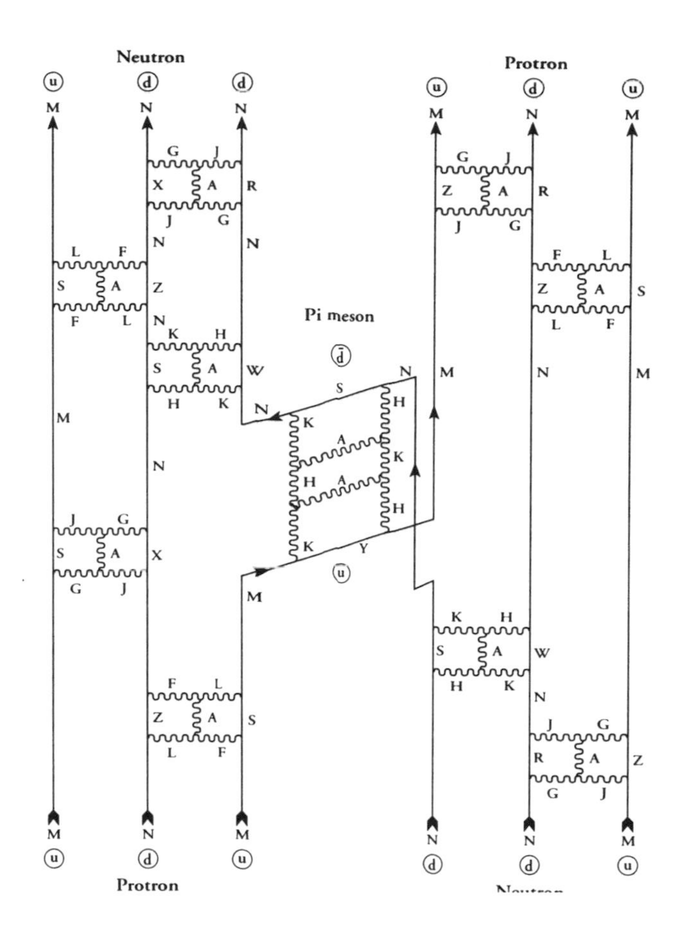

McKay group, or both. In fact, we will argue that the McKay-group algebra corresponds to the physical aspect of the world, while the Lie algebra corresponds to the mental aspect. The reflection space exists in the intersection of the McKay-group algebra and the Lie algebra (actually the infinite-dimensional Lie algebra defined by the extended Coxeter graph). Since this intersection space mediates between the two algebras, and since it is so active, I propose to identify it with *universal consciousness.*

It might be supposed that the idea of consciousness as reflection space is simply a bad pun. After all, mathematicians have taken an ordinary word *reflection* and given it a technical meaning which is different from, but similar to, the usual meaning. The ordinary meaning has also generated the figurative meaning, by which reflection is a kind of thinking. It may be, however, that there is something deeper in this figure of speech. Perhaps thinking of all kinds is a kind of reflection. And perhaps this kind of reflection is akin to the mathematician's use of the word. Of course, there is more to consciousness than thinking. But there is also much more to reflection space than reflection. It is often assumed that a theory of consciousness is impossible because consciousness is so complicated. Perhaps the few aspects of reflection space which I have so far described in this paper will suggest that mathematical complexities may be rich enough to describe so rich an experience as consciousness.

Figures of speech have a habit of cutting two ways. We say that certain molecules have a "memory" if they return to an original shape

FIG. A14

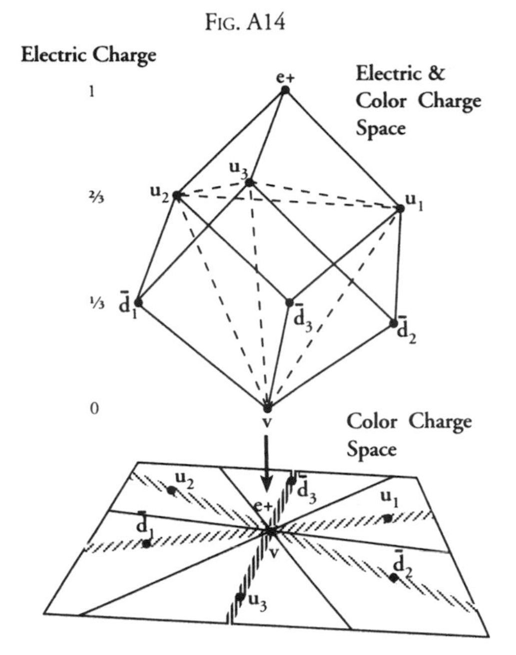

THE A_3 REFLECTION SPACE projects down the A_2 reflection space, so that a tetrahedron (inscribed in a cube) is projected down to a triangle. The vertices of the tetrahedron carry the labels of three colored up quarks u_1, u_2, and u_3, and a neutrino *v*. A reciprocal tetrahedron carries the anti-down quark labels $\bar{d}_1$, $\bar{d}_2$, and $\bar{d}_3$, and the anti-electron e^+.

Thus, the eight vertices of the cube in A_3 reflection space encode the charge and color eigenvalues of a fermion family.

Note that the dotted lines in the A_2 reflection space correspond to mirrors, while the solid lines correspond to roots (or mirror vectors).
Cf. Fig. A4.

after being deformed. This is a figure of speech (in fact, a gross anthropomorphism), of course. However, it is not entirely foolish to imagine that our own memories may, in part, be describable by using molecules with a "memory."

In the philosophical debates over the mind-body problem, this scheme of consciousness as reflection space could be considered either dualistic or monistic, since we are proposing that the world is a single space consisting of two different (but intersecting) algebras. This intersection is identified as universal consciousness. Thus the scheme is monistic geometrically, but dualistic algebraically.

Moreover, there is no problem here with the notion of the mental realm acting on the physical real and vice versa, because both realms in this theory are spaces. In fact, the action of each of these realms upon the other is mediated by the overlapping space — the reflection space.

The question now becomes: is there a reflection space with the appropriate qualities to fulfill this very demanding role? And does it have further properties which confirm it in this role?

Because of the theoretical successes of the $E_8 \times E_8$ superstring theory, we might first try out the E_8 reflection space. However, an examination of the McKay groups *ID, OD* and *TD* of the exceptional Lie algebras makes E_7 with its McKay group *OD* more plausible.

Remember that the balance numbers of the E_7 graph are 1, 2, 3, 4, 2, 3, 2, 1, which are the dimensions of the iirep vector spaces on which *OD* acts. According to the theory of complex group algebras, each *n*-dimensional iirep space corresponds to a unitary Lie group *U(n)* embedded in the group algebra. This, along with other facts of group theory, implies that the maximal compact subspace of the E_7 McKay-group algebra, which we label **C**[*OD*], is the 48-d unitary Lie group (Note: $1^2 + 2^2 + 3^2 + 4^2 + 2^2 + 3^2 + 2^2 + 1^2 = 48$):

$$P = U(1) \times U(2) \times U(3) \times U(4) \times U(2) \times U(3) \times U(2) \times U(1)$$

which we call *P* because it plays the role of the *principal fiber bundle* of our scheme.

A *fiber bundle* is a total space *B* which projects down to a subspace *X* (called the base space) in such a way that all the points of *B* which project to a point of *X* constitute a fiber *F;* moreover, there is a Lie group which acts on each fiber *F* as a symmetry group. In the case of a principal fiber bundle, the fibers are all copies of the symmetry group, so that the action of the symmetry group on a fiber is the action of the Lie group on itself. For any fiber bundle, we can write $b = f + x$, where b, f and x are dimensions of B, F and X, respectively. In fact each point of X can be considered to be a copy of the Lie group F (see Fig. A15).

Every unified field theory is specified, in part, by constructing a principal fiber bundle whose base space is spacetime, and whose fiber is the symmetry group of the world (cf. Bleecker, 1981; Duff, 1986).

For example, the principal fiber bundle of the $E_8 \times E_8$ superstring theory would be a

FIG. A15

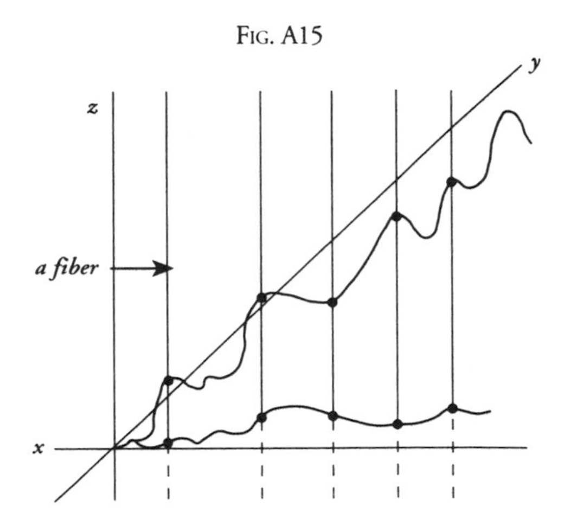

x axis = time
y axis = space
z axis = internal space

A FIBER BUNDLE with 2–d spacetime as base space, and a 1–d fiber. More complicated fiber bundles are not picturable.

If the 1–d fibers were replaced by circles of unit radius — e.g. by connecting the ends of the 1–d fibers — we would have an example of a principal fiber bundle, because a unit circle is equivalent to the U(1) group, and thus each fiber would be a copy of U(1).

The bundle pictured above is a vector bundle, because each fiber is a vector space (1–d, in each case).

The path in the bundle is projected onto the path in the base space via the bundle projection, which projects each fiber onto a point of the base space.

506-d bundle, with a 10-d base space and the 496-d fiber, $G(E_8 \times E_8)$ — i.e., the Lie group generated by the Lie algebra $E_8 \times E_8$. This fiber bundle is a rather large structure, and much ingenuity has been expended in masking most of it in order to make contact with ordinary (low-energy) particle physics which corresponds to the product Lie group $U(1) \times SU(2) \times SU(3)$.

In the scheme of this paper, E_7 is the high-energy symmetry group, but because of the McKay correspondence between E_7 and OD, the unitary Lie group P in $\mathbf{C}[OD]$ contains the low-energy symmetry group. P is a 48-d principal fiber bundle projecting down to a 10-d base space S, each point of which is a copy of the 38-d fiber G. We write these spaces out as the following Lie groups:

$S = U(2) \times T^6$;

$G = U(1) \times SU(2) \times SU(3) \times SU(4) \times SU(2)$

where $U(2)$ is a 4-d spacetime called *conformally compactified Minkowski space;* T^6 is a 6-d torus = $U(1) \times U(1) \times U(1) \times U(1) \times U(1) \times U(1)$.

We consider S as the 10-d spacetime of superstring theory and G as the symmetry group of the following six forces, which we identify in sequence as: electromagnetism, weak, strong (color), hyperweak, gravity and perhaps the feeble force.

We can write: $P = S \times G$. Since $U(2) = U(1) \times SU(2)$, we can rearrange S as $S = SU(2) \times T^7$. We regard $SU(2)$, .i.e., *spherical 3-space* S^3, as the space of cosmology — the space in which we as macroscopic bodies appear to live. Every point of a macroscopic body is a point of S^3. Thus, if we view S as a fiber bundle, every point of a macroscopic body is actually a 7-d space which is a copy of T^7. Note that the 7-torus T^7 incorporates the factor $U(1)$ from the $U(2)$ spacetime, and thus includes time.

Now T^7 corresponds (via McKay's theorem) to the 7-d reflection space of E_7 as follows:

$T^7 = \mathbf{R}^7/L_r$

where $\mathbf{R}^7$ is the *real* part of the E_7 complex reflection space $\mathbf{C}^7$, and L_r is the E_7 root lattice. This means that all the points of the lattice are identified as a single point, the identity element of T^7, and every other point of T^7 is a copy of L_r.

FIG. A16

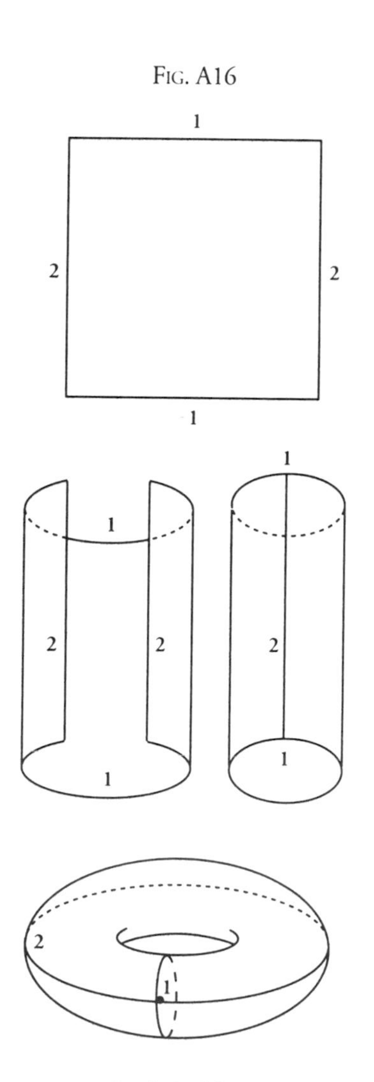

A TORUS can be derived from a square by connecting up the sides labeled 2, and then the sides labeled 1. Note that all four corners of the square become a single point of the torus.

To see how this works, you can generate a 2-torus (the surface of a donut) from a rectangular grid on a 2-d plane. First take one square of this grid, e.g., a sheet of paper, and glue together two opposite edges. You now have a tube. Now connect the ends of the tube to each other, and you will have a 2-torus. Notice that the four corners of the sheet of paper at the same point of the 2-torus (see Fig. A16).

If the sheet of paper with a rectangular grid were infinite in extent, you could still get a 2-torus by rolling the paper an infinity of times along one grid direction and then doing the same along an orthogonal grid direction. In doing this, all the vertices of the grid (the lattice points) become identified as one point in T^2. In fact, since the positioning of the grid of paper is arbitrary, we can consider every point of T^2 as a copy of all the grid lattice points (see Fig. A17).

This procedure for generating tori works even if the grid is not rectangular and is hyperdimensional, as is the case of the E_7 root-lattice L_r.

Since, by our fiber bundle construction, each point of the base space S^3 is a copy of T^7, we can consider each point of a macroscopic body to be a copy of $\mathbf{R}^7/L_r$. Since $\mathbf{R}^7$ is the (real) E_7 reflection space, it is the home of the fundamental eigenvalues of our unification scheme. If we identify consciousness with $\mathbf{C}^7$ space, it would appear that every point of any macroscopic body has access to consciousness. This is why we must consider $\mathbf{C}^7$ as space of universal consciousness. From this point of view, the long evolution of larger and larger brains is the evolution of richer and richer access to the universal consciousness.

Our experience suggests that consciousness is also causal. For example, I can choose to raise my arm. Even though most of what I do occurs unconsciously, these unconscious

FIG. A17

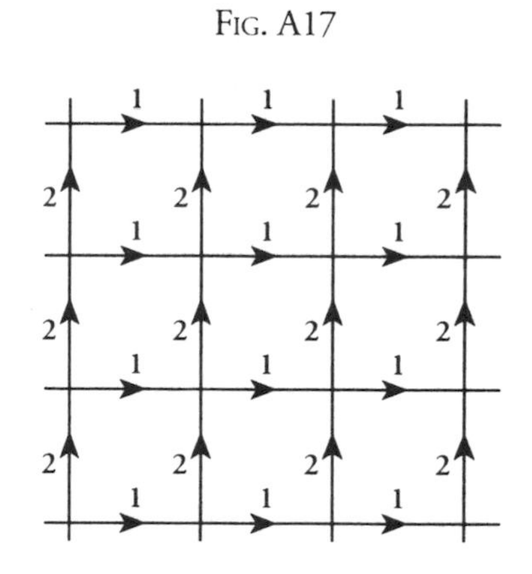

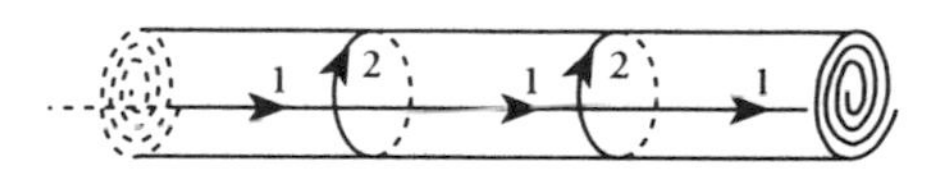

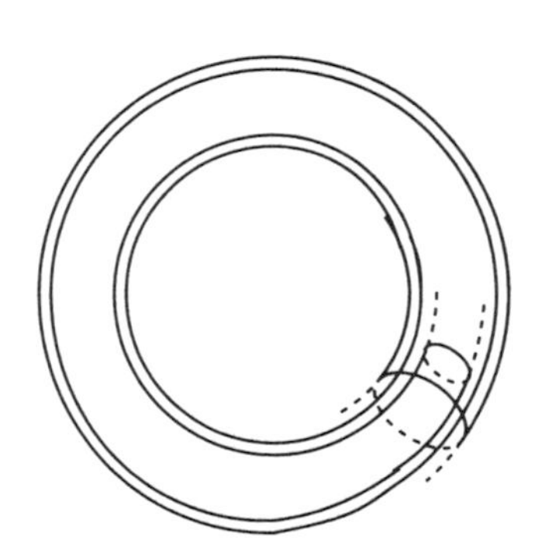

A TORUS can be derived from a lattice on a plane by rolling the plane an infinity of times in one direction so that an infinite tube (of circumference equal to the lattice spacing 2) is produced. Then wrap the tube an infinity of times around a circle of circumference equal to the lattice spacing 1. (*Drawing after Stillwell*)

happenings seem to be ultimately under my control as a conscious entity. Raising my arm, after all, requires a host of activities only some of which are understood by physiologists, cell biologists, molecular biologists, chemists and physicists. Nevertheless, all this unconscious activity is coordinated by my desire to life my arm.

I believe that the causal aspect of consciousness derives also from our access to the E_7 reflection space $\mathbf{C}^7$. In this case it must be viewed as the *critical value space* of a catastrophe (see Figs. A18 and A19).

A *catastrophe* is a large change in a dynamic system caused by a small change in the parameter space on which the system depends. The mathematician Rene Thom (1975) invented catastrophe theory in order to provide a framework for a theory of biological development. Such a theory has not yet emerged. However, catastrophe theory has been successful in describing dynamical systems in the physical sciences. Certainly the mathematics of catastrophe theory is sound and very fruitful. In fact, the Russian mathematician Vladimir Arnold (1981) has been able to define Thom's

FIG. A18

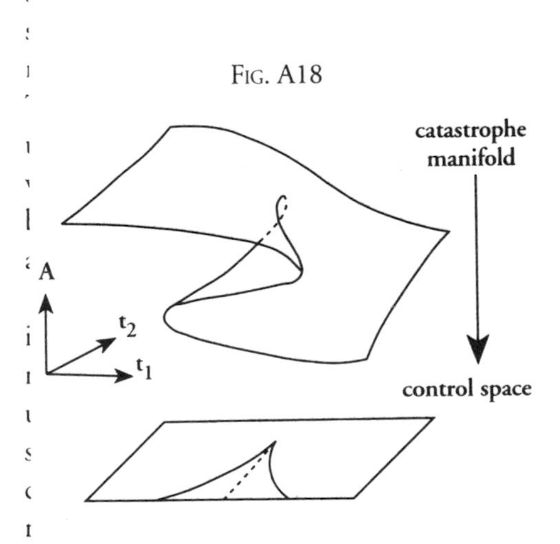

THE A_3 CATASTROPHE is pictured via the real catastrophe manifold, which is the net of critical points of the A_3 catastrophe polynomial:

$$K = A^4 + t_1A^2 + t_2A$$

The catastrophe manifold is projected down to the control space, here 2–d, and the folds of the manifold become cusp lines in the control space.

Such a control space diagram is called a separatrix. The same separatrix occurs if we add any number of squared terms to K.

Fig. A19

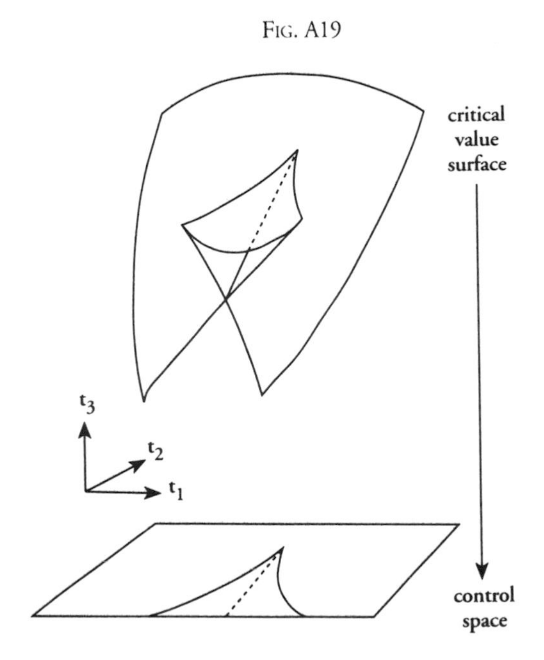

THE A_3 CATASTROPHE is pictured via the real critical value surface embedded in the 3–d base space R^3 (with parameters t_1, t_2, t_3) of the A_3 catastrophe bundle B^5. B^5 is the zero set of the A_3 polynomial:

$$K = A^4 + B^2 + C^2 + t_1A^2 + t_2A + t_3$$

which is the value of K on the catastrophe manifold of critical points of K (Cf. Fig. A18).

K is a complex polynomial. The corresponding polynomial for E_7 is:

$$K = A^3 + AB^3 + C^2 + t_1C^2 + t_2B^3 + t_2B^4 + t_3B^4 + t_4AB + t_5AB^2 + t_6A^2 + t_7$$

seven elementary catastrophes via the Lie algebra reflection spaces. Moreover, Arnold showed that there is an infinite hierarchy of catastrophes such that there is a 1:1 correspondence between the simple catastrophes and the *A-D-E* complex reflection spaces.

Note: A *complex catastrophe* is modeled by a catastrophe bundle, a fiber bundle which is a complex manifold — i.e., each small piece of the manifold looks like a complex vector space of dimension equal to that of the manifold. For example, S^2 can be regarded as a 1-d complex manifold, because each small piece of S^2 looks like a 2-d real plane which can be viewed as the plane of complex numbers, called a 1-d complex plane. Since a complex number is an ordered pair of real numbers, an *n*-dimensional complex space can be considered a $2n$ dimensional real space (see Fig. A20).

The *A-D-E* classification is very deep since it also classifies simple singularities of maps, degenerate critical points of functions, caustics, wave fronts, quivers, crystallographic reflection groups, all the finite subgroups of *SU(2)* and therefore the regular polyhedra. In fact, Arnold (1986) hypothesizes that the *A-D-E* scheme classifies all "simple" objects in mathematics.

Fig. A20

The Manifold S^2: The 2–sphere

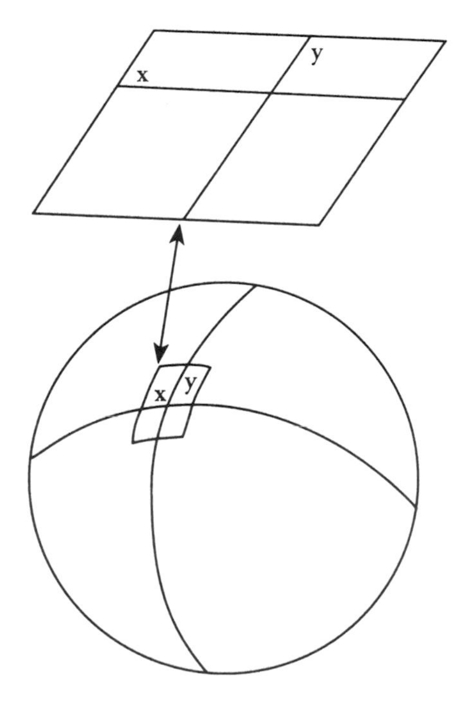

A small piece of S^2 looks like a flat plane. S^2 can be regarded as a 2–d real manifold or as a 1–d complex manifold. In this case, we relabel the y-axis on the plane as *i*y, where $i = \sqrt{-1}$

In other words, there is a complex hierarchical structure, specified by the hierarchy of the *A-D-E* Coxeter graphs. All the mathematical structures listed by Arnold (and probably many yet to be discovered) are different ways of viewing this structure. The advantage of this discovery is that what is almost impossible to see in one view is easy in another view. By using all these views, a deep understanding of this structure will emerge. Arnold calls this field of mathematics by the suggestive name *Platonics* (cf. Arnold, 1986).

The correspondence between reflection spaces and catastrophes is via the *catastrophe germs* listed in Figure A6.

As we will see, for a given Coxeter graph, the actions of the McKay group and the Lie algebra interact in a truly marvelous way in the structure of a catastrophe.

In the case of E_7, we generate the following construction:

$$E_7 \Rightarrow M^9 \Rightarrow \mathbf{C}^7/W \Leftarrow \mathbf{C}T^7 \Leftarrow \mathbf{C}[OD]$$

This means that the 133-d complex Lie algebra E_7 projects down to the 9-d complex-catastrophe bundle M^9, which in turns down to a 7-d *complex vector space* $\mathbf{C}^7/W$, the final base space, which is the *critical value space* for M^9.

(Note: W is the E_7 *Weyl reflection group*, and $\mathbf{C}^7/W$ is the *orbit* space consisting of the set of W orbits in the E_7 reflection space $\mathbf{C}^7$.)

Since the reflection group W acts on complex as well as real spaces, $\mathbf{C}^7$ can be regarded as the complex reflection space of E_7. Remember that $T^7 = \mathbf{R}^7/L_r$, where T^7 is the real E_7 torus, $\mathbf{R}^7$ is the real reflection space of E_7 and L_r is the E_7 root lattice. Moreover, $\mathbf{C}^7$ is the intersection of the Lie algebra E_7 and $\mathbf{C}[OD]$.

Let us focus on the *catastrophe projection:* $M^9 \to \mathbf{C}^7/W$. This is a *fiber-bundle projection* where each point of the base space $\mathbf{C}^7/W$ is a space with a *singularity* (i.e., a point which is not smooth, such as a cusp) (cf. Arnold, 1981). This singularity space F_0 is the *zero set* of the E_7 *catastrophe germ:* $g = A^3 + AB^3 + C^2$. Thus F_0 is the set of solutions derived by setting this polynomial g equal to zero. This polynomial is called a catastrophe germ because each fiber of the bundle M^9 is a *deformation* of the fiber F_0. In fact, we can write out the full E_7 catastrophe bundle polynomial K as:

$$K = A^3 + AB^3 + C^2 + t_1B^2 + t_2B^3 + t_3B^4 + t_4AB + t_5AB^2 + t_6A^2 + t_7.$$

It is interesting that the *degrees* of those *monomials* in K which are *coefficients* of the t-parameters (i.e., 2, 3, 4, 2, 3, 2 and 0 for the constant 1) match the balance numbers on the extended Coxeter graph — if we exclude the node marked with a star (see Fig. 6) — and we have used this fact to make the indexing of the t-parameters match the indexing of the Coxeter graph (see Figs. A7 and A10).

There are 10 complex variables in this polynomial: $A, B, C, t_1,\dots,t_7$. Thus the zero set is a 9-d complex space, i.e., the *complex manifold* M^9.

(Note: in general, the zero set of an n-variable polynomial is an $(n-1)$-dimensional space (cf. Kendig, 1977).

If we substitute a complex number for each of the parameters t_1 through t_7, the solution set of the equation, $K = 0$, is a complex space. Note: Once we specify the seven parameters, there remain only the three variables: A, B, C; (and $3 - 1 = 2$). This 2-d solution set is the fiber attached to the base $\mathbf{C}^7/W$ at the point specified by the seven complex coordinate numbers. This is why we call $\mathbf{C}^7/W$ a parameter space. Note that if we set all seven parameters to zero, i.e., choose the origin of $\mathbf{C}^7/W$, all the terms beyond the germ become zero. Thus the fiber at the origin of $\mathbf{C}^7/W$ is the zero set of the germ itself. There is a smooth *deformation* (or *unfolding*) of the fiber as one traces a path in C^7/W. This smoothness does not preclude "rapid" changes in the fiber corresponding to "slow" changes in the parameter space. This is what is meant by the word *catastrophe* (see Fig. A21).

In the language of catastrophe theory, the six parameters $t_1,...,t_6$ are the parameters of the *control space* of the E_7 catastrophe. This implies that the E_7 control space is a subspace of the E_7 base space C^7/W. This base space is a *critical value space* in the following sense:

> Every mapping has a set of *critical points,* i.e., points on the graph of the mapping which locate qualitative changes. For example, $Y= X^2$ is a mapping from X to Y (each 1-d spaces); the graph of this map is a *parabola* with the lowest point at the origin 0,0) of X, Y space. This point is the only *critical point* of the mapping, since only at this point does the graph change direction from going down to going up. The value of the graph at the critical point is called the *critical value* (e.g., the critical value of the parabola is zero) (see Fig. A22).

FIG. A21

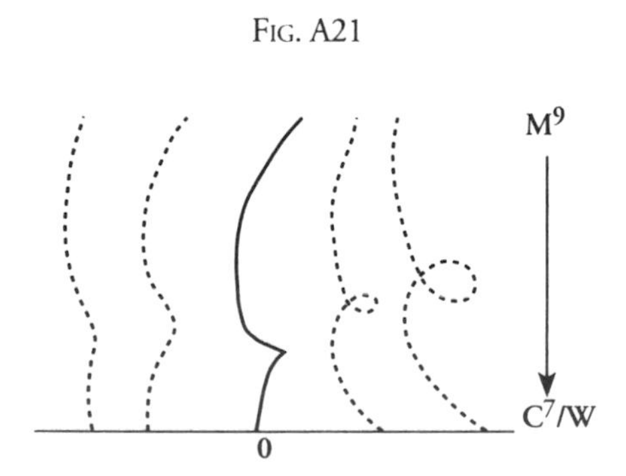

A 2–d fiber bundle analogous to the 9–d complex fiber bundle M^9. The fiber at the origin 0 of the 1–d base space has a cusp singularity. The fiber at the origin of the 7–d complex base space C^7/W is the 2–d complex space C^2/OD. As we pick fibers further from the origin, the singularity becomes more benign (to the right), and even disappears (to the left). The change in the singularity structure of the fiber C^2/OD is determined by the critical surface Σ in C^7/W.

FIG. A22

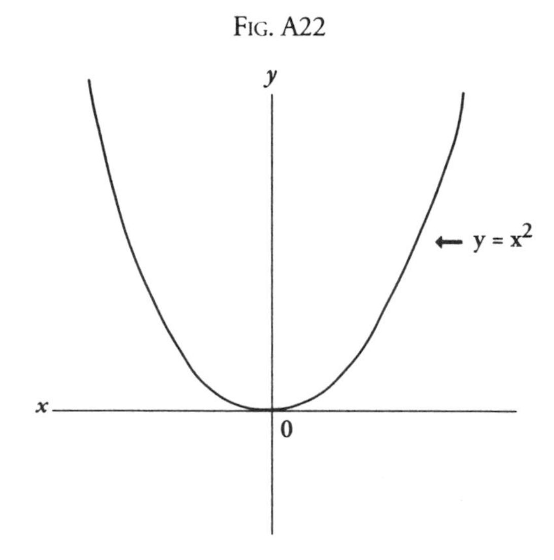

A PARABOLA is the graph of the equation $y = x^2$. The point at the origin 0 is a critical point.

In the much more complicated case of a mapping from C^{10} to C^1, as described by the E_7 catastrophe polynomial K, the set of all values of the t parameters $\{t_1,...,t_7\}$ for which K has zero as a critical value forms a 6-d hypersurface Σ in C^7/W. This set of critical values is called the *critical value hypersurface* (see Fig. A19).

The hypersurface Σ contains a great deal of information about the *singularities* of the fibers attached to C^7/W. Any point of Σ in C^7/W is attached to a fiber which has a singularity. The type of singularity is determined by the nested Coxeter graphs. The most severe singularity occurs at the origin and is described by the E_7 graph. This singular point when *resolved* (i.e., pulled apart to form a non-singular structure) looks like a "bouquet" of seven 2-spheres just touching each other according to the *adjacency pattern* of the nodes of the E_7 graph (see Fig. A23).

As one moves along a path away from the origin in the critical value surface, one picks out fibers whose singularity structures becomes simpler and simpler just as their Coxeter graphs become simpler and simpler. Since Σ is a 6-d space, it is "thin" within the 7-d space $\mathbf{C}^7/W$— just as a 2-d surface is "thin" within the 3-d space. A path crossing this thin Σ will correspond to a rapid change in fiber. Thus a small change along this crossing path corresponds to a large change in the fiber (and therefore in the catastrophe manifold; see Fig. A21).

This kind of activity is the essence of

Fig. A23

The Resolution of the Singular Point in C^2/OD

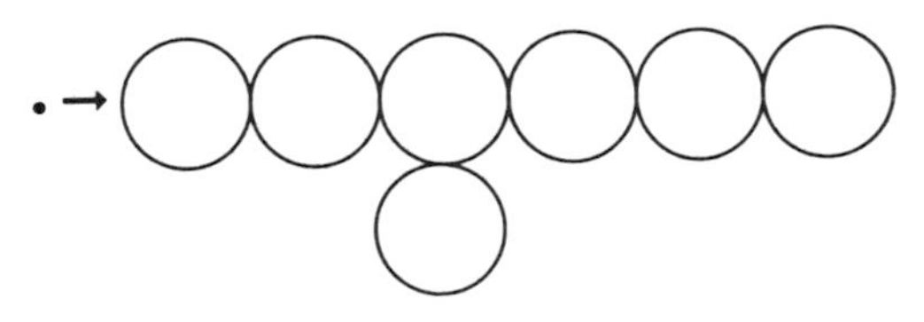

The seven 2–spheres are in contact in accordance with E_7 Coxeter graph.

mental control of the body: a small change in mental activity corresponds to a large change in bodily activity. But how is the fiber in this catastrophe bundle related to bodily activity?

Remember that E_7 corresponds to the mental world (both conscious and unconscious), whereas $\mathbf{C}[OD]$ corresponds to the physical world, the overlap $\mathbf{C}^7$ being universal consciousness.

Amazingly, one can also describe the singular fiber F_0 at the origin of $\mathbf{C}^7/W$ as $\mathbf{C}^2/OD$. In other words, each point of F_0 is actually a set of 48 points and is a copy of OD. This is because OD acts on $\mathbf{C}^2$ by its representation via 2-by-2 matrices. In this action certain polynomials on $\mathbf{C}^2$ remain invariant.

According to the mathematics of *invariant theory* (Springer, 1977) there are three fundamental invariant polynomials of the group OD acting on $\mathbf{C}^2$. This means that any polynomial which is invariant under OD is a polynomial of these fundamental polynomials — i.e., any OD-invariant polynomial can be written as a polynomial whose variables are the three fundamental invariant polynomials. These three polynomials are not independent, but form a relation called a *syzygy*. The OD syzygy can be written in the form $A^3 + AB^3 + C^2 = 0$. Thus A, B and C are not simply variables. They are also invariant polynomials of OD. In order to describe A, B and C, we first describe the fundamental invariant polynomials of the *tetrahedral* group TD, which is a subgroup of OD.

There are three fundamental invariant polynomials of TD:

$f = x^5y - xy^5$; ($f = 0$: 6 vertices of octahedron)

$h = x^8 + 14x^4y^4 + y^8$; ($h = 0$: 8 vertices of cube)

$j = x^{12} - 33x^8y^4 - 33x^4y^8 + y^{12}$; ($j = 0$: 12 vertices of cuboctahedron)

where x and y are complex variables, so that the elements of TD act on (x,y)-vectors.

Note: A vector in an n-dimensional space is an ordered set of n numbers. Here the vector space is $\mathbf{C}^2$, so a vector is an ordered pair of complex numbers.

Using the rules of matrix multiplication, the action of c, one particular element of TD, on the general vector $\mathbf{V} = (x,y)$ is $c\mathbf{V} = \mathbf{V}'$:

$$\begin{matrix} 0 & -1 \\ 1 & 0 \end{matrix} \begin{bmatrix} x \\ y \end{bmatrix} = \begin{bmatrix} -y \\ x \end{bmatrix}$$

Then by substituting $-y$ for x and x for y in the polynomial f, we have:

$f' = (-y)^5x - (-y)x^5 =$

$-y^5x + yx^5 =$

$x^5y - xy^5 = f$

Thus f is invariant under the action of element c. Similarly, f is invariant under all 12 elements of TD. In the same way, h and j are invariant under TD.

Since TD is a subgroup of OD, we expect a close relationship between the fundamental invariant polynomials f, h and j of the group TD and the fundamental invariant polynomials A, B and C of the group OD. In fact, A, B, C are constructed from f,h,j as follows:

$A = f^2;$

$B = h;$

$C = fj.$

The syzygy between the fundamental OD invariants A, B and C, expressed by $A^3 + AB^3 + C^2 = 0$, has the consequence that $\mathbf{C}^2/OD$ is a space which is equivalent to the zero set of $A^3 + AB^3 + C^2$.

Furthermore, the syzygy between A, B and C makes the *OD-invariant algebra* finite dimensional. In fact, this invariant algebra, the set of all OD-invariant polynomials $\mathbf{C}[A,B,C]$ — i.e., the set of polynomials in A, B and C — is also the *coordinate algebra* of $\mathbf{C}^2/OD$. This coordinate algebra is 7-dimensional and has as basis elements the seven *coefficients* of the seven parameters $t_1,\ldots,t_7$. These coefficients are listed in the catastrophe polynomial K and are: B^2, B^3, B^4, AB, AB^2, A^2, 1.

Note: the *coordinate algebra* of a space is the set of all distinct polynomials which live on that space. Thus $\mathbf{C}[A,B,C]$ is the set of distinct polynomials living on $\mathbf{C}^2/OD$.

In *string theory*, the spacetime in which the string vibrates is generated as the set of *scalar fields* living on the 2-d string manifold. Since a *scalar field* on a space is a continuous assignment of a number (a scalar) to each point of the space, a polynomial is a good example of a scalar field. This suggests that we should interpret $\mathbf{C}[A,B,C]$ as a set of scalar fields living on $\mathbf{C}^2/OD$. In other words: the deformations of $\mathbf{C}^2/OD$ are analogous to the vibrations of the string surface.

An important difference between our theory and string theory, however, is that a string surface is a 2-d real space, whereas $\mathbf{C}^2/OD$ is a 2-d complex space, which suggests that our theory entails a complexification of string theory.

Thus we start with 10 complex variables: A, B, C, t_1, ...,t_7. There is a polynomial K in these variables. The zero set of K is a complex space M^9, called the catastrophe manifold. There is a fiber bundle projection, $M^9 \Rightarrow \mathbf{C}^7/W$, such that the fiber F_0 at the origin of $\mathbf{C}^7/W$ is a singular space $\mathbf{C}^2/OD$ (see Fig. A21).

Remember: $\mathbf{C}^7/W$ is the set of points (an orbit) reached by acting on one point of $\mathbf{C}^7$ with all the elements of the E_7 reflection group W acting on $\mathbf{C}^7$.

A, B and C are the fundamental invariants of the OD group acting on $\mathbf{C}^2$; $t_1,\ldots,t_7$ are fundamental invariants of the $E7$ reflection group W acting on $\mathbf{C}^7$.

The actions of OD and W are very different because OD is not a reflection group, whereas W is a reflection group. The action of a finite *nonreflection* group on a complex vector space is to create a space of the same dimension but different *topology*. The action of a *reflection* group on a complex vector space is to create another copy of that space — i.e., we regard $\mathbf{C}^7/W$ as a copy of $\mathbf{C}^7$ because they are isomorphic vector spaces.

The action of OD on $\mathbf{C}^2$ creates the complex string surface $\mathbf{C}^2/OD$ which is the zero set of the germ $A^3 + AB^3 + C^2$. The deformations of $\mathbf{C}^2/OD$ are parameterized by $t_1,\ldots,t_7$. Thus for each point t of $\mathbf{C}^7/W$ a different perturbation of $\mathbf{C}^2/OD$ is selected.

Since the action of W of $\mathbf{C}^7$ creates another copy of $\mathbf{C}^7$, $\mathbf{C}^7/W$ is also a 7-d complex vector space. However, $\mathbf{C}^7/W$ is related to $\mathbf{C}^2/OD$ in a wonderful and useful way:

C^7/W contains the E_7 critical value surface Σ, which is a 6-d hypersurface cutting through itself in a complicated way and thus dividing the reflection space into many regions. Every point of Σ picks out a deformation of C^2/OD which contains a singularity. Since C^7/W is the set of orbits of W in C^7, W creates Σ in C^7/W in the sense that Σ is the set of *nonregular* W orbits in C^7.

Note: W has 288 x 7! x 2 = 2,903,040 elements. If the number of distinct points in an orbit of W is equal to this element number, the orbit is called *regular;* otherwise it is called *nonregular.* As it turns out, each point of Σ in C^7/W is a nonregular element — i.e. consists of fewer than 2,903,040 distinct points of C^7.

In case the number 288 x 7! x 2 looks too abstract, we can remember that 288 = 1 x 2 x 3 x 4 x 3 x 2 x 1 x 2, which are the node numbers appearing on the extended E_7 Dynkin diagram and are thus also the dimensions of the iireps of the OD group. (This is one more way in which the structure of OD and W intertwine). The 7! (7-factorial) is 7 x 6 x 5 x 4 x 3 x 2 x 1, which is the number of elements in the symmetric-7 group S_7, which is the group of all permutations of seven objects (e.g., the basic mirror plans of C^7), and the factor 2 corresponds to the bilateral symmetry of the extended E_7 Coxeter graph itself. Thus we have a geometric meaning for all of the factors in the number of W elements 288 x 7! x 2.

We can use geometric reasoning to clarify further the orbit mapping $C^7 \Rightarrow C^7/W$ as follows (cf. Bott, 1979):

> Consider that the 63 mirror planes in C^7 cut this reflection space into 2,903,040 Coxeter chambers, since each chamber can be reached from the fundamental chamber by an action of one of the elements of the Weyl group, each of whose elements is a series of reflections. Thus any point inside any of these Coxeter chambers is copied by reflections into all the other Coxeter chambers, thereby generating a regular orbit of that point. Moreover, in the orbit mapping C^7 fi C^7/W, this regular orbit of the Coxeter chamber point is mapped onto a point outside the critical value surface Σ.

In contrast to this regular orbit mapping, if a point belonging to one of the 63 mirrors of C^7 is chosen and acted on by reflections, that mirror point will be reflected only onto other mirrors. Thus the biggest mirror orbit will be a reflection onto all 63 mirrors. These orbits of mirror points are clearly nonregular orbits. In effect, the orbit mapping $C^7 \Rightarrow C^7/W$ transforms a mirror point into a point of the critical surface Σ, and a chamber point into a point out side this critical surface.

To summarize: Within E_7 there is a fiber bundle projection

$M^9 \Rightarrow C^7/W$

where M^9 is a compact, complex manifold, which is the zero set of a polynomial K in the 10 variables $A, B, C, t_1, \ldots, t_7$. The fiber at the origin O_7 of C^7/W is C^2/OD, which is the zero set of $A^3 + AB^3 + C^2$. Each point of C^7/W, parameterized by $t_1, \ldots, t_7$, picks out a perturbation of C^2/OD — i.e., a different fiber in M^9.

A, B and *C* are fundamental invariants of OD acting on C^2.

$t_1, \ldots, t_7$ are fundamental invariants of OD acting on C^7.

OD is a nonreflection group acting on C^2; and thus this action creates an orbit space C^2/OD which has a singularity: the zero set of $A^3 + AB^3 + C^2$, which is a 2-d closed, complex surface.

W is a reflection group on C^7; thus this action creates an orbit space C^7/W, which is isomorphic to C^7. Under this orbit mapping, the 63 mirrors in $\mathbf{C}^7$ get mapped onto the critical value surface Σ in $\mathbf{C}^7/W$, and the space between the mirrors (the Coxeter chambers) gets mapped onto the space outside the critical surface Σ.

Σ is called a critical value surface because crossing it corresponds to undergoing a rapid change in the catastrophe bundle M^9. Moreover, the points of Σ pick out singularity-containing perturbations of $\mathbf{C}^2/OD$.

All of this activity is going on inside the 133-d Lie algebra E_7, because M^9 is a subspace of E_7; W is the Weyl reflection-group of E_7; and $\mathbf{C}^7/W$ is the action of W on the largest commutative subalgebra $\mathbf{C}^7$ of E_7.

With this summary of the E_7 side of the theory, we must now look more closely at the $\mathbf{C}[OD]$ side.

Since $\mathbf{C}^2/OD$ is the identity fiber in M^9, we are reminded that *OD* plays a fundamental role within the E_7 algebra; in fact, $\mathbf{C}^7$ is the intersection of E_7 with $\mathbf{C}[OD]$. As we have seen, $\mathbf{C}^7$ (universal consciousness) in the form of $\mathbf{C}^7/W$ controls the deformation structure of $\mathbf{C}^2/OD$ in the space E_7, which we have identified with universal mind.

FIG. A24

THE 24–CELL

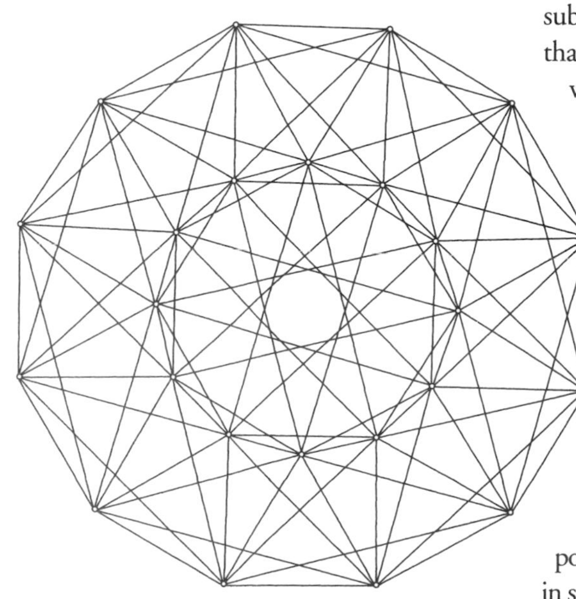

This is a 2–d projection of a 4–d figure which has 24 vertices and 24 three–d faces (called cells), each of which is an octahedron. (Coxeter, 1973)

We have yet to identify $\mathbf{C}^2/OD$ and its deformation with anything other than the fibers of the E_7 catastrophe bundle M^9. Having identified $\mathbf{C}[OD]$ with the physical world, we expect $\mathbf{C}^2/OD$ to have an intimate connection with the physical world. One could say that $\mathbf{C}^2/OD$ and its perturbations are mental images of the physical world. This is because the deformations of $\mathbf{C}^2/OD$ are controlled by the critical space $\mathbf{C}^7/W$, which is the set of W orbits in $\mathbf{C}^7$, the intersection of the mental world E_7 and the physical world $\mathbf{C}[OD]$. In effect, $\mathbf{C}^2/OD$ is a counterpart to ordinary physical space — i.e., cosmic space S^3. This is suggested by the mathematical properties of these spaces:

$\mathbf{C}^2$ can be viewed as the set of quaternions $\mathbf{Q}$, since a quaternion is an ordered pair of complex numbers. S^3 can be viewed as the set of unit quaternions. *OD* is a finite subgroup of S^3. And in fact, it is well known that S^3/OD = the intersection of $\mathbf{C}^2/OD$ with S^5 (the unit sphere in $\mathbf{C}^3$, the space parameterized by *A*, *B* and *C*, the fundamental invariants of *OD*) (cf. Milnor, 1968,1975). In other words, S^3/OD provides a good approximation to the topology of $\mathbf{C}^2/OD$ near the singular point.

Incidentally, *OD* is a very symmetric set of 48 points in S^3. OD is two copies of the *24-cell* (each reciprocal to the other). One 24-cell is the subgroup *TD*.

Note: The 24-cell is a unique regular polytope in $\mathbf{R}^4$ space, since there is no analog in spaces of higher or lower dimension. The 24 vertices of the 24-cell (i.e., the points *TD* in S^3) are the points at which the 24 S^3-spheres touch the central S^3 sphere in the most efficient *sphere packing* in $\mathbf{R}^4$ (see Fig. A24).

Thus the 24-cell as a sphere-packing figure has analogs in other spaces, such as $\mathbf{R}^7$, where the 126 sphere-packing points are the E_7 roots. Because of the hierarchy of the reflection space structures, the 24-cell is a substructure of the 126 vertex E_7 root figure V_{126}.

This suggests that the 24-cell is a structure mediating between V_{126} and the V_{48}, the double 24-cell or *OD*.

Moreover, all these considerations suggest that the cosmic space S^3 as embedded in $\mathbf{C}[OD]$ must be a fiber similar to the fiber $\mathbf{C}^2/OD$.

In order to describe the relationship of $\mathbf{C}^2/OD$ to S^3, we can display Figure A25, where (1) is a fiber bundle projection with identity fiber $\mathbf{C}^2/OD$, and (2) is a fiber bundle projection with identity fiber $S^3 \times S^3$. (Remember that $\mathbf{C}^7/L_r$ is the complex 7-torus $\mathbf{C}T^7$.) Therefore the *spliced bundle* $SB^{181} = E_7 \times \mathbf{C}[OD]$ contains the *spliced sub-bundle* $M^9 \times [T^7 \times S^3]^{\mathbf{C}}$ with the projection:

$$M^9 \times [T^7 \times S^3]^{\mathbf{C}} \Rightarrow \mathbf{C}T^7/W$$

with identity fiber $(\mathbf{C}^2/OD) \times S^3 \times S^3$. As different points along a path in $\mathbf{C}T^7/W$ are chosen, the identity fiber undergoes perturbations. Since $\mathbf{C}T^7/W$ is the Lie group version of $\mathbf{C}^7/W$, these perturbations are described by the critical value surface Σ in $\mathbf{C}^7/W$.

Thus we have identified the physical counterpart of $\mathbf{C}^2/OD$ with $S^3 \times S^3$. Notice that both of these structures are complex, and that the real part of $\mathbf{C}^2/OD$ is a 2-d surface while the real part of $S^3 \times S^3$ is the spherical cosmic space S^3. We can regard $S^3 \times S^3$ as complexified cosmic space.

This implies that every deformation of $\mathbf{C}^2/OD$ is linked to a deformation of the cosmic space S^3. We can view each point of S^3 as a copy of $\mathbf{C}^2/OD$, and each point of a perturbed S^3 as a copy of a perturbed $\mathbf{C}^2/OD$. The perturbations are controlled by the points of $\mathbf{C}^7/W$. Thus each path through $\mathbf{C}^7/W$ corresponds to a different evolution of the fiber $\mathbf{C}^2/OD \times S^3 \times S^3$.

What about *time?* Remember that S^3 is the space part of *cosmic spacetime* $U(2)$, because $U(2) = T^1 \times S^3$. The one-torus (a circle) T^1 is part of the T^7 in the *superstring spacetime* $S^3 \times T^7$. Thus time becomes complexified as a parameter t_1 in the seven-parameter basis of $\mathbf{C}^7/W$. This implies that *time is a part of universal consciousness.*

The fact that we must identify t_1 with time raises the question: What is the interpretation of the fibration $M^9 \Rightarrow \mathbf{C}^7/W$ such that t_1 as a deformation parameter of $\mathbf{C}^2/OD$ can be considered as time?

We can consider the critical surface Σ in $\mathbf{C}^7/W$ to be a family of *6-d wave fronts* evolving in time; under this interpretation t_1 does indeed become a time parameter, so that $\mathbf{C}^7/W$ is a complex 7-d spacetime. The singularities of the wave fronts are described by the E_7 singularity structure (cf. Arnold, Gusein-Zade, Varchenko, 1985).

Note: The *same* time parameter t_1 plays three roles:

1. *Cosmic time* in the 4-d spacetime: $U(1=2) = T^1 \times S^3$
2. *Supertime* in 10-d superstring spacetime: $T^7 \times S^3$
3. *Complex time* in the 7-d consciousness spacetime $\mathbf{C}^7/W$.

Fig. A25
The Spliced Bundle (SB[181]) Mappings

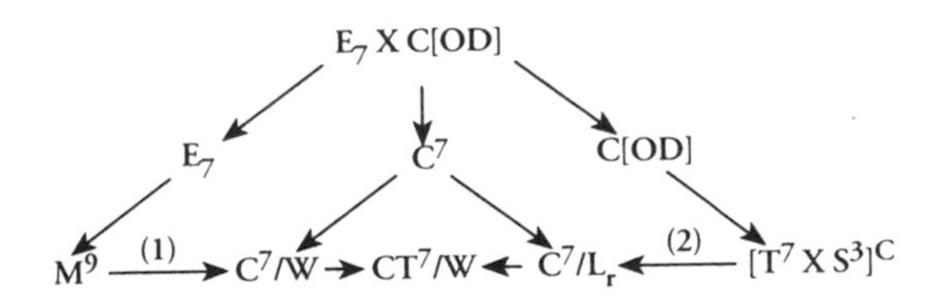

The picture we now have is that (corresponding respectively to the three roles of the time parameter) every spacetime path in $\mathbf{C}^7/W$ synchronizes three evolutions:

1. An evolution of cosmic space S^3

2. A series of deformations of the $\mathbf{C}^2/OD$ fiber

3. An evolution of wave fronts in $\mathbf{C}^7/W$.

Because of the fact that $\mathbf{C}^7/W$ is the base space of the SB^{181} fiber bundle, $\mathbf{C}^7/W$ contains an *image* of all the most essential structure of the entire SB^{181} space. Remembering that $\mathbf{C}^7/W$ is the Lie group version of the Lie algebra space $\mathbf{C}^7/W$, we can say that the paths in $\mathbf{C}^7/W$ — i.e., the set of evolutions of wave fronts in $\mathbf{C}^7/W$ — are the images of the perturbations of $\mathbf{C}^2/OD$, as well as the evolutions of S^3.

Every point of S^3 is a location in the space of the macroscopic world. However, the identity-subfiber s_0 in SB^{181} is the direct product of $(\mathbf{C}^2/OD) \times S^3 \times S^3$. Thus every point of S^3 can be regarded as a copy of $\mathbf{C}^2/OD$. This view is especially appropriate in the present state of the universe — i.e., S^3 is of "cosmic" size, approximately 10^{28} cm in radius. Thus, as we look at smaller and smaller regions of S^3, we will see the structure of the 4-(real)-dimensional $\mathbf{C}^2/OD$ emerge, rather than a mere 0-d point.

Remember that every point along a path in $\mathbf{C}^7/W$ selects a different deformation (or unfolding) of $\mathbf{C}^2/OD$. The critical value surface Σ in $\mathbf{C}^7/W$ (i.e., the family of wave fronts) determines what happens along the path (once the path is chosen). The zero point O_7 (the origin) of $\mathbf{C}^7/W$ selects $\mathbf{C}^2/OD$ itself. This a complicated space with one singular point (rather like a line crossing itself three times at one point). As one moves along a chosen path away from O_7, the deformations of $\mathbf{C}^2/OD$ change this 2-d fiber so that it becomes less and less complicated. If the path selects a fiber outside Σ, the singularity in the fiber goes away.

It is natural to suppose that O_7 corresponds to the "big bang" singularity of cosmology. The cosmological singularity, however, refers to a condition of infinite density of the cosmos. This is a different meaning of "singularity" than that associated with $\mathbf{C}^2/OD$.

The study of spaces such as $\mathbf{C}^2/OD$ is a branch of mathematics called *singularity theory* (cf. Arnold, 1981), and it deals with "singular" points such as cusps and nodes, where the space becomes unsmooth. (For the sake of reader who have studied calculus, let me say that at such unsmooth points on a curve the derivative becomes undefined, and we say that the curve is undifferentiable at a singularity; analogous statements about partial derivatives hold for higher dimensional unsmooth spaces, which are undifferentiable at a singular point.)

It is striking, however, that the curve which represents the *scale factor R* of the universe as a function of time has a *cusp singularity* at $t = 0$, just when the *cosmological singularity* occurs. For a *compact* universe (e.g., S^3) this curve is the *cycloid curve* (i.e., the curve traced out by a point on the rim of a rolling wheel); there is also a cusp singularity corresponding to the cosmic singularity of the "big crunch" (see Fig. A26).

This cosmic cycloid curve would be a subspace of the parameter space $\mathbf{C}^7/W$, which includes time. It is possible to regard the scale factor R as a measure of a variable gravitational "constant" G (cf. Sirag, 1983). In other words, G is effectively one of the seven parameters in $\mathbf{C}^7/W$.

More correctly, we should regard G as a *toroidal* parameter of $\mathbf{C}^7/W$. As we have said, $\mathbf{C}^7/W$ is the Lie group version of the Lie algebra space $\mathbf{C}^7/W$. The relationship between a Lie group *LG* and Lie algebra *LA* can be expressed via the *exponential map:*

$$LG = \exp(LA)$$

This implies that in our case the parameter g in $\mathbf{C}^7/W$ which corresponds to G in T^7 has the following relationships:

$e^g = G;$

$e^0 = 1;$

$e^{-0.69} = 1/2.$

where e is the base of the *natural logarithms:* 2.71828...etc.

If we set $G = 1$ (i.e., $g = 0$ at the time $t = 0$, then as the cosmos S^3 expands to its maximum R, the gravitational parameter G goes to 1/2 (i.e., g goes to –0.69). Similarly, as S^3 contracts to its minimum size, $G = 1/2$ ($g = -0.69$) goes back to $G = 1$ ($g = 0$). In other words, as G oscillates between 1 and 1/2 (i.e., g oscillates between 0 and –0.69), S^3 oscillates between $R = 1$ and $R = 10^{41}$. Thus a small change in the parameter space $\mathbf{C}^7/W$ corresponds to a large change in the fiber S^3 (cf. Sirag, 1983; Marciano, 1984; Appelquist et al., 1985).

Since each point of S^3 is a copy of $\mathbf{C}^2/OD$, we should also consider the changes in the radius l of $\mathbf{C}^2/OD$. According to string theory arguments, l is approximately 10^{-30} cm, which in natural units is $137G^{1/2}$. Thus, by the construction considered here, as G goes from 1 to 1/2, l goes from 10^{-30} cm to $.707(10^{-30})$ cm. This small change in d (less than one order of magnitude) corresponds to a change of 41 orders of magnitude in R, radius of S^3. Moreover, the changes go in the opposite direction: d gets smaller as R gets bigger, and vice versa.

We are in a position to describe the relationship of mind to body. There is a *universal body,* $\mathbf{C}/OD$. There is a *universal mind,* E_7. The intersection of the two — $\mathbf{C}^7$ — is *universal consciousness.*

There is a *universal geometric entity,* the spliced bundle SB^{181}, which is the direct product $E_7 \times \mathbf{C}/[OD]$.

Notice that from the point of view of SB^{181}, the system is monistic; but from the point of view of the constituent algebras E_7 and $\mathbf{C}[OD]$, the system is dualistic.

SB^{181} projects down to the base space $\mathbf{C}^7/W$. The subfiber attached to the identity element of $\mathbf{C}^7/W$ is $S_0 = (\mathbf{C}^2/OD) \times S^3 \times S^3$.

Every observer corresponds to a different path in $\mathbf{C}^7/W$— i.e., each observe has his own cosmos, his own copy of s_0 carried along the $\mathbf{C}^7/W$ path. Each point along the path is, in effect, a new observation since a new copy of s_0 is selected for each point of the path.

In one sense, each observer is separate, having his own path in reflection space $\mathbf{C}^7$. However, in $\mathbf{C}^7/W$ many coincide part of the way. The coincidence of two or more paths over any distance is called the *contact* between the paths. Remember, these are not paths in spacetime, but paths in critical space $\mathbf{C}^7/W$ (which includes complexified time). Moreover, since we have identified the reflection space $\mathbf{C}^7$ with universal consciousness, the contact structure of $\mathbf{C}^7/W$ must

FIG. A26

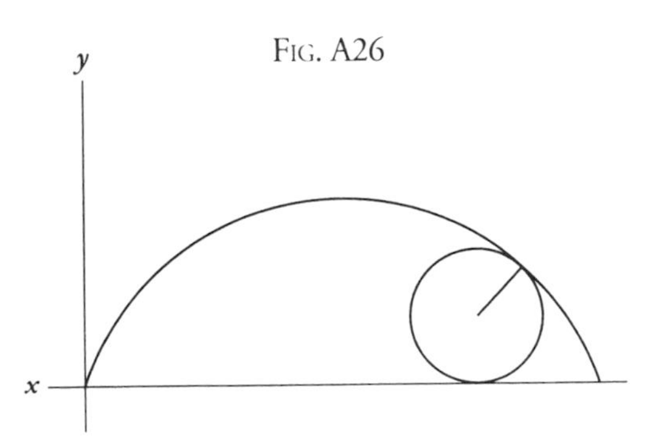

THE CYCLOID CURVE is the line traveled by a point on the rim of a rolling wheel.

Applied to cosmology:

x = time

y = the radius of a spherical universe S^3

play an important role in coordinating the observations of separate observers. For example, even though two separate observers live in separate "universes" via paths in $\mathbf{C}^7$, they can communicate with each other meaningfully if they are in contact via paths in $\mathbf{C}^7/W$— i.e., while the paths are in contact, the observers see the same universe, because they experience the same eigenvalues.

In fact, since eigenvalues are observations, and the reflection space has been identified with eigenvalue space, we can make the following postulate:

> Within $\mathbf{C}^7$, point a is aware of point b if and only if point a is in contact with point b in $\mathbf{C}^7/W$— i.e., a and b in $\mathbf{C}^7$ are mapped onto the same point of $\mathbf{C}^7/W$.

This postulate justifies, in a fundamental way, our calling the space $\mathbf{C}^7$ consciousness (cf. Culbertson, 1982).

Notice that we are not explaining consciousness here. Rather we are defining the conditions under which awareness (an aspect of consciousness) exists between two points. We are assuming the existence of consciousness as fundamental. As some philosophers (e.g., Descartes) have emphasized, the existence of consciousness is the safest starting point for any theory of reality — i.e., if the philosopher knows nothing certain about the world, he at least knows that he is aware of his uncertainty, so he can begin his speculations on reality by assuming his own awareness.

Given the above postulate, we can see that the "purpose" of the projection from $\mathbf{C}^7$ onto $\mathbf{C}^7/W$ is to make points which are separate in $\mathbf{C}^7$ identical in $\mathbf{C}^7/W$— i.e., bring the points into contact and thus into mutual awareness.

Moreover, it becomes clear that a brain must be a set of points in cosmic space S^3 attached to a set of paths in $\mathbf{C}^7/W$ in such a way as to increase its contact structure. The entire evolution of biological entities is in general an evolution of increasing the richness of contact (and thus awareness).

The nature of the contact structure of all the paths in $\mathbf{C}^7/W$ is determined by the orbit structure of the E_7 reflection group W acting on $\mathbf{C}^7$. Remember that a single point of $\mathbf{C}^7/W$ is an orbit, which is a set of (as many as 2,903,040) points in $\mathbf{C}^7$. Thus the reflection group W weaves together sets of points in $\mathbf{C}^7$ to create points in $\mathbf{C}^7/W$, so that paths in contact in $\mathbf{C}^7/W$ can be lifted to separate paths in $\mathbf{C}^7$, while separate paths in $\mathbf{C}^7$ can be *lowered* to paths in contact in $\mathbf{C}^7/W$. (Note: Lifting and lowering are the standard mathematical terms for these mappings.)

Moreover, there is a close relationship between the contact structure of $\mathbf{C}^7/W$ and the wave front structure Σ, because Σ in $\mathbf{C}^7/W$ is the set of nonregular W orbits in $\mathbf{C}^7$.

Remember: The 63 mirror planes in $\mathbf{C}^7$ are mapped onto Σ in $\mathbf{C}^7/W$; the point where all the mirrors intersect is the origin of $\mathbf{C}^7/W$, to which is attached the identity fiber $\mathbf{C}^2/OD$ (cf. Bott, 1979).

Because of the relationship between the contact structure and the wave front structure in $\mathbf{C}^7/W$, we can see that there must be a close relationship between the contact structure and the causal properties of both the *mental fiber* $\mathbf{C}^2/OD$ and the *physical fiber* S^3. This means that there is an intimate relationship between *awareness* (contact) and *volition* (wave fronts).

At this point the reader may feel that the choice of $\mathbf{C}^7$ as the space of primary consciousness is arbitrary. Why not $\mathbf{C}^8$ or $\mathbf{C}^9$ or infinite dimensional C-space?

The answer is that we have identified $\mathbf{C}^7$ as the space of consciousness because it is the intersection of E_7 with $\mathbf{C}[OD]$; in turn, these algebras are brought into the picture by the structure of a unified field theory which unites the forces as we know them in the physical world. The physical world is decisive in this discussion because we know that physical factors affect consciousness (e.g., close your eyes) and that consciousness affects the physical world (e.g., you can decide to close, or not to close, your eyes).

But from a purely mathematical point of view, our choice of $\mathbf{C}^7$ is arbitrary. We could have chosen any of infinity of finite subgroups of *SU(2)* as our starting point. Call that finite subgroup (a McKay group) g; then we could write:

$\mathbf{C}^n$ is the intersection of X_n with $\mathbf{C}[g]$

where X_n is some Lie algebra: A_n, D_n, or E_6, E_7, E_8. And $\mathbf{C}[g]$ is the group algebra of (McKay) group g. Thus we can form a spliced-bundle projection:

$$X_n \times \mathbf{C}[g] \Rightarrow \mathbf{C}^n$$

In fact, as we have pointed out early in this paper, there is a hierarchy of these *A-D-E* algebras, so that, for instance, we have the projections:

$$E_8 \Rightarrow E_7 \Rightarrow E_6 \Rightarrow D_5 \Rightarrow D_4 \Rightarrow A_3 \Rightarrow A_2 \Rightarrow A_1$$

The structure of this hierarchy derives from the structure of the Coxeter graph of these algebras. The subscripts are the ranks of the algebras and correspond to the number of nodes in the algebra's Coxeter graph. Not only is the algebra of lower rank a subalgebra, but, more important, these projections entail a hierarchy of singularity structure — i.e., the perturbations of the higher rank singularity contains the deformations of the lower rank singularity.

The difference between any of the *A-D-E* spliced-bundle schemes and the scheme of this paper is that quite different unified field theories would be entailed — i.e., different from the physical forces we know about.

It is conceivable that the entire *A-D-E* set of schemes is active, and that we have here described only one scheme E_7. If this were so, there would be an infinite set of consciousness structures *(realms)*, all intimately tied together by the hierarchical structure of the *A-D-E* classification based on the Coxeter graphs (see Fig. A27).

There are many possible applications of such a hierarchy. Since the actual existence of such a hierarchy of realms is speculative, we shall content ourselves with the observation that it is suggested by the mathematical structure of our unification scheme E_7.

This is the same position that Maxwell found himself in when (in 1864) he unified electricity and magnetism and discovered the electromagnetic theory of light. The mathematics of his unification suggested to him the speculative idea that visible light is only a small part of an infinite spectrum of light frequencies. In due course, radio waves, X rays and other forms of light verified his speculation.

It may be that there is an infinite set of consciousness realms, hierarchically organized according to the *A-D-E abutment* scheme as depicted in Figure A27. It should be mentioned that the hierarchy of abutments corresponds as well to hierarchy of control in the sense that higher catastrophe structures embed and control lower catastrophe structures (cf. Gilmore, 1981).

This means that, if the hierarchy is actual (not merely mathematical), there is a

Fig. A27

The A–D–E Hierarchy

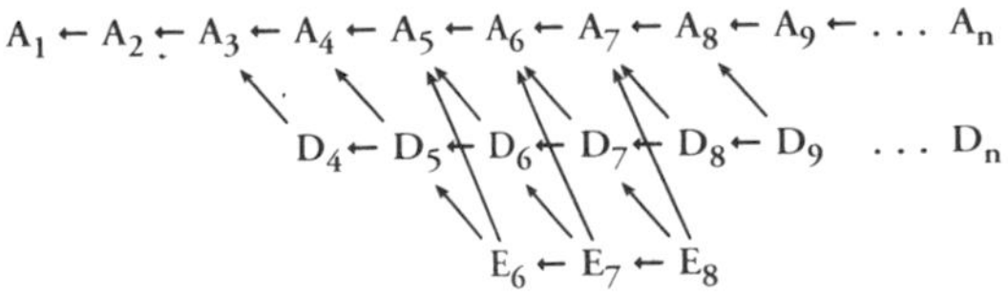

$A_1 \leftarrow A_2 \leftarrow A_3 \leftarrow A_4 \leftarrow A_5 \leftarrow A_6 \leftarrow A_7 \leftarrow A_8 \leftarrow A_9 \leftarrow \ldots A_n$

$D_4 \leftarrow D_5 \leftarrow D_6 \leftarrow D_7 \leftarrow D_8 \leftarrow D_9 \quad \ldots D_n$

$E_6 \leftarrow E_7 \leftarrow E_8$

whole hyperphysical world with its own set of forces above us, the E_8 realm, and also a world directly below us, the E_6 realm.

The connections between these separate realms is via the spaces of consciousness:

$$C^8 \Rightarrow C^7 \Rightarrow C^6$$

because the Dynkin diagrams are nested according to this hierarchy and the lattices and Weyl groups are embedded according to this hierarchy.

At this point it is useful to mention that there is indeed something special about the three algebras E_8, E_7, E_6. For one thing, the corresponding McKay groups are symmetry groups of the Platonic solids: icosahedron (and dodecahedron), octahedron (and cube), tetrahedron. Perhaps more important for a theory of consciousness is the fact that the root lattices corresponding to these three algebras are generated by error-correcting codes (cf. Conway and Sloane, 1988).

This is a large subject with much connection to other aspects of the theory presented here (including unified field theory). I will just point out that the E_7 lattice is generated by the Hamming-7 code and that the E_8 lattice is generated by the Hamming-8 code. Moreover, *h,* a fundamental invariant polynomial of the *OD* group, is the *weight polynomial* of the Hamming-8 code. Thus there is a connection between E_7 and E_8 via coding theory, in addition to the ones we have already considered. Coding theory is an application of *information* theory, so that it is natural to suppose a connection between coding theory and cognitive aspects consciousness. Needless to say, this is an active area of my own research (cf. Sirag, 1984, 1986).

There is much room for further speculation and comparison with philosophical and mystical ideas. I will mention only one such mystical idea (cf. Chang, 1971):

Fa Tsang (643–712), the Chinese master of Hwa Yen Buddhism, prepared for the Empress Wu an octagonal room completely covered with mirrors, including the floor and ceiling. In the center he placed an image of the Buddha with a burning torch. He brought the Empress into this room and said (in part):

Your Majesty, this is a demonstration of Totality in the Dharmadhatu. In each and every mirror within this room you will find the reflections of all the other mirrors with the Buddha's image in them. And in each and every reflection of any mirror you will find all the reflections of all the other mirrors, together with the specific Buddha image in each, without omission or misplacement. The principle of interpenetration and containment is clearly shown by this demonstration. Right here we see an example of one in all and all in one — the mystery of realm embracing realm ad infinitum is thus revealed. The principle of simultaneous arising of different realms is so obvious here that no explanation is necessary. These infinite reflections of different realms now simultaneously arise without the slightest effort; they just naturally do so in a perfectly harmonious way....

As for the principle of the nonobstruction of space, it can be demonstrated in this manner....(saying which, he took a crystal ball from his sleeve and placed it in the palm of his hand). Your Majesty, now we see all the mirrors and their reflections within this small crystal ball. Here we have an example of the small containing the large, as well as of the large containing the small. This is a demonstration of the nonobstruction of "sizes," or space.

As for the nonobstruction of times, the past entering the future and the future entering the past cannot be shown in this demonstration, because this is, after all, a static one, lacking the dynamic quality of the temporal elements. A demonstration of the nonobstruction of times, and of time and space, is indeed difficult to arrange by ordinary means. One must reach a different level to be capable of witnessing a "demonstration" such as that. But in any case, your Majesty, I hope this simple demonstration has served its purpose to your satisfaction.

Notice that the suggestion of a need for a "different level" to explain the temporal aspects of the mystical experience is a suggestion that reality is hyperdimensional. Moreover, we have found in the catastrophe and wave front evolution structure a dynamical description of this hyperdimensional realm.

Let us quote from the *Hwa Yen Sutra* itself (cf. Chang, 1971):

> The Indescribable-Indescribable
> Turning permeates what cannot be described....
> It would take eternity to count
> All the Buddha's universes.
> In each dust-mote of these worlds
> Are countless worlds and Buddhas....
> An excellent mathematician could not enumerate them
> But a Bodhisattva can clearly explain them all....

Perhaps a seventh-century mathematician could not cope with the mystic vision, but Fa Tsang appears to have had a method that can be enlarged upon significantly by today's mathematics. Moreover, the mathematics of singularity theory and the *A-D-E* hierarchy is rather recent, and therefore many aspects of this hierarchy remain to be discovered. I will close with a speculation of one of the most perceptive mathematicians who has contributed to the study of the *A-D-E* hierarchy, V. I. Arnold (1986):

> At first glance, functions, quivers, caustics, wave fronts and regular polyhedra have no connection with each other. But in fact, corresponding objects bear the same label not just by chance: for example, from the icosahedron one can construct the function $x^2 + y^3 + z^5$, and from it the diagram E_8 and also the caustic and wave front of the same name.
>
> To easily checked properties of one of a set of associated objects correspond properties of the others which need not be evident at all. Thus the relations between all the *A, D, E* classifications can be used for the simultaneous study of all simple objects, in spite of the fact that the origin of many of these relations (for example, of the connections between functions and quivers) remains an unexplained manifestation of the mysterious unity of all things.

Coda

In order to clarify the plethora of algebraic and geometrical relationships entailed in our overall scheme, let us lay out Figure A28.

Note that the diagram forms a box divided into two compartments with the following features:

1. The back panel maps algebras to each other
2. The front panel maps groups to each other
3. The upper panel is projected down to the lower panel
4. The left compartment contains the Lie (algebra and group) structures

FIG. A28

E_7 MAPPING BOX DIAGRAM

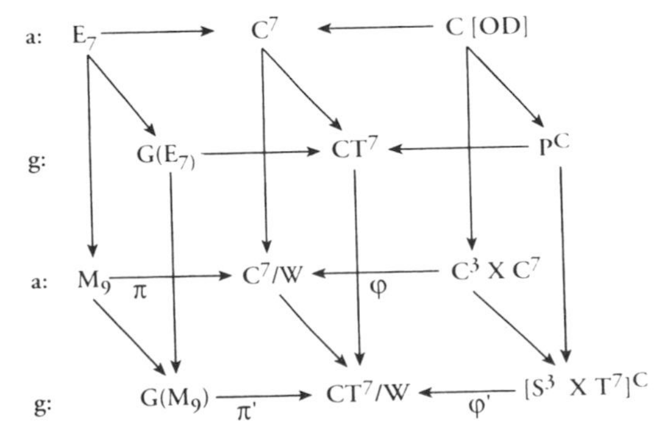

5. The right compartment contains the finite group-algebra structures

6. The panel dividing the box into the two compartments contains the intersections between the Lie structures and the finite group-algebra structures.

In particular:

Algebra Structures

E_7 is the 133-d (complex) Lie algebra.

$\mathbf{C}[OD]$ is the 48-d (complex) OD group algebra.

$\mathbf{C}^7$ is the intersection of E_7 with $\mathbf{C}[OD]$.

$\mathbf{C}^7$ is also the Cartan subalgebra (maximal commutative subalgebra) of E_7.

$\mathbf{C}^7$ is also a subalgebra of $\mathbf{C}^8$, which is the center of $\mathbf{C}[OD]$.

M^9 is the 9-d (complex) E_7 catastrophe bundle.

$\mathbf{C}^7/W$ is the critical space of the E_7 catastrophe bundle.

$\mathbf{C}^7 \times \mathbf{C}^3$ is the $\mathbf{C}[OD]$ whose Lie group is $[T^7 \times S^3]^{\mathbf{C}}$.

Group Structures

$G(E_7)$ is the 133-d (complex) Lie group whose Lie algebra is E_7.

$P^{\mathbf{C}}$ is the 48-d (complex) Lie group which consists of the set of invertible elements of $\mathbf{C}[OD]$.

P is the 48-d (real) manifold consisting of all unitary elements of $\mathbf{C}[OD]$ — i.e., P is the unitary Lie group in $\mathbf{C}[OD]$.

P is also the maximal compact subspace of $\mathbf{C}[OD]$.

$\mathbf{C}T^7$ is the intersection of $G(E_7)$ and $P^{\mathbf{C}}$.

$\mathbf{C}T^7$ is also the Cartan subgroup of $G(E_7)$.

$\mathbf{C}T^7$ is also a (complex) 7-torus, which is a maximal torus in $G(E_7)$ — i.e., the maximal commutative subgroup of $G(E_7)$.

$G(M^9)$ is the (complex) 9-d catastrophe bundle embedded in the Lie Group $G(E_7)$.

$\mathbf{C}T^7/W$ is the Lie-group view of the critical space $\mathbf{C}^7/W$ of the E_7 catastrophe bundle.

$[T^7 \times S^3]^{\mathbf{C}}$ is the (complex) 10-d superstring spacetime, which is a base space of the (complex) principal fiber bundle $P^{\mathbf{C}}$.

Mappings

$G = \exp(A)$ is the *exponential map* which transforms Lie algebra structures A into Lie group structures G.

$A = \log(G)$ is the *logarithmic map* which transforms Lie group structures G into Lie algebra structures A.

π = the fiber bundle projection from M^9 to $\mathbf{C}^7/W$, with $\mathbf{C}^2/OD$ as the fiber at the origin of $\mathbf{C}^7/W$.

π' = the fiber bundle projection from $G(M^9)$ to $\mathbf{C}T^7/W$, also with $\mathbf{C}^2/OD$ as the fiber at the origin of $\mathbf{C}T^7/W$.

Φ= the fiber bundle projection from $\mathbf{C}^7 \times \mathbf{C}^3$ to $\mathbf{C}^7$, with $\mathbf{C}^3$ as the fiber at the origin of $\mathbf{C}^7$.

Φ' = the fiber bundle projection from $[T^7 \times S^3]^{\mathbf{C}}$ to $\mathbf{C}T^7$, with $(S^3)^{\mathbf{C}}$ as fiber. (Note: $(S^3)^{\mathbf{C}} = S^3 \times S^3$.)

References

Appelquist, T., Chodos, A and Freund, P. G. O. *Modern Kaluza-Klein Theories.* Addison-Wesley, 1985.

Arnold, V. I. *Catastrophe Theory* 2nd ed. Springer-Verlag, 1986.

Arnold, V. I., Gusein-Zade, S. M. and Varchenko, A. N. *Singularities of Differentiable Maps,* Vol. 1. Birkhauser, 1985.

Bleecker, David. *Gauge Theory and Variational Principles.* Addison-Wesley, 1981.

Bott, Roul. "The Geometry and Representation Theory of Compact Lie Groups," pp. 65–90 in *Representation Theory of Lie Groups,* M. F. Atiyah, ed. Cambridge University Press, 1979.

Broad, C.D. *Scientific Thought.* Routledge and Kegan Paul, 1923.

Bröcker, T. and Dieck T. t. *Representations of Compact Lie Groups.* Springer-Verlag, 1985.

Chang, Garma C. C. *The Buddhist Teaching of Totality.* Pennsylvania State University Press, 1971.

Conway, J. H. and Sloane, N. J. A. *Sphere Packings, Lattices and Groups.* Springer-Verlag, 1988.

Coxeter, H. S. M. *Regular Polytopes,* 3rd ed. Dover, 1973.

Coxeter, H. S. M. *Complex Regular Polytopes.* 2nd ed. Cambridge University Press, 1991.

Culbertson, James T. *Consciousness:Natural and Artificial.* Libra, 1982.

d'Espagnat, Bernard. *Conceptual Foundations of Quantum Mechanics,* 2nd ed. Benjamin, 1976.

d'Espagnat, Bernard. "The Quantum Theory and Reality," *Scientific American,* 241:5, 158–181. November 1979.

d'Espagnat, Bernard. *In Searach of Reality.* Springer-Verlag, 1983.

Duff, Michael J. "Recent Results in Extra Dimensions" in *Physics in Higher Dimensions,* T. Piran and S. Weinberg, eds. World Scientific, 1986.

Du Val, Patrick. *Homographies, Quaternions and Rotations.* Oxford, 1964.

Elliot, J. P. and Dawber, P. G. *Symmetry in Physics,* Vol. I and II. Oxford, 1979.

Gardiner, Martin. *The Whys of a Philosophical Scrivener.* Quill, 1983.

Georgi, Howard. "A Unified Theory of Elementary Particles and Forces." *Scientific American,* 244:4, 48–63. April 1981.

Georgi, Howard. *Lie Algebras in Particle Physics.* Benjamin/Cummings, 1982.

Gilmore, Robert. *Lie Groups, Lie Algebras, and Some of Their Applications.* Wiley, 1974.

Gilmore, Robert. *Catastrophe Theory for Scientists and Engineers.* Wiley, 1981.

Green, Michael B. "Superstrings," *Scientific American,* 255:3, 48–60. September 1986.

Grove, L. C. and Benson, C. T. *Finite Reflection Groups,* 2nd ed. Springer-Verlag, 1985.

Harnad, J., Shnider, S. and Vinet, L. "The Yang-Mills Equation in Compactified Minkowski Space," *J. Math Phys.,* 20:5, 931. 1979.

Heisenberg, Werner. *Physics and Philosophy.* Harper, 1958.

Herbert, Nick. *Quantum Reality.* Doubleday, 1985.

Hinton, C. Howard. *The Fourth Dimension.* Allen and Unwin. 1904.

Hinton, Charles H. *Speculations on the Fourth Dimension,* Rudolf v.B. Rucker. Dover, 1980.

Humphreys, James E. *Introduction to Lie Algebra and Representation Theory.* Springer-Verlag, 1972.

Kendig, Keith. *Elementary Algebraic Geometry.* Springer-Verlag, 1977.

Lang, Serge. *Algebra,* 2nd ed. Addison-Wesley, 1984.

Marciano, William J. "Time Variation of the Fundamental 'Constants' and Kaluza-Klein Theories," *Phys. Rev. Lett.* 52, 489–491. 1984.

McKay, John. "Graphs, Singularities, and Finite Groups," in *Procedings of Symposia in Pure Mathematics,* Vol. 37, 183–186. 1980.

Milnor, John. *Singular Points of Complex Hypersurfaces.* Princeton, 1968.

Penrose, Roger. "The Geometry of the Universe," in *Mathematics Today,* L. A. Steen, ed. Springer-Verlag, 1978.

Poor, Walter A. *Differential Geometric Structures.* McGraw-Hill, 1981.

Rucker, Rudy. *The 4th Dimension.* Houghton Mifflin, 1984.

Russell, Bertrand. *Analysis of Matter.* Dover, 1954.

Schwartz, J. W. "Lectures on Superstring Theory," in *Physics in Higher Dimensions,* T. Piran and S. Weinberg, eds. World Scientific, 1986.

Sirag, Saul-Paul. "Why There Are Three Fermion Families," *Bulletin of the Amer. Phys. Soc.,* 27:1, 31. 1982.

Sirag, Saul-Paul. "Physical Constants as Cosmological Constraints," *International Journal of Theoretical Physics,* Vol. 22, 1067–1089. 1983.

Sirag, Saul-Paul. "A Discrete Approach to Unified Field Theory," *Proceedings of the 1st Annual Western Regional ANPA Meeting, Stanford University.* Alternative Natural Philosophy Association, 1985.

Sirag, Saul-Paul. "An E_7 Unification Scheme Via the Octahedral Double Group," contributed paper at the XXIII International Conference of High Energy Physics. Berkeley, California, 1986.

Sirag, Saul-Paul. "A Finite Group Algebra Unification Scheme," *Bulletin of the Amer. Phys. Soc.,* 34:1. 1989.

Slodowy, P. "Platonic Solids, Kleinian Singularities, and Lie Groups," *Lecture Notes in Math: 1008, Algebraic Geometry.* I. Dolgachev, ed., pp. 102–137. Springer-Verlag, 1983.

Springer, T. S. *Invariant Theory.* Springer-Verlag, 1977.

Thom, Rene. *Structural Stability and Morphogenesis.* Benjamin, 1975.

Weinberg, Steven. "Towards the Final Laws of Physics," in *Elementary Particles and the Laws of Physics, the 1986 Dirac Memorial Lectures.* Cambridge University Press, 1987.

Weyl, Hermann. *The Theory of Groups and Quantum Mechanics.* Dover, 1949.

Weyl, Hermann. *Symmetry.* Princeton University Press, 1952.

Wheeler, J. A. and Zurek, W. H. *Quantum Theory and Measurement.* Princeton University Press, 1983.

Wigner, Eugene. *Symmetries and Reflections.* Indiana, 1967, republished by Ox Bow, 1979.

Zee, A. *Fearful Symmetry.* Macmillan, 1986.

Glossary

Algebra. the study of the structure of numbers and the systems of entities abstracted from and generalized upon this structure. The most basic structure in this study is the *group* From groups one can define *rings,* and from rings and groups one can define *modules,* and *vector spaces* and *associative algebras.* We can construct the following hierarchy of structures:

Semigroup. {*S*,•} closure and associativity defined on S with composition of elements under Σ.

Group. {G,•} identity element and inverses also defined on *G.*

Commutative group. {*G*, +} commutativity ($a + b = b + a$) defined on *G.* Note: We conventionally regard composition of elements of a commutative as addition.

Ring. additive group *{R, +}* multiplicative semigroup (i.e., closure and associativity) {*R*, •}

Module. scalar ring {*R*, •}, i.e., only the multiplicative aspect of the ring is used; vector commutative group {V, +}; scalar multiplication distributes over vector addition : $r\bullet\ (v_1 + v_2) = r\bullet v_1 + r\bullet v_2$.

Vector space. the multiplicative structure of the ring (with zero omitted becomes a *commutative group* {*R* – 0, •}; vector commutative group {V, +}.

Associative algebra. multiplicative semigroup (with composition denoted by •) in the vector space {V, +, •}. Note: Mathematicians used to call *algebra* as defined here *hypercomplex numbers,* or *abstract algebra.* Now they simply call it algebra. *High school algebra* is essentially the study of the associative algebras **R** (the set of real numbers) **C** = **R** + √–1 **R** (the set of complex numbers) because every *polynomial* in one unknown has a root (or solution) in **C**. The multiplicative structure of **C** (if we leave out 0) is a commutative group. For these reasons, **C** is a very special case of an algebra.

Both **R** and **C** are examples of commutative algebras. The first noncommutative algebra was discovered by Hamilton in 1842 and was called by him the *quaternions,* which we designate **Q**.

An associative algebra (also called a *linear algebra*). A mathematical structure *A* which closely imitates the essential features of the real number line **R**. These features are: an additive structure on *A*, *a multiplicative structure on A*, and a distributive relationship between these two structures. The additive structure is provided by requiring *A* to be a *vector space*, which by definition is a commutative group. The multiplicative structure is not necessarily a group, but must obey the properties of closure and associativity. Thus not every element of *A* has a multiplicative inverse, nor does the multiplicative structure necessarily have an identity element. Moreover, the additive identity element, *zero*, when used multiplicatively has the effect of making any element of *A* into zero: $a0 = 0a = 0$, for all elements *a* of *A*.

Examples of associative algebras are: the real numbers **R**; the complex numbers **C**; the *total matrix algebra M(n)*, the set of all $n \times n$ matrices. These examples are all *associative algebras;* i.e., the multiplicative structure obeys the associative law: $a(bc) = (ab)c$.

Any associative algebra can be made into a *Lie algebra* by defining a new multiplicative structure based on the underlying associative multiplicative structure. Conventionally, this new structure is called the *Lie bracket:* $[a,b] = ab - ba$, where *ab* and *ba* are defined by the underlying associative multiplication structure. This is usually done via matrix representations of the underlying associative algebra. It is a basic theorem of algebra that any finite dimensional associative algebra is equivalent to a matrix algebra and thus can be faithfully represented by a matrix algebra. The dimension of the algebra is, of course, the dimension of the vector space which provides the additive structure for the algebra. For example, the dimension of the matrix algebra *M(n)* is n^2, the number of components of an $n \times n$ matrix. This is obvious to anyone who remembers that two matrices are added by componentwise addition.

Basis. a set **B** of vectors in a vector space such that any vector in the space is equivalent to a sum over **B**. The number of vectors in **B** is equal to the *dimension* of the vector space. Thus, in an *n*-dimensional vector space **V**, if **B** is the set $\{b_1, b_2, \ldots, b_n\}$, then any vector **v** of **V** can be written as $V = c_1 b_1 + c_2 b_2 + \ldots + c_n b_n$, where the set $\{c_1, c_2, \ldots, c_n\}$ are numbers which are called *coordinates.* Thus a basis provides a *coordinate system* for the vector space.

Note: When a vector is regarded as an ordered set of numbers, these numbers are the coordinates and are usually called the *components* of the vector.

Bell's Theorem. a statement proved by John S. Bell (1964) that if a quantum state vector corresponds to some objective reality, that reality must have nonlocal effects.

Consciousness. awareness — not necessarily self-awareness. Awareness entails *perceptual* and *cognitive* aspects of reality. Self-awareness entails also the *volitional* aspect. (For example, in a state of muscular paralysis due to anesthesia, it is possible to have awareness without self-awareness.)

Component. see Basis.

Coordinate system. see Basis.

Deformation. a transformation of a mathematical structure which shrinks, twists or otherwise changes the structure without tearing.

Dimension. the number of degrees of freedom in a space. The number of coordinates necessary to locate a point in a space. The coordinates are not necessarily rectangular, e.g., the surface of the earth is 2-dimensional; customarily it has two coordinates called longitude and latitude.

Dual space. the set of *linear functions* of a vector space. An *n*-dimensional space has an *n*-dimensional dual space. The dual space of a dual space is the original vector space.

Eigenvalue. the numerical solution *x* to operator equation:

$$Av = xv$$

where *A* is the operator, and y is a vector called the *eigenvector* solution to this operator equation. Geometrically, this means that *A* is acting upon the vector *v* does not change its direction but only its length; this length change is by the factor *x*. E.g., if *x* is 2, *v* is doubled in length by *A*; if *x* is 1/2, *v* is cut in half by *A;* if *x* is 1, *v* doesn't change at all.

Usually A is represented by an $n \times n$ matrix where n is a dimension of v, i.e., the number of components in v. In many cases, the solutions can be found by transforming the matrix M representing A into *diagonal form;* i.e., M is transformed into diag(M) = gMg^{-1} (where g is an $n \times n$ matrix representing an element of a group, and g^{-1} is the inverse of g). In diag(M) all components are zero except for the components of the primary diagonal. These components are the eigenvalues of M. It should be noted that M and gMg^{-1} represent the same transformation, but with differing coordinate systems, on the vector space in which the v live. In fact, the *eigenvectors* v are the basis vectors of the coordinate system in which diag(M) is defined.

Eigenvector. a vector solution v to the operator equation described under *eigenvalue.* In general, there are several eigenvector solutions to an operator equation. The number n of solutions is the dimensionality of the space the vectors live in. The matrix representing the operator is an $n \times n$ matrix. If there are n eigenvectors belonging to an operator acting on an n-dimensional space, the n eigenvectors form a *basis* for this space.

Field. a smooth assignment of some type mathematical object to each point of some space; e.g., a *scalar* field assigns a scalar (a number) to each point of a space; a *vector* field assigns a vector to each point of a space.

Force field. (also called a *gauge field*) a smooth assignment of an element of a Lie algebra to each point of spacetime. The standard correspondence of Lie algebras to forces is as follows (using Lie group labels): $U(1)$, electromagnetism; $SU(2)$, weak force; $SU(3)$, strong (color) force.

Function. a mapping from an n-dimensional space to a 1-dimensional space. This means that n numbers go into the function and one number comes out; e.g., given the equation $z = x^2 + 3y$, we can say that z is a function of x and y (see Mapping).

A *linear function* is an additive entity in the sense that the sum of two linear functions is also a linear function on the space on which the function is defined.

Hyperspace. a space of more than three dimensions.

Lie algebra. a nonassociative *algebra* which obeys two extra rules:

1. Anticommutativity: $a \bullet b = -b \bullet a$

2. The Jacobi identity: $a \bullet (b \bullet c) = (a \bullet b) \bullet c + b \bullet (a \bullet c)$, which replaces the associative law $a(bc) = (ab)c$.

Since most Lie algebras can be faithfully represented by sets of matrices, the Lie product $a \bullet b$ can be faithfully represented via the Lie bracket: i.e., $a \bullet b = [a,b] = ab - ba$, where the product ab is simply the ordinary product of two matrices. In fact, any matrix algebra is an associative algebra which can be made into a Lie algebra by defining Lie brackets on the underlying matrix algebra.

The building blocks of Lie algebras are called *simple Lie algebras.*

Mapping. from an n-dimensions space N to an x-dimensional space X is a rule which assigns to each element of N a unique element of X. note that N and X may be the same space. A *function* is a special case of a mapping, in which X the space being mapped to is 1-dimensional.

Matter field. a smooth assignment of a complex vector (of appropriate dimension) to each point of spacetime. The vector space is acted upon by the Lie group associated with the force field or fields involved.

Mind. the realm of all mental events (such as beliefs, memories, images, thoughts) including *subconscious* events.

Nonlocality. independence from spatial and temporal constraints. A nonlocal effect is instantaneous and undiminished over distance. Note: A nonlocal effect does not necessarily transfer energy or information from one point to another. In particular, transfer of energy or information is not implicit in the nonlocality referred to in *Bell's Theorem.* The understanding of a nonlocal effect which does not transfer energy or information is an open problem in physics.

Orbit. the set of points in a space selected by the action of all the elements of a group on a single point of the space. If the group is continuous (a Lie group), the orbit will be continuous; e.g., *SO(3)* acting on a single point of ordinary 3-d space creates an ordinary orbit — a circle. If the group is discrete, the orbit of a point will be a discrete set of points; e.g., the orbit of a reflection group acting on a point between the mirrors (i.e., in a reflection chamber) is a point in each reflection chamber; thus the number of points in such an orbit is equal to the number of elements in the reflection group.

Operator. a mathematical entity that transforms a space; e.g., on a *vector space,* an operator may rotate, stretch, translate or perform some combination of these transformations on the vectors of the space. An operator which acts on an n-dimensional space is represented by an $n \times n$ *matrix.*

Parameter. a variable which selects members of a family of structures. For example, a and b are parameters in the equation for a straight line: $y = a + bx$; a selects the point where the line intercepts the y axis; b selects the slope of the line. In this paper, $t_1,...,t_7$ are parameters which select a fiber of the catastrophe manifold X^9; these fibers are called the *deformations* of the identity fiber C^2/OD (see Unfolding).

Quaternions. an ordered pair of complex numbers, or equivalently, an ordered quartet of real numbers. Although quaternions are not commutative, they do have multiplicative inverses, which means that the set of quaternions (with zero omitted) forms a group under multiplication. If we think of the quaternions as an ordered quartet of real numbers, we will regard the quaternions as forming a 4-d vector space. The set of all quaternions of unit length form a sphere S^3, which is also a Lie group with the usual label *SU(2).*

Realm. reality structure. In this paper we use the word realm to refer to the combination of a universal mind with a universal body. Thus there is a hierarchical spectrum of realms corresponding to the *A-D-E* hierarchy.

Space. a set with a *continuous infinity* of elements. These elements are called *points.* The simplest space is a line.

Unfolding. a family of functions F which contains a particular function f is called an unfolding of f. If F contains all the functions close to f, the folding is called a *universal unfolding* (see Deformation).

Unified field theory. a theory combining two or more force fields into a single *force field.* The theory must also give an adequate description of the *matter fields* which feel the force fields entailed in the unified theory, because matter fields interact with each other by exchanging force fields.

Universal body. the physical realm in all its aspects. Individual bodies are substructures of the universal body. In this paper, the universal body is identified with the group algebra, C[*OD*]. It is possible that there is a universal body corresponding to each of the McKay-group algebras, so that there would be a hierarchical spectrum of universal bodies matching the spectrum of universal minds. One could view the entire spectrum of universal bodies as a Supreme Body.

Universal consciousness. that consciousness of which individual consciousnesses are substructures. In this paper, universal consciousness is identified with C^7, the intersection of the Lie algebra E_7 with the group algebra C[*OD*].

Universal mind. the mental realm in all its aspects, including the *subconscious.* Individual minds are viewed as substructures within the universal structure. In this paper, the universal mind is described as the 133-dimensional E_7 group. It may be necessary to extend this description to the infinite-dimensional Kac-Moody group $\wedge E_7$. Also, it is possible that there is a universal mind corresponding to each of the *A-D-E* Lie groups, so that there would be a hierarchical spectrum of universal minds comprising a Supreme Mind.

Vector. a set of ordered numbers called components. The number of these components is the dimension of the vector. A real vector has real numbers as components; a complex vector has complex numbaers as components. Since a complex number is an ordered pair of real numbers, an *n*-dimensional complex vector can be regarded as a 2*n*-dimensional real vector.

Geometrically, a vector is an entity having magnitude and direction. We can recover the algebraic definition, if we imagine the vector tied to the origin of a coordinate system. Then the components of the vector are the coordinates of the tip of the vector.

Vector space. a space whose elements, called vectors v, form a commutative group, which by convention is considered additive so that the 0-vector is the identity element. It must also be possible to multiply v by a scalar *s* (i.e., a number); a law of *distribution* holds for this scalar multiplication:

$s v_1 + s v_2 = s(v_1 + v_2)$.

We say that a vector space is defined over the set of scalars. If this set is the *real numbers,* we call the space a real vector space. If this set is the *complex numbers,* we call the space a complex vector space. Since a complex number is an ordered pair of real numbers, it is always possible to consider an *n*-dimensional complex vector space as a 2*n*-dimensional real vector space.

Wave function. a function ψ which provides a description of a quantum system. It is a function of the observables of the system in such a way that the square of the amplitude of the wavefunction is the probability for seeing a particular value of an observable. For example, if we consider only the observable of position *x*, then the square of the amplitude of $\psi(x)$ is the probability of seeing the system (say, a particle) at *x*. The word "wavefunction" is really a misnomer in the sense that most wavefunctions are not wavelike. Like waves, however, wavefunctions are additive: i.e., two wavefunctions (for the same system) can be added to produce another wave function. An alternative view of the wavefunction is a vector called the **state vector** of the system. Whenever a system is observed a particular aspect of the wavefunction, called an eigenfunction, is observed; alternatively a projection of the state vector, called an eigenvector, corresponds to an observed state.

NOTES

Introduction

1. Paradoxically, one might also say that the *reverse* is true: the only thing that we know objectively and directly is our own consciousness. The rest is all secondary. The great physicist, Sir Arthur Eddington, put it this way: *Primarily the sphere of objective law is the interplay of thoughts, emotions, memories and volitions in consciousness.* The resolution of this paradox, that consciousness might be both objective and subjective, will be the focus of a detailed discussion in the Appendix.

2. Michael J. Mahoney, *Scientist as Subject: The Psychological Imperative.* Cambridge, MA: Ballinger, 1976.

3. Harry M. Pinch and Trevor C. Collins, *Frames of Meaning.* London: Routledge and Kegan Paul, 1982.

4. Paul Feyerabend, *Science in a Free Society.* London: NLB, 1978.

5. Sigmund Koch, "The Nature and Limits of Psychological Knowledge. Lessons of a Century Qua 'Science,'" *American Psychologist, 36*(3), March 1981, p. 260.

6. *Rig Veda* 10.129 in W. T. de Bary (ed.), *Sources of Indian Tradition, Vol. 1.* New York: Columbia University Press, 1961.

7. Arthur M. Young, *The Reflexive Universe.* New York: Delacorte, 1975. Young's theory, which is outlined in more detail in Section IV, suggest a mathematical process linking science and mythology.

Section I

1. Heinz Werner, *Comparative Psychology of Mental Development.* New York: Science Editions, 1961. This is a classic text of cross-cultural psychological development.

2. Alexander Marshack, *The Roots of Civilization.* New York: McGraw-Hill, 1971. Scientific detective work of quality — well illustrated, quite technical, yet very readable.

3. Charles Muses, "A New Way of Altering Consciousness: Manual of Dynamic Resonance Meditation," *The Journal for the Study of Consciousness, 5*(2), 141-164. An interdisciplinary scholar, Muses edited and contributed to this now defunct journal.

4. Gerald S. Hawkins, *Stonehenge Decoded.* New York: Dell, 1965. This solution to an ancient mystery is now the basis for planetarium exhibits throughout the world.

5. Mircea Eliade, *Shamanism.* Princeton, NJ: Princeton University Press, 1972.

6. Michael J. Harner, *The Way of the Shaman* (#S059) in *Living Traditions* (#Q304), videotapes available from Thinking Allowed Productions. For further information write to 2560 Ninth Street, Suite 123, Berkeley, CA 94710 or phone (510) 548-4415.

7. Michael J. Harner, (ed.), *Hallucinogens and Shamanism.* New York: Oxford University Press, 1973. Harner is currently active in training Westerners in shamanistic approaches.

8. Many readers will undoubtedly be familiar with the marvelous accounts of shamanic initiation written by Carlos Casteneda. Fewer, however, are aware that scholars have uncovered so many inconsistencies in Casteneda's writings that it is now generally acknowledged that his work is most properly viewed as fiction in the *magical realism* genre than as legitimate anthropological inquiry. (See Richard DeMille, *Casteneda's Journey.* Santa Barbara, CA: Capra Press, 1976. Also see Richard DeMille (ed.), *The Don Juan Papers.* Santa Barbara, CA: Ross-Erikson, 1980.) Similar criticism is now emerging suggesting that the accounts of Lynn Anderson, another popular writer on shamanic initiation, may also be best viewed as fiction.

9. Adrian K. Boshier, "African Apprenticeship," *Parapsychology Review, 5*(5), July-August 1974.

10. A. Leo Oppenheim, "Mantic Dreams in the Ancient Near East," in G. E. Grunebaum and Roger Callois (eds.), *The Dream and Human Societies.* Berkeley: University of California Press, 1966. A volume touching on the psychological as well as sociological nature of dreams.

11. Kurt Seligman, *Magic, Supernaturalism and Religion.* New York: Random House, 1948, pp. 1-11.

12. Roger Callois, "Logical and Philosophical Problems of the Dream," in *The Dream and Human Societies, op. cit.*

13. A. L. Basham, *The Wonder That Was India.* New York: Grove Press, 1959. A fascinating and scholarly account of ancient India.

14. The English word *yoke* is derived from the same proto-Indo-European root word as *yoga.* Thus we can see the close linguistic connection between spiritual freedom and physical bondage.

15. Rammurti S. Mishra, *Yoga Sutras.* Garden City, New York: Doubleday, 1973. Linguist, yogi, physician, Mishra has been one of the most articulate representatives of the Hindu tradition in the West.

16. Eugene Taylor, "Asian Interpretations: Transcending the Stream of Consciousness." In Kenneth S. Pope and Jerome L. Singer (eds.), *The Stream of Consciousness.* New York: Plenum Press, 1978.

17. Arthur Waley, *The Way and Its Power.* New York: Grove Press, 1958, p. 197.

18. *The Secret of the Golden Flower,* trans. by Richard Wilhelm. New York: Harcourt, Brace and World, 1962.

19. Manley Palmer Hall, *The Secret Teachings of All Ages.* Los Angeles: Philosophical Research Society, 1973. A well-illustrated volume which may be somewhat distorted by Hall's tendency to romanticize about esoteric traditions.

20. Ceasare de Vesme, *A History of Experimental Spiritualism, Vol 1,* translated by Stanley de Brath. London: Rider and Co., 1931. This scholarly work quotes Herodotus (i, 47-8).

21. The best available collection of ancient source materials related to Pythagoras is D. R. Fideler (ed.), *The Pythagorean Sourcebook and Library: An Anthology of Ancient Writings which Relate to Pythagoras and Pythagorean Philosophy.* K. S. Guthrie (trans.), Grand Rapids, MI: Phanes Press, 1987.

22. The myth of Orpheus traveling into the realm of the dead has been beautifully portrayed in two modern films, *Orpheus* and *Black Orpheus.*

23. For a detailed explication of Pythagorean number lore, see R. Waterfield (trans.), *The Theology of Arithmetic: On the Mystical, Mathematical and Cosmological Symbolism of the First Ten Numbers, Attributed to Iamblichus.* Grand Rapids, MI: Phanes Press, 1988.

24. Walter Buckert, *Lore and Science in Ancient Pythagoreanism,* translated by Edwin J. Minar, Jr. Cambridge, MA: Harvard University Press, 1972, p. 357.

25. E. R. Dodds, "Supernormal Phenomena in Classical Antiquity," *Proceedings of the Society for Psychical Research,* March 1971, p. 194. Since its inception, the SPR in Britain has been graced with gifted classical scholars, Dodds being among the most recent.

26. I. F. Stone, *The Trial of Socrates.* Boston: Little, Brown, 1988.

27. Jacob Needleman, *Spirituality and the Intellect* (#S032), in *Living Philosophically* (#Q374), videotapes available from Thinking Allowed Productions, Berkeley, CA. (510) 548-4415.

28. Plato, *The Republic, Book VII*, translated by B. Jowett. New York: Vintage.

29. Ernest G. McClain, *The Pythagorean Plato: Prelude to the Song Itself.* York Beach, ME: Nicholas-Hays, 1978.

30. Plato, *The Republic, Book X*, translated by B. Jowett. New York: Vintage.

31. Frank Thilly, *A History of Philosophy*, revised by Ledger Wood. New York: Henry Holt and Company, 1957.

32. *Iamblichus on The Mysteries of the Egyptians, Chaldeans and Assyrians*, translated by Thomas Taylor. London: Bertram Dobell, 1895.

33. Tacitus, *Historia, Lib. IV*, p. 81. This passage was quoted in Ceasare de Vesme, *A History of Experimental Spiritualism, op. cit.*

34. Lynn Thorndike, *A History of Magic and Experimental Science, Vol. I.* New York: Columbia University Press, 1923. This magnificent treatise is the authoritative work in its field. It contains eight volumes which were written over thirty-five years. If not otherwise noted, much of the material from this chapter through Newton, can be traced to Thorndike.

35. Violet MacDermot, *The Cult of the Seer in the Ancient Middle East.* Berkeley: University of California Press, 1971. An entertaining and most thorough scholarly examination.

36. Charles Ponce, *Kabbalah.* San Francisco: Straight Arrow, 1973, p. 35. This book is most valuable for its integration of many Hebrew and Christian systems.

37. Matthew 22:37-40.

38. Morton Smith, *Clement of Alexandria and a Secret Gospel of Mark.* Cambridge, MA: Harvard University Press, 1972, p. 357.

39. Mark 16:17-18.

40. I Corinthians 12:8-10.

41. I Corinthians 12:12-13. Many conventional Christian ministers will disagree with my inclusive interpretation. However, contemporary theologians generally agree that the community of the spirit within Christ is not limited to professed believers. As an example, see the Thinking Allowed interview with David Steindl-Rast, Ph.D., a Benedictine monk, titled *Human and Divine Nature.*

42. I Corinthians 13:1-3.

43. I Corinthians 13:8-10.

44. I Corinthians 13:13.

45. I Corinthians 14:1.

46. Evelyn Underhill, *Mysticism.* New York: E. P. Dutton, 1961, p. 331. Augustine here is quoted in a volume first published in 1911 which still remains the authoritative analysis of mysticism in Western culture. The source is given as St. Augustine, *Book vii*, cap. xvu.

47. Violet MacDermot, *op. cit.*

48. Francis Yates, *Giordano Bruno and The Hermetic Tradition.* New York: Random House, 1969, p. 150. Yates has been one of the most active and knowledgeable scholars exploring the occult roots in renaissance culture.

49. Wayne Shumaker, *The Occult Sciences in the Renaissance.* Berkeley: University of California Press, 1972, p. 225. A useful reference volume lacking both condescension and credulousness.

50. Aldous Huxley, *The Perennial Philosophy.* New York: Harper and Row, 1945.

51. Huston Smith, *The Primordial Tradition* (#S048), in *Living Philosophically* (#Q374), videotapes available from Thinking Allowed Productions, Berkeley, CA. (510) 548-4415.

52. Steven T. Katz, "Language, Epistemology and Mysticism," in Steven T. Katz (ed.), *Mysticism and Philosophical Analysis.* New York: Oxford University Press, 1978, p. 66.

53. Lynn Thorndike, *op. cit.*, Vol. I, p. 643.

54. *Ibid.*

55. *Ibid.*, pp. 517-592.

56. Sayed Idries Shah, *The Secret Lore of Magic*, New York: The Citadel Press, 1970, pp. 22-29. Shah is quoting from Turner's eighteenth-century translation of Agrippa.

57. Lynn Thorndike, *op. cit.*, Vol. V., pp. 617-651.

58. Jolande Jacobi (ed.), *Paracelsus, Selected Writings*. Princeton, NJ: Princeton University Press, 1951.

59. Peter J. French, *John Dee*. London: Routledge and Kegan Paul, 1972. This book has helped to reestablish Dee's once-forgotten importance as a renaissance figure.

60. Francis A. Yates, *The Rosicrucian Enlightenment*. London: Routledge and Kegan Paul, 1972. A fascinating, yet scholarly, history regarding the controversial origins of the famous Rosicrucians.

61. "Under the shadow of thy wings, Jehova."

62. Paul M. Allen (ed.), *A Christian Rosenkreutz Anthology*. Blauvelt, NY: Rudolf Steiner Publications, 1974, p. 179.

63. Francis Bacon, "Selections From New Atlantis," in Edward A. Tiryakian (ed.), *On the Margin of the Visible*. New York: John Wiley and Sons, 1974, p. 143. This volume is the first product of a wave of scholars who are attempting sociological analyses of esoteric movements, modern and historical.

64. Francis Yates, *op. cit.*

65. Thomas Vaughan, "The Holy Mountain, A Rosicrucian Allegory," in *A Christian Rosenkreutz Anthology*. For a modern version of the same story see Rene Daumal, *Mt. Analogue*. San Francisco: City Lights Books, 1968.

66. Lynn Thorndike, *op. cit.*, Vol. VII, pp. 544-566.

67. Frank Thilly, *op. cit.*, pp. 384-399.

68. Lynn Thorndike, *op. cit.*, Vol. VIII, p. 591. Thorndike here is quoting Lord Keynes.

69. *Ibid.*

70. *Ibid.*

Section II

1. Theodore Roszak, *Why Astrology Endures: The Science of Superstition and the Superstition of Science*. San Francisco: Robert Briggs Associates, 1980.

2. Paul Feyerabend, *Science in a Free Society*. London: NLB, 1978.

3. *Ibid.*, p. 96.

4. Lynn Thorndike, *op. cit.*, Vol. VII, pp. 11-32.

5. Wolfgang Pauli, "The Influence of Archetypal Ideas on The Scientific Theories of Kepler," in *The Interpretation of Nature and The Psyche*. New York: Pantheon, 1955. Pauli, a Nobel laureate physicist, was one of the first to inquire into the quantum physics of consciousness. He was known throughout Europe for the poltergeist-like effects which he seemed to have on laboratory equipment (the "Pauli effect"). Pauli studied with Carl Jung who contributed his important essay on *synchronicity* to this volume.

6. Michel Gauquelin, *Birthtimes*. New York: Hill and Wang, 1983.

7. Dennis Rawlins, "sTarbaby," *Fate*, October 1981, pp. 67-98.

8. Michel Gauquelin, *Cosmic Influences on Human Behavior*, translated by Joyce E. Clemow. New York: Stein and Day, 1973.

9. Suitbert Ertel. "Scientific quality and progressive dynamics within the Gauquelin paradigm," *Zeitschrift fur Parapsychologie und Grenzgebiete der Psychologie, 28*(1/2), 1986, pp. 104-135.

10. Michel Gauquelin, *Birthtimes, op. cit.*, p. 141.

11. Carl Gustav Jung, "Synchronicity: An Acausal Connecting Principle," in *The Interpretation of Nature and The Psyche*. New York: Pantheon, 1955, pp. 60-94. Jung's arduous discussion of an important philosophical concept is illustrated by his astrology experiment.

12. Walter Boer, Peter Niehenke and Ulrich Timm, "Can 'Accident-Prone Persons' Be Diagnosed in Terms of Astrology? An Exploratory Experiment," *Zeitschrift fur Parapsychologie und Grenzgebiete der Psychologie, 28*(1/2), 1986, p. 65.

13. John A. West and Jan G. Toonder, *The Case for Astrology.* Baltimore: Penguin Books, 1973, pp. 204-209.

14. J. E. Vidmar, "Astrological Discrimination Between Authentic and Spurious Birthdates," *Cosmology Bulletin,* 8/9, 1979.

15. Nona Press, "Suicide in New York," *Journal of the National Council of Geocosmic Research, 3,* 1978.

16. J. Mayo, O. White, and H. J. Eysenck, "An empirical study of the relation between astrological factors and personality," *Journal of Social Psychology, 105,* 1978, pp. 229-36.

17. Hans Jurgen Eysenck, "Scientific Research in Astrology and the Demand for 'Naive' Subjects," *Zeitschrift fur Parapsychologie und Grenzgebiete der Psychologie, 23*(2), 1981, pp. 89-93.

18. Geoffrey A. Dean, "Planets and Personality Extremes," *Correlation, 1*(2), pp. 15-18.

19. P. Niehenke, "The Whole Is More than the Sum of Its Parts," *Astro-Psychological Problems, 1*(2), pp. 33-37.

20. Michael Startup, "The Astrological Doctrine of 'Aspects': A Failure to Validate With Personality Measures," *British Journal of Social Psychology,* 24, 1985, pp. 307-315.

21. Louise Lacey, *Lunaception: The New Revolutionary Natural Way to Control Your Body, Your Life and Your Fertility.* New York: Warner Books, 1986.

22. Gordon W. Russell and Jane P. de Graaf, "Lunar Cycles and Human Aggression: A Replication," *Social Behavior and Personality, 13*(2), 1985, pp. 143-146.

23. George O. Abell and Bennett Greenspan, "The Moon and the Maternity Ward," *Skeptical Inquirer, III*(4), Summer 1979, pp. 17-25. This article challenges the notion that birthrates correlate with the phases of the moon and also offers additional evidence countering the alleged correlation of suicide and homicide rates with lunar cycles.

24. Shawn Carlson, "A Double Blind Test of Astrology," *Nature, 318,* December 5, 1985, pp. 419-425.

25. D. H. Saklofske, I. W. Kelly, and D. W. McKerracher. "An Empirical Study of Personality and Astrological Factors," *Journal of Psychology, 11,* 1982, pp. 275-280. An examination of certain hypothesized relationships between zodiac signs and personality among 214 students finds none. "There were no significant differences between subjects classified according to odd vs. even sign and the personality dimensions of extraversion, neuroticism, and psychoticism. Neuroticism scores were not significantly different between subjects classified according to water and nonwater signs."

26. Michel Gauquelin, "Zodiac and Personality: An Empirical Study," *Skeptical Inquirer, VI*(3), Spring 1982, pp. 57-65.

27. Douglas P. Lackey, "A Controlled Test of Perceived Horoscope Accuracy," *Skeptical Inquirer, VI*(1), Fall 1981, pp. 29-31. People rate "placebo" horoscopes to be as accurate as their own.

28. David Lester, "Astrologers and Psychics as Therapists," *American Journal of Psychotherapy, 36*(1), January 1982, pp. 56-66.

29. Shawn Carlson, private communication, October 6, 1989.

30. Michel Gauquelin, *The Scientific Basis of Astrology.* New York: Stein and Day, 1969. pp. 198-211. Tchijewski's work is not, to my knowledge, available in English.

31. *Ibid.,* pp. 211-221. Further information is available in G. Piccardi, *The Chemical Basis of Medical Climatology.* Springfield, IL: Charles C. Thomas, 1963.

32. *Ibid.*

33. *Ibid.,* pp. 222-231. For further reference Gauquelin cites Takata's article in *Helvetica Medica Acta,* 1950.

34. *Ibid.,* pp. 183-184.

35. F. A. Brown, J. Woodland Hastings, and John D. Palmer, *The Biological Clock—Two Views.* New York: Academic Press, 1970. In addition to discussing Brown's evidence of

biological rhythms being tied to astronomical cycles, this book discusses the theory of an internally controlled timing mechanism. Both views are necessary for an overall understanding of bio-rhythms.

36. G. Edgar Folk, *Environmental Physiology.* Philadelphia: Lea and Febiger, 1966, p. 62.

37. Michel Gauquelin, *op. cit.*, p. 48. Gauquelin refers to J. H. Heller and A. A. Teixeira-Pinto, "A New Physical Method of Creating Chromosomal Aberrations," *Nature*, No. 4645, 1959.

38. Charles Muses, *Time and Destiny* (#S460), in *New Pathways in Science* (#Q134), videotapes available from Thinking Allowed Productions. For further information write to 2560 Ninth Street, Suite 123, Berkeley, CA 94710 or phone (510) 548-4415.

39. Arthur M. Young, *The Geometry of Meaning.* New York: Delacorte, 1975.

40. In the early 1970s, Arthur M. Young and Charles Muses worked together as editors of *The Journal for the Study of Consciousness* and the anthology *Consciousness and Reality.*

41. Sergio Bernardi, "Shamanism and Parapsychology," in Betty Shapin and Lisette Coly (eds.), *Parapsychology, Philosophy and Religious Concepts.* New York: Parapsychology Foundation, 1987, pp. 41-54.

42. Dean Sheils, "A Cross-Cultural Study of Beliefs in Out-Of-The-Body Experiences, Waking and Sleeping," *Journal of the Society for Psychical Research, 49*(775), March 1978, pp. 697-741.

43. E. A. Wallis Budge, *The Egyptian Book of the Dead.* New York: Dover, 1967, lviii-lxx.

44. D. Scott Rogo, "Astral projection in Tibetan Buddhist literature," *International Journal of Parapsychology, 10*(3), August 1968, pp. 277-284.

45. Carl B. Becker, "The Centrality of Near-Death Experiences in Chinese Pure Land Buddhism," *Anabiosis: The Journal for Near-Death Studies, 1*(2), December 1981, pp. 154-171.

46. Ganymede was a Trojan boy of great beauty who was caught up by love's eagle and made a cupbearer to the gods.

47. Dante Alighieri, *Purgatorio*, translated by John D. Sinclair. New York: Oxford University Press, 1968. Canto ix.

48. Apparently Ramacharaka is neglecting the extra-terrestrial possibilities here.

49. Yogi Ramacharaka, *Yogi Philosophy and Oriental Occultism.* Chicago: Yogi Publication Society, 1931, pp. 192-199.

50. Sylvan Muldoon and Hereward Carrington, *The Projection of the Astral Body.* New York: Samuel Weiser, 1970, pp. 61-2. Originally published in 1929, as a collaborative effort between an excellent researcher and an unusual psychic, this book remains a standard reference in the OBE literature.

51. Frederick Myers, *Human Personality and Its Survival of Bodily Death*, Vol. I, New York: Longmans, Green, 1954, p. 682 ff.

52. Muldoon and Carrington, *op. cit.*, p. 164.

53. *Ibid.*, p. 80.

54. Oliver Fox, *Astral Projection.* New Hyde Park, NY: University Books, 1962.

55. Roy Ald, *The Man Who Took Trips.* New York: Delacorte, 1971.

56. Robert Monroe, *Journeys Out of the Body.* New York: Doubleday, 1971. The book contains a preface by Charles Tart describing research with Monroe.

57. Robert Crookall, *The Study and Practice of Astral Projection.* New York: University Books, 1966.

58. Robert Crookall, *Out-of-the-Body Experiences.* New York: University Books, 1970, pp. 1-5.

59. *Ibid.*

60. Carlos S. Alvarado, "ESP During Out-Of-Body Experiences: A Review of Experimental Studies," *Journal of Parapsychology, 46*(3), September 1982, pp. 209-230.

61. Carlos S. Alvarado, "Research on Spontane-

ous Out-Of-Body Experiences: A Review of Modern Developments, 1960-1984," in Betty Shapin and Lisette Coly (eds.), *Current Trends in Psi Research.* New York: Parapsychology Foundation, 1986, pp. 140-174.

62. Susan J. Blackmore, "A Postal Survey of OBEs and Other Experiences," *Journal of the Society for Psychical Research, 52*(796), February 1984, pp. 225-244.

63. Gertrude R. Schmeidler, "Interpreting Reports of Out-Of-Body Experiences," *Journal of the Society for Psychical Research, 52*(794), June 1983, pp. 102-104.

64. Arnold Mindell, *Working with the Dreaming Body.* Boston: Routledge and Kegan Paul, 1985.

65. F. Gordon Greene, "Multiple Mind/Body Perspectives and The Out-Of-Body Experience," *Anabiosis: The Journal for Near-Death Studies, 3*(1), June 1983, pp. 39-62.

66. J. H. M. Whiteman, "Whiteman Replies to Chari," *Parapsychological Journal of South Africa, 4*(2), December 1983, pp. 144-149.

67. Susan Blackmore, "The Aspirations of Psychical Research," in D. H. Weiner and R. L. Morris (eds.), *Research in Parapsychology 1987.* Metuchen, NJ: Scarecrow Press, 1988, pp. 135-139.

68. Stanley Krippner, *Psychic and Spiritual Healing* (#S334) in *Perspectives on Healing* (#Q284), videotapes available from Thinking Allowed Productions, Berkeley, CA. (510) 548-4415.

69. Charles Muses, "Trance Induction Techniques in Ancient Egypt," in *Consciousness and Reality*, pp. 9-17.

70. Martin Rossman, *Healing Yourself With Mental Imagery.* New York: Walker and Company, 1987.

71. Martin Rossman, *Healing Yourself With Mental Imagery* (#W352), an *InnerWork* videotape available from Thinking Allowed Productions, Berkeley, CA. (510) 548-4415.

72. As Nero's unofficial chief minister, Seneca literally administered the Roman empire during five prosperous and peaceful years which were later described by the emperor Trajan as the finest period in the history of imperial Rome. He was the leading intellectual of his time and elevated the Graeco-Roman philosophy of Stoicism to a new height of humanistic synthesis. He stood alone in opposing the slaughter of gladiators and slaves in the Coliseum. His spiritual writings served as a great inspiration to the early Christian fathers who claimed him as one of their own. His drama was a major influence on virtually all of the great renaissance playwrights. And his influence apparently still lives. Ordered to commit suicide by Nero in 65 A.D., Seneca's last words to his beloved friends and family were said to be, "Study my life." One historian has written that in all of history, no one (with the exception of Socrates) was more prepared to face death than was Seneca.

73. Seneca, *Letters from a Stoic*, translated by Robin Campbell. London: Penguin, 1969, Letter LXXVIII.

74. Rabbi Nathan of Nemirov, *Rabbi Nachman's Wisdom*, translated by Rabbi Aryeh Kaplan. Brooklyn: Leonard Kaplan, 1973.

75. Benjamin Franklin, Lavoisier, Bailly, Guillotin, et al., "Secret Report on Mesmerism or Animal Magnetism," in Ronald E. Shor and Martin T. Orne (eds.), *The Nature of Hypnosis.* New York: Holt, Rinehart and Winston, 1965, p. 37.

76. Suzy Smith, *ESP and Hypnosis.* New York: Macmillan, 1972, p. 27.

77. Franz Anton Mesmer, *Memoire sur la Decouverte du Magnetisme.* Quoted in ibid.

78. The word *pass* is common to all magnetizers: it signifies all the movements made by the hand in *passing* over the body, whether by slightly touching, or at a distance.

79. J. P. F. Deleuze, "Rules of Magnetising," in *The Nature of Hypnosis*, pp. 24-29.

80. Charles Richet, *Thirty Years of Psychical Research*, translated by Stanley de Brath. New York: Macmillan, 1923, p. 22. Richet is a French physiologist, also a Nobel Laureate, who figures

prominently in the early years of psychical research.

81. Frank Podmore, *From Mesmer to Christian Science.* New Hyde Park, NY: University Books, 1965.

82. In this context, it is somewhat understandable that an International Study Group on Unorthodox Healing sponsored by the Parapsychology Foundation in 1954 concluded that it would be premature to consider an alleged psychic influence in the multifarious types of mental healing before the whole field had been investigated in regard to its normal aspects. It would be fair to maintain that this task is far from complete.

83. Rene Beck and Erik Peper, "Healer-Healee Interactions and Beliefs in Therapeutic Touch: Some Observations and Suggestions," in M.D. Borelli and P. Heidt (eds.). *Therapeutic Touch: A Book of Readings.* New York: Springer, 1981, pp. 129-137.

84. Harry Edwards, "The Organization Behind the Healing Intelligence," *Journal of Pastoral Counseling, 6*(2), Fall-Winter 1971/1972, pp. 15-20.

85. Peter Tompkins and Christopher Bird, *The Secret Life of Plants.* New York: Harper and Row, 1973, pp. 317-360.

86. A. R. G. Owen, *Psychic Mysteries of the North.* New York: Harper and Row, 1975, p. 125. Owen briefly describes his research with a radionics operator.

87. Vernon D. Wethered, *An Introduction to Medical Radiesthesia and Radionics.* Ashingdon, Rochford, Essex, England: C. W. Daniel Company, 1957, p. 75.

88. Tompkins and Bird, *op. cit.*, p. 333.

89. Arthur M. Young, "Reflections." Transcribed from a seminar at the Institute for the Study of Consciousness, Berkeley, California.

90. Francis K. Farrelly, "The Enigmatic Status of Radionics in the United States," First International Conference on Psychotronics, Prague, 1973.

91. William A. Tiller, "Radionics, Radiesthesia and Physics," *The Varieties of Healing Experience.* Los Altos, CA: Academy of Parapsychology and Medicine, 1971.

92. Tompkins and Bird, *op. cit.*, pp. 350-351.

93. Arthur M. Young, unpublished study.

94. Tompkins and Bird, *op. cit.*, 351.

95. James Randi, "Edgar Cayce: The Slipping Prophet," *Skeptical Inquirer, IV*(1), Fall 1979, pp. 51-57.

96. Stanley Krippner and Alberto Villoldo, *The Realms of Healing* (third edition). Berkeley: Celestial Arts, 1986, p. 24.

97. *Ibid.*

98. Andrija Puharich, M.D., "Some Biophysical Aspects of Healing," *Dimensions of Healing,* Los Altos, CA: Academy of Parapsychology and Medicine, 1973.

99. *Ibid.*

100. Anne Dooley, *Every Wall a Door.* New York: E. P. Dutton, 1974, p. 144. Dooley here is quoting Puharich.

101. Under these circumstances, such a high degree of reported accuracy would lead most observers to question the diligence of those reporting such a claim. In the realm of healing 100% accuracy claims are almost inevitably the mark of a true believer and *not* an independent observer.

102. Puharich, *op. cit.*

103. Anne Dooley, *op. cit.*

104. Tom Valentine, *Psychic Surgery.* Chicago: Henry Regnery, 1973.

105. Jesus B. Lava and Antonio S. Araneta, *Faith Healing and Psychic Surgery in the Philippines.* Manila: The Philippine Society for Psychical Research Foundation, 1983.

106. *Ibid.*, p. 10.

107. *Ibid.*, p. 6.

108. James Randi, *The Faith Healers.* Buffalo, NY: Prometheus Books, 1989.

109. Astronomer Carl Sagan, in his forward to Randi's *The Faith Healers*, maintains that "the book can properly be described as a tirade." He further characterizes Randi as "rambling, anecdotal, crotchety and ecumenically offensive." Yet, in spite of his many weaknesses, I recommend Randi's writings. The enormity of the problems in these areas, and the timidity of more reasonable individuals to critique such outrageous (yet influential) claims, has created a cultural vacuum which Randi, in his unique way, fills.

110. *Ibid.*, p. 199.

111. Brendon O'Regan, *The Inner Mechanisms of Healing* (#S238) in *Perspectives on Healing* (#Q284), videotapes available from Thinking Allowed Productions, Berkeley, CA. (510) 548-4415.

112. Donald J. West, *Eleven Lourdes Miracles.* New York: Garrett/Helix, 1957. p. viii. As an objective presentation of the difficulties involved in assessing the efficacy of faith healing, this book is valuable for anyone with an interest in the subject. It contains an appendix of cures accepted by the Lourdes Medical Bureau in the period 1925-50.

113. James Randi, *op. cit.*

114. Jeanne Achterberg, *Imagery in Healing: Shamanism and Modern Medicine.* Boston: New Science Library, 1985.

115. Anees A. Sheikh (ed.), *Imagination and Healing: Imagery and Human Development Series.* Farmingdale, NY: Baywood Publishing, 1984.

116. F. Papentin, "Self-purification of the Organism and Transcendental Meditation: A Pilot Study," in D. W. Orme-Johnson, L. Domash, and J. Farrow (eds.), *Scientific Research on Transcendental Meditation: Collected Papers, Vol. 1.* Los Angeles: MIU Press, 1975.

117. A. H. Schmale Jr, M.D. and P. Iker, Ph.D., "The Affect of Hopelessness and the Development of Cancer," *Psychosomatic Medicine, 28*, 1966, pp. 714-721.

118. Jean Shinoda Bolen, M.D., "Meditation in the Treatment of Cancer," *Psychic Magazine*, August 1973, p. 20.

119. O. Carl Simonton, M.D., "Management of the Emotional Aspects of Malignancy," Symposium of the State of Florida, Department of Health and Rehabilitative Service, June 1974.

120. Steven Locke and Douglas Colligan, *The Healer Within: The New Medicine of Mind and Body.* New York: E.P. Dutton, 1986.

121. Frances Vaughan, *The Inward Arc: Healing and Wholeness in Psychotherapy and Spirituality.* Boston: New Science Library (Shambhala), 1986, p. 7.

122. Inge Strauch, "Medical Aspects of Mental Healing," *International Journal of Parapsychology, 5*, 1963, pp. 140-141.

123. Interestingly enough, another survey indicates that two-thirds of the psychic healers in Britain are men. Sally Hammond, "What the Healers Say," *Psychic Magazine*, August 1973.

124. Inge Strauch, *op. cit.*

125. Yogi Ramacharaka, *op. cit.* pp. 154-163.

126. Yogi Ramacharaka, *The Hindu-Yogi Science of Breath.* Chicago: Yogi Publishing Company, 1905, pp. 68-72.

127. *Ibid.*

128. Stephen Levine, *Conscious Living/Conscious Dying* (#W343), an *InnerWork* videotape available from Thinking Allowed Productions, Berkeley, CA. (510) 548-4415.

129. Joseph Campbell, *Understanding Mythology* (#S075), in *The Roots of Consciousness* (#Q154), videotapes available from Thinking Allowed Productions, Berkeley, CA. (510) 548-4415.

130. Annie Besant and C. W. Leadbeater, *Thought Forms.* Wheaton, IL: Theosophical Publishing House, 1971, pp. 8-17. Originally published in 1901. Besant, a former mistress of George Bernard Shaw, became head of the international theosophical movement after the death of Madame Blavatsky.

131. Yogi Ramacharaka, *Yogi Philosophy and Oriental Occultism*, *op. cit.*, pp. 73-90. Originally published in 1903.

132. Eileen Garrett, *Awareness.* New York: Helix Press, 1943, pp. 99-100. An eloquent, lyrical testimony by the founder of the Parapsychology Foundation.

133. Yogi Ramacharaka, *Yogi Philosophy and Oriental Occultism, op. cit.*, pp. 64-66.

134. Ray Stanford, "On Viewing the Aura," KPFA-FM and the lnterdisciplinary Parapsychology Program of the University of California, Berkeley, Parapsychology Symposium, February 1974.

135. Charles T. Tart, "Concerning the Scientific Study of the Human Aura," *Journal of the Society for Psychical Research, 46*(751), March 1972.

136. A. R. G. Owen, "Generation of an 'Aura': A New Parapsychological Phenomenon," *New Horizons, I*(1), Summer 1972, pp. 11-13.

137. E. N. Santini, *Photographie des Effluves Humains.* Paris, 1896.

138. Max Heindel, *The Rosicrucian Cosmo-Conception.* Oceanside, CA: The Rosicrucian Fellowship, 1969, pp. 59-64. Originally published in 1909.

139. A. R. G. Owen and G. A. V. Morgan, "The 'Rim Aura': An Optical Illusion — A Genuine but Non-psychic Perception," *New Horizons I*(3), January 1974, pp. 19-31.

140. A. R. G. Owen, "Generation of an 'Aura'," *op. cit.*, pp. 14-23.

141. C. W. Leadbeater, *The Chakras.* Wheaton, IL: Theosophical Publishing House, 1972, p. l. Originally published in India in 1927, this book contains a perplexing combination of allegedly firsthand clairvoyant reports and Theosophical dogma. There are a number of illustrations.

142. The *etheric body* is essentially a pseudoscientific term which became popular at a time when many scientists still supposed that an unknown substance called the *ether* permeated the entire universe and mediated the transmission of electromagnetic waves. Einstein's theory of relativity has subsequently superseded this view in science; however, occultists who simplify the teachings of earlier generations still sometimes retain this terminology.

143. C. W. Leadbeater, *op. cit.*, 45.

144. *Ibid.*, pp. 19-20. According to Leadbeater, *Theosophica Practica* was originally issued in 1696. The illustrations to the book were apparently added about 1720. A French translation, used by Leadbeater, was published in 1897 in the Bibliotheque Rosicrucienne (No. 4) by the Bibliotheque Charcornac, Paris.

145. Shafica Karagula, *Breakthrough to Creativity.* Santa Monica, CA: De Vorss, 1967.

146. Shafica Karagula, "Higher Sense Perception and New Dimensions of Creativity," American Psychiatric Association Convention, May 1974.

147. William A. Tiller, "Radionics, Radiesthesia and Physics," in *The Varieties of Healing Experience,* Academy of Parapsychology and Medicine, October 1971, pp. 72-78.

148. Lee Sannella, *Kundalini: Psychosis or Transcendence.* San Francisco: H. S. Dakin, 1976.

149. Dr. Hiroshi Motoyama, *Chaka, Nadi of Yoga and Meridian Points of Acupuncture.* Tokyo: Institute of Religious Psychology, 1972, pp. 15-17.

150. Rammurti S. Mishra, *Yoga Sutras.* New York: Doubleday, 1973, pp. 295-296.

151. Lama Anagarika Govinda, *Foundations of Tibetan Mysticism.* New York: Samuel Weiser, 1969, p. 135. An exposition of the esoteric teachings underlying the mantra, OM MANI PADME HUM.

152. John F. Thie and Mary Marks, *Touch for Health.* Santa Monica, CA: DeVorss, 1973.

153. Jacques de Langre, *The First Book of Do-In.* Hollywood: Happiness Press, 1971.

154. J. F. Chaves and T. X. Barber, "Acupuncture Analgesia: A Six-Factor Theory," *Psychoenergetic Systems, 1*, 1974, pp. 11-21.

155. Joan Steen Wilentz, *The Senses of Man.* New York: Thomas Crowell, 1968, pp. 89-109.

156. Felix Mann, "The Probable Neurophysiological Mechanism of Acupuncture," *Transcript of the Acupuncture Symposium.* Los Altos, CA:

Academy of Parapsychology and Medicine, 1972, pp. 23-31.

157. *Ibid.*

158. Wilhelm Reich, *The Discovery of The Orgone, Vol. II: The Cancer Biopathy.* New York: Farrar, Straus and Giroux, 1973, p. 1521. This book, the major volume in which Reich describes his orgone research, contains more than seventy microphotographs.

159. Today, Reich's claims are considered, by most, comparable to the mysterious "N-rays" that French physicists once thought they had discovered. See Section III for a further discussion of this cognitive error.

160. *Ibid.*, pp. 108-142.

161. Wilhelm Reich, *History of the Discovery of Life Energy — The Einstein Affair.* Rangeley, ME: Orgone Institute Press, 1953. A documentation of the original correspondence between Reich and Einstein.

162. W. Edward Mann, *Orgone, Reich and Eros.* New York: Simon and Schuster, 1973. Mann, a retired professor of sociology from York University in Toronto, Ontario, is one of Canada's foremost sociologists. His book is a major document tracing the scientific impact of Reich's work within a sociological framework which includes other research into life energies.

163. Wilhelm Reich, *The Cancer Biopathy, op. cit.*, pp. 23-25.

164. Thelma Moss, et al., "Bioenergetics and Radiation Photography," First International Conference on Psychotronics, Prague, 1973.

165. Bernard Grad, "Orgone Treatment of Cancerous Rats," Esalen Institute Symposium on Reich and Orgone, San Francisco, August 1974.

166. V. M. Inyushin, "Biological Plasma of Human and Animal Organisms," *Symposium of Psychotronics*, Prague, September 1970. Published by the Paraphysical Laboratory, Downton, Wiltshire, England.

167. *Ibid.*

168. Thelma Moss has retired from UCLA, and research in high-voltage photography no longer continues there. The general consensus among researchers is that there were too many uncontrolled extraneous variables in virtually all of the high-voltage photography studies to enable any conclusions to be drawn of a psychological nature.

169. Thelma Moss, John Hubacher, and Francis Saba, "Visual Evidence of Bioenergetic Interactions Between People?" American Psychological Association Convention, New Orleans, May 1974.

170. Sheila Ostrander and Lynn Schroeder, *Psychic Discoveries Behind the Iron Curtain.* New York: Prentice Hall, 1970, p. 223.

171. E. Douglas Dean, "High-voltage Radiation Photography of a Healer's Fingers," in S. Krippner and D. Rubin (eds.), *The Kirlian Aura.* New York: Doubleday, 1974.

172. Jeffrey Mishlove and Douglas Dean, "From Parapsychology to Paraphysics," *Mind's Ear* radio program, broadcast on KPFA-FM in Berkeley, California, July 5, 1973.

173. Thelma Moss, Kendall Johnson, Jack Grey, John Hubacher, Roger MacDonald, and Francis Saba, "Bioenergetics and Radiation Photography," First International Conference on Psychotronics, Prague, 1973.

174. *Ibid.*

175. *Ibid.*

176. *Ibid.*

177. Many individuals used the title "Dr." when referring to Olga Worrall as a token of respect, although she lacked formal academic or medical credentials.

178. Arleen J. Watkins and William S. Bickel, "The Kirlian Technique: Controlling the Wild Cards," *Skeptical Inquirer, 13*(2), Winter 1989, pp. 172-184.

179. Thelma Moss, John Hubacher, and Francis Saba, "Visual Evidence of Bioenergetic Interactions Between People?" *op. cit.*

180. Thelma Moss, John Hubacher, Francis

Saba, and Kendall Johnson, "Kirlian Photography: An Electrical Artifact?" American Psychological Association, August 1974.

181. Thelma Moss, John Hubacher, and Francis Saba, "Anomalies in Kirlian Photography: Interactions Between People Reveal Curious 'Disappearances' and 'Merging' Phenomena," Second International Psychotronics Conference, Monaco, 1975.

182. William A. Tiller, "Energy Fields and the Human Body: Part I," A.R.E. Medical Symposium on Mind Body Relationships in the Disease Process, Phoenix, Arizona, January 1972.

183. William A. Tiller, "Some Energy Field Observations of Man and Nature," *The Kirlian Aura, op. cit.*, p. 122.

184. John Hubacher and Thelma Moss, *The "Phantom Leaf Effect" as Revealed Through Kirlian Photography.* UCLA Center for the Health Sciences, 1974.

185. Thelma Moss, Ph.D., *The Probability of the Impossible.* Los Angeles: J. P. Tarcher, 1974. pp. 54-58. This lively book documents the many research activities of Dr. Moss, a former Broadway actress. It is valuable for her firsthand accounts of her travels in the former Soviet Union and Czechoslovakia as well as the inside story of her own laboratory activities.

186. Clark Dugger, John Hubacher, Thelma Moss, and Francis Saba, "'The Phantom Leaf,' Acupuncture, and Altered States of Consciousness," Second International Psychotronics Conference, Monaco, 1975.

187. Larry Burton, William Joines, and Brad Stevens, "Kirlian Photography and Its Relevance to Parapsychological Research," Parapsychological Association Convention, New York, 1974. Also presented before the Symposium of the Institute of Electronic and Electrical Engineers, November 1974.

188. C. S. Hall and G. Lindzey, *Theories of Personality.* New York: John Wiley and Sons, 1957, pp. 296-335. This book provides summaries of many psychological theories. Of particular interest is the classification of each theory according to its emphasis along each of eighteen different parameters (p. 548).

189. Stanley Krippner and Sally Ann Drucker, "Field Theory and Kirlian Photography: An Old Map for a New Territory," in *The Kirlian Aura, op. cit.*

190. Kenneth Demarest, "The Winged Power," in Charles Muses and Arthur M. Young (eds.), *Consciousness and Reality.* New York: Avon, 1972, p. 351. The accurate recording of the event referred to is quite veiled as the only preserved records come from Syncellus in Greek and Eusebius in Latin, both quoting the Greek chronicler Alexander Polyhistor, who is quoting from Berosus, who is in turn quoting from more ancient texts. This article tracing the esoteric symbology of the winged gods appears in an anthology by the editors of *The Journal for the Study of Consciousness.*

191. D. D. Home, *Lights and Shadows of Spiritualism.* London: Virtue and Company, 1878, p. 77. Actually Home is quoting directly from Augustine, but neglects to acknowledge the specific source. It's interesting to note that the nineteenth-century medium placed himself in the same tradition as the saint.

192. Idries Shah, *Oriental Magic.* New York: Philosophical Library, 1956, pp. 61-2. Hundreds of these delightful Sufi tales have been recorded and translated by Shah, who is considered a spiritual father to story lovers throughout the world. It is said that one can develop inwardly by merely listening to these "teaching stories."

193. Irina Tweedie, *Spiritual Training* (#S058), in *Personal and Spiritual Development* (#Q184), videotapes available from Thinking Allowed Productions, Berkeley, CA. (510) 548-4415.

194. Robert Frager, *Common Threads in Mysticism* (#S050) in *Mystical Paths* (#Q244), videotapes available from Thinking Allowed Productions, Berkeley, CA. (510) 548-4415.

195. William Rodarmor, "The Secret Life of Swami Muktinanda: Abuses of Power in the Ashmam of the 'Guru's Guru,'" *Co-Evolution Quarterly,* Winter 1983, pp. 104-111.

196. Katy Butler, "Events Are the Teacher: A Buddhist Community Asks Its Leader to Stop," *Co-Evolution Quarterly,* Winter 1983, pp. 112-123.

197. Unrestrained endorsements of Love-Ananda's teachings and books have come from Ken Wilbur, Willis Harman, Charles Tart, Fred Alan Wolf, Ken Wilbur, Irina Tweedie, Lee Sannella, and Larry Dossey among others.

198. Heart-Master Da Love-Ananda, *The Basket of Tolerance: A Guide to Perfect Understanding of the One and Great Tradition of Mankind.* Second prepublication edition. Clearlake, CA: Free Daist Communion, 1989.

199. Heart-Master Da Free John, *The Dawn Horse Testament.* San Rafael, CA: Dawn Horse Press, 1985.

200. Da Free John, *The Knee of Listening.* Clearlake, CA: Dawn Horse, Press, 1972, 1988.

201. Franklin Jones, *The Method of the Siddhas.* Clearlake, CA: Dawn Horse Press, 1973, 1988.

202. Also spelled *Kabala, Kabbalah, Qabala,* etc.

203. Charles Ponce, *Kabbalah.* San Francisco: Straight Arrow, 1973.

204. MERU Foundation, "The MERU Project: A Geometric Metaphor for Transcendence, 1989 (P. O. Box 1738, San Anselmo, CA 94960).

205. Emmanuel Swedenborg, *The Heavenly Arcana, Vol. II.* New York: American Swedenborg Publishing Society, 1873, pp. 114-121.

206. Wilson Van Dusen, *The Presence of Other Worlds.* New York: Harper and Row, 1974. A sensitive biography of Swedenborg. Formerly the chief psychologist at a mental hospital in California, Van Dusen began to treat the hallucinations of his patients as if they were spirits. The experiment worked. In fact, he hypothesized that he was encountering the very same world of spirits as described in the encyclopedic writings of Swedenborg.

207. C. D. Broad, *Religion, Philosophy and Psychical Research.* New York: Harcourt Brace, 1953, pp. 115-116. The case for Swedenborg's clairvoyance is presented by Emmanuel Kant in a letter sent to Fraulein von Knobloch. Later, in an anonymously published manifesto, Kant essentially retracted his support of the claims for Swedenborg. The evidence one way or another is quite shaky. Broad, an eminent philosopher of this century, demonstrated that Kant, undoubtedly a great thinker, was, nevertheless, a careless psychical researcher. This book is useful for its examination of the philosophical implications of psi.

208. G. Stanley Hall, *Founders of Modern Psychology.* New York: D. Appleton and Company, 1924, pp. 129-130. An innovative American psychologist, Hall studied with many of the German pioneers, such as Fechner, about whom he writes. This book touches on the human side of psychology's birth as a science.

209. *Ibid.,* pp. 148-149.

210. *Ibid.,* p. 151.

211. Richard Hodgson, "Report of the Committee Appointed to Investigate Phenomena Connected with the Theosophical Society," *Proceedings of the Society for Psychical Research,* Part IX, December 1885.

212. Victor A. Endersby, *The Hall of Magic Mirrors.* New York: Carlton Press, 1969. Delightfully written with a consistent sense of humor, this book is now difficult to obtain.

213. Judith Skutch Whitson, *A Course in Miracles* (#S360), in *Channels and Channeling* (#Q214), videotapes available from Thinking Allowed Productions, Berkeley, CA. (510) 548-4415.

214. J. Allen Hynek, *The UFO Experience: A Scientific Inquiry.* Chicago: Henry Regnery, 1972.

215. Shafica Karagula, *Breakthrough to Creativity.* Santa Monica, CA: De Vorss, 1967, pp. 110-113. Dr. Karagula is a neuropsychiatrist. However, her book lacks the detail and precision necessary for a scientific evaluation.

216. Peter Dawkins, *Francis Bacon and Western Mysticism* (#S077), in *Living Traditions* (#Q304), videotapes available from Thinking Allowed Productions, Berkeley, CA. (510) 548-4415.

217. Sara Grey Thomason, "'Entities' in the Linguistic Minefield," *Skeptical Inquirer, 13*(4), Summer 1989, pp. 391-396.

218. Ray Stanford, *Fatima Prophecy.* Austin, TX: Association for the Understanding of Man, 1972, pp. 3-24.

219. *lbid.*, pp. 43-50.

220. C. J. Ducasse, *Paranormal Phenomena, Science and Life After Death.* New York: Parapsychology Foundation, 1969. Ducasse is quoting from Abraham Cummings, *Immortality Proved by the Testimony of Sense*, Bath, Maine, 1826.

221. C. G. Jung, *Flying Saucers.* New York: Harcourt, Brace and Co., 1959.

222. Edward J. Ruppelt, *The Report on Unidentified Flying Objects.* New York: Doubleday, 1956. An inside story of the Air Force UFO Investigations.

223. Maj. Donald E. Keyhoe, *Aliens From Space.* New York: Doubleday, 1973. Keyhoe's stance is generally quite critical of the Air Force investigations.

224. Andrija Puharich, *Uri.* New York: Doubleday, 1974.

225. *Ibid.*, pp. 151-152.

226. *Ibid.*, p. 152.

227. An interesting coincidence further connecting Ray Stanford to the archetypal Horus intelligence was evidenced in the January 1974 issue of *Analog* science-fiction magazine. The magazine cover, illustrating a story called "The Horus Errand" depicted a hawk-shaped vehicle, and a "psychonaut" named Stanford whose duty was to guide the souls of departed citizens of a futuristic pyramid-shaped city safely into their next chosen incarnation. Further inquiry revealed the coincidence that the cover artist, Kelly Freas (who also designed the official Skylab uniform patches) had known Ray Stanford and had had a psychic reading from him some fifteen years earlier.

228. Berthold E. Schwarz, "Stella Lansing's UFO Motion Pictures," *Flying Saucer Review, 18*(1), January-February 1972, p. 312. *Flying Saucer Review*, published in England, generally provides the best international coverage of UFO sightings.

229. Berthold E. Schwarz, "Stella Lansing's Movies of Four Entities and Possible UFO," *Flying Saucer Review*, Special Issue No. 5, *UFO Encounters.*

230. Berthold E. Schwarz, "Stella Lansing's Clocklike UFO Patterns," *Flying Saucer Review, 20*(4), January 1975, p. 39.

231. Berthold E. Schwarz, "Stella Lansing's Clocklike UFO Patterns. Part II," *Flying Saucer Review, 20*(5), March 1975, pp. 20-27.

232. Dwight Connelly and Joseph M. Brill, "Rhodesian Case Involves Occupants, Transportation of Auto," *Skylook*, No. 89, March 1975, p. 39. This is the monthly publication of the Mutual UFO Network.

233. Ray Stanford, "Uri: The 'Geller Effect,'" and "The 'Geller Effect' Part Two," *Journal of the Association for the Understanding of Man*, 1974. This journal is available from the Association for the Understanding of Man, P.O. Box 5310, Austin, TX 78763.

234. Aime Michel, "The Strange Case of Dr. X," *Flying Saucer Review*, Special Issue No. 3, September 1969, p. 316.

235. Jacques Vallee, "UFOs: The Psychic Component," *Psychic Magazine*, February 1974, pp. 12-17.

236. Ralph Blum and Judy Blum, *Beyond Earth: Man's Contact With UFOs.* New York: Bantam, 1974.

237. *Ibid.*, pp. 143-145.

238. Benjamin Simon, M.D., "Introduction," in John G. Fuller, *The Interrupted Journey*, New York: Dial Press, 1966.

239. Terrence Dickinson, "The Zeta Reticuli Incident," *Astronomy, 2*(12), December 1974, pp. 4-19.

240. Phillip J. Klass, *UFO Abductions: A Dangerous Game.* Buffalo: Prometheus, 1989. Klass' book is a serious challenge to current claims regarding abductions, particularly those of Whitley Strieber and Bud Hopkins.

241. Ralph Blum, *op. cit.*, pp. 107-120. Blum's account contains transcripts of hypnotic sessions with Schirmer.

242. Jeffrey Mishlove, Leo Sprinkle, Herbert Schirmer, "Nebraska UFO Contact," radio interview broadcast on *Mindspace*, KSAN-FM in San Francisco, November 1974.

243. Jeffrey Mishlove and Robert Monroe, "UFOs and Out-of-Body Experience," radio interview broadcast on *Mindspace*, March 1975.

244. Robert Monroe, *Journeys Out of The Body.* New York: Doubleday, 1971, p. 153.

245. "Few and Far Between," in Charles Bowen (ed.), *The Humanoids.* Chicago: Henry Regnery, 1969, pp. 20-22.

246. "Interview with Ray Stanford," *Psychic Magazine*, April 1974, p. 11.

247. Jeffrey Mishlove and Ray Stanford, "UFOs and Psychic Phenomena," *Mindspace* radio broadcast, December 1974.

248. "Interview with Ray Stanford," *op. cit.*, p. 10. While Stanford clearly distinguishes his psychic experiences from his UFO contact, he also maintains that the "visitation of UFOs to earth is archetypically a counterpart of the higher being coming down and contacting man through the traditional third eye and, perhaps through the pineal and pituitary glands." This applies, he claims, to UFOs that are experienced in dreams.

249. Jacques and Janine Vallee, *Challenge to Science: The UFO Enigma.* Chicago: Henry Regnery, 1966.

250. J. Allen Hynek, "Twenty-one Years of UFO Reports," in Carl Sagan and Thornton Page (eds.), *UFOs, A Scientific Debate.* New York: W. W. Norton, 1972. This book contains the proceedings of a symposium on UFOs at the convention of the American Academy for the Advancement of Science in Boston, 1969.

251. E. U. Condon, *Scientific Study of Unidentified Flying Objects.* New York: Bantam, 1969.

252. J. Allen Hynek, *The UFO Experience: A Scientific Inquiry, op. cit.*

253. Bruce C. Murray, "Reopening the Question. Review of The UFO Experience by J. Allen Hynek," *Science, 177*(4050), August 25, 1972, pp. 688-689.

254. James Harder, *Extra-Terrestrial Intelligence* (#S430), in *Life in the Universe* (#Q364), videotapes available from Thinking Allowed Productions, Berkeley, CA. (510) 548-4415.

255. Jacques Vallee, *The Implications of UFO Phenomena* (#S440), in *Life in the Universe* (#Q364), videotapes available from Thinking Allowed Productions. For information, write to 2560 Ninth Street, Suite 123, Berkeley, CA 94710, or phone (510) 548-4415.

256. E. A. Wallis Budge, *The Egyptian Book of the Dead.* New York: Dover, 1967, pp. lviii-lxx. A classic.

257. E. A. Wallis Budge, *Osiris: the Egyptian Religion of Resurrection.* New Hyde Park, New York: University Books, 1961. First published in 1911.

258. W. Y. Evans-Wentz, *The Tibetan Book of the Dead.* London: Oxford University Press, 1960. The text contains commentaries by Carl Jung, Lama Anagarika Govinda, and Sir John Woodroffe.

259. *Ibid.*, p. 103.

260. *Ibid.*, pp. 165-166.

261. Timothy Leary, Richard Alpert, and Ralph Metzner, *The Psychedelic Experience.*

262. W. Y. Evans-Wentz, *Tibet's Great Yogi, Milarepa.* New York: Oxford University Press, 1958.

263. *Ibid.*, pp. 133-136.

264. Sir William Crookes, "Notes of an Inquiry into the Phenomena called Spiritual," in R. G. Medhurst (ed.), *Crookes and the Spirit World.* New York: Taplinger, 1972. This volume contains descriptions of Crookes' experiments as well as his replies to his critics.

265. Nandor Fodor, *Encyclopedia of Psychic Science.* London: Arthur's Press, 1933, p. 95. Crookes is quoted by Fodor.

266. Charles Richet, *Thirty Years of Psychical Research*, translated by Stanley de Brath. New York: Macmillan, 1923.

267. Camille Flammarion, *Mysterious Psychic Forces.* Boston: Small, Maynard and Company, 1907. The famous French astronomer recounts his own personal experiences with Kardec.

268. Pedro McGregor, *The Moon and Two Mountains.* London: Souvenir Press, 1966. A firsthand account of spiritualism in Brazil.

269. *Proceedings of the Society for Psychical Research, I,* 1882, p. 34.

270. F. W. H. Myers, *The Human Personality and Its Survival of Bodily Death.* New York: Longmans, Green and Company, 1954. Originally published in 1903, this book is possibly the greatest classic of psychical research.

271. *Ibid.*, p. 333. Myers is quoting from Wordworth's "Prelude," Book VI.

272. *Ibid.*, pp. 360-366. Myers quotes the report published by Dr. E. W. Stevens in the *Religio-Philosophical Journal*, Chicago, 1879.

273. *Ibid.*, 367-368. Myers quotes Hodgson's report published in the *Religio-Philosophical Journal* for December 20, 1890.

274. Edmund Gurney, F. W. H. Myers, and Frank Podmore, *Phantasms of the Living.* Gainesville, Florida: Scholars' Facsimiles and Reprints, 1970, pp. 163-164. This passage is quoted in Alan Gauld, *The Founders of Psychical Research*, London: Routledge and Kegan Paul, 1968, pp. 165-166.

275. "Notes on the Evidence Collected by the Society for Phantasms of the Dead," *Proceedings of the Society for Psychical Research, IIII,* 1885, pp. 69-150. This information is cited in Gauld, *op. cit.*

276. G. N. M. Tyrell, *Apparitions.* New Hyde Park, NY: University Books, 1961, pp. 69-70. Originally published in 1953.

277. *Ibid.*, pp. 77-80.

278. *Ibid.*, pp. 132-133. The original material comes from H. M. Wesermann, *Der Magnetismus und die Allgemeine Weltsprache*, 1822.

279. Karlis Osis, *Deathbed Observations by Physicians and Nurses.* New York: Parapsychology Foundation, 1961.

280. Raymond Moody, *Life After Life* (#W417), an *InnerWork* videotape available from Thinking Allowed Productions, 2560 9th Street, #123, Berkeley, CA 94710.

281. Alan Gauld, *The Founders of Psychical Research, op. cit.*, p. 253. Gauld here is quoting William James from *The Proceedings of the Society for Psychical Research*, 1886.

282. Nandor Fodor, *Encyclopedia of Psychic Science.* London: Arthur's Press, 1933, p. 170.

283. G. N. M. Tyrell, *Science and Psychical Phenomena.* New Hyde Park, NY: University Books, 1961, pp. 175-179. The actual quote is from William James in *Proceedings of the Society for Psychical Research, 28*, pp. 117-121.

284. Sir Oliver Lodge, *Raymond or Life and Death.* New York: George H. Doran, 1916, p. 90.

285. *Ibid.*, pp. 98-99.

286. *Ibid.*, p. 100.

287. *Ibid.*, p. 102.

288. Rosalind Heywood, *Beyond the Reach of Sense.* New York: E. P. Dutton, 1974, p. 118. A researcher as well as a sensitive, Heywood was one of the grand ladies of psychical research.

289. G. N. M. Tyrell, *op. cit.*, pp. 230-250.

290. Nandor Fodor, *op. cit.* pp. 71-72.

291. *Ibid.*, pp. 67-68.

292. Many of the historical quotes cited under this heading were collected in J. Head and S. L. Cranston (eds.), *Reincarnation: The Phoenix Fire Mystery.* New York: Julian Press, 1977.

293. Jonathan Venn, "Hypnosis and the Reincarnation Hypothesis: A Critical Review and Intensive Case Study," *Journal of the American Society for Psychical Research,* October 1986, *80*(4), pp. 409-426.

294. Ian Stevenson, *Twenty Cases Suggestive of Reincarnation.* New York: American Society for Psychical Research, 1966.

295. Ian Stevenson, "Xenoglossy: A Review and Report of a Case," *Proceedings of the American Society for Psychical Research, 31*, February 1974.

296. Stephen E. Braude, *The Limits of Influence: Psychokinesis and the Philosophy of Science.* New York: Routledge and Kegan Paul, 1986, pp. ix-xii.

297. Carl. G. Jung, *Memories, Dreams and Reflections.* London: Routledge and Kegan Paul, 1963, p. 152.

298. Charles Richet, *op. cit.*, pp. 407-408.

299. Sir William Crookes, "Experimental Investigation of a New Force," *Crookes and The Spirit World, op. cit.*, p. 24.

300. *Ibid.*, p. 26.

301. D. D. Home, *op. cit.*

302. Sir William Crookes, "The Last of Katie King," in *Crookes and the Spirit World, op. cit.*, p. 138. A poignant, yet comical story.

303. Sir William Crookes, "Spirit Forms," in *Crookes and the Spirit World, op. cit.*, pp. 135-136.

304. Harry Price, *Fifty Years of Psychical Research.* London: Longmans, Green and Company, 1939.

305. Charles Richet, *op. cit.*, pp. 506-508.

306. *Ibid.*, p. 543.

307. *lbid.*, pp. 543-544.

308. Soji Otani, "Past and Present Situation of Parapsychology in Japan," *Parapsychology Today: A Geographic View*, pp. 34-35.

309. J. Gaither Pratt, *ESP Research Today.* Metuchen, NJ: The Scarecrow Press, 1973, pp. 108-109. An insider's view of developments in psychic research.

310. Jule Eisenbud, *The World of Ted Serios.* New York: William Morrow, 1967, p. 332.

311. J. Gaither Pratt, *op. cit.*, p. 114.

312. Sheila Ostrander and Lynn Schroeder, *Psychic Discoveries Behind the Iron Curtain.* New York: Prentice-Hall, 1969. p. 84.

313. *Ibid.*, pp. 60-61.

314. *Ibid.*, p. 407.

315. J. Gaither Pratt and H. H. J. Keil, "First-hand Observations of Nina S. Kulagina Suggestive of PK Upon Static Objects," Parapsychological Association Convention, Charlottesville, Virginia, 1973.

316. H. H. J. Keil and Jarl Fahler, "Nina S. Kulagina: A Strong Case for PK Involving Directly Observable Movements of Objects Recorded on Cine Film," Parapsychological Association Convention, New York, 1974.

317. Montague Ullman, "Report on Nina Kulagina," Parapsychological Association Convention, 1973.

318. Benson Herbert, "Report on Nina Kulagina," *Journal of Paraphysics*, 1970, Nos. 1, 3, 5.

319. Lecture presented by Stanley Krippner at the University of California, Davis, 1973.

320. Andrija Puharich, *Beyond Telepathy.* New York: Doubleday, 1972.

321. Russell Targ and Harold Puthoff, "Experiments with Uri Geller," Parapsychological Association Convention, 1973.

322. H. H. J. Keil and Scott Hill, "Mini-Geller PK Cases," Parapsychological Association Convention, 1974.

323. Uri Geller, *My Story.* New York: Praeger, 1975. Geller's own account of his worldwide spoon-bending stir.

324. A. R. G. Owen, "Editorial," *New Horizons, 2*(1), April 1975, p. 1.

325. Wilbur Franklin, "Fracture Surface Physics Indicating Teleneural Interaction," *New Horizons, 2*(1), April 1975, p. 813.

326. W. G. Roll, "Poltergeists," in Richard Cavandish (ed.), *Encyclopedia of the Unexplained.* New York: McGraw-Hill, 1974, p. 200.

327. *Ibid.*

328. A. R. G. Owen, *Can We Explain the Poltergeist?* New York: Taplinger, 1964.

329. Matthew Manning, *The Link.* New York: Holt, Rinehart and Winston, 1975.

330. *Matthew Manning: Study of a Psychic.* This movie, made on location in England, shows how Matthew, an English schoolboy, developed ostensible powers of clairvoyance and psychokinesis and brought them under voluntary control. The film may be rented or purchased from George Ritter Films Limited, Toronto, Ontario, Canada.

331. Peter Bander, "Introduction," *The Link.* New York: Holt, Rinehart and Winston, 1975.

332. Brian Josephson, "Possible Relations Between Psychic Fields and Conventional Physics," and "Possible Connections Between Psychic Phenomena and Quantum Mechanics," *New Horizons, 1*(5), January 1975.

333. A. R. G. Owen, "A Preliminary Report on Matthew Manning's Physical Phenomena," *New Horizons, 1*(4), July 1974, pp. 172-173.

334. Joel L. Whitton, "'Ramp Functions' in EEG Power Spectra during Actual or Attempted Paranormal Events," *New Horizons,* July 1974, pp. 173-186.

335. Iris M. Owen and Margaret H. Sparrow, "Generation of Paranormal Physical Phenomena in Connection with an Imaginary Communicator," *New Horizons, 1*(3), January 1974, pp. 6-13.

336. K. J. Batcheldor, "Report on a Case of Table Levitation and Associated Phenomena," *Journal of the Society for Psychical Research, 43*(729), September 1966, pp. 339-356.

337. C. Brookes-Smith and D. W. Hunt, "Some Experiments in Psychokinesis," *Journal of the Society for Psychical Research, 45*(744), June 1970, pp. 265-281.

338. C. Brookes-Smith, "Data-tape Recorded Experimental PK Phenomena," *Journal of the Society for Psychical Research, 47*(756), June 1973, pp. 68-89.

339. *Philip, The Imaginary Ghost.* This film may be rented or purchased from George Ritter Films Limited in Toronto, Canada.

340. Iris M. Owen, "'Philip's' Story Continued," *New Horizons, 2*(1), April 1975.

341. Joel L. Whitten, "Qualitative Time-domain Analysis of Acoustic Envelopes of Psychokinetic Table Rappings," *New Horizons,* April 1975.

342. Giovanni Ianuzzo, "'Fire Immunity': Psi Ability or Phychophysiological Phenomenon," *Psi Research, 2*(4), December 1983, pp. 68-74.

343. Ruth Inge Heinze, "'Walking on Flowers' in Singapore," *Psi Research, 4*(2), June 1985, pp. 46-50.

344. Meyn Reid Coe, Jr., "Fire-Walking and Related Behaviors," *The Psychological Record, 7*(2), April 1957, p. 107.

345. Larissa Vilenskaya, "An Eyewitness Report: Firewalking in Portland, Oregon," *Psi Research, 2*(4), December 1983, pp. 85-97.

346. Larissa Vilenskaya, "Firewalking: a New Fad, a Scientific Riddle, an Excellent Tool for Healing, Spiritual Growth and Psychological Development?" *Psi Research, 3*(2), June 1984, pp. 102-118.

347. J. Doherty, "Hot Feat: Firewalkers of the World," *Science Digest, 66,* August 1982, pp. 67-71.

348. Harry Price, "A Report on Two Experimental Firewalks," *Bulletin II.* London: University of London Council for Psychical Investigation, 1936.

349. Harry Price, *Fifty Years of Psychical Research, op. cit.*, pp. 250-262.

350. Earl of Dunraven, *Experiences in Spiritualism with D.D. Home.* Glasgow: Robert Maclehose and Company, 1924. Introduction by Sir Oliver Lodge.

351. William Crookes, "Notes of Seances with D. D. Home," *Proceedings of the Society for Psychical Research, 6,* 1889-1890.

352. Elmer Green and Alyce Green, "The Ins and Outs of Mind-Body Energy," *Science Year*

1974, World Book Science Annual. Chicago: Field Enterprises Educational Company, 1973, p. 146.

353. Berthold E. Schwarz, "Ordeal By Serpents, Fire and Strychnine," *Psychiatric Quarterty, 34,* July 1960, pp. 405-429.

354. George Egely, "Why Have I Failed With Calculations?" *Psi Research, 3*(2), June 1984, pp. 94-101.

355. Julianne Blake, "Attribution of Power and the Transformation of Fear: An Empirical Study of Firewalking," *Psi Research, 4*(2), June 1985, pp. 62-88.

356. Loyd M. Auerbach, "Psi-fi: Psi in Science Fiction," *Applied Psi,* 3-4, Spring 1984.

357. Louisa E. Rhine, *PSI.* New York: Harper and Row, 1975.

358. Ronald Rose, *Primitive Psychic Power.* New York: New American Library, 1956.

359. Joan Halifax-Grof, "Hex Death," in Alan Angoff and D. Barth (eds.), *Parapsychology and Anthropology: Proceedings of an International Conference.* New York: Parapsychology Foundation, 1974.

360. Leonid I. Vasiliev, *Experiments in Distant Influence.* New York: E. P. Dutton, 1976.

361. Sun Tzu, *The Art of War,* translated by S. B. Griffith. New York: Oxford University Press, 1963.

362. Zdenek Rejdak, "Parapsychology — War Menace or Total Peace Weapon?" in S. Ostrander and L. Schroeder (eds.), *The ESP Papers: Scientists Speak Out from Behind the Iron Curtain.* New York: Bantam, 1976.

363. J. H. Brennan, *The Occult Reich.* New York: New American Library, 1974.

364. C. L. Linedecker, *The Psychic Spy.* New York: Doubleday, 1976.

365. Leonid L. Vasiliev, *op. cit.*

366. Michael Rossman, *New Age Blues.* New York: E. P. Dutton, 1979.

367. E. Read, "In Vietnam Life Can Depend On a Dowsing Rod," *Fate, 21*(4), April 1968, pp. 52-59.

368. SRI report

369. SRI report

370. A letter from Dr. Ray Hyman reporting his salient observations of Uri Geller during those meetings refered to the penetrating "blue eyes" of Uri Geller. Geller's eyes are brown.

371. *Paraphysics Research and Development — Warsaw Pact.* Defense Intelligence Agency, Washington, DC, March 30, 1978.

372. Charles T. Tart, "A Survey of Negative Uses, Government Interest and Funding of Psi," *Psi News, 1*(2), 1978, p. 3.

373. Barbara Honegger, who received an M.S. degree in parapsychology at John F. Kennedy University, worked in the White House as an assistant to Martin Anderson, Ronald Reagan's domestic policy advisor. She co-authored, with me, a paper titled "National Security Implications of Psi" for the *Applied Psi Newsletter, 1*(5), November-December 1982, which is the basis for much that is reported here of government interest in psi. She has also recently published a book, *October Surprise* (New York: Tudor Publishing, 1989), detailing behind-the-scenes activities relating to Iranian arms sales during the Reagan administration.

374. Robert C. Beck, "Extreme Low Frequency Magnetic Fields Entrainment: A Psychotronic Warfare Possibility?" *Association for Humanistic Psychology Newsletter,* April 1978.

375. Lt. Col. John B. Anderson, "The New Mental Battlefield," *Military Review,* December 1980, pp. 47-54.

376. Thomas E. Bearden, *Excalibur Briefing.* San Francisco: Walnut Hill, 1980.

377. Thomas E. Bearden, "Soviet Psychotronic Weapons: A Condensed Background," *Specula,* March-June 1978.

378. Candice Borland and G. Landrith, *Improved Quality of City Life: Decreased Crime Rate. MERU Report 7502.* Weggis, Switzerland: Department of Sociology, Center for the Study of Higher States of Consciousness, Maharishi European Research University, 1975.

379. M.C. Dillbeck, T.W. Bauer, and S.I. Seferovich, *The Transcendental Meditation Program as a Predictor of Crime Rate Changes in the Kansas City Metropolitan Area.* Unpublished paper (1978). Available from the International Center for Scientific Research, Maharishi International University, Fairfield, Iowa 52556.

380. S. Giles, "Analysis of Crime Trend in 56 Major U.S. Cities," *Scientific Research on the Transcendental Meditation Program: Collected Papers,* Vol II. Rheinweiler, West Germany, nd.

381. G. Hatchard, *Influence of Transcendental Meditation Program on Crime Rate in Suburban Cleveland,* Unpublished paper (1978). Available from the International Center for Scientific Research, Maharishi International University, Fairfield, Iowa 52556.

382. G. Landrith, *The Maharishi Effect and Invincibility: Crime, Automobile Accidents and Fires.* Unpublished paper (1978). Available from the International Center for Scientific Research, Maharishi International University, Fairfield, Iowa 52556.

383. Franklin D. Trumpy, "An Investigation of the Reported Effect of Transcendental Meditation on the Weather," *Skeptical Inquirer, VIII*(2), Winter 1983-84, pp. 143-148.

384. Joseph Banks Rhine, "Some Exploratory Tests in Dowsing," *Journal of Parapsychology, 14*(4), December 1950, pp. 278-286.

385. Kenneth Roberts, *Henry Gross and His Divining Rod.* New York: Doubleday, 1951.

386. Kenneth Roberts, *The Seventh Sense.* New York: Doubleday, 1953.

387. Kenneth Roberts, *Water Unlimited.* New York: Doubleday, 1957.

388. Berthold E. Schwarz, *A Psychiatrist Looks at ESP.* New York: New American Library, 1965. Originally published as *Psycho-Dynamics,* the book contains studies of three different psychics.

389. A. G. Bakirov, "The Geological Possibilities of the Biophysical Method," translated by C. Muromcew and C. Bird, *The American Dowser,* August 1974, pp. 110-112.

390. A useful skeptical account of dowsing is Evon Z. Vogt and Ray Hyman, *Water Witching U.S.A.* Chicago: University of Chicago Press, 1959.

391. Jonsson, incidentally, is the same psychic who participated with astronaut Edgar Mitchell in a less than successful ESP experiment from outer-space. See Edgar D. Mitchell, "An ESP Test from Apollo 14," *Journal of Parapsychology, 35*(2), 1971.

392. "News Ambit," *Psychic Magazine,* December 1974.

393. Ostrander and Schroeder, *op. cit.*

394. In May of 1975, Lozanov spoke at several conferences of American educators interested in utilizing his techniques.

395. Milan Ryzl, *ESP in the Modern World.* San Jose: Milan Ryzl, 1972. Available from the author, P.O. Box 9459, Westgate Station, San Jose, CA. 95117.

396. Frederick Bligh Bond and T. S. Lea. *The Gate of Remembrance: The Story of the Psychological Experiment Which Resulted in the Discovery of Edgar Chapel at Glastonbury.* 2nd and 4th editions revised. Oxford: Basil Blackwell, 1918, 1921.

397. J. N. Emerson, "Intuitive Archeology: A Psychic Approach," *New Horizons, 1*(3), January 1974.

398. J. N. Emerson, "Intuitive Archeology: The Argillite Carving," Department of Anthropology, University of Toronto, March 1974.

399. J. N. Emerson, "Intuitive Archeology: A Developing Approach," Department of Anthropology, University of Toronto, November 1974.

400. Chris Bird and Jeffrey Mishlove, "Soviet Parapsychology," *Mindspace* radio broadcast, May 1975.

401. Stephen A. Schwartz, *The Secret Vaults of Time.* New York: Grosset and Dunlap, 1978.

402. Stephen A. Schwartz and H. E. Edgerton, *A Preliminary Survey of the Eastern Harbour, Alexandria, Egypt, Combining both Technical and*

Extended Sensing Remote Sensor Exploration. Los Angeles: The Mobius Group, 1980.

403. Stephan A. Schwartz and Rand de Mattei, "The Discovery of an American Brig: Fieldwork Involving Applied Archaeological Remote Viewing," in Linda A. Henkel and Rick E. Berger (eds.), *Research in Parapsychology 1988.* Metuchen, NJ: Scarecrow Press, 1989.

404. A major critic of this report has been Keith Harary, a psi researcher who has served as a Mobius respondent.

405. Paul Tabori, *Crime and the Occult.* New York: Taplinger, 1974.

406. *Ibid.*

407. *Ibid.*

408. M. B. Dykshorn and Russell H. Felton, *My Passport Says Clairvoyant.* New York: Hawthorne, 1974.

409. "Interview: Irene Hughes," *Psychic Magazine,* December 1971.

410. Tabori, *op. cit.*

411. *Ibid.*

412. Piet Hein Hoebens, "Gerard Croiset: Investigation of the Mozart of 'Psychic Sleuths' — Part I," *Skeptical Inquirer, VI*(1), Fall 1981, pp. 17-28.

413. Piet Hein Hoebens, "Croiset and Professor Tenhaeff: Discrepancies in Claims of Clairvoyance," *Skeptical Inquirer, VI*(2), Winter 1981-2, pp. 32-40.

414. Kathlyn Rhea, *Mind Sense.* Berkeley, CA: Celestial Arts, 1988.

415. Kathlyn Rhea, *The Psychic Is You* (2nd ed.). Berkeley, CA: Celestial Arts, 1988.

416. Pat Michaels, "Pat Michaels, Clairvoyant Newsman," *Fate,* August 1966, pp. 442-53.

417. William H. Kautz, "Rosemary Case of Alleged Egyptian Xenoglossy," *Theta, 10*(2), 1982, pp. 26-30.

418. Jeffrey Goodman, *Psychic Archeology.* New York: G. P. Putnam, 1977.

419. Douglas Dean, John Mihalsky, Shiela Ostrander and Lynn Schroeder, *Executive ESP.* Englewood Cliffs, NJ: Prentice-Hall, 1974.

420. Eric Mishara, "Psychic Stock Market Analyst," *Omni,* February 1983.

421. Uri Geller and Guy Lyon Playfair, *The Geller Effect.* New York: Henry Holt, 1986.

422. J. Cook, "Closing the Psychic Gap," *Forbes,* 1984, *133,* 12, pp. 90-95.

423. *Ibid.*

424. J. C. Barker, "Premonitions of the Aberfan Disaster," *Journal of the Society for Psychical Research, 44,* 1967, pp. 169-180.

425. E. W. Cox, "Precognition: An Analysis, I," *Journal of the American Society for Psychical Research, 50,* 1956, pp. 47-58.

426. E. W. Cox, "Precognition: An Analysis, II," *Journal of the American Society for Psychical Research, 50,* 1956, pp. 97-107.

427. Alan Vaughan, "Applying Precognition to Space Systems," *Applied Psi Newsletter, 1*(2), May/June 1982, pp. 1-6.

428. Alan Vaughan, *Patterns of Prophecy.* New York: Hawthorne, 1973, pp. 37-55. Vaughan is also a researcher and theorist. His book provides a detailed discussion of the archetypal patterns of time.

429. For an interesting research study involving Vaughan and his students, see Gertrude R. Schmeidler and J. Goldberg, "Evidence for Selective Telepathy in Group Psychometry," in W. G. Roll, R. L. Morris, and J. D. Morris (eds.), *Research in Parapsychology 1973.* Metuchen, NJ: Scarecrow Press, 1974.

430. For a skeptical account of Alan Vaughan's abilities, see James Randi, "'Superpsychic' Vaughan: Claims Versus the Record," *Skeptical Inquirer, V*(4), Summer 1981, pp. 19-21.

431. *Ibid.*, pp. 66-69, 101-102.

432. Carol Liaros, "Psi Faculties in the Blind," *Parapsychology Review, 5*(6), November-December 1974, pp. 25-26.

433. Yvonne Duplessis, *The Paranormal Perception of Color.* New York: Parapsychology Foundation, 1975.

434. Jeffrey Mishlove, *Psi Development Systems.* New York: Ballantine, 1988.

435. W. V. Rauscher, "Beethoven Lives Through Rosemary Brown," In Marton Ebon (ed.), *The Psychic Scene.* New York: New American Library, 1974, pp. 140-150.

436. E. W. Russell, "Radionics — Science of the Future," in John White and Stanley Krippner (eds.), *Future Science.* New York: Anchor Books, 1977.

437. Michael Murphy and Rhea A. White, *The Psychic Side of Sports.* Reading, MA: Addison-Wesley, 1979.

438. Michael Murphy, *Transforming the Human Body* (#S320). Videotape available from Thinking Allowed Productions, 2560 9th Street, #123, Berkeley, CA 94710.

439. Patric V. Geisler, "Parapsychological Anthropology: Multi-Method Approaches to the Study of Psi in the Field Setting," in W. G. Roll, J. Beloff, and R. A. White (eds.), *Research in Parapsychology 1982.* Metuchen, NJ: Scarecrow Press, 1983, pp. 241-244.

440. Annie Besant and C. W. Leadbeater, *Occult Chemistry.* London: Theosophical Publishing House, 1919.

441. Z. W. Wolkowski, "A Case of Apparent Direct Observation of Matter at the Molecular and Sub-Atomic Level," *Proceedings of the Second International Congress on Psychotronic Research,* 1975, Monte Carlo, pp. 288-290.

442. Janet Mitchell, "A Psychic Probe of the Planet Mercury," *Psychic, 6*(3), May/June 1975, pp. 16-21.

443. Jeffrey Mishlove, *Preliminary Investigation of Events Which Suggest the Applied Psi Abilities of Mr. Ted Owens.* San Francisco: Washington Research Center, 1977, 1978.

444. T. J. Constable, "Orgone Energy Weather Engineering Through the Cloudbuster," in John White and Stanley Krippner (eds.), *Future Science.* New York: Anchor Press, 1977.

445. Sir Arthur Grimble, *We Chose the Islands.* New York: William Morrow, 1952.

446. Dr. Charles Tart (who at that time served on my Ph.D. advisory committee in Berkeley) described a similar state in an article titled, "Mutual Hypnotic Induction," published in his classic anthology, *Altered States of Consciousness,* New York: Doubleday, 1971.

447. Wade Doak, *Dolphin Dolphin.* New York: Sheridan House, 1988. While this most interesting book, and others by Doak, may be hard to find in bookstores, they can be ordered directly from the publisher at 146 Palisades, Dobbs Ferry, New York.

448. Wade Doak, *Encounters With Whales and Dolphins.* Auckland: Hodder and Stoughton, 1988.

449. William H. Kautz, "Intuitive Consensus: A Novel Approach to the Solution of Difficult Scientific and Technical Problems." Brochure published by the Center for Applied Intuition.

450. Published in San Francisco by Harper and Row in 1987 and 1989 respectively.

451. E. G. Boring, "The Present Status of Parapsychology," *American Scientist, 43,* 1955, pp. 108-116.

452. P. W. Bridgeman, "Probability, Logic and ESP," *Science, 123,* 1956, pp. 15-17.

Section III

1. Trevor Pinch and Harry Collins, *Frames of Meaning.* London: Routledge and Kegan Paul, 1985.

2. William James, *Psychology.* New York: Henry Holt, 1982, p. 152.

3. Jack R. Strange, "The Search for Sources of the Stream of Consciousness," in Kenneth S. Pope and Jerome L. Singer (eds.), *The Stream of Consciousness.* New York: Plenum Press, 1978.

4. William James, *Essays in Radical Empiricism.* New York: Longmans, Green, 1912, p. 3

5. *Ibid.*, p. 4.

6. E. B. Titchener, *A Text-Book of Psychology.* New York: Macmillan, 1909, p. 6.

7. James B. Watson, *Behaviorism.* New York: Norton, 1924, p. 3.

8. Philosopher Michael Grosso, author of *The Final Choice: Playing the Survival Game* (Waltham, MA: Stackpole, 1987) maintains that this academic denial of the psyche is an expression of *Thanatos,* the Freudian death instinct. In the following excerpt from my *Thinking Allowed* interview with him, he states:

There are many aspects of our culture that do seem to express a death instinct. One illustration is that many contemporary academic philosophers, scientists and psychologists deny the existence of consciousness. They have gotten to the point of denying the existence of mind. That impresses me as a kind of a suicide, a denial of our very livingness. In other words, we use our intellect to deny our livingness. I think that we can see this process taking place on many different levels of our culture: the way we are destroying the environment, the way we use our intelligence to create weapons that threaten us with mass extinction. I see that as evidence for a death instinct.

9. Howard Gardner, *The Mind's New Science: A History of the Cognitive Revolution.* New York: Basic Books, 1985.

10. Karl H. Pribram, "Mind, Brain and Consciousness: The Organization of Competence and Conduct," in Julian M. Davidson and Richard J. Davidson (eds.), *The Psychobiology of Consciousness.* New York: Plenum Press, 1980, p. 47.

11. Indeed, the very name *parapsychology* reflects the peculiar twists of history. I would argue that, if the data of psi research has validity, it probably belongs within *psychology.* Yet, J. B. Rhine felt constrained to use the German term parapsychology to describe his research. It may have been a useful maneuver at the time. Rhine's *Journal of Parapsychology* has been in continuous existence for over fifty years (an indication of some success for this discipline). I believe that the name *parapsychology* denotes an artificial distinction which presently creates more problems than it solves.

12.The material presented under this heading is largely based upon a summary of research prepared by Stanford psychologist Dale Griffin, titled "Intuitive Judgment and the Evaluation of Evidence," commissioned by the National Academy of Science's *Committee on Techniques for the Enhancement of Human Performance.* Washington, DC: 1987.

13. P. C. Wason, "On the Failure to Eliminate Hypotheses in a Conceptual Task," *Quarterly Journal of Experimental Psychology, 12,* 1960, pp. 129-140.

14. A. Koriat, S. Lichtenstein, and B. Fischoff, "Reasons for Confidence," *Journal of Experimental Psychology: Learning and Memory,* 1980, pp. 107-118.

15. Robert K. Merton, "The Self-Fulfilling Prophecy," *Antioch Review, 8,* 1948, pp. 193-210.

16. R. Rosenthal and L. Jacobson, *Pygmalion in the Classroom: Teacher Expectation and Pupil's Intellectual Development.* Hillsdale, NJ: Holt, Rinehart and Winston, 1968.

17. A meta-analysis is the quantitative combination of the results of a group of studies on a given topic.

18. A. Hastorf and H. Cantril, "They Saw A Game: A Case Study," *Journal of Abnormal and Social Psychology, 49,* 1967, pp. 129-134.

19. R. E. Nisbett and L. Ross, *Human Inference: Strategies and Shortcomings of Social Judgment.* Englewood Cliffs, NJ: Prentice-Hall, 1980.

20. D. Kahneman and A. Tversky, "On the Study of Statistical Intuitions," in *Judgment Under Uncertainty, op. cit.,* pp. 463-492.

21. B. F. Skinner, "Superstition in the Pigeon," *Journal of Experimental Psychology, 38,* 1948, pp. 168-172.

22. T. Gilovich, B. Vallone, and A. Tversky, "The Hot Hand in Basketball: On the Misperception of Random Sequences," *Cognitive Psychology, 17,* 1986, pp. 295-314.

23. R. Falk, "On Coincidences," *Skeptical Inquirer, 6,* 1981-82, pp. 18-31.

24. R. E. Nisbett and T. D. Wilson, "Cognitive Manipulation of Pain," *Journal of Experimental Social Psychology, 21,* 1966, pp. 227-236.

25. G. A. Quattrone and A. Tversky, "Causal Versus Diagnostic Contingencies: On Self-Deception and the Votor's Illusion," *Journal of Personality and Social Psychology, 46,* 1984, pp. 237-248.

26. Dale Griffin, *op. cit.*

27. E. J. Langer, "The Illusion of Control," in *Judgment Under Uncertainty, op. cit.*, pp. 32-47.

28. Erving Goffman, *Interaction Ritual.* New York: Anchor, 1967.

29. L. H. Strickland, R. J. Lewicki, and A. M. Katz, "Temporal Orientation and Perceived Control as Determinants of Risk Taking," *Journal of Experimental Social Psychology, 3,* 1966, pp. 143-151.

30. F. Ayeroff and R. P. Abelson, "ESP and ESB: Belief in Personal Success at Mental Telepathy," *Journal of Personality and Social Psychology, 34,* pp. 240-247.

31. Victor A. Benassi, P. D. Sweeney, and G. E. Drevno, "Mind Over Matter: Perceived Success at Psychokinesis," *Journal of Personality and Social Psychology, 37,* 1979, pp. 1377-1386.

32. Leon Festinger, *A Theory of Cognitive Dissonance.* Stanford, CA: Stanford University Press, 1957.

33. Leon Festinger, H. W. Riecken, and S. Schachter, *When Prophecy Fails.* Minneapolis, MN: University of Minnesota Press, 1956.

34. Actually, when one reads closely Festinger's classic book, *When Prophecy Fails,* it seems that the social psychologists are as much in error as the UFO cultists under study. Most members of the UFO group actually did give up their cult beliefs and drifted away — contrary to Festinger's predictions. Nevertheless, the theory of *cognitive dissonance* has survived. Festinger's own behavior in supporting his theory in spite of the evidence may be an example of the *confirmation bias.* Ironically, this error is often overlooked by social psychologists.

35. C. D. Batson, "Rational Processing or Rationalization: The Effect of Disconfirming Information on Stated Religious Belief," *Journal of Personality and Social Psychology, 32,* 1975, pp. 176-184.

36. D. Druckman and J. A. Swets (eds.). *Enhancing Human Performance: Issues, Theories and Techniques.* Washington, DC: National Academy Press, 1988.

37. The chairman of the NRC committee asked Harvard researchers Monica J. Harris and Edward Rosenthal to withdraw their favorable conclusions regarding psi research. When they refused to do so, the final report of the committee ignored their presentation. See John A. Palmer, Charles Honorton, and Jessica Utts, *Reply to the National Research Council Study on Parapsychology.* Research Triangle Park, NC: Parapsychological Association, 1988.

38. Rene Prosper Blondlot, *"N" Rays: A Collection of Papers Communicated to the Academy of Sciences,* translated by J. Garcin. London: Longmans, Green, 1905.

39. Robert Wood. "The n-rays." *Nature, 70,* 1904, pp. 530-531.

40. J. P. Gilbert, B. McPeek, and F. Mosteller, "How Frequently Do Innovations Succeed in Surgery and Anesthesia?" in J. Tanur, F. Mosteller, W. H. Kruskal, R. F. Link, R. S. Pieters, and G. R. Rising (eds.), *Statistics: A Guide to the Unknown.* San Francisco: Holden-Day, 1978, pp. 45-58.

41. F. J. Roethlisberger and W. J. Dickson, *Management and the Worker.* New York: Wiley, 1939.

42. James E. Alcock and Laura P. Otis, "Critical Thinking and Belief in the Paranormal," *Psychological Reports, 46,* April 1980, pp. 479-482.

43. A survey published in *New Scientist,* on January 25, 1973, indicated that 25% of scientists polled considered extrasensorimotor phenomena "an established fact." Another 42%

opted for "a likely possibility."

44. Harry Price, *Fifty Years of Psychical Research.* London: Longmans, Green and Company, 1939, pp. 73-74. Price, who founded the National Laboratory of Psychical Research in London, was involved in exposing many fraudulent "psychics."

45. Joseph Banks Rhine, *Extra-Sensory Perception.* Boston: Society for Psychical Research, 1933, pp. 73-74.

46. B. H. Camp, [Statement in notes.] *Journal of Parapsychology, 1,* 1937, p. 305.

47. J. Gaither Pratt, Joseph Banks Rhine, et al., *Extra Sensory Perception After Sixty Years.* New York: Henry Holt and Company, 1940. This book was a Bible, in its day, for card-guessing researchers.

48. Rhine and his associates borrowed a German term and designated their experimental work *parapsychology.* This was done both to distinguish it from the earlier term *psychical research,* which was generally a non-experimental field, and to denote an inquiry which was closely related to psychology. Although, I myself hold a doctoral diploma in *parapsychology* (from the University of California at Berkeley, 1980), I am refraining from using that term in this book, along with the related term *paranormal.* Such terms may have caused damage to a field whose subject matter is, in my view, properly conceived of as *normal* and *psychological.* The term *psi* is simply defined as "interactions between organisms and their environment which are not mediated by recognized sensorimotor functions."

49. Joseph Banks Rhine, *Extra-Sensory Perception, op. cit.*

50. George R. Price, "Science and the Supernatural," *Science, 122,* pp. 359-367.

51. C. E. M. Hansel, *ESP: A Scientific Evaluation.* New York: Scribner's, 1966.

52. Ian Stevenson, "An Antagonist's View of Parapsychology. A Review of Professor Hansel's ESP: A Scientific Evaluation," *Journal of the American Society for Psychical Research, 61,* July 1967, pp. 254-267. Stevenson points out that Hansel based his conclusions on an inaccurate diagram of Pratt's office.

53. Betty Marwick, "The Soal-Goldney Experiments with Basil Shackleton: New Evidence of Manipulation," *Proceedings of the Society for Psychical Research, 56,* p. 211.

54. In the absence of experimental consistency and theoretical underpinnings, some psychic investigators feel that it is premature to claim that even the best experiments support a *psi* hypothesis. Perhaps, in the future, researchers and critics working together will uncover conventional explanations for the existing data. Therefore they prefer to refer to the existing data of psi research as *anomalies.* See John Palmer, "Have We Established Psi?" *Journal of the American Society for Psychical Research, 81,* 1987, pp. 111-123; K. Ramakrishna Rao and John Palmer, "The Anomaly Called Psi: Recent Research and Criticism," *Behavioral and Brain Sciences, 10,* 1987, pp. 539-551.

55. E. Douglas Dean, "The Plethysmograph as an Indicator of ESP," *Journal of the Society for Psychical Research, 41,* 1962, pp. 351-353.

56. E. Douglas Dean and Carroll B. Nash, "Plethysmograph Results Under Strict Conditions," Sixth Annual Convention of the Parapsychological Association, New York, 1963.

57. Charles T. Tart, "Possible Physiological Correlates of Psi Cognition," *International Journal of Parapsychology, 5,* 1963, pp. 375-386.

58. Montague Ullman, Stanley Krippner, and Alan Vaughan, *Dream Telepathy.* New York: Macmillan, 1973. A valuable feature of this book is that, as in *ESP After Sixty Years,* the authors invited contributions from known critics of their work.

59. Naturally these findings caused some scientists to echo the thought of Shakespeare that "we are the stuff that dreams are made of." This notion may eventually take on some rather precise physical and mathematical coloring, as the Pythagorean tradition finds renewal mathematical theorists (see Appendix).

60. Stanley Krippner, Charles Honorton, and

Montague Ullman, "An Experiment in Dream Telepathy with The Grateful Dead," *Journal of the American Society of Psychosomatic Dentistry and Medicine, 20*(1), 1973.

61. John Palmer, *An Evaluative Report on the Current Status of Parapsychology*. Alexandria, VA: U.S. Army Research Institute for the Behavioral and Social Sciences, 1985.

62. Irvin L. Child, "Psychology and Anomalous Observations: The Question of ESP in Dreams," *American Psychologist, 40*(11), November 1985, pp. 1219-1229.

63. Milan Ryzl, "A Method of Training in ESP," *International Journal of Parapsychology, 8*(4), Autumn 1966.

64. Charles Honorton, "Significant Factors in Hypnotically-Induced Clairvoyant Dreams," *Journal of The American Society for Psychical Research, 66*(1), January 1972, pp. 86-102.

65. Edward A. Charlesworth, "Psi and the Imaginary Dream," Seventeenth Annual Convention of the Parapsychological Association, New York, 1974.

66. Gertrude R. Schmeidler, "High ESP Scores After a Swami's Brief Instruction in Meditation and Breathing," *Journal of The American Society for Psychical Research, 64*(1), January 1970, pp. 101-103.

67. Karlis Osis and Edwin Bokert, "ESP and Changed States of Consciousness Induced by Meditation," *Journal of The American Society for Psychical Research, 65*(1), January 1971, pp. 17-65.

68. Emile Boirac, *Our Hidden Forces*, London: Rider, 1918.

69. D. Scott Rogo, *Parapsychology: A Century of Inquiry*. New York: Taplinger, 1975, p. 238.

70. Emile Boirac, *op. cit.*

71. *Ibid.*

72. Shiela Ostrander and Lynn Schroeder, *Psychic Discoveries Behind The Iron Curtain,* Englewood Cliffs, NJ: Prentice-Hall, 1970. pp. 37-40.

73. Charles Honorton and Stanley Krippner, "Hypnosis and ESP: A Review of the Experimental Literature," *Journal of The American Society for Psychical Research, 63*, 1969, pp. 214-252.

74. Ephriam I. Schechter, "Hypnotic Induction vs. Control Conditions: Illustrating an Approach to the Evaluation of Replicability in Parapsychological Data," *Journal of the American Society for Psychical Research, 78,* 1984, pp. 1-27.

75. Ephriam I. Schechter, personal communication, September 12, 1989.

76. Rex G. Stanford, "Altered Internal States and Parapsychological Research: Retrospect and Prospect," in D. H. Weiner and D. I. Radin (eds.), *Research in Parapsychology 1985.* Metuchen, NJ: Scarecrow Press, 1986, pp. 128-131.

77. J. Gaither Pratt, *ESP Research Today,* Metuchen, NJ: Scarecrow Press, 1973. pp. 84-100.

78. Martin Gardner, *How Not to Test a Psychic: Fads and Fallacies in the Name of Science.* Buffalo, NY: Prometheus, 1989.

79. H. Kanthamani and E. F. Kelly, "Awareness of Success in an Exceptional Subject," *Journal of Parapsychology, 38*(4), December 1974, pp. 355-382.

80. Persi Diaconis, "Statistical Problems in ESP Research," *Science, 201,* 1978, pp. 131-136.

81. Stanford Research Institute, news release, October 1974. See also Harold Puthoff and Russell Targ, "Information Transmission Under Conditions of Sensory Shielding," *Nature,* October 18, 1974.

82. Martin Gardner, "How Not to Test a Psychic: The Great SRI Die Mystery," *Skeptical Inquirer, VII*(2), Winter 1982-83, pp. 33-39.

83. Charles Honorton and James C. Terry, "Psi-mediated Imagery and Ideation in the Ganzfeld: A Confirmatory Study," Seventeenth Annual Convention of the Parapsychological Association, New York, 1974.

84. Lendell W. Braud and William G. Braud, "The Psi Conducive Syndrome: Free Response

GESP Performance During an Experimental Hypnagogic State Induced by Visual and Acoustic Ganzfeld Techniques," Parapsychological Association Convention, New York, 1974.

85. Charles Honorton, "Meta Analysis of Psi Ganzfeld Research: A Response to Hyman," *Journal of Parapsychology, 49,* 1985, pp. 51-91.

86. Susan Blackmore, "The Extent of Selective Reporting of ESP Ganzfeld Studies," *European Journal of Parapsychology, 3,* 1980, pp. 213-219.

87. Monica J. Harris and Robert Rosenthal, *Interpersonal Expectancy Effects and Human Performance Research.* Washington, DC: National Academy Press, 1988.

88. Susan Blackmore, "A Report of a Visit to Carl Sargent's Laboratory," *Journal of the Society for Psychical Research, 54*(808), July 1987, pp. 186-198.

89. Adrian Parker and Nils Wiklund, "The Ganzfeld Experiments: Towards an Assessment," *Journal of the Society for Psychical Research, 54*(809), October 1987, pp. 261-265.

90. Ray Hyman, "The Ganzfeld/Psi Experiment: A Critical Appraisal," *Journal of Parapsychology, 49,* 1985, pp. 3-49.

91. Charles C. Honorton, Rick E. Berger, Mario P. Varvoglis, M. Quant, P. Derr, George P. Hansen, Ephriam Schechter, D. C. Ferrari, "Psi Ganzfeld Experiments Using an Automated Testing System: An Update and Comparison with a Meta-Analysis of Earlier Studies," *Proceedings of Presented Papers, the Parapsychological Association 32nd Annual Convention,* San Diego, August 1989, pp. 93-109.

92. If the participant choose not to bring a friend, a Psychophysics Research Laboratory staff member served as sender.

93. These courtesies and considerations may seem either obvious or trivial. Experience suggests, however, that they should not be taken for granted. The emotional tone is noticeably different where researchers are hostile to the possibility of positive psi results and are suspicious that subjects will engage in fraud.

94. This statement, cited in Honorton, et al., *op. cit.,* 1989 is attributed to correspondence received in May 1989.

95. This quote is cited in a news brief titled "Psychologist for Psi," in *Parapsychology Review, 20*(5), September-October 1989, p. 14.

96. National Research Council, *Enhancing Human Performance: Issues, Theories, and Techniques.* Washington, DC: National Academy Press, 1988, p. 175.

97. Ray Hyman and Charles Honorton, "A Joint Communique: The Psi Ganzfeld Controversy," *Journal of Parapsychology, 50,* 1984, pp. 353-354.

98. J. Gaither Pratt and M. Price, "The Experimenter-Subject Relationship in Tests for ESP," *Journal of Parapsychology,* 1938, pp. 84-94.

99. Charles Honorton, M. Ramsey, and C. Cabibbo, "Experimenter Effects in Extrasensory Perception," *Journal of the American Society for Psychical Research, 69,* 1975, pp. 135-149.

100. Judith L. Taddonio, "The Relationship of Experimenter Expectancy to Performance on ESP Tasks," *Journal of Parapsychology, 40,* 1976, pp. 107-114.

101. Adrian Parker, "A Pilot Study of the Influence of Experimenter Expectancy on ESP Scores," Parapsychological Association Convention, New York, 1974.

102. John Beloff and I. Mandelberg, "An Attempted Validation of the 'Ryzl Technique' for Training ESP Subjects," *Journal of the Society for Psychical Research, 43,* 1966, pp. 229-249.

103. John Beloff and J. Bate, "An Attempt to Replicate the Schmidt Findings," *Journal of the Society for Psychical Research, 46,* 1971, pp. 21-30.

104. John Beloff, "The 'Sweethearts' Experiment," *Journal of the Society for Psychical Research, 45,* 1969, pp. 1-7.

105. Gertrude Schmeidler, *Parapsychology and Psychology.* Jefferson, NC: McFarland, 1988.

106. H. C. Berendt, "Parapsychology in Israel," in Allan Angoff and Betty Shapin (eds.),

Parapsychology Today: A Geographic View. New York: Parapsychology Foundation, 1973, p. 68.

107. Gertrude R. Schmeidler and Robert A. McConnell, *ESP and Personality Patterns.* New Haven: Yale University Press, 1958.

108. John A. Palmer, "Scoring in ESP Tests as a Function of Belief in ESP. Part I. The Sheep-Goat Effect," *Journal of The American Society for Psychical Research, 65,* 1971, pp. 373-408.

109. John A. Palmer, "Scoring in ESP Tests as a Function of Belief in ESP. Part I: The Sheep-Goat Effect," *Journal of the American Society for Psychical Research, 65,* 1971, pp. 373-408.

110. Gertrude R. Schmeidler, personal communication, September 18, 1989.

111. J. E. Crandall, "Effects of Favorable and Unfavorable Conditions on the Psi-Missing Displacement Effect," *Journal of the American Society for Psychical Research, 79,* 1985, pp. 27-38.

112. K. Ramakrishna Rao, "The Bidirectionality of Psi," *Journal of Parapsychology, 29,* 1965, pp. 230-250.

113. Harvey J. Irwin, *An Introduction to Parapsychology.* Jefferson, NC: McFarland, 1989. This is an introductory text, suitable for college classes. In particular, see the discussion on "The Bidirectionality of ESP: Psi-Missing."

114. B. K. Kanthamani and K. R. Rao, "Personality Characteristics of ESP Subjects," *Journal of Parapsychology, 36,* 1972, pp. 56-70.

115. John A. Palmer. "Attitudes and Personality Traits in Experimental ESP Research," in B. B. Wolman (ed.), *Handbook of Parapsychology.* New York: Van Nostrand Reinhold, 1977, pp. 175-201.

116. Gertrude Schmeidler, *Parapsychology and Psychology.* Jefferson, NC: McFarland, 1988.

117. Robert L. Morris, "The Concept of the Target," in L. A. Henkel and R. E. Berger, *Research in Parapsychology 1988.* Metuchen, NJ: Scarecrow Press, 1989, pp. 89-91.

118. Martin Johnson, "A New Technique of Testing ESP in a Real-Life, High Motivational Context," *Journal of Parapsychology, 37,* 1973, pp. 210-217. This study, however, was not actually designed to test Stanford's PMIR model.

119. Rex G. Stanford and Gary Thompson, "Unconscious Psi-mediated Instrumental Response and Its Relation to Conscious ESP Performance," Parapsychological Association Convention, Charlottesville, Virginia, 1973.

120. John A. Palmer, *op. cit.*

121. Rex G. Stanford, "Toward Reinterpreting Psi Events," *Journal of the American Society for Psychical Research, 72,* 1978, pp. 197-214.

122. A. A. Foster, "Is ESP Diametric?" *Journal of Parapsychology, 4,* 1940, pp. 325-328.

123. This works both ways as many PK experiments can be interpreted as evidence of precognition.

124. Helmut Schmidt, "A Quantum Process in Psi Testing," in J. B. Rhine (ed.), *Progress in Parapsychology.* Durham, NC: Parapsychology Press, 1973, pp. 28-35.

125. Helmut Schmidt, "A Quantum Mechanical Random Number Generator for Psi Tests," *Journal of Parapsychology, 34,* 1970, pp. 219-224.

126. Helmut Schmidt, "Precognition of a Quantum Process," *Journal of Parapsychology, 33,* 1969, pp. 99-108.

127. Helmut Schmidt, "PK Tests with a High-Speed Random Number Generator," *Journal of Parapsychology,* December 1973, pp. 105-118.

128. C. E. M. Hansel, "Critical Analysis of Schmidt's PK Experiments," *Skeptical Inquirer, V*(3), Spring 1981, pp. 26-33.

129. Ray Hyman, "Further Comments on Schmidt's PK Experiments," *Skeptical Inquirer, V*(3), Spring 1981, p. 39.

130. J. E. Alcock, *A Comprehensive Review of Major Empirical Studies in Parapsychology Involving Random Event Generators or Remote Viewing.* Washington, DC: National Academy Press, 1988.

131. Charles Honorton and Diane C. Ferrari,

"Future Telling — A Meta-Analysis of Forced Choice Precognition Experiments, 1935-1987," *Proceedings of Presented Papers, the Parapsychological Association 32nd Annual Convention,* San Diego, August 1989, pp. 110-121.

132. Charles T. Tart, "Information Acquisition Rates in Forced-Choice ESP Experiments: Precognition Does Not Work as Well as Present-Time ESP," *Journal of the American Society for Psychical Research, 77*(4), October 1983, pp. 293-310.

133. Charles Honorton, "Precognition and Real-Time ESP Performance in a Computer Task with an Exceptional Subject," *Journal of Parapsychology, 51*(4), December 1987, pp. 291-320.

134. Dean I. Radin. "Precognition of Probable Versus Actual Futures: Exploring Futures That Will Never Be," in D. H. Weiner and R. L. Morris (eds.), *Research in Parapsychology 1987.* Metuchen, NJ: Scarecrow Press, 1988, pp. 1-5.

135. Harold E. Puthoff and Russell Targ, "A Perceptual Channel for Information Transfer over Kilometer Distances: Historical Perspective and Recent Research," *Proceedings of the Institute of Electrical and Electronics Engineers, 64,* 1976, pp. 329-354.

136. Brenda J. Dunne, York H. Dobyns, and S. M. Intner, *Precognitive Remote Perception III: Complete Binary Data Base with Analytical Refinements.* (Technical Note PEAR 89002.) Princeton, NJ: Princeton University School of Engineering and Applied Sciences, 1989.

137. *Ibid.*, pp. i-ii.

138. Louisa E. Rhine, *ESP in Life and Lab — Tracing Hidden Channels.* New York: Macmillan, 1967, pp. 166-168.

139. Joseph Banks Rhine, "Psychokinesis," in R. Cavandish (ed.), *Man, Myth and Magic.* New York: Marshall Cavandish, 1970, pp. 2285-2291.

140. Joseph Banks Rhine and Louisa E. Rhine, "The Psychokinetic Effect. l. The First Experiment," *Journal of Parapsychology, 7,* 1943, p. 2043.

141. Louisa E. Rhine, *op. cit.*, p. 175.

142. Helmut Schmidt, Robert L. Morris, and Luther Rudolph. "Channeling Psi Evidence to Critical Observers," in W. G. Roll, R. L. Morris, and R. A. White (eds.), *Research in Parapsychology 1981.* Metuchen, NJ: Scarecrow Press, 1982, pp. 136-138.

143. Helmut Schmidt, Robert L. Morris, and Luther Rudolph. "Channeling Evidence for a PK Effect to Independent Observers," *Journal of Parapsychology,* 1986, 50, pp. 1-16.

144. J. E. Alcock. *A Comprehensive Review of Major Empirical Studies in Parapsychology Involving Random Event Generators or Remote Viewing.* Washington, DC: National Academy Press, 1988, pp. 99-102.

145. According to a personal communication from Robert Morris, October 1989, a replication of this study did not yield significant data.

146. Robert G. Jahn, Brenda J. Dunne, and Roger D. Nelson, "Engineering Anomalies Research," *Journal of Scientific Exploration, 1*(1), 1987, pp. 21-50. The researchers at the Princeton Engineering Anomalies Research program have offered to make full details of the design of this equipment available upon request.

147. Roger D. Nelson, Brenda J. Dunne, and Robert G. Jahn, *An REG Experiment With Large Database Capability, III: Operator Related Anomalies* (Technical Note PEAR 84003. Princeton Engineering Anomalies Research). Princeton, NJ: Princeton University School of Engineering/Applied Science, 1984, p. 10.

148. Robert G. Jahn and Brenda Dunne, *Margins of Reality: The Role of Consciousness in the Physical World.* New York: Harcourt Brace Jovanovich, 1987. As this research is ongoing, the database is growing.

149. *Ibid.*, p. 25

150. John A. Palmer, *op. cit.*

151. James E. Alcock, *A Comprehensive Review of Major Empirical Studies in Parapsychology Involving Random Event Generators or Remote Viewing.* Washington, DC: National Academy Press, 1988.

152. Roger D. Nelson, G. John Bradish, and York H. Dobyns, *Random Event Generator Qualification, Calibration, and Analysis.* (Technical Note PEAR 89001.) Princeton, NJ: Princeton University School of Engineering/Applied Sciences, 1989.

153. In a personal communication, October 16, 1989, Brenda Dunne informed me that several skeptics including James Randi and Phillip Klass have made detailed inquiries about the experimental setup and have not, to her knowledge, uncovered ways in which data-tampering could have occurred.

154. Brenda J. Dunne, Roger D. Nelson, Y. H. Dobyns, and Robert G. Jahn, "Individual Operator Contributions in Large Data Base Anomalies Experiments." (Technical Note PEAR 88002.) Princeton, NJ: Princeton University School of Engineering and Applied Science, 1988.

155. Brenda Dunne, personal communication, October 16, 1989.

156. Dean I. Radin and Roger D. Nelson. *Replication in Random Event Generator Experiments: A Meta-Analysis and Quality Assessment.* (Technical Report 87001). Princeton, NJ: Princeton University Human Information Processing Group, 1987.

157. For example, with a ten hidden-node network, trained for four thousand passes, the difference between the mean number of correctly identified individuals obtained with the transfer dataset versus the same mean using the random dataset resulted in t(198 df) = 3.02; the t test between the transfer dataset and scrambled dataset produced t = 4.01; and the t test between the random and scrambled datasets was t = 0.98. Other tests using different network configurations showed similar results. See Dean I. Radin, "Searching for 'Signatures' in Human-Machine Interaction Data: A Neural Network Approach," in Linda A. Henkel and Rick E. Berger (eds.), *Research in Parapsychology 1988.* Metuchen, NJ: Scarecrow Press, 1989.

158. Brenda J. Dunne, Roger D. Nelson, and Robert G. Jahn, "Operator-Related Anomalies in a Random Mechanical Cascade," *Journal of Scientific Exploration, 2*(2), 1988, pp. 155-179.

159. Chen Hsin and Mai Lei, "Study of the Extraordinary Function of the Human Body in China," in W. G. Roll, J. Beloff, and R. White (eds.), *Research in Parapsychology 1982.* Metuchen, NJ: Scarecrow Press, 1983, pp. 278-282.

160. L. L. Haft, "Abstracts of Chinese Reports on Parapsychology," *European Journal of Parapsychology, 4,* 1982, pp. 399-402.

161. Harold E. Puthoff, "Report on Investigations Into 'Exceptional Human Body Function' in the People's Republic of China," in W. G. Roll, J. Beloff, and R. White (eds.), *Research in Parapsychology 1982.* Metuchen, NJ: Scarecrow Press, 1983, pp. 275-278.

162. Chinese Academy of Sciences, "Exceptional Human Body Radiation," *Psi Research, 1*(2), 1982, pp. 16-25.

163. Wu Xiaoping, "Report of a Chinese Psychic's Pill-Bottle Demonstration," *Skeptical Inquirer, 13*(2), Winter 1989, pp. 168-171.

164. G. Scott Hubbard, Edwin C. May, and Harold E. Puthoff, "Possible Production of Photons During a Remote Viewing Task: Preliminary Results," in D. H. Weiner and D. I. Radin (eds.), *Research in Parapsychology 1985.* Metuchen, NJ: Scarecrow Press, 1986. pp. 66-70.

165. John Hasted, *The Metal Benders.* London: Routledge & Kegan Paul, 1980.

166. John A. Palmer, *op. cit.*

167. Hasted, *op. cit.*

168. Charles Crussard and J. Bouvaist, "Experiences Psychocinetiques Sur Eprouvettes Metalliques [Psychokinetic Experiments with Metal Test Samples]," *Memoires Scientifiques de la Revue de Metallurgie,* 1978, pp. 13-23.

169. John A. Palmer, *op. cit.*

170. N. Richmond, "Two Series of PK Tests on Paramecia," *Journal of the American Society for Psychical Research, 46,* 1952, pp. 577-587.

171. L. Passidomo, "PK Effects on the Course Direction of Eurycereus, Lamaellatus, Eruycerine," *New Realities, 1*(1), 1977, pp. 40-44.

172. Pierre Duval, "Exploratory Experiments with Ants," *Journal of Parapsychology, 35,* 1971, p. 58. This study was actually conducted by the famous French biologist, Prof. Remy Chauvin of the Sorbonne, who often published his psi research studies using the pseudonym of P. Duval. For an overview of his perspective see Remy Chauvin, *Parapsychology: When the Irrational Rejoins Science,* translated by K. M. Banham. Jefferson, NC: McFarland, 1985.

173. Carroll B. Nash, "Psychokinetic Control of Bacterial Growth," *Journal of the American Society for Psychical Research, 51,* 1982, pp. 217-331.

174. J. Barry, "General and Comparative Study of the Psychokinetic Effect on a Fungus Culture," *Journal of Parapsychology, 32,* 1968, pp. 237-243.

175. William H. Tedder and M. L. Monty, "Exploration of Long-Distance PK: A Conceptual Replication of the Influence on a Biological System," in *Research in Parapsychology 1980.* Metuchen, NJ: Scarecrow Press, 1981, pp. 90-93.

176. Elizabeth A. Rauscher and Beverly A. Rubik, "Human Volitional Effects on a Model Bacterial System," *Psi Research, 3*(3/4), 1984, pp. 26-41.

177. Carroll B. Nash, "Test of Psychokinetic Control of Bacterial Mutation," *Journal of the American Society for Psychical Research, 78,* 1984, pp. 145-152.

178. Charles M. Pleass and N. Dean Day, "Using the Doppler Effect to Study Behavioral Responses of Motile Marine Algae to Psi Stimuli," in D. H. Weiner and D. I. Radin (eds.), *Research in Parapsychology 1985.* Metuchen, NJ: Scarecrow Press, 1986, pp. 70-73.

179. Loyd M. Auerbach, "Psi-fi: Psi in Science Fiction," *Applied Psi,* Spring 1984, pp. 3-4.

180. Andrew Neher, "Comments on 'The Schism within Parapsychology' by Jeffrey Mishlove," *Zetetic Scholar, 8,* 1981, pp. 94-96.

181. Gertrude Schmeidler, "Comments on 'The Schism within Parapsychology' by Jeffrey Mishlove," *Zetetic Scholar, 8,* 1981, pp. 99-102.

182. Robert A. McConnell, private correspondence, January 1981.

183. Robert Brier and Walter V. Tyminski, "Psi Application," in Joseph Banks Rhine (ed.), *Progress in Parapsychology.* Durham, NC: Parapsychology Press, 1973.

184. J. C. Carpenter. *Toward the Effective Utilization of Enhanced Weak-Signal ESP Effects.* Paper presented at the Meeting of the American Association for the Advancement of Science, New York, NY, January 1975.

185. Harold E. Puthoff, Edwin C. May, and M. J. Thomson, "Calculator-Assisted Psi Amplification II: Use of the Sequential-Sampling Technique as a Variable-Length Majority Vote Code," in D. H. Weiner and D. I. Radin (eds.), *Research in Parapsychology 1985.* Metuchen, NJ: Scarecrow Press, 1986, pp. 73-76.

186. Harold E. Puthoff, "ARV (Associational Remote Viewing) Applications," in *Research in Parapsychology 1984.* Metuchen, NJ: Scarecrow Press, 1985.

187. Bernard Grad, R. J. Cadoret, and G. I. Paul, "The Influence of an Unorthodox Method of Treatment on Wound Healing of Mice," *International Journal of Parapsychology, 3,* 1961, pp. 5-24.

188. Bernard Grad, "A Telekinetic Effect on Plant Growth," *International Journal of Parapsychology, 6,* 1964, p. 473.

189. Bernard Grad, "The 'Laying on of Hands': Implications for Psychotherapy, Gentling, and the Placebo Effect," *Journal of The American Society for Psychical Research, 61*(4), October 1967, pp. 286-305.

190. Bernard Grad, "The Biological Effects of the 'Laying on of Hands' on Animals and Plants: Implications for Biology," in G. R. Schmeidler (ed.), *Parapsychology: Its Relation to Physics, Biology, Psychology and Psychiatry.* Metuchen, NJ: Scarecrow Press, 1976.

191. Daniel P. Wirth, "Unorthodox Healing: The Effect of Noncontact Therapeutic Touch on the Healing Rate of Full Thickness Dermal Wounds," *Proceedings of Presented Papers, the Parapsychological Association 32nd Annual Convention,* San Diego, August 1989, pp. 251-268.

192. Carroll B. Nash and C. S. Nash, "The Effect of Paranormally Conditioned Solution on Yeast Fermentation," *Journal of Parapsychology, 31,* 1967, p. 314.

193. M. Justa Smith, "Effect of Magnetic Fields on Enzyme Reactivity," in Madeleine F. Barnothy (ed.), *Biological Effects of Magnetic Fields.* New York: Plenum Press, 1969.

194. M. Justa Smith, "The Influence on Enzyme Growth By the 'Laying on of Hands,'" *Dimensions of Healing.* Los Altos, CA: Academy of Parapsychology and Medicine, 1973.

195. Douglas Dean, "The Effects of Healers on Biologically Significant Molecules," *New Horizons, 1*(5), January 1975, pp. 215-219. This issue contains the Proceedings of the First Canadian Conference on Psychokinesis and Related Phenomena, June 1974.

196. Douglas Dean, *An Examination of Infra-Red and Ultra-Violet Techniques to Test for Changes in Water Following the Laying-On of Hands.* Ph.D. dissertation. Saybrook Institute, 1983.

197. Stephen A. Schwartz, Randall J. De Mattei, Edward G. Brame, Jr., and James P. Spottiswoode, *Infrared Spectra Alteration in Water Proximate to the Palms of Therapeutic Practitioners. Final Report.* Los Angeles: The Mobius Society, 1986.

198. Graham K. Watkins and Anita M. Watkins, "Possible PK Influence on the Resuscitation of Anesthetized Mice," *Journal of Parapsychology, 35*(4), December 1971, pp. 257-272.

199. Graham K. Watkins and Roger Wells, "Linger Effects in Several PK Experiments," Parapsychological Association Convention, Charlottesville, Virginia, 1973.

200. Roger Wells and Judith Klein, "A Replication of a 'Psychic Healing' Paradigm," *Journal of Parapsychology, 36*(2), June 1972, pp. 144-149.

201. William Braud and Marilyn Schlitz, "A Methodology for the Objective Study of Transpersonal Imagery," *Journal of Scientific Exploration, 3*(1), 1989.

202. Jean Achterberg, *Imagery in Healing.* Boston: New Science Library (Shambala), 1985, p. 5.

203. Much of the material included under this heading is based upon a paper by psychologist Charles Akers titled, "Methodological Criticisms of Parapsychology," in Stanley Kripper (Ed.), *Advances in Parapsychological Research, Vol. 4.* Jefferson, NC: McFarland, 1984, pp. 112-164.

204. W. K. Feller, "Statistical Aspects of ESP," *Journal of Parapsychology, 4,* 1940, pp. 271-298.

205. L. Zusne and W. H. Jones, *Anomalistic Psychology: A Study of Extraordinary Phenomena of Behavior and Experience.* Hillsdale, NJ: Erlbaum, 1982.

206. This constructive suggestion was made by C. E. M. Hansel in a paper commented upon by fellow skeptic Ray Hyman. See Hansel, "A Critical Analysis of H. Schmidt's Psychokinesis Experiments," *Skeptical Inquirer, 5,* 1981, pp. 26-33.

207. R. Wilson, "Deviations from Random in ESP Experiments," *International Journal of Parapsychology, 8,* 1966, pp. 387-395.

208. James W. Davis and Charles Akers, "Randomization and Tests for Randomness," *Journal of Parapsychology, 42,* 1974, pp. 393-407.

209. F. C. C. Hansen and A. Lehmann, "Ueber Unwillkurliches Flustern," *Philosophische Studien, 11,* 1895, pp. 471-530.

210. J. L. Kennedy, "Experiments on 'Unconscious Whispering,'" *Psychological Bulletin, 35,* 1938, p. 526 (Abstract).

211. J. L. Kennedy, "A Methodological Review of Extra-Sensory Perception," *Psychological Bulletin, 36,* 1939, pp. 59-103.

212. A. Moll, *Hypnotism* (4th ed.). New York: Scribner, 1898.

213. G. M. Stratton, "The Control of Another Person by Obscure Signs," *Psychological Review, 28,* 1921, pp. 301-314.

214. L. Warner and M. Raible, "Telepathy in the Psychophysical Laboratory," *Journal of Parapsychology, 1,* 1937, pp. 44-51.

215. Robert Rosenthal, "Clever Hans: A Case Study of Scientific Method," in O. Pfungst, *Clever Hans.* New York: Holt, Rinehart and Winston, 1965.

216. Robert Rosenthal, *Experimenter Effects in Behavioral Research.* New York: Appleton-Century-Croft, 1966.

217. George R. Price, "Science and the Supernatural," *Science, 122,* 1955, pp. 359-367.

218. James W. Davis, "A Developmental Program for the Computer Based Extension of Parapsychological Research and Methodology," *Journal of Parapsychology, 38,* 1974, pp. 69-84.

219. Deborah Delanoy, "Work With a Fraudulent PK Metal-Bending Subject," in D. H. Weiner and R. L. Morris (eds.), *Research in Parapsychology 1987.* Metuchen, NJ: Scarecrow Press, 1988, pp. 102-105.

220. George H. Estabrooks, *Spiritism.* New York: E. P. Dutton, 1947.

221. For an authoritative overview and annotated bibliography, see Earle J. Coleman, *Magic: A Reference Guide.* New York: Greenwood Press, 1987.

222. J. L. Kennedy and H. F. Uphoff, "Experiments on the Nature of Extra-Sensory Perception: III. The Recording Error Criticism of Extra-Chance Scores," *Journal of Parapsychology, 3,* 1939, pp. 226-245.

223. F. D. Sheffield, R. S. Kaufman, and J. B. Rhine, "A PK Experiment at Yale Starts a Controversy," *Journal of the American Society for Psychical Research, 46,* 1952, pp. 111-117.

224. Robert Rosenthal, "How Often Are Our Numbers Wrong?" *American Psychologist, 33,* 1978, pp. 1005-1008.

225. James W. Davis, "The Stacking Effect: Its Practical Significance in Parapsychology," *Journal of Parapsychology, 42,* 1978, p. 67.

226. Betty M. Humphrey, "Further Work With Dr. Stuart on Interest Ratings and ESP," *Journal of Parapsychology, 13,* 1949, pp. 151-165.

227. Brenda J. Dunne, et al., *op. cit.*

228. Jessica Utts, "Successful Replication Versus Statistical Significance," in L. A. Henkel and R. E. Berger (eds.), *Research in Parapsychology 1988.* Metuchen, NJ: Scarecrow Press, 1989, pp. 44-48.

229. Martin Gardner, "Quantum Theory or Quack Theory," *New York Review,* May 17, 1979. Here journalist and skeptic Gardner reports on the (failed) effort of noted physicist John Archibald Wheeler to persuade the American Association for the Advancement of Science (AAAS) to consider a process for disaffiliation with the Parapsychological Association. Gardner incorporates Wheeler's essay titled "Drive the Pseudos Out of the Workshop of Science."

230. Trevor Pinch, "Some Suggestions from the Sociology of Science to Advance the Psi Debate," *Brain and Behavior, 10*(4), December 1987, pp. 603-605. This issue is devoted to many articles regarding the controversy among psi researchers and their critics. It offers a window into the rhetoric of the debate.

231. *Skeptical Inquirer*

232. John A. Palmer, Charles Honorton, and Jessica Utts, *Reply to the National Research Council Study on Parapsychology.* Research Triangle Park, NC: Parapsychological Association, 1988.

233. John T. Sanders, "Are There Any 'Communications Anomalies'?" *Brain and Behavior, 10*(4), December 1987, p. 608.

234. Jeffrey Mishlove, "Parapsychology Research: Interview with Ray Hyman," *Skeptical Inquirer, V*(1), Fall 1980, pp. 63-67. Hyman described his attendance at the 1979 conference of the Parapsychological Association as follows: "Coming here as a skeptic, I would say that my overwhelming impression is the high quality of the research I have heard reported and the

impressive insights and awareness of the problems demonstrated by the people I have met." While not accepting the existence of ESP or PK, Hyman acknowledged that "most of the criticism of current parapsychological research is uninformed and misrepresents what is actually taking place."

235. I myself received a $30,000 settlement in a libel suit resulting from an October 1980 article in *Psychology Today* which claimed that my dissertation work at Berkeley was "incompetent" and that I possibly did not receive or deserve my doctoral diploma in parapsychology.

236. Monica J. Harris and Robert Rosenthal, *Human Performance Research: An Overview.* Washington, DC: National Academy Press, 1988. This paper compares psi ganzfeld research favorably in quality with other areas of human performance research.

237. Psychologist Donald Hebb made the following statement in his presidential address to the American Psychological Association: "Why do we not accept E.S.P. as a psychological fact? Rhine has offered us enough evidence to have convinced us on any other issue....I cannot see what other basis my colleagues have for rejecting it....My own rejection of [Rhine's] views is in a literal sense prejudice." See Harry M. Collins and Trevor J. Pinch, "The Construction of the Paranormal," in R. Wallace (ed.), *On the Margins of Science.* Sociological Review Monograph 27, University of Keele, 1979.

238. William James, "What Psychical Research Has Accomplished," in Gardner Murphy and Robert O. Ballou (eds.), *William James on Psychical Research.* London: Chatto and Windus, 1961. Originally published in *The Will to Believe and Other Essays,* 1897.

239. Eric J. Dingwall, "Responsibility in Parapsychology," in A. Angoff and B. Shapin (eds.), *A Century of Psychical Research: The Continuing Doubts and Affirmations.* New York: Parapsychology Foundation, 1971.

240. William James, "The Final Impressions of a Psychical Researcher," *The American Magazine,* October 1909. Reprinted in Gardner Murphy and Robert O. Ballou (eds.), *William James on Psychical Research.* London: Chatto and Windus, 1961.

241. Harry M. Collins, *Changing Order.* Newbury Park, CA: Sage, 1985.

242. S. Shapin and S. Schaffer, *Leviathan and the Air Pump: Hobbes, Boyle and the Experimental Life.* Princeton, NJ: Princeton University Press, 1985.

243. Trevor J. Pinch, "Theory Testing in Science — The Case of Solar Neutrinos: Do Crucial Experiments Test Theories or Theorists?" *Philosophy of the Social Sciences, 15,* 1985, pp. 167-187.

244. William James, "What Psychical Research Has Accomplished," *op. cit.*

Section IV

1. Michael Scriven, "The Frontiers of Psychology: Psychoanalysis and Parapsychology," in R. G. Colodny (ed.), *Frontiers of Science and Philosophy.* Pittsburgh, PA: University of Pittsburgh Press, 1962, pp. 78-129.

2. H. J. Irwin, *Introduction to Parapsychology.* Jefferson, NC: McFarland, 1989, p. 149.

3. Douglas Stokes, "Theoretical Parapsychology," in S. Krippner (ed.), *Advances in Parapsychological Research 5.* Jefferson, NC: McFarland, 1989, p. 189.

4. Gertrude Schmeidler, *Parapsychology and Psychology.* Jefferson, NC: McFarland, 1989.

5. Since there are no pain receptors in the brain itself, only the scalp needs to be anesthetized. While there are regions of the brain that seem to elicit pain when stimulated, these "pain centers" (i.e., in the limbic region) are simply the parts of the brain activated by the pain receptors of the body.

6. Richard F. Thompson, *Foundations of Physiological Psychology.* New York: Harper and Row, 1967.

7. Joseph E. Bogen, "The Other Side of the Brain: An Appositional Mind," in Robert Ornstein (ed.), *The Nature of Human Conscious-*

ness. San Francisco: W. H. Freeman, 1973. pp. 101-125. An anthology of scientific, philosophical, and literary material.

8. A. T. W. Simeons, M.D., *Man's Presumptuous Brain.* New York: E. P. Dutton, 1961. One of the most enjoyable and knowledgeable studies of brain science. Even in an age of information explosion, still worth reading.

9. Barbara Brown, *New Body, New Mind.* New York: Harper and Row, 1974. The story of biofeedback research seen through the eyes of one of the pioneer investigators.

10. The last two drugs are derived in the body directly from serotonin — and bufotenine is also the active ingredient in the toads that are proverbially used in witches' brews.

11. Frank X. Barron, Murray Jarvik, and Sterling Bunnell, Jr., "The Hallucinogenic Drugs," *Contemporary Psychology Readings From Scientific American.* San Francisco: W. H. Freeman, 1971, p. 305.

12. A. P. Krueger and S. Kotaka, "The Effects of Air Ions on Brain Levels of Serotonin in Mice," *International Journal of Biometeorology, 13*(1), 1969, p. 27.

13. Angela Longo, "'To Sleep; Perchance to Dream?' A Neurochemical Study of the States of Sleep." Unpublished paper, 1971.

14. John N. Bliebtrau, *The Parable of the Beast.* New York: Macmillan Company, 1968, p. 74.

15. Phillip Handler, ed., *Biology and the Future of Man.* New York: Oxford University Press, 1970, pp. 59-60. A survey of the life sciences sponsored by the National Academy of Sciences.

16. Frank Barr, "Melanin: The Organizing Molecule," *Medical Hypotheses, 11*(1), 1983, pp. 1-140.

17. Frank Barr, *What is Melanin?* Berkeley, CA: Institute for the Study of Consciousness, 1983.

18. L. R. Squire, "Mechanisms of Memory," *Science, 232,* 1986, pp. 1612-1619.

19. Michael A. Persinger, *The Paranormal: Part I. The Patterns.* New York: MSS Information, 1974.

20. Michael A. Persinger, "Psi Phenomena and Temporal Lobe Activity: The Geomagnetic Factor," in L. A. Henkel and R. E. Berger (eds.), *Research in Parapsychology 1988.* Metuchen, NJ: Scarecrow Press, 1989. pp. 124-125.

21. P. Gloor, A. Olivier, L. F. Quensey, F. Andermann, and S. Horowitz, "The Role of the Limbic System in Experiential Phenomena of Temporal Lobe Epilepsy," *Annals of Neurology, 12,* 1982, pp. 129-144.

22. P. Gloor, "Role of the Human Limbic System in Perception, Memory and Affect: Lessons from Temporal Lobe Epilepsy," in B. K. Doane and K. E. Livingston (eds.), *The Limbic System.* New York: Raven, 1986, pp. 159-169.

23. P. Gloor, et al., *op. cit.*

24. As a person who has experienced seizures, I am interested in understanding the extent to which one's orientation to the field of psi research and the philosophical issues related to consciousness may be shaped by neurological propensities.

25. D. M. Bear and P. Fedio, "Quantitative Analysis of Interictal Behavior in Temporal Lobe Epilepsy," *Archives of Neurology, 34,* 1977, pp. 454-467.

26. W. P. Spratling, *Epilepsy and Its Treatment.* Philadelphia: W. B. Saunders, 1904.

27. Edmund Gurney, F. W. H. Myers, and Frank Podmore, *Phantasms of the Living.* London: Trubner, 1886.

28. Eleanor Sidgewick, "Phantasms of the Living," *Proceedings of the Society for Psychical Research, 33,* 1922, pp. 23-429.

29. Michael Persinger and G. B. Schaut, "Geomagnetic Factors in Subjective Telepathic, Precognitive, and Post-Mortem Experiences," *Journal of the American Society for Psychical Research, 82,* 1988, pp. 217-235.

30. Michael A. Persinger, "Psi Phenomena and Temporal Lobe Activity," *op. cit.*, p. 127.

31. J. R. Stevens, "Sleep Is for Seizures: A New Interpretation of the Role of Phasic Ocular

Events in Sleep and Wakefulness," in M. B. Sterman and M. N. Shouse (eds.), *Sleep and Epilepsy.* New York: Academic Press, 1982, pp. 249-264.

32. Michael A. Persinger and K. Makarec, "Temporal Lobe Epilepsy Signs and Correlative Behaviors Displayed by Normal Populations," *Journal of General Psychology, 114,* 1987, pp. 179-195.

33. Michael A. Persinger, "Psi Phenomena and Temporal Lobe Activity," *op. cit.*, p. 131.

34. Albert Krueger, "Are Negative Ions Good for You?" *New Scientist,* June 14, 1973, p. 668.

35. Albert P. Krueger, "Preliminary Consideration of the Biological Significance of Air Ions," *Scientia,* September 1969.

36. A. P. Krueger and S. Kotaka, "The Effects of Air Ions on Brain Levels of Serotonin in Mice," *International Journal of Biometeorology, 13*(1), 1969, pp. 31-44.

37. Albert P. Krueger, P. C. Andriese, and S. Kotaka, "Small Air Ions: Their Effect on Blood Levels of Serotonin in Terms of Modern Physical Theory," *International Journal of Biometeorology, 12*(3), pp. 225-239.

38. A. P. Krueger and S. Kotaka, "The Effects," *op. cit.*, p. 33.

39. A. P. Krueger, "Are Negative Ions Good for You?" *op. cit.*

40. N. Robinson and F. S. Dirnfield, "The Ionization of the Atmosphere As a Functioning of Meterological Elements and of Various Sources of Ions," *International Journal of Biometeorology, 3*(2), March 1963.

41. A. Danon and F. G. Sulman, "Ionizing Effect of Winds of Ill Repute and Serotonin Metabolism," *Proceedings of the Fifth International Biometerological Congress,* September 1969.

42. Albert P. Krueger, "Biological Effects of Ionization of the Air," in S. W. Tromp (ed.), *Progress in Biometerology.* Amsterdam: Swets and Zeitlinger, 1974, p. 32.

43. Albert P. Krueger, personal communication to the author.

44. Walter M. Elsasser, "The Earth as Dynamo," *Scientific American,* May 1958. This article provides a basic explanation of the earth's magnetic field.

45. James B. Beal, "The Emergence of Paraphysics: Research and Applications," in E. D. Mitchell and J. White (eds.), *Psychic Explorations.* New York: Putmans, 1974.

46. T. Dull and B. Dull, "Uber die abhangigkeit des Gesundheitszustandes von plotzlichen Eruptionen auf der Sonne und die Existenz einer 27 taigigen Periode in den Sterbefillen," *Virschows* Archiv, No. 293, 1934. This study is summarized in Michel Gauquelin, *The Scientific Basis of Astrology.* New York: Stein and Day, 1969.

47. Howard Friedman, Robert O. Becker, and Charles Bachman, "Geomagnetic Parameters and Psychiatric Hospital Admissions," *Nature, 200,* November 16, 1963, pp. 620-628.

48. Michael A. Persinger, "ELF Waves and ESP," *New Horizons, 1*(5), January 1975, pp. 232-235.

49. Michael A. Persinger, *The Paranormal. Part II: Mechanisms and Models.* New York: M. S. S. Information Corp., 1974.

50. Selco Tromp, "Review of the Possible Physiological Causes of Dowsing," *International Journal of Parapsychology, 10*(4), 1968. Tromp, a Dutch researcher, has been the executive editor of the *International Journal of Biometerology.*

51. Y. Rocard, "Actions of a Very Weak Magnetic Gradient: The Reflex of the Dowser," in Madeleine F. Barnothy (ed.), *Biological Effects of Magnetic Fields.* New York: Plenum Press, 1969.

52. *Ibid.*, p. 281.

53. A double blind is a basic experimental technique in which neither the subject nor the experimenter knows whether a particular condition is part of the control or the test group, i.e., whether the magnetic field is on or off.

54. Selco Tromp, *op. cit.*

55. Yuri A. Kholodov, "Electromagnetic Fields

and the Brain," *Impact of Science on Society, 24*(4), October 1974, pp. 291-297. Kholodov is one of the Soviet researchers in the area of biomagnetic interactions. This issue of *Impact*, published by UNESCO, was devoted to the international developments in the "parasciences."

56. Yuri A. Kholodov, "The Brain and the Magnetic Field," *Journal of Paraphysics, 6*(4), 1972, pp. 144-147. This article provides a more detailed description of Kholodov's experiments. Several other articles in this issue of the *Journal of Paraphysics*, published in England by Benson Herbert, deal with bio-magnetics.

57. A. S. Presman, *Electromagnetic Fields and Life*, translated by F. L. Sinclair, edited by F. A. Brown. New York: Plenum Press, 1970. This volume is a compendium of the Soviet work in bio-magnetics. Presman is on the biophysics faculty at Moscow University.

58. Victor Yagodinsky, "The Magnetic Memory of the Virus," *Journal of Paraphysics, 6*(4), 1972, p. 141. Translated from the Russian.

59. M. Justa Smith, "The Influence on Enzyme Growth by the 'Laying on of Hands,'" in *Dimensions of Healing.* Los Altos, CA: Academy of Parapsychology and Medicine, 1973.

60. Svetlana Vinokurava, "Life in a Magnetic Web," *Journal of Paraphysics, 5*(4), 1971, p. 135.

61. Homer Jensen, "The Airborn Magnetometer," *Scientific American, 202*(6), June 1961, p. 152.

62. Pedro McGregor, *The Moon and Two Mountains.* London: Souvenir Press, 1966. This book offers an unusual balance of emotional involvement and sociological objectivity. The author, an educated journalist, is also the founder of a spiritist church which is attempting to synthesize the many conflicting strains of Brazilian magical tradition.

63. Charles C. Conley, "Effects of Near-Zero Magnetic Fields on Biological Systems," *Biological Effects of Magnetic Fields*, Vol. 2., *op. cit.*

64. R. R. Koegler, S. M. Hicks, L. Rogers, and J. H. Barger, "A Preliminary Study in the Use of Electrosleep Therapy in Clinical Psychiatry," in Norman L. Wulfson (ed.), *The Nervous System and Electric Currents.* New York: Plenum Press, 1970, pp. 137-143. Not satisfied with the quality of the European work, these American researchers conducted their own study with encouraging results.

65. Gay Gaer Luce, *Biological Rhythms in Human and Animal Physiology.* New York: Dover, 1971, pp. 120-132. This is an unabridged version of a report originally prepared for the National Institute of Mental Health.

66. John N. Ott, *Health and Light.* Old Greenwich, CT: Devin-Adair, 1973. Using the techniques of time-lapse photography, this volume demonstrates the effects of light variations on plants, and points to similar responses in animals and people.

67. Robert O. Becker, "The Effect of Magnetic Fields Upon the Central Nervous System," *Biological Effects of Magnetic Fields*, Vol. 2, pp. 207-214.

68. Howard Friedman, Robert O. Becker, and Charles H. Bachman, "Psychiatric Ward Behavior and Geophysical Parameters," *Nature, 205*, March 13, 1965, pp. 1050-1052.

69. Howard Friedman, Robert O. Becker, and Charles H. Bachman, "Effect of Magnetic Fields on Reaction Time Performance," *Nature, 213*, March 4, 1967, pp. 949-950.

70. Wilder Penfield, *Mystery of the Mind.* Princeton: Princeton University Press, 1975, p. 114.

71. Charles Sherrington, *Man On His Nature.* New York: Macmillan, 1941.

72. John C. Eccles, *Facing Reality.* New York: Springer-Verlag, 1970.

73. Roger W. Sperry, "Mental Phenomena as Causal Determinants in Brain Function," in G. G. Globus, G. Maxwell, and I. Savodnik (eds.), *Consciousness and the Brain.* New York: Plenum, 1976.

74. Karl R. Popper and John C. Eccles, *The Self and Its Brain.* London: Routledge and Kegan Paul, 1983, p. 98.

75. Bertrand Russell, "Mind and Matter," in *Portraits From Memory.* New York: Simon and Schuster, 1956, p. 145.

76. Gertrude R. Schmeidler, "Respice, Adspice and Prospice," in W. G. Roll, R. L. Morris, and J. D. Morris (eds.), *Proceedings of the Parapsychological Association, No. 8, 1971.* Durham, NC: Parapsychological Association, 1972.

77. John Archibald Wheeler, *Geometrodynamics.* New York: Academic Press, 1962.

78. Bob Toben and Fred Alan Wolf, *Space-Time and Beyond.* New York: Bantam Books, 1982.

79. Fred Alan Wolf, *The Body Quantum.* New York: Macmillan, 1986.

80. Fred Alan Wolf, *Star Wave: Mind, Consciousness, and Quantum Physics.* New York: Macmillan, 1984.

81. Fred Alan Wolf, "Trans-World I-ness: Quantum Physics and the Enlightened Condition," in *Humor Suddenly Returns: Essays on the Spiritual Teaching of Master Da Free John.* Clearlake, CA: Dawnhorse Press, 1984.

82. Fred Alan Wolf, "The Quantum Physics of Consciousness: Towards a New Psychology," *Integrative Psychiatry, 3*(4), December 1985, p. 236.

83. Fred Alan Wolf, *Parallel Universes.* New York: Simon and Schuster, 1988.

84. Russell Targ, Harold E. Puthoff, and Edwin C. May, "Direct Perception of Remote Geographical Locations," in C. T. Tart, H. E. Puthoff, and R. Targ (eds.), *Mind At Large.* New York: Praeger, 1979, pp. 78-106. The authors state that this work was in conjunction with physicist Gerald Feinberg — who is well known for his postulation of the existence of tachyons, particles that travel faster than light.

85. Physicist Evan Harris Walker ("Review of *Mind At Large,*" *Journal of Parapsychology, 45,* 1981, pp. 184-191) has observed, however, that if we retain the inverse-square law for gravity, the effect of four extra dimensions on planetary trajectories should have been observed.

86. Elizabeth A. Rauscher, "Some Physical Models Potentially Applicable to Remote Perception," in A. Puharich (ed.), *The Iceland Papers.* Amherst, WI: Essential: 1979, pp. 50-93.

87. Elizabeth A. Rauscher, "The Physics of Psi Phenomena in Space and Time. Part I. Major Principles of Physics, Psychic Phenomena, and Some Physical Models," *Psi Research, 2*(2), 1983, pp. 64-88.

88. Elizabeth A. Rauscher, "The Physics of Psi Phenomena in Space and Time. Part II. Multidimensional Geographic Models," *Psi Research, 2*(3), 1983, pp. 93-120.

89. C. Ramon and Elizabeth A. Rauscher, "Superluminal Transformations in Complex Minkowski Spaces," *Foundations of Physics, 10,* 1980, pp. 661-669.

90. John S. Bell, "On the Einstein Podolsky Rosen Paradox," *Physics, 1*(3), 1964, pp. 195-200.

91. Nick Herbert, "Crytographic approach to hidden variables," *American Journal of Physics,* Vol. 43, No. 4, April 1975, pp. 315-316. This paper presents a proof of Bell's theorem by considering error rates in binary message sequences. It also speculates about the possibility of faster-than-light signaling.

92. Nick Herbert, *Faster Than Light.* New York: New American Library, 1988.

93. J. S. Bell, *Nature, 248,* March 22, 1974, p. 297.

94. S. D. Drell, "Electron-Positron Annihilation and the New Particles," *Scientific American,* June 1975.

95. John F. Clauser and Abner Shimony, "Bell's Theorem: Experimental Tests and Implications," *Reports on Progress in Physics 41,* 1978, p. 1881.

96. Alain Aspect, Jean Dalibard, and Gerard Roger, "Experimental Test of Bell's Inequalities Using Time-varying Analyzers," *Physical Review Letters 49,* 1982, p. 1804.

97. E. Wigner, *Symmetries and Reflections.* Indiana University, 1967, and Cambridge, MA: M.I.T. Press paperback edition, 1970.

98. Brian Josephson, "Possible Connections Between Psychic Phenomena and Quantum

Mechanics," *New Horizons,* January 1975, pp. 224-226.

99. P.A. Schilpp, *Albert Einstein Philosopher Scientist.* New York: Harper Torchbook, 1959, p. 85. Einstein uses the German word "telepathisch" in the original version.

100. I. J. Good, "Speculations Concerning Precognition," in I. J. Good (ed.), *The Scientist Speculates.* New York: Basic Books, 1962, pp. 151-157.

101. Nick Herbert, *Quantum Reality.* New York: Doubleday, 1985.

102. David Bohm, *Wholeness and the Implicate Order.* London: Routledge and Kegan Paul, 1980.

103. Evan Harris Walker, "Foundations of Paraphysical and Parapsychological Phenomena," in L. Oteri (ed.), *Quantum Physics and Parapsychology.* New York: Parapsychology Foundation, 1975, pp. 1-53.

104. Evan Harris Walker, "A Review of Criticisms of the Quantum Mechanical Theory of Psi Phenomena," *Journal of Parapsychology, 48,* 1984, pp. 277-332.

105. Evan Harris Walker, "Measurement in Quantum Physics Revisited: A Response to Phillips' Criticism of the Quantum Mechanical Theory of Psi," *Journal of the American Society for Psychical Research,* October 1987, *81*(4), pp. 333-369.

106. Richard D. Mattuck, "Random Fluctuation Theory of Psychokinesis: Thermal Noise Model," in J. D. Morris, W. G. Roll, and R. L. Morris (eds.), *Research in Parapsychology 1976.* Metuchen, NJ: Scarecrow Press, 1977, pp. 191-195.

107. Richard D. Mattuck, "A Model of the Interaction Between Consciousness and Matter using Bohm-Bub Hidden Variables," in W. G. Roll, R. L. Morris, and R. A. White (eds.), *Research in Parapsychology 1981.* Metuchen, NJ: Scarecrow Press, 1982, pp. 146-147.

108. Saul-Paul Sirag, "A Combinatorial Derivation of the Proton-Electron Mass Ratio," *Nature, 268,* July 7, 1977, p. 254.

109. Saul-Paul Sirag, "Physical Constants as Cosmological Constraints," *International Journal of Theoretical Physics, 22,* 1983, pp. 1067-1089.

110. Saul-Paul Sirag, "Why There Are Three Fermion Families," *Bulletin of the American Physics Society, 27*(1), 1982, p. 31.

111. G. Abrams, et al., "Initial Measurements of the Z Boson Resonance, *Physical Review Letters, 63*(7), August 14, 1989, pp. 724-727.

112. "Experts Focus on the Birth of the Universe," *San Francisco Examiner,* October 13, 1989, p. 2.

113. Douglas M. Stokes, "Theoretical Parapsychology," in Stanley Krippner (ed.), *Advances in Parapsychological Research, Vol. 5.* Jefferson, NC: McFarland, 1987.

114. Ken Wilbur, "Introduction," in K. Wilbur (ed.), *Quantum Questions: Mystical Writings of the World's Great Physicists.* Boston: Shambhala, 1984.

115. In fact, *Beyond Science* is the name of a television series, produced by Arthur Bloch, based on the work of Arthur Young.

116. Action has the measure formula Mass x $Length^2$/Time and is always an integral multiple of *h*, Planck's constant (in MKS units, 6.63×10^{-34} Joule-seconds). The smallest whole unit of action is equivalent to h, which is the quantum. While energy is proportional to frequency, action is a constant of the proportion between energy and frequency ($E = hv$) and comes in wholes. Gravity, the strong force, and the weak force can all be expressed in terms of action.

117. Max Planck, *Scientific Autobiography and Other Papers,* translated by Frank Gaynor. New York: Philosophical Library, 1949, p. 178. Quoted by Arthur Young in *The Reflexive Universe,* New York: Delacorte, 1975.

118. Alfred North Whitehead, *The Function of Reason.* Princeton, NJ: Princeton University Press, 1929. Quoted by Arthur Young in *The Reflexive Universe, op. cit.*

119. Arthur M. Young, *The Reflexive Universe, op. cit.*

PERMISSIONS

The publisher wishes to thank the following for permission to reprint selections in this book. Any inadvertent omission will be corrected in future printings on notification to the publisher.

John Wiley and Sons for material from Kilton Stewart, "Dream Theory in Malaya," in C. Tart (ed.), *Altered States of Consciousness.*

Parapsychology Review for material taken from Adrian Boshier, "African Apprenticeship," July 1974.

Columbia University Press for material from William de Bary (ed.), *Sources of Indian Tradition,* 1958.

Oxford University Press for material from W. Y. Evan-Wentz, *The Tibetan Book of the Dead.*

The Regents of the University of California for material from Wayne Schumaker, *The Occult Sciences in the Renaissance,* originally published by the University of California Press.

Manley Palmer Hall for material from *The Secret Teachings of All Ages.*

Idries Shah for material from *Oriental Magic,* published by E. P. Dutton and Company, Inc., New York, 1973 edition, pp. 61-2.

Harper and Row for material from Aldous Huxley, *The Perennial Philosophy.*

The Journal of Parapsychology for material from Helmut Schmidt, "PK Tests with a High Speed Random Number Generator," December 1973; "A Quantum Mechanical Random Number Generator for Psi Tests," 1970; and "A Quantum Process in Psi Testing," *Progress in Parapsychology.*

Parapsychology Foundation for material from H. C. Berendt, "Parapsychology in Israel," in A. Angoff and B. Shapin (eds.), *Parapsychology Today: A Geographic View,* 1973; and Inge Strauch, "Medical Aspects of 'Mental Healing,'" *International Journal of Parapsychology, V,* 1963. Copyright Parapsychology Foundation, 1963.

C. W. Daniel Company, Ltd., London, for material from Vernon Wethered, *Medical Radiesthesia and Radionics.*

Psychic Magazine for material from Robert N. Miller, "The Positive Effect of Prayer on Plants," April 1972; "Interview with Ray Stanford," April 1974; and Jean Shinoda Bolen, M. D., "Meditation in the Treatment of Cancer," August 1973.

William Morrow and Company for material from Jule Eisenbud, *The World of Ted Serios,* copyright 1966, 1967 by Jule Eisenbud.

E. P. Dutton and Company for material from Anne Dooley, *Every Wall a Door,* copyright 1973 by Anne Dooley. First published in 1974 by E. P. Dutton.

Harper and Row for material from Peter Tompkins and Chris Bird, *The Secret Life of Plants*; Ambrose and Olga Worrall, *Explore Your Psychic World*; and Louisa E. Rhine, *Psi,* 1975.

The Estate of Eileen J. Garrett for material from Eileen Garrett, *Awareness,* copyright 1943.

Doubleday and Company for material from Andrija Puharich, *Uri: A Journal of the Mystery of Uri Geller,* copyright 1974 by Lab Nine, Ltd.

Gordon and Breach for material from S. Krippner and D. Rubin (eds.), *Galaxies of Life,* 1973.

Thelma Moss for material from "Bioenergetics and Radiation Photography," International Psychotronics Conference, Prague, 1973.

Theosophical Publishing House, Adyar, Madras, India, for material from C. W. Leadbeater, *The Chakras.*

From the book, *Psychic Discoveries Behind the Iron Curtain,* by Ostrander and Schroeder, copyright 1970 by Sheila Ostrander and Lynn Schroeder. Published by Prentice-Hall, Inc., Englewood Cliffs, New Jersey.

From the book, *The Seth Material* by Jane Roberts, copyright 1970 by Jane Roberts. Published by Prentice-Hall, Inc., Englewood Cliffs, New Jersey.

From William G. Roll, "Poltergeists," in Richard Cavandish (ed.), *Encyclopedia of the*

Unexplained, copyright 1974 by Rainbird Reference Books Ltd. Used with permission of McGraw-Hill Book Company.

Rudolf Steiner, *Goethe As Scientist.*

Julian Press for material from Rammurti S. Mishra, *The Textbook of Yoga Psychology.*

New Scientist for material from Dr. Albert Krueger, "Are Negative Ions Good for You?" June 14, 1973.

Samuel Weiser, Inc. for material from Muldoon and Carrington, *The Projection of the Astral Body*, and Lama Anagarika Govinda, *Foundations of Tibetan Mysticism.*

Regents of the University of California for material from G. Grunebaum and Roger Callois (eds.), *The Dream and Human Societies.*

C

D

G

H

I

J

K

L

O

P

S

T

ABOUT THE AUTHOR

JEFFREY MISHLOVE, PH.D., currently serves as Director of the Intuition Network, an organization of thousands of individuals in business, government, health, science and education who are interested in cultivating and applying intuitive skills. The Intuition Network operates a computer conference and also sponsors conferences, seminars, tours, research and publications. For more information, please write to Intuition Network, 369-B Third Street, #161, San Rafael, CA 94901 or send E-mail to intuition.network@intuition.org.

As host of the weekly, national public television series *Thinking Allowed,* Jeffrey Mishlove has pioneered the introduction of deep, authentic and thoughtful discussions in a broadcast medium noted mostly for soundbites and sensationalism. In this capacity he conducts interviews with leading figures in science, philosophy, psychology, health and spirituality. *Thinking Allowed* programs have been released every week since April 1986 — making it one of the longest running programs on national television. The *Thinking Allowed* website address is www.thinkingallowed.com.

In addition to The *Roots of Consciousness,* Dr. Mishlove is also author of *Psi Development Systems,* a book evaluating dozens of methods for training extrasensory abilities. Another book, *Thinking Allowed* consists of edited transcripts from the television series. And, in 1996, an anthology titled *Intuition At Work,* with contributions by Dr. Mishlove and others, was published by New Leaders Press in cooperation with the Intuition Network.

Dr. Mishlove has the distinction of holding a doctoral diploma in "Parapsychology" which is the only such diploma ever awarded by an accredited, American University. He received this degree in 1980 from the University of California at Berkeley. He is also past-president of the California Society for Psychical Study and past-vice-president of the Association for Human Psychology.

THINKING ALLOWED AND *INNERWORK*

Readers interested in further exploration of the subjects covered herein may wish to become acquainted with the *Thinking Allowed* and *InnerWork* videotape collections. This growing library now contains well over a hundred intimate conversations between Jeffrey Mishlove and the leading consciousness explorers of our generation. The video format allows for a vitality approaching that of being with these individuals in your own home.

Many of the *Thinking Allowed* and *InnerWork* programs have been cited in the footnotes and picture captions as directly relevant to the text of this book.

A complete catalog of the *Thinking Allowed* and *InnerWork* video collections may be obtained by contacting Thinking Allowed Productions, 2560 Ninth Street, Suite 123-RC, Berkeley, California 94710, phone 1-800-999-4415.